T0203899

The Froehlich / Kent

ENCYCLOPEDIA OF TELECOMMUNICATIONS

VOLUME 15

The Froehlich/Kent
ENCYCLOPEDIA OF
TELECOMMUNICATIONS

Editor-in-Chief

Fritz E. Froehlich, Ph.D.

Professor of Telecommunications
University of Pittsburgh
Pittsburgh, Pennsylvania

Co-Editor

Allen Kent

Distinguished Service Professor of Information Science
University of Pittsburgh
Pittsburgh, Pennsylvania

Administrative Editor

Carolyn M. Hall

Arlington, Texas

VOLUME 15

RADIO ASTRONOMY to
SUBMARINE CABLE SYSTEMS

CRC Press
Taylor & Francis Group
Boca Raton London New York

CRC Press is an imprint of the
Taylor & Francis Group, an **informa** business

CRC Press
Taylor & Francis Group
6000 Broken Sound Parkway NW, Suite 300
Boca Raton, FL 33487-2742

First issued in paperback 2019

© 1998 by Taylor & Francis Group, LLC
CRC Press is an imprint of Taylor & Francis Group, an Informa business

No claim to original U.S. Government works

ISBN-13: 978-0-8247-2913-4 (hbk)
ISBN-13: 978-0-367-40083-5 (pbk)

Library of Congress Cataloging-in-Publication Data

The Froehlich/Kent Encyclopedia of Telecommunications / editor-in-chief, Fritz E. Froehlich ; co-editor, Allen Kent.
    p.   cm.
    Includes bibliographical references and indexes.
    ISBN 0-8247-2902-1 (v. 1 : alk. paper)
    1. Telecommunication—Encyclopedias.   I. Froehlich, Fritz E.,
Kent, Allen.
    TK5102.E646   1990
    384′.03—dc20

90-3966
CIP

Visit the Taylor & Francis Web site at
http://www.taylorandfrancis.com

and the CRC Press Web site at
http://www.crcpress.com

CONTENTS OF VOLUME 15

CONTRIBUTORS TO VOLUME 15

Ashok K. Agrawala, Ph.D.   Professor, Department of Computer Science, University of Maryland, College Park, Maryland: *Real-Time Communication*

Cleo D. Anderson, M.E.E.   Consultant, Tyco Submarine Systems, Holmdel, New Jersey; AT&T Bell Laboratories (Retired), Holmdel, New Jersey: *Submarine Cable Systems*

Gregory M. Bubel, M.S.M.E.   Director, Tyco Submarine Systems, Holmdel, New Jersey: *Submarine Cable Systems*

Aubrey M. Bush, Sc.D. Deputy Director, Division of Networking and Communications Research and Infrastructure, National Science Foundation, Arlington, Virginia: *Status of the Internet*

Armando A. Cabrera, M.S.M.E.   Distinguished Member of Technical Staff, Tyco Submarine Systems, Holmdel, New Jersey: *Submarine Cable Systems*

Ardas Cilingiroglu   Graduate Student, University of Maryland, College Park, Maryland: *Real-Time Communication*

Marc Davidson, B.A.   Editor, News-Journal Center Online Services, Daytona Beach News-Journal, Daytona Beach, Florida: *Sarnoff, David*

Sudhir S. Dixit, Ph.D.   Principal Research Engineer, Nokia Research Center, Burlington, Massachusetts: *Services and Network Interworking in Wide-Area Networks*

William G. Duff, DSc.E.E.   Senior Consulting Engineer, Computer Sciences Corporation, Springfield, Virginia: *Review of Electromagnetic Compatibility in Telecommunications*

James T. Ellis, A.M.   Member of the Technical Staff, CERT Coordination Center, Software Engineering Institute, Carnegie Mellon University, Pittsburgh, Pennsylvania: *Security of the Internet*

Darrel T. Emerson, Ph.D.   Assistant Director for Arizona Operations, National Radio Astronomy Observatory, Tucson, Arizona: *Radio Astronomy*

Lewis E. Franks, Ph.D.   Professor Emeritus, Department of Electrical and Computer Engineering, University of Massachusetts, Amherst, Massachusetts: *Signal Theory*

Vijay K. Garg, Ph.D.   Principal Staff Engineer, Cellular Infrastructure Group, Motorola, Arlington Heights, Illinois: *Standard Interfaces for Wireless Communications*

Alex Gillespie, Ph.D.   Senior Member of Professional Staff, BT Laboratories, Ipswich, England; Member IEEE; Associate Member IEE: *Review of Management Modeling for Telecommunications*

Zorach R. Glaser, Ph.D.   Senior Scientific Consultant, Division of Standards Development, U.S. Pharmacopeia, Rockville, Maryland; Adjunct Associate Professor of Environmental Health Engineering, Department of Environmental Health Sciences, School of Public Health, Johns Hopkins University, Baltimore, Maryland; President, Glaser and Associates, Biomedical Consultants, Laurel, Maryland: *Radio-Frequency and Microwave Radiation Biological Effects: An Overview*

Robert F. Gleason, Ph.D.   Director, Tyco Submarine Systems, Holmdel, New Jersey: *Submarine Cable Systems*

Roch H. Glitho, M.Sc.   Principal System Engineer, Ericsson Research Canada, Montreal, Quebec, Canada: *Routing in Telecommunications Networks*

Mark A. Gordon, Ph.D.   Scientist, National Radio Astronomy Observatory, Tucson, Arizona: *Radio Astronomy*

Shawn V. Hernan   Member of the Technical Staff, CERT Coordination Center, Software Engineering Institute, Carnegie Mellon University, Pittsburgh, Pennsylvania: *Security of the Internet*

Robert W. Klessig, Ph.D.   Director of Systems Marketing, 3Com Corporation, Santa Clara, California; ATM Forum, Vice Chair of Technical Committee: *SMDS— Switched Multimegabit Data Service*

Sung Lee   Graduate Student, University of Maryland, College Park, Maryland: *Real-Time Communication*

Howard F. Lipson, Ph.D.   Member of the Technical Staff, CERT Coordination Center, Software Engineering Institute, Carnegie Mellon University, Pittsburgh, Pennsylvania: *Security of the Internet*

Thomas A. Longstaff, Ph.D.   Member of the Technical Staff, CERT Coordination Center, Software Engineering Institute, Carnegie Mellon University, Pittsburgh, Pennsylvania: *Security of the Internet*

Robert L. Lynch, M.E.E.   Director, Tyco Submarine Systems, Holmdel, New Jersey: *Submarine Cable Systems*

Robert D. McMillan, B.App.Sc.   Member of the Technical Staff, CERT Coordination Center, Software Engineering Institute, Carnegie Mellon University, Pittsburgh, Pennsylvania: *Security of the Internet*

Michael I. Meyerson, J.D.   Professor of Law, University of Baltimore School of Law, Baltimore, Maryland: *Regulation of Telecommunications by the U.S. Federal Government—A Guide to the 1996 Telecommunications Act*

Linda Hutz Pesante, M.A.   Member of the Technical Staff, CERT Coordination Center, Software Engineering Institute, Carnegie Mellon University, Pittsburgh, Pennsylvania: *Security of the Internet*

Michael B. Pursley, Ph.D.   Holcombe Professor of Electrical and Computer Engineering, Clemson University, Clemson, South Carolina: *Spread-Spectrum Communications*

Bruce O. Rein   Director, Tyco Submarine Systems, Morristown, New Jersey: *Submarine Cable Systems*

Marylin K. Sheddan, M.S.   Senior Research Writer, Embry-Riddle Aeronautical University, Daytona Beach, Florida: *Shockley, William; Steinmetz, Charles Proteus*

Derek Simmel, M.S.   Member of the Technical Staff, CERT Coordination Center, Software Engineering Institute, Carnegie Mellon University, Pittsburgh, Pennsylvania: *Security of the Internet*

William F. Sirocky, M.S.M.E.   Director, Tyco Submarine Systems, Morristown, New Jersey: *Submarine Cable Systems*

Elysia C. Tan, B.Sc.   Principal Technical Staff Member, AT&T Bell Laboratories, Holmdel, New Jersey: *SMDS — Switched Multimegabit Data Service*

Kaj Tesink, M.S.   Director of Advanced Internetworking, Bell Communications Research, Red Bank, New Jersey: *SMDS — Switched Multimegabit Data Service*

Paul E. Teske, Ph.D.   Associate Professor, Department of Political Science, State University of New York at Stony Brook, Stony Brook, New York: *Regulation of Telecommunications at the State Level in the United States*

Mark D. Tremblay, M.S.E.E.   Technical Director, Tyco Submarine Systems, Holmdel, New Jersey: *Submarine Cable Systems*

Kwei Tu, Ph.D.   Corporate Engineer, LinCom Corporation, Houston, Texas: *Space Communications*

Joseph E. Wilkes, Ph.D.   Senior Research Scientist, Bell Communications Research, Inc., Red Bank, New Jersey: *Standard Interfaces for Wireless Communications*

The Froehlich / Kent

ENCYCLOPEDIA OF TELECOMMUNICATIONS

VOLUME 15

Radio Astronomy

What Is Radio Astronomy?

Radio astronomy refers to astronomy using radio waves naturally generated by cosmic bodies. The term excludes *radar astronomy*, which consists of studying radar echoes from objects in the solar system.

Today, astronomy (or, better, astrophysics) research is generally independent of wavelength or frequency. Modern astronomers combine observations from many wavelength ranges to study particular astronomical phenomena.

For astronomers, *radio* is an imprecise term that today might include the electromagnetic spectrum between λ 300 μm (micrometers) and λ 30,000 m (meters), a range of 10^8. Overlaps with other named bands occur. The far infrared band includes submillimeter wavelengths. The International Telecommunication Union specifies the use of radio bands from 9 kilohertz (kHz) (λ 33,333 m) to 275 GHz (gigahertz) (λ 1.1 millimeters [mm]) to minimize conflict of radio services.

Conducting astronomy research at radio wavelengths requires understanding astrophysics in general and, specifically, the physical mechanisms that generate cosmic radio emissions. Observationally, it means understanding the specialized equipment designed to respond quantitatively to cosmic radio waves.

Early Years

Radio astronomy began in the first half of this century. Thomas Edison suggested observations of radio waves from the sun around 1890. Sir Oliver Lodge unsuccessfully searched for radio emissions from the sun between 1897 and 1900. Karl G. Jansky is considered the father of radio astronomy through his discovery of cosmic radio emission at a frequency of 20.5 MHz from the Milky Way (our galaxy) in 1932 (1). To honor his discovery, the fundamental unit of radio flux density is the Jansky, and the United States' National Radio Astronomy Observatory awards the Jansky Lectureship annually to someone who has made significant contributions to the field.

Following Jansky's detection, in 1938 Grote Reber built a 30-ft diameter parabolic reflector in the back yard of his Wheaton, Illinois, house and a radio receiver to investigate the Milky Way emission at a frequency of 3300 MHz (2).

1

Unable to confirm Jansky's results, he decreased his observing frequency to 910 MHz, still without success. Finally, reducing the frequency to 162 MHz, he was able to confirm Jansky's detections. These observations were the first indications that the cosmic radiation emission was nonthermal. In the 1940s, Reber surveyed the sky at 160 MHz with greatly improved electronics. This time, in addition to the Milky Way, he detected the supernova remnant Cassiopeia A and the sun. Changing frequency to 480 MHz, he again detected the Milky Way but also variable radio emission from the sun.

The great improvements in radio and radar technology driven by World War II contributed to the emergence of radio astronomy (3). Many radar and radio operators detected signals of unknown origin in the frequency range 55 to 80 MHz. Generally, these reports were classified. One proved to be variable emission from sunspots, later reported by James S. Hey in 1946 (4). Another was the 1942 detection of the emission from the quiet sun at 9.4 GHz, later published by G. C. Southworth in 1945 (5).

At least one theoretical study carried out during World War II proved especially important to astronomers. In occupied Netherlands, H. C. van de Hulst considered the possibility of detecting line emission at radio frequencies from the most abundant constituent of cosmic gases, hydrogen. Publishing this paper in 1945 after the war (6), he predicted that astronomers should be able to detect line emission from atomic hydrogen at λ 21 cm (centimeters) (1420 MHz). In 1951, the line was detected on March 25 by H. I. Ewen and E. M. Purcell at Harvard University (Cambridge, Mass.) (7), on May 11 by C. A. Muller and J. H. Oort at Leiden (Netherlands) (8), and on July 12 by W. N. Christiansen and J. V. Hindman in Sydney, Australia (8). The availability of radio line radiation allowed the study of cold gases from which stars form. The line frequency uniquely identified hydrogen as the emitter, thereby facilitating measurements of line-of-sight velocities of cold cosmic gas through the Doppler effect.

Following these detections and others, radio astronomy grew rapidly in many countries, attracting the attention of both theorists and observers. Several countries formed national centers for radio astronomy to accommodate astronomers wanting to observe in the radio domain. These observatories, sometimes associated with particular universities, are usually open to astronomers everywhere on a competitive basis.

These centers made radio astronomy available to nonspecialists. Their engineers provide state-of-the-art instrumentation. When needed, staff astronomers and telescope operators assist visiting astronomers with observing. The centers free astronomers to concentrate on astrophysical problems without having to know the technical details of the instrumentation. In this way, radio astronomy evolved from a peculiar research specialty into the full field of general astronomy that it is today.

Technical Aspects

Specific aspects distinguish modern radio astronomy from telecommunications: sensitivity, angular resolution, observing techniques, and the transmission of the earth's atmosphere This article considers these rather than the general technology of receivers and antennas also used in radio astronomy and discussed elsewhere in this encyclopedia.

Coupling of Radiation to Antennas

Cosmic radio emission is often weak compared to optical emission. The brightness or specific intensity (9), I_ν, emitted by cosmic sources at any frequency has units of

$$[I_\nu] = \text{energy/time/area/unit frequency interval/solid angle} \tag{1}$$

as it passes through a unit area in space and confined to a solid angle. There is a wavelength form of specific intensity,

$$I_\lambda d\lambda = I_\nu d\nu \tag{2}$$

The specific intensity radiated by black bodies, that is, in thermal equilibrium, is

$$I_\nu = B_\nu \tag{3}$$

$$= \frac{2h\nu^3}{c^2} \frac{1}{\exp(h\nu/kT) - 1} \tag{4}$$

where B_ν is the frequency form of the Planck function for a body with a temperature T. At the long wavelengths characteristic of much of radio astronomy,

$$B_\nu \approx \frac{2kT}{\lambda^2}, \qquad h\nu \ll kT \tag{5}$$

which is the Rayleigh–Jeans approximation for the red end of the spectrum. Note that this approximation often does not hold for cold gases radiating at millimeter and submillimeter wavelengths.

Much of astrophysics involves determining and applying the appropriate form of I_ν. Many cosmic processes emit nonthermal radiation, only near-thermal radiation, or a mixture of both. Here, B_ν would not apply.

Astronomical telescopes detect spectral flux density, a quantity independent of solid angle. The spectral flux density, F_ν, detected by an antenna in a particular direction

$$F_\nu = \int_{\text{source}} I_\nu(\Omega)\, d\Omega \tag{6}$$

where I_ν is the specific intensity. The units of F_ν are Janskys, 10^{-26} W/m²/Hz.

The detected signal power per unit bandwidth, P_ν, in that direction is

$$P_\nu = F_\nu A \tag{7}$$

where A is the capture area of the antenna, which may not be the same as the physical collecting area (10). Detection causes an effective increase of the antenna temperature T_A due to the source at that frequency and in that direction:

$$T_A = P_\nu / k \tag{8}$$

$$= F_\nu A / 2k \tag{9}$$

where the factor of 2 arises because usually only one polarization and, hence, only half the power can be detected. In practice, there are additional factors depending upon whether the source is larger or smaller than the antenna beam and upon the shape of the antenna beam (10,11). The situation is more complex at low frequencies because of the ionosphere and at millimeter and submillimeter wavelengths because of atmospheric extinction (12).

There are alternative ways to include the effective receiving pattern of the antenna. These lead to different forms of the equations above. However, the principles and quantitative results remain the same.

Sensitivity

Astronomical signals at radio wavelengths are generally weak when observed with conventional radio telescopes. Many are measured in units of milliKelvins or even microKelvins with telescopes with large collecting areas. There are exceptions. At long wavelengths, cosmic emission can be strong. Even at centimeter wavelengths, some "point" sources have large antenna temperatures when seen with sufficient angular resolution.

To enhance sensitivity, astronomers use antennas with large collecting areas and extremely low noise receivers. At centimeter wavelengths, parabolic antennas often have collecting areas measured in thousands of square meters. At meter wavelengths, phased-array telescopes can have even greater collecting areas. Receiver elements are often cooled to temperatures of 4 K or lower. Today, effective receiver noise temperatures at centimeter wavelengths are characterized by the rule-of-thumb 1 K/GHz or better. Receiver noise temperatures at a frequency of 10 GHz may be approximately 10 K and at 30 GHz, 30 K. Typical receiver noise temperatures at 100 GHz are 50 K, including all of the optics; at 230 GHz, they are 70 K.

Averaging observations over long time periods can enhance the sensitivities. Thermal receiver noise dominates most observations. The rms (root-mean-square) amplitude of this noise σT depends on the relationship

$$\sigma T = \text{const} \frac{T_{sys}}{\sqrt{B_w t}} \tag{10}$$

where B_w is the detected bandwidth and t is the length of the integration. The "system temperature" T_{sys} characterizes all sources of noise associated with the telescope, including the receiver, antenna, and the atmosphere. The constant is of the order 1 to 2, depending on the details of the observing technique. For differential observing in which signals from two parts of the sky—for example, a source and a reference position—are subtracted, the constant can be greater than or equal to 2.

Typically, radio astronomy observations range from minutes to hours. Exceptions would include detection of time-varying bursts from, say, the sun or pulsars involving very short integration times or from searches for extremely weak spectral lines involving integration times measured in days. Long integrations and careful calibrations produce precise, accurate measures of T_{sys}, making possible measurements of small increases (i.e., a few times ΔT) in antenna temperatures due to a cosmic source.

While increasing the bandwidth B_w in Eq. (10) will also increase sensitivity, in practice this is not always possible. For example, radio spectral lines from atoms and molecules are important sources of information for astronomers. Their full widths at half intensity rarely exceed 10^{-4} of their rest frequencies and are usually very much less. The shapes of the lines convey important information regarding the temperature environment and the velocity field of the emitters. Expanding the bandwidth wider than the lines would destroy much of the value of the observations. This problem can also exist for "continuum" sources, for which wide bandwidths can enhance detection but mask inflections in the continuum emission essential for understanding the emission mechanisms.

Transmission of the Atmosphere

The earth's atmosphere limits the cosmic radiation available to ground-based astronomers. Figure 1 shows the zenith transmission through the earth's atmosphere with total precipitable water vapor of 1 and 5 mm, values typical of arid and merely dry sites, respectively. Atmospheric molecules, principally O_2 and H_2O, form wide absorption bands with "windows" between them. The transparency of the windows becomes increasingly sensitive to water vapor toward high frequencies. Not shown is absorption below several (e.g., 30) megahertz by the ionosphere.

These windows limit ground-based astronomy. First, only cosmic radiation within the windows can reach ground-based telescopes. Second, the opacity of the warm atmosphere radiates noise that increases T_{sys}, making detection of weak cosmic signals more difficult. Third, observations of cosmic radio sources are usually made in directions other than the zenith, thereby increasing the path length of signals through the atmosphere and, correspondingly, the effective absorption and radiation of the atmosphere. Astronomers working at high frequencies situate their telescopes at high altitudes in arid climates to facilitate observations.

Interference

An enormous problem for astronomy at radio wavelengths is interference from man-made signals. The great sensitivities of the receivers render them particu-

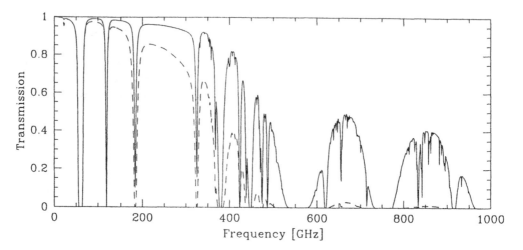

FIG. 1   The zenith transmission of the earth's atmosphere as a function of frequency. The solid
line corresponds to precipitable water vapor of 1 mm, typical of a high-altitude dry site
like Mauna Kea, Hawaii; the broken line, for precipitable water vapor of 5 mm, is
typical of a moderately dry site.

larly vulnerable, although the angular selectivity of the telescopes gives some
protection. The explosion of artificial radio signals from cellular telephones,
pagers, satellite communication links, navigational radars, arcing electrical
fences, and television and radio transmitters is creating an increasingly hostile
environment for astronomy. These artificial signals consist of nonrandom emissions that add to T_{sys} but do not decrease with integration, drastically decreasing
the sensitivity of radio telescopes. An analogy is the effect of city lights on
optical telescopes.

Linked to the physical laws of nature, cosmic radio signals have immutable
frequencies with respect to an astronomer. Spectral lines radiate at specific
frequencies determined by the physical characteristics of the emitter. "Continuum" emission involves emission best observed at specific wavelengths. It may
be impossible to choose different frequencies to study these astronomical phenomena.

Angular Resolution

Angular resolution has always been a concern of astronomers, particularly at
radio wavelengths. Most radio telescopes are diffraction limited; the angular
resolution improves as the diameter of the telescope in wavelengths increases.
The precise beam width in radians obtained with a parabolic dish is a function
of the degree of illumination taper but is usually around $1.2\lambda/D$. To be effective, radio telescopes need to be large because radio wavelengths are so much
larger than optical wavelengths. To obtain an angular resolution comparable to

the eye at optical wavelengths, a radio telescope operating at a wavelength of 1 meter has to be a mile or more wide.

Today's radio telescopes are divided between filled-aperture instruments such as a parabolic dish and interferometric designs. With a filled aperture such as a parabolic dish, low side lobe performance may be more important to a radio astronomer than the highest antenna gain or lowest possible beam size. Stray radiation such as thermal radiation from the ground enters the receiving system through side lobes, thereby degrading the ratio of telescope gain to system temperature (G/T) that ultimately defines the system sensitivity. Side lobes may also prevent the study of weak celestial sources near strong emitters and may make the telescope more vulnerable to terrestrial interference. With a parabolic dish, tapering the illumination of the surface toward the edge of the dish reduces side lobes. Minimizing aperture blockage such as that caused by feed support legs or instrumentation at the prime focus of the telescope also reduces side lobes. One approach to reducing aperture blockage and obtaining the lowest possible side lobe level is the offset parabola, such as the 7-m telescope of AT&T Bell Laboratories in Holmdel, New Jersey (13), and the 100-m Green Bank Telescope (GBT) of the National Radio Astronomy Observatory now being constructed in Green Bank, West Virginia.

Interferometry provides a way to achieve much higher angular resolution. In 1946, astronomers used antennas situated on a high cliff near the sea to generate interference patterns. The fringe pattern resulted from the interference of the direct waves from the cosmic radio source with those reflected from the sea; this is the radio analog of Lloyd's mirror (14). The narrow interference lobes allowed more precise determination of source positions and angular sizes.

Astronomers generalized this principle by building antennas separated perhaps by several miles but bringing the signals together to a central site. A breakthrough in technique was the introduction of the phase-switching interferometer (15). Especially noteworthy was the development of aperture synthesis with connected-element interferometers at the Mullard Radio Astronomy Observatory of the University of Cambridge, England, which led to the Nobel Prize in Physics being awarded in 1974 to Sir Martin Ryle.

Today, connected-element arrays used for aperture synthesis radio telescopes are common. Perhaps the most spectacular is the Very Large Array near Magdalena, New Mexico, which consists of 27 parabolic antennas 25 meters wide located on railroad tracks laid out in the shape of a Y. The combination of centimeter wavelengths and the 35-km (kilometer) size of the Y produces radio images with angular resolutions of a few hundredths of an arc second, better than the best images produced from ground-based optical telescopes. The movable antennas allow astronomers to select the combination of angular resolution and field of view most appropriate to their investigations.

The extreme of aperture synthesis is very large baseline interferometry (VLBI). In this technique, wideband tape recorders and atomic clocks replace the cables connecting the elements of the usual radio interferometer. Observations are made by pointing each antenna—perhaps separated from its closest neighbor by thousands of kilometers—to the same cosmic radio source and recording the received signals on magnetic tape. Later, cross-correlation of the tapes against each other, using special-purpose hardware and computers, provides the visibility information needed to construct an astronomical image.

Imaging in Radio Astronomy

For imaging in radio astronomy, radio telescopes are capable of producing detailed images of objects. With a filled-aperture telescope like a parabolic dish, the image of a celestial object may be built up point by point, by physically moving the antenna each time to point its single beam at different parts of the source, following some regular grid. This technique requires spending sufficient time integrating on each pixel to satisfy the sensitivity requirements of a given observation. For spectral line observations, a complete spectral map might require 1000 frequency points at each pixel. This becomes a three-dimensional imaging problem: two celestial coordinates plus the frequency domain.

Using multiple receivers may speed up spectral mapping so that integration occurs simultaneously at several pixels of the image. For example, the submillimeter common-user bolometer array (SCUBA) system on the James Clerk Maxwell Telescope on Mauna Kea, Hawaii (16), contains 37 separate feeds at $\lambda 850$ μm and 91 at $\lambda 450$ μm, speeding up the mapping process by the same factors, respectively.

There are other methods of forming images in radio astronomy. For a filled-aperture telescope using a phased array as a feed, simultaneous multiple beams may be formed electronically by amplifying, splitting, and phasing the signals from the much smaller elements making up the complete array. However, it is not necessary to have a fully filled aperture. A sparsely populated array can be used, with pairs of elements connected as interferometers. The potential angular resolution is then set by the maximum separation, in wavelengths, of the outer elements of the array, rather than the size of individual elements.

Too few elements result in inadequate images. Early attempts at imaging with simple interferometers produced too few "visibility points" (see below) to create unambiguous radio images of the sky. Using various assumptions, astronomers produced crude images by fitting models to these data. Nevertheless, these images often were sufficient to place useful constraints on the distribution of emission from a given object, such as total angular extent, even though a unique image could not always be formed.

Interferometers and Aperture Synthesis

The complex visibility function is the fundamental data produced by a radio interferometer. When two antennas are connected to form a coherent interferometer, the output signals are electronically multiplied to give both in-phase and in-quadrature terms. The hardware performing the multiplication is called a correlator, and the correlated signal is usually known as the (complex) visibility function. McCready, Pawsey, and Payne-Scott in 1947 were the first to point out the intensity distribution across the source is the Fourier transform of correlated power received by an interferometer (14).

A given pair of antennas records one Fourier component of the angular source intensity distribution. If the relative position of one antenna is moved with respect to the other, the system responds to a different Fourier component. If enough relative antenna positions are used, it is possible to build an adequate description of the source in terms of Fourier components of its brightness distribution. A simple Fourier transform then yields the angular source distribution. This process is known as *aperture synthesis*. The potential angular resolution is set by the maximum separation of antennas, and it is not essential to measure the Fourier components simultaneously.

Using an interferometer with many antenna elements and using the rotation of the earth, astronomers can synthesize a large, single telescope. Seen from a distant cosmic source, a two-element interferometer with fixed spacing will appear to rotate and to change spacing as the earth turns. Several hours of observing will create a track of source visibilities in the two-dimensional, complex visibility U-V plane, often signified by $f(U,V)$. Changing the linear separation of antennas will produce a different visibility track. Combining multiple tracks will fill the U-V plane with a distribution of visibilities. A Fourier transform of these data will produce an image of the source $f(\theta,\phi)$ similar to what would have been observed by a large, filled-aperture radio telescope, that is, the two domains are related by (17)

$$f(U,V) \supset f(\theta,\phi), \tag{11}$$

where the operator indicates a two-dimensional Fourier transform and the symbols and θ and ϕ indicate orthogonal angular dimensions on the sky.

In 1962, the One Mile Telescope of the Mullard Radio Astronomy Observatory of the University of Cambridge, England, was constructed to use these principles. The first data from this telescope were obtained in 1963. The image quality was equivalent to what might have been obtained from a one-mile diameter paraboloid with a beam that was scanned point by point across the source.

The number of antenna elements determines the number of visibilities that can be measured instantaneously. An aperture synthesis telescope with N elements gives $N(N-1)/2$ possible pairs of antennas that can be correlated together at any instant to give the same number of visibilities. The Very Large Array (VLA) telescope near Magdalena, New Mexico, has 27 antennas, giving 351 instantaneous visibilities. Depending on the complexity of the astronomical source being studied, with a suitable configuration of the 27 antennas on the ground, an adequate, almost instantaneous image (known as a "snapshot") may be obtained. By continuing to take data as the earth rotates, each of the 351 pairs of antennas contributes a track of visibilities.

Details of the sampling of the U-V plane depend on the geometry of the antenna elements, the range of time over which the data are taken, and the celestial coordinates of the source. Depending on how irregularly sampled the U-V data are, a straightforward Fourier transform of the visibility data may

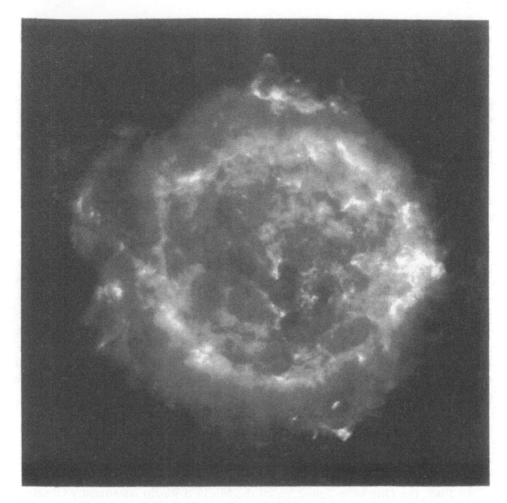

FIG. 2 Radio image of Cassiopeia A. The "A" designates the strongest radio source in the constellation Cassiopeia (the name of a mythical Ethiopian queen). The image was made with the very large array synthesis telescope at λ20 cm at a resolution of 1".3. The false color image shows nonthermal radio emission emitted by relativistic electrons trapped in magnetic fields, called synchrotron emission. This emission is the remains of a supernova, a star in our Milky Way galaxy that exploded in the year 1572. The filaments move at speeds exceeding 4000 km s^{-1} with respect to each other. Supernovas typically occur in spiral galaxies at a rate of a few per hundred years, enriching the interstellar gas from which other stars will form. (Courtesy of the National Radio Astronomy Observatory, Ref. 22.)

yield a poorly synthesized beam shape with high side lobes. The measurement defects may contribute artificial features to the computed image.

Powerful techniques have been developed to eliminate this artificial structure. Most involve a system for rationally guessing the information missing from the observations of the visibility function. One of the most widely used is CLEAN (18). This technique is equivalent to iteratively constructing a model image of the source that, when convolved with the true point spread function with its high side lobe level, is consistent with the measured visibility points. The model distribution is conventionally convolved with a clean beam such as a two-dimensional Gaussian distribution of appropriate half-power width. This best "CLEAN map" is then presumed to be the best estimate of the true source image within the limits of angular resolution of the observations. The more independent visibility points that are measured, the closer the CLEAN map result is to a unique image of the radio source.

Other image-processing algorithms are routinely used to enhance radio astronomical images. A particularly powerful one is the maximum entropy method (MEM) originally developed for seismic images in geosciences (19). It uses principles from information theory to minimize image distortions arising from flaws and omissions in observations of the visibility function.

The earth's atmosphere also limits the quality of an aperture synthesis image (20). Fluctuations in the electrical path length through the troposphere can introduce phase errors in the complex visibility function measured between antennas. These errors are analogous to optical "seeing," which limits the angular resolution attainable from the earth's surface. A powerful algorithm, SELF-CAL, developed in 1980 (21), often eliminates these atmospheric problems. The technique rests on the fact that calibration errors caused by tropospheric fluctuations are antenna based, while the number $N(N-1)/2$ of instantaneous visibilities measured on the source is proportional approximately to the square of the number of antennas. With sufficient antennas, with very weak assumptions about the source distribution or using the redundancy available when more than one pair of antennas has measured the same source visibility function, it is possible to solve for these antenna-based calibration errors and to construct an image almost as if there had been no perturbations from the atmosphere. This technique is now used routinely, often with some spectacular results (see Fig. 2) (22).

The Future

The future will undoubtedly bring substantive improvements in radio astronomy techniques. These might include placing antennas on earth satellites to extend interferometer baselines to enhance angular resolution greatly. Orbiting antennas would also eliminate absorption by the earth's atmosphere and allow observations at much higher frequencies. This would extend our knowledge of the cosmic information radiated in atomic and molecular spectral lines. In orbit the absence of local gravity would permit construction of truly giant parabolic antennas impossible to build on earth to gain sensitivity. Observatories on the far side of the moon would be shielded from the growing radio interference from earth.

Bibliography

Bracewell, R. M., *The Fourier Transform and Its Applications*, McGraw-Hill, New York, 1965.

Kraus, J. D., *Radio Astronomy*, Cygnus-Quasar Books, Powell, OH, 1982.

Sullivan, W. T., III, *Classics in Radio Astronomy*, Kluwer, Boston, 1982.

Verschuur, G. L., and Kellermann, K. I. (eds.), *Galactic and Extragalactic Radio Astronomy*, 2d ed., Springer-Verlag, New York, 1988.

References

1. Jansky, K. G., Directional Studies of Atmospherics at High Frequencies, *Proc. IRE*, 20:1920 (1932).
2. Reber, G., Early Radio Astronomy at Wheaton, Illinois, *Proc. IRE*, 46:15 (1958).
3. Buderi, R., *The Invention That Changed the World: How a Small Group of Radar Pioneers Won the Second World War and Launched a Technological Revolution*, Simon and Schuster, New York, 1996.
4. Hey, J. S., Solar Radiation in the 4–6 Metre Radio Wave-Length Band, *Nature*, 157 (1946).
5. Southworth, G. C., Microwave Radiation from the Sun, *J. Franklin Inst.*, 239:285 (1945).
6. van de Hulst, H. C., The Origin of Radio Waves from Space, *Ned. Tijd. v. Natuurkunde* (in Dutch), 11 (1945).
7. Ewen, H. I., and Purcell, E. M., Radiation from Galactic Hydrogen at 1420 mc/s, *Nature*, 168 (1951).
8. Muller, C. A., and Oort, J. H., The Interstellar Hydrogen Line at 1420 mc/s and an Estimate of Galactic Rotation, *Nature*, 168 (1951).
9. Chandrasekhar, S., *Radiation Transfer*, Dover, New York, 1996.
10. Kraus, J. D., *Radio Astronomy*, McGraw-Hill, New York, 1966.
11. Baars, J. W. M., The Measurement of Large Antennas with Cosmic Radio Sources, *IEEE Trans. Ant. Prop.*, AP-21:461–474 (1973).
12. Gordon, M. A., Baars, J. W. M., and Cocke, W. J., Observations of Radio Lines from Unresolved Source: Telescope Coupling, Doppler Effects, and Cosmological Corrections, *Astron. Astrophys.*, 264:337–344 (1992).
13. Chu, T. S., Wilson, R. W., England, R. W., Gray, D. A., and Legg, W. E., The Crawford Hill 7-Meter Millimeter Wave Antenna, *Bell Sys. Tech. J.*, 57:1257–1288 (1978).
14. McCready, L. L., Pawsey, J. L., and Payne-Scott, R., Solar Radiation at Radio Frequencies and Its Relation to Sunspots, *Proc. Roy. Soc.*, A-190:357–375 (1947).
15. Ryle, M., A New Radio Interferometer and Its Application to the Observation of Weak Radio Stars. *Proc. Roy. Soc.*, A-211:351–375 (1952).
16. Gear, W. K., and Cunningham, C. R., SCUBA: A Camera for the James Clerk Maxwell Telescope. In D. T. Emerson and J. M. Payne (eds.), *Multi-Feed Systems for Radio Telescopes*, Vol. 75, *Conf. Ser.* Ast. Soc. Pacific, San Francisco, CA, 1995, pp. 215–221.
17. Bracewell, R. M., *The Fourier Transform and Its Applications*, McGraw-Hill, New York, 1965.
18. Högbom, J. A., Aperture Synthesis with a Non-Regular Distribution of Interferometer Baselines, *Astron. Astrophys. Suppl. Ser.*, 15:417–426 (1974).
19. Ables, J. G., Maximum Entropy Spectral Analysis, *Astron. Astrophys. Suppl. Ser.*, 15:383–393 (1974).
20. Hinder, R., and Ryle, M., Atmospheric Limitations to Angular Resolution of

Aperture Synthesis Radio Telescopes, *Mon. Not. R. Astron. Soc.*, 154:229-253 (1971).

21. Schwab, F. R., Adaptive Calibration of Radio Interferometry Data, *Proc. Soc. Phot-opt Instrum. Eng.*, 231:18-25 (1980).

22. Condon, J. J., and Wells, D. (eds.), *Images from the Radio Universe*, National Radio Astronomy Observatory, Charlottesville, VA, 1992. (CD-ROM of FITS radio images)

MARK A. GORDON
DARREL T. EMERSON

Radio Systems—Digital   (see Digital Radio Systems)

Radio Wave Propagation
(see Electromagnetic Signal Transmission)

Radiocommunication Sector of the ITU
(see The International Telecommunication Union—Radiocommunication Sector)

Radio-Frequency and Microwave Radiation Biological Effects: An Overview

Introduction

This article briefly introduces the physical description of the applicable portion of the electromagnetic (EM) spectrum that includes the nonionizing radio-frequency (RF) and microwave energy (also termed "RF and microwave radiation"). Such a description involves the wavelengths and frequencies, and the associated quantum energies. The electric (E) and magnetic (H or B) field components are described, as are some of the representative sources and uses of the energy. These uses include communications applications, especially the cellular and personal communications services (PCS) "wireless" applications of RF energy. The concepts of exposure, incident energy, dose, dose rate, and specific absorption rate (SAR) are briefly discussed (in general, nontechnical terms). Comments are made relating to the potential exposure of various populations associated with the use of this energy.

In connection with concern for the potential for risk to the health of personnel exposed to RF energy, an overview is presented of the biological effects and clinical responses (as contained in the published scientific, engineering, and biomedical literature) to the EM energy at these frequencies. Attempts are made to note the magnitude (i.e., the intensity and/or field strengths) of the energy necessary to produce the observed biological effect ("bioeffect") or "biological change." The issue of the reversibility of the change produced is also noted, with emphasis on the alleged adverse effects thought to be of concern to humans. Examples of the problems, deficiencies, and criticisms of some of the published scientific and engineering studies are pointed out, as are the major implications from the applicable studies. Some of the recognized effects of EM fields on implanted medical (and related) devices, and on sensitive electrical equipment are also noted. Recommendations are made relating to needed research in these areas. The basis for the human exposure standards, recommendations, and guidelines that have been developed for this portion of the RF spectrum, is briefly discussed.

Radio-Frequency Electromagnetic Energy

General Characteristics

Significant increases in the development and use of equipment that produces nonionizing radiant energy have occurred in the past 25 years. Many questions have been raised as to whether adequate measures have been, and are being, taken to protect the user, the patient, and the public from possible adverse health effects that may be associated with exposure to such energy. Nonionizing

15

EM energy (in contrast to ionizing EM energy, or ionizing radiation) is of longer wavelength and therefore lower frequency, and is intrinsically less energetic (i.e., lower photon or quantum energy) in its interaction with biological tissue. Consequently, nonionizing radiation generally does *not* produce ions in biological materials as it interacts with the tissue.

The term *nonionizing radiation* refers to the group of electromagnetic radiations with energies less than about 10 electron volts (eV), corresponding to wavelengths in the near ultraviolet (UV), visible (VIS), infrared (IR), and RF/microwave portions of the spectrum (Fig. 1). This includes some of the energy often referred to as "optical" radiation, that which is termed "light," and most coherent laser/maser energy. In addition, ultrasound (which is due to pressure variations and mechanical vibration) is also included under the heading of non-

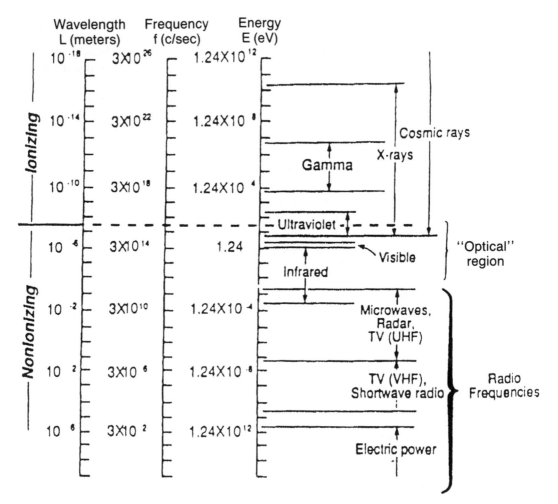

FIG. 1 Comparison of the wavelength, frequency, and energy of the electromagnetic spectrum. Nonionizing radiation is that portion of the electromagnetic spectrum with energies less than about 1–10 electron volts (UHF = ultra-high frequency; VHF = very high frequency). (From Ref. 1.)

ionizing radiation. The reader is reminded that ultrasonic energy is *non-electromagnetic* energy. The RF region of the EM spectrum includes millimeter waves, radio and TV broadcast, power transmission, and low-frequency electric and magnetic fields.

The portion of the EM spectrum referred to as the RF and/or microwave region is the subject of this article. The RF region usually applies to the spectral region having frequencies between 300 megahertz (MHz; million cycles/second) and DC (i.e., direct current; 0 cycles/second), but often is generalized to include the microwave region, and is described as extending to a frequency of 300 gigahertz (GHz; billion cycles/sec).

Devices and processes that use or generate RF nonionizing energy, and its various applications, are widely found in industry; medicine, dentistry, and veterinary science; telecommunications and the entertainment industry; many research laboratories; building and construction; navigation; various military applications; educational applications; geodesy; transportation; advertising; commercial food preparation; various leisure activities; and in the home. Non-ionizing energy in its various forms generally pervades our entire environment, exposing all (to various degrees) and, except for the narrow band of visible energy to which the retina of the eye responds, is not perceived by any of the human senses* unless its intensity becomes so great that it is felt as heat.

The depth of penetration and the sites of absorption of RF energy by the human body depend (to a great extent) on the wavelength of the energy, and consequently vary widely. Absorption of RF energy refers to the conversion of EM energy into other forms of energy, and usually results in the attenuation of the wave energy as it passes into an absorptive medium (i.e., a "lossy" dielectric). Many questions remain regarding the immediate and long-term consequences of acute and/or chronic exposure to various intensities and types of RF energy. These questions include considerations of potential occupational risks, public health hazards, risks to patients (and others), and certain environmental issues (including the exposure of large populations to the generally low exposures associated with power lines, and power transmission and distribution). Continued examination of exposure standards, protection strategies, and their application and enforcement is under way to ensure the safest possible use of RF energy.

It is also important to recognize that certain nonionizing radiation-producing devices utilize or generate ionizing radiation as a by-product, or as a requirement for their operation. Considerations for the protection of personnel from this energy are also necessary.

Historical Perspectives and Important Contributors

Faraday, Oersted, Ampere, James Clerk Maxwell (1864), Heinrich Hertz (1886), and Guglielmo Marconi (1901) are names familiar to many students of physics.

---

*It is recognized that under certain circumstances, some individuals can "hear" (i.e., can perceive via auditory or other cues) pulsed RF energy of the correct pulse width, pulse shape, intensity, and modulation.

(The years noted here are those of significant discovery, invention, or event.) These scientists and engineers all made significant contributions to the experimental observations and theoretical descriptions of the (at that time) newly discovered high-frequency electrical "energy." They performed early studies and experimental work with radio-frequency energy, wave propagation, and high-frequency energy detection. The frequency unit (i.e., the term) for cycles per second, the *hertz* (Hz), has been established in honor of the many contributions of that scientist. The implications of these findings were clear to researchers such as Marconi and deForest, who engineered the development of wireless communication using the newly recognized RF energy.

During and following World War II (1940–1945), many significant developments were made in the areas of radio communications, navigation, radar, and other uses of the RF spectrum. During this period, investigation also began into the study of the biological effects resulting from exposure to RF energy. This evolved, in part, due to some accidental exposures (i.e., "overexposures") that occurred with some of the investigators and their technicians. The true nature of the very real occupational exposure problem was not fully appreciated until relatively recently, however.

Other technical developments followed, leading to the microwave oven, precision navigation and radar detection, long-range radio communication using tropospheric reflection and scatter, radio and television transmission ("broadcast"), induction furnaces and dielectric heaters, communications using earth-orbiting satellites, "plasma" generation, cellular phones, PCS, and other so-called wireless applications of RF energy. (These latter items have been termed elements in the "wireless revolution" taking place in communications.)

Basic Physical (and Related) Principles of Radio-Frequency Energy

Physical Description, Characteristics, and Biological Interactions

Figure 2 portrays schematically the relationship between the electric (E) field, and the magnetic (H or B) field components (vector quantities) of a propagating EM wave. Intensities of the E and H fields are represented by the amplitude (magnitude) of the x and y axes, respectively. *Field strengths* (the measure of the intensities of the E and H [or B] fields) are expressed in units of volts/m (for E), and amps/m (for H). *Power density*, the time-averaged energy flow, or the power incident on a cross-sectional area perpendicular to the direction of propagation of the wave front, are described in units of watts/square meter (W/m^2), or milliwatts/square centimeter (mW/cm^2). *Current density* refers to the current flow (in amps/m^2) through a given cross-sectional area (such as in biological tissue), and often is used for interspecies comparison of experimental bioeffects results.

Early in the development of RF communications, D'Arsonval, a physician, used low-frequency and RF electrical energy to investigate the electrophysiological properties of nerves and muscles. He detected distinct differences in the effects of low- and high-frequency RF EM fields on tissue preparations; one significant difference was that high-frequency fields induced tissue heating. Be-

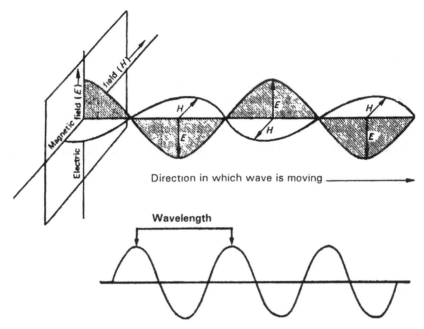

FIG. 2 An electromagnetic monochromatic wave. Electromagnetic waves consist of electrical and magnetic forces that move in consistent wavelike patterns at right angles to one another.

lieving that this heating was therapeutically beneficial, D'Arsonval (and others) applied RF fields to the treatment of various human ailments, including cancer. Diathermy (the general term used to describe tissue heating) has been the principal therapeutic application of RF and microwave energy.

Electrocautery, which uses RF energy to control bleeding following surgical procedures, has been widely used since the mid-1930s. Diagnostic medical applications, as well as other exciting therapeutic applications (including bone and wound repair, and pain control), of RF energy are now also being investigated.

A frequency continuum exists from static fields (at 0 Hz), through the regions referred to as extremely low frequency (ELF) and very low frequency (VLF) fields, medium frequency (MF), high frequency (HF), very high frequency (VHF), ultra-high frequency (UHF), super-high frequency (SHF), the microwave and millimeter wave regions, and so on.

The relationship between the wavelength (L) of the energy, and the frequency (f), within the EM spectrum is shown in Fig. 1. The relationships between the quantum energies (E) and f and L are also shown.

Some typical sources and applications of RF energy were mentioned earlier.

Exposure Considerations and Dosimetry. *Exposure* is the irradiation or contact of part or all of the target (i.e., the animal or human body) with E or H fields or with electric currents induced by those fields. Exposure can occur in an "open" system, such as from a horn or dipole (or other) antenna element transmitting RF energy, or within a "closed system," such as within a microwave oven cavity. *Incident energy* is the sum of the total energy impinging on the target (the animal or human). *Dosimetry* is a quantitative measure of the amount of incident energy actually deposited within (or absorbed by) the exposed body (i.e., the "dose"). A number of phenomena can occur that result in the incident energy being significantly different from the energy actually absorbed. These phenomena include (but are not limited to) reflection, refraction, diffraction, and transmission of the incident energy. The RF energy absorbed by the target is usually converted to heat, especially by water molecules (of which animals and the human body are largely composed). Absorbed RF energy may be reradiated at different wavelengths (e.g., within the infrared portion of the spectrum).

Specific Absorption Rate.   The SAR is the mass-normalized deposition of power. It is used as the criterion by which comparison is possible for biological effects studies and for determination of exposure criteria.

*Other Considerations Relating to the Biological Responses to
Radio-Frequency Exposure*

Many considerations and factors can have an influence on an animal's or individual's response to exposure to EM energy. Among these are the dose, temporal and spatial considerations (i.e., issues relating to the timing of the exposure, its delivery, and the portion[s] of the body directly [and/or indirectly] exposed), and the possibility of concomitant exposure to other physical agents, chemicals, and drugs (those that are prescribed, as well as the possibility of "street drugs"), biological agents, psychological stress, and other factors. Rarely in "real life" does exposure occur to an isolated, single stress or agent. This likely exposure to multiple stresses often complicates the interpretation of the observed biological effects following exposure to RF energy.

Other important considerations that can affect an individual's response to exposure to EM fields include age, gender, height, weight, body type, heredity, health and nutritional status, fitness/physical conditioning, presence of certain medications in the body, diet, smoking history, and more. In addition, many variables associated with the EM exposure are possible and require consideration. These include whether the RF energy is continuous wave (CW) or pulsed; if pulsed, the pulse shape, its pulse width, its peak power, and so on need to be considered.

Effect of Environmental Factors on Human and Animal Responses to Electromagnetic Exposure.   A number of factors (e.g., ambient temperature, humidity, task or work rate, fluid intake and electrolyte balance, ventilation considerations, type and amount of clothing present, whether protective equipment is being worn, etc.) also can significantly influence the response of a biological organism (such as an animal or human) to RF exposure. Since much (but not all) of the response to the exposure to an external source of RF energy has been demonstrated to be the well known physiologic response to a source of heat, factors that adversely affect the organism's ability to cope with the added thermal stress can or will influence the animal's physiologic, chemical, and behavioral responses.

Exposure Criteria and Type. *Acute exposure* often refers to a high dose and/or to the immediate response (if any) following such exposure. *Chronic exposure* often refers to a low or lower dose and/or to long-duration (i.e., low-level) exposure, and sometimes to the long-term consequences of such exposures. Exposures may be fractionated (i.e., delivered in small increments) as opposed to given all at once. It has frequently been shown that a fractionated exposure often allows the repair mechanisms and processes to occur between the exposure segments, and thereby enabling many organisms to receive more total energy compared to the exposure situation in which all the energy is delivered at one time. This is often the case with exposure to ionizing radiation.

Accidental exposures to high doses of RF EM energy, and the resulting adverse biological effects, began to be recognized as an occupational risk within the group of physicists and electrical engineers who were developing the early RF communications and radar systems. Such exposures were generally of short duration and at intensities sufficient to cause significant tissue heating.

The hypothesis emerged that the biological effects of RF energy were principally indirect or nonspecific thermal effects. It was assumed that by limiting the intensity (or power density) of incident RF fields to values that did not induce significant tissue heating (i.e., a temperature increase of 1°C or less), adverse biological effects would be averted. This reasoning, together with data from experimental studies (of the effects of exposure to RF energy) conducted on laboratory animals, resulted in the early recommendations for a maximum safe exposure intensity for humans of 10 milliwatts per square centimeter (mW/cm^2, independent of the RF frequency (2,3).

Later exposure recommendations for workers and the general public were based on much more extensive (and sophisticated) experimental research (often using animals "higher" in the evolutionary chain, isolated organs, *in vitro* preparations [i.e., in test tubes], and more sensitive indicators of exposure), on computer modeling and theoretical predictions, on "accident" and overexposure reports, and on the limited quantity of data involving human exposures to RF energy. It should be noted that broad extrapolation (often across decades of the frequency spectrum), and from small animals to humans, were necessary to arrive at the next generation of human exposure recommendations. This has been a constant (but probably appropriate) criticism of the "standards" for human exposure to RF energy. Additional comments are provided below relating to the limitations of the existing biological effects studies.

Interactions of Radio-Frequency Energy with Matter

Radio-frequency energy interacts with biological tissue at the atomic, molecular, cellular, and whole-body levels. The microscopic interactions are usually averaged on the macroscopic level (i.e., on the scale of the live, intact animal), and usually manifest themselves as responses to applied electric and/or magnetic fields. Mechanisms for the interaction include (at the molecular level) alignment of polar molecules, molecular rotation and bond vibration, and the transfer of kinetic energy to electrons and ions. As molecules vibrate and rotate, they meet with resistive forces and experience the interactions associated with neighboring molecules, which result in frictional heating within the material.

The oscillation of electrons and ions has important consequences for the interaction and functional activity of biological macromolecules and related materials. As noted above, the absorption of RF energy by a biological material is generally accompanied by reflection, refraction, diffraction, reradiation, and scattering of the incident energy. The actual situation is quite complex because the target animal or human body is nonhomogeneous; composed of layers of tissues having different dielectric characteristics. These differences produce multiple interfacial reflections and standing waves between the various tissue layers and at the air-tissue interface.

Factors important in the absorption of RF energy include the relationship between the size (i.e., dimensions) of the target material and the wavelength of the incident energy; the wave polarization relative to the geometrical length (i.e., the long axis) and girth (or axial ratios) of the target; the dielectric properties of the target material; and the presence of conductive and/or reflective surfaces in the local environment. In addition to these considerations, other factors have an important role in influencing the absorption of RF energy by a biological material. These include the issue of geometrical resonance(s) in objects of a certain size and shape, and oriented in specific ways within the EM field, and the possible presence of a ground plane on which the target is placed. The SAR can vary significantly depending on the orientation of an object in the exposure field, and the frequency at which maximum absorption of energy occurs can be significantly altered by the presence of a ground plane. As noted above, the possibility of the generation of induced and/or contact currents also enters into the complex exposure considerations.

Physical Parameters That Determine Energy Transfer

Physical parameters that determine energy transfer (especially into the human body) include conduction, coupling, and the various absorption mechanisms. All are dependent on the wavelength (L) of the energy, and on the body's distance (d) from the RF-emitting or RF-radiating source.

The distance relative to wavelengths L and the corresponding interaction mechanisms are

    0 L = Conduction (contact)
0–0.2 L = Coupling (direct transfer of a charge)
 >0.2 L = Absorption (conversion to internal heat at frequencies greater than
           about 1000 MHz)

Conduction. Conduction occurs when the body makes contact with an RF source (e.g., an antenna element or an exposed transmission line). The resulting detrimental effects include electrical shock, burn, and an involuntary reflex (i.e., nervous system activation of muscle) resulting in a "jerking" action, which can result in additional injury. Falls from elevated locations, such as metallic ladders or platforms, are possible as the individual making contact may lose grip and/or footing as a result of the shock/burn.

At frequencies above 100 kHz, most of the energy delivered through contact with an RF source will be absorbed within a few millimeters of the RF current's travel through the tissue. In these cases, the SARs involved may be significant if a small volume of tissue absorbs a large amount of energy.

The exposure limits (ELs), guidelines, and standards developed to prevent RF shock and burns are intended to limit induced RF current flow through the body for frequencies less than about 100 MHz.

Coupling.   An individual can be exposed to the stored energy fields (at frequencies lower than about 1000 MHz) that are present close to antenna elements or transmission lines. The body absorbs this energy through capacitive or inductive coupling, which is the direct transfer of a charge from one conductor to the body. Physical contact need not actually occur, and the SARs are difficult to predict and to measure. When a conducting object (such as a human) is placed in an RF radiative field, the object will absorb three- to fivefold more RF energy due to coupling if the field is at the absorbing object's resonant frequency.

Resonance occurs when an object's dimensions approximate one-half the wavelength of the incident energy. The human body standing in a vertically polarized field is resonant in the frequency band between 30 and 100 MHz; that is, a person 175 cm in height would be resonant at a frequency of 85 MHz (with 175 cm = 0.5 of the wavelength L at this frequency). This resonance exists because the body acts as an antenna. As body size decreases, the frequency for resonance increases (Fig. 3). A number of other factors also influence the resonance frequency to a varying extent.

The permissible exposure level (PEL) in the resonance frequency region for humans is reduced slightly from the PEL in the nonresonant region. The SAR still remains 0.4 watts per kilogram (W/kg), and the effects are still thermally induced. Only the PEL, which is derived from the SAR, changes.

Absorption.   Absorption is the principle mechanism for energy transfer at frequencies greater than about 1000 MHz. At lower frequencies, RF energy transfer occurs through a combination of radiation conduction, coupling, and absorption.

The PEL at frequencies greater than about 1000 MHz is still based on a whole-body SAR of 0.4 W/kg. This SAR corresponds to the absorption of 100% of the RF energy incident on the body, and equates to a measured power density of about 10 mW/cm. For energy absorption at frequencies greater than about 1000 MHz, the biological effects are thought to be induced primarily thermally.

Biological Effects and/or Responses to Exposure to Radio-Frequency Energy

The absorption of energy is the primary mechanism by which RF EM fields affect living cells. Electric and magnetic fields are induced within a biological

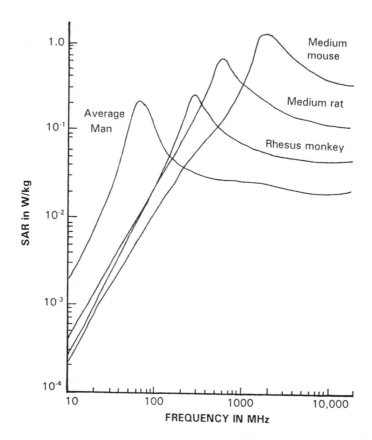

FIG. 3 Whole-body-averaged specific absorption rate of prolate spheroidal models of average human, rhesus monkey, medium rat, and medium mouse for E polarization at incident plane-wave power density of 1 mW/cm². (From Durney, C. H., Johnson, C. C., Barber, P. W., et al., *Radiofrequency Radiation Dosimetry Handbook*, 2d ed., Publication No. SAM-TR-78-22, U.S.A.F. School of Aerospace Medicine [RZP], Brooks Air Force Base, Texas, 1978.)

system exposed to microwave or RF energy. As noted earlier, such energy is transformed from these electric and magnetic fields into one or more types of energy modes (similar to translational, vibrational, rotational, and other modes) in the target material (i.e., the biological tissue). When translational modes are excited, the ambient cell temperature rises due to the heat generated by these modes. If the temperature rise is sufficient, proteins denature and their internal bonding becomes disrupted, which results in a change in their three-dimensional shape, resulting in their inability to participate in certain biochemical reactions.

Depositing RF energy into the body increases its overall thermal load. The body's thermoregulatory system responds to the increased thermal load by transfer of energy to the surrounding environment through convection, evaporation of body water, and reradiation (primarily of infrared [IR] energy). When the absorption of RF energy causes localized heating of certain organs, such as the eye or the testes, prolonged exposure to this thermal stress can result in serious damage (usually irreversible) to that organ.

To understand the resulting biological effects (upon irradiation with RF energy), it is necessary to determine the induced field strengths at various internal points within the system. Knowing the electrical and geometrical characteristics of the irradiated object and the external exposure conditions, it is possible, in principle, to calculate the rate at which energy is absorbed throughout the interior of the irradiated object.

The magnitude of interior and exterior scattered and reflected fields associated with an irradiated object depends on many factors: the frequency and configuration of the incident field, the electrical properties of the various layers (i.e., biological tissues) of which the irradiated system is composed, the shape, the size relative to the wavelength, and the relative orientation of the system. Biological systems are usually of complex exterior and interior geometry, and consist of several layers with various electrical properties (complex permitivity). As a result, the internal energy deposition in biological systems is usually of a nonuniform nature. Depending on the thermal properties and blood flow of tissues, there can be marked differences in the magnitude and the rate of increase in temperature, and thermal gradients within the exposed object can result.

The intensity of the internal electric field, or the amount of energy absorbed per unit time per unit mass (i.e., the SAR) are both used in RF and microwave dosimetry. The most generally used units for SAR are W/kg and mW/g.

Measurements of internal electric fields within various dielectric media are possible if a small, insulated dipole array is used. Such a device has been developed in miniature form and has been used to measure the internal electric fields in "phantoms" (i.e., models that exhibit the desired physical dimensions and the correct electrical [e.g., dielectric] properties) and in various animals (including carcasses, live anesthetized animals, and unrestrained animals).

It is important to note that many of the early (and even relatively recent) studies of the biological effects of exposure to RF EM energy have implied that the effects are primarily "thermal" in nature, and are an indirect consequence of RF energy absorption. Data from a number of more recent studies (including those demonstrating a number of cellular alterations) have provided strong evidence of possible direct RF field effects on biological systems. Such effects are "potentially significant in assessing bioeffects because they introduce the possibility of low-field intensity thresholds, frequency specificity, and dependence on the instantaneous induced electric or magnetic field strength per se" (4). Additional studies in these areas are under way, and the results must be examined very carefully.

Biological Effects of Radio-Frequency and Microwave Energy

A number of excellent descriptions and thorough discussions exist on the interaction of this nonionizing EM radiation with biological tissue. The resulting biological effects and, in some cases, the hazards, are also described in many well-written books and articles (including some that are scholarly and some written for the lay reader). It is beyond the scope of this article to include much discussion of this specific topic); however, some references to this literature are provided in the bibliography. One should note that, in most cases, the data available regarding human exposure are limited, and may (in part) result from accidental exposure and/or overexposure incidents, and may (to some degree) not involve careful and/or reproducible dosimetry.

Effects from Low-Level Radio-Frequency Exposure

Much controversy exists regarding the possibility (and the extent and/or the reversibility) of biological effects at low levels (i.e., low doses) of RF exposure. Low-level effects are often not easily observable. Research studies have included controlled exposures of cells, organs, and/or animals; computer simulations and modeling; evaluation of accident/overexposure incident reports involving humans; and sophisticated epidemiologic analysis (including retrospective and concurrent studies) of animal and human exposures. The point (or, more usually, the range of exposures) at which "effects" demonstrated in various organisms and in animals represent a "hazard" to humans has been widely debated.

Summary of the RF Bioeffects Literature

Biological effects and clinical responses to RF exposures discussed (as in the literature) can be classified into a number of categories:

- Acute (immediate response) and/or chronic (long-term) responses (as noted above, the term *acute* is also sometimes used to refer to a short-term, high-dose exposure, and *chronic* has been used to imply an exposure of long duration but low dose).
- Molecular and cellular studies (so-called *in vitro* studies because they often are conducted in test tubes, not actually on intact [live or otherwise] vertebrate animals). Includes studies of macromolecules, cell membranes, enzyme activity, and mitochondria; studies of carbohydrate, lipid, and protein metabolism; studies of clinical chemistry (involving blood), serum proteins, and electrolytes; studies of various microorganisms, such as bacteria, fungi, and viruses; studies of protozoa and other unicellular organisms; and studies on invertebrates.
- Vertebrate animal studies and studies conducted on isolated biological tissue and organs (so-called *in vivo* studies conducted on intact, live [but often restrained and/or anesthetized] animals); includes thermal- and pain-perception studies. Problems (such as the effect on the study results of the anesthetic agent) exist with some of these data.
- Experimental studies involving deliberate exposure to RF energy of humans are mostly absent; accident reports and case histories do exist, but the data

are generally not of reliable quality and generally do not have accurate dosimetric data.

- Human and animal studies on thermoregulatory responses, metabolism, and modeling. These include studies on the physiologic regulation of temperature by higher animals, hyperthermia, and cell kinetics; adaptation of animals to thermal stress; acute lethality to whole-body exposure of animals to RF energy; response to localized exposure to RF energy; comparison of RF exposure of animals versus exposure of the animals to infrared energy; and therapeutic application of RF energy (such as diathermy).

- "Perception" of RF energy by such mechanisms as the "RF hearing" phenomena.

- Studies of shock and burns (i.e., "contact" currents).

- Studies demonstrating that "induced currents" can be produced in the body (i.e., so-called "foot current").

- Behavioral effects and physiology; includes effects on the performance of various learned tasks following exposure of the animal to RF energy, behavioral "baseline" studies, and studies demonstrating altered sensitivity to certain drugs following exposure to RF energy.

- Eye effects/ophthalmic responses; demonstrates that cataracts can be produced on exposure to high doses of RF energy (the cataract occurs in the lens of the irradiated subject), effects produced on other structures of the eye (including the retina and the cornea), and biochemical changes produced in the eye. Frequent reports of the "dry eye" phenomena.

- Effect of exposure to RF energy on certain "biorhythms" (including sleep, ovulation/menstruation).

- Endocrine and neuroendocrine responses, including studies demonstrating changes in the concentration of growth hormone following exposure to RF energy, the hypothalamic-hypophysial-adrenal (HT-HP-adrenal) response to RF exposure; the HT-HP-thyroid response to exposure, and neuroendocrine activity and cardiovascular function.

- Immunologic responses to RF exposure have been demonstrated. The immune system is a physiological defense of the body against a large spectrum of pathogens, including bacteria, viruses, fungi, parasites, tumors, toxins from organisms, and miscellaneous chemical substances. Since there often is considerable adaptability and redundancy in the immune system, many of the demonstrated perturbations of the immune system may not have clinical significance.

- Neural effects and nervous system/neurologic responses, including electroencephalographic (EEG) changes, biochemical changes, histopathologic changes, changes in the influence of drugs on the body, central nervous system structural alterations, and blood-brain barrier studies.

- Hematologic responses, including suggested changes, following RF exposure, in the blood and blood-forming system, in hematopoiesis and the production of hemocytological shifts, and in the blood chemistry of exposed subjects.

- Cardiovascular and cerebrovascular responses to RF exposure have included changes in blood flow and pharmacodynamics, and reports of atherosclerosis (i.e., plaque formation).

- Epidemiologic data and related evidence includes nervous system and cardiovascular studies, ocular effects, fertility and sterility studies, growth and developmental changes, and studies of cancer production/causation. Some problems and criticism include the difficulty of determining (especially in retrospective studies) who actually was exposed, and determining dose and duration also often is very difficult.
- Reproduction, development, growth, and genetic effects include studies of embryonic development; fertility; spermatogenesis and egg production, fertilization, and implantation; reproductive efficiency; sex ratios (per litter) of the newborn; embryo and fetal toxicity; numbers of stillbirths; production of sterility; mutagenicity and teratogenicity (i.e., production of mutations and birth defects).

Other Effects Resulting from Exposure to Radio-Frequency Energy

Other effects resulting from exposure to RF energy include the effects of EM fields on implanted medical/electrical devices, and effects of EM fields on sensitive electrical equipment/devices (not implanted).

Radio-frequency energy can also pose a threat to the health of personnel through EM interference (EMI, also termed RF interference or RFI) with sensitive electronic devices, especially medical equipment (such as cardiac monitors and some other types of patient monitors) by disrupting their normal operation. The EMI can also affect the normal operation of certain implanted devices, such as electronic cardiac pacemakers, nerve stimulators, implanted drug delivery pumps, and the like. Interaction of EMI with some video display terminals (VDTs) also has been demonstrated. Fortunately, methods have been developed (including shielding, redesign, and other techniques) to make many of these devices and equipment less susceptible to EMI/RFI.

Other exposure concerns include combined stress caused by exposure to RF/microwave energy *plus* exposure to certain chemicals/drugs, biological compounds, and/or other physical agents has been demonstrated to often result in synergistic responses (i.e., heightened response to the combined stresses, compared to exposure to each stress alone). There also appears to be a psychological stress contribution, as if the individual (or animal) is aware that exposure to RF energy is taking place.

Standards/Guidelines for Human Exposure to Nonionizing Radiation

Basis for the Standards/Guidelines

Some of the standards and/or applicable guidelines relating to exposure to various modalities of nonionizing radiation of equipment users/operators (i.e., occupational exposure), patients, and/or members of the general public (i.e., environmental exposure) are very simple and straightforward. Other standards/ guidelines are much more complicated and vary with frequency/wavelength of the energy, the duration and/or repeatability of the exposures, and/or a number of other considerations. The reader is reminded that exposures of patients (therapeutic and/or diagnostic) are performed under the direction and control of licensed practitioners of the healing arts following consideration by the practitioner of the risks of exposure and the benefits to the patient. The reader of this article is referred to some of the applicable standards/guidelines (for human nonionizing radiation exposure) cited in the reference section.

The reader is alerted that some of the standards/guidelines are presently undergoing revision or updating due to the evaluation of new bioeffects data. Consequently, the reader is cautioned to ensure that the standard or guideline applicable to the particular exposure situation (i.e., occupational, environmental, or medical) is the most recent.

Other Considerations

Other considerations relating to protection from RF exposure include the measurement and quantitation of RF/microwave E and/or H fields to determine the risk for exposure of personnel. This is very important, but discussion is not included here.

Also, limitation of exposure and exposure control(s) and protection (including engineering and administrative controls and protective equipment) are very important, but a detailed treatment is beyond the scope of this article. However, the topic includes

- Health and safety (i.e., exposure) standards and guidelines (relating to the absorption of energy) for occupational groups or for the general public (the population at large) or relating to medical exposures or for environmental (or other) exposure(s).
- Performance criteria (i.e., emission standards) for specific types or items of RF-emitting equipment (e.g., microwave ovens).
- Shielding (often can be placed between the personnel and the source of RF energy to reduce exposure of individuals).
- Distance of personnel from the source of RF energy. Usually, the exposure decreases as the distance from the source of energy increases.
- Exposure time limitations. Generally, nonpreventable exposures to RF energy should be kept as short as possible.
- Beam access limitations. Access of personnel to intense beams of RF energy, or to other areas in which intense RF energy is present, often can be prevented using a variety of techniques, including locked enclosures, interlocked doors, key control of power supplies, floor and wall markings, alarms, recorded warning messages, and so on.
- Personal protective equipment (PPE) can be utilized to reduce (or eliminate) exposure to RF energy. Such PPE includes goggles, protective clothing, and the like.

Bibliography

Radio-Frequency Bioeffects Literature

Adey, W. R., Joint Actions of Environmental Nonionizing Electromagnetic Fields and Chemical Pollution in Cancer Production, *Environmental Health Perspectives*, 86: 297–305, (1990).

American Conference of Governmental Industrial Hygienists (ACGIH), Threshold Limit Values for Radiofrequency and Microwave Radiation. In *Threshold Limit Values (TLV) and Biological Exposure Indices (BEI) Book*, ACGIH, Cincinnati, OH, 1996.

Dodge, C. H., and Glaser, Z. R., Trends in Nonionizing Electromagnetic Radiation Bioeffects Research and Related Occupational Health Aspects, *J. Microwave Power*, IMPI Symposium issue, 12(4):319–334, 385 (1977).

Gandhi, O. P. (ed.), *Biological Effects and Medical Applications of Electromagnetic Energy*, Prentice-Hall, Englewood Cliffs, NJ, 1990.

Glaser, Z. R., Non-Ionizing Radiation, Electromagnetic (EM) and Non-EM (Including Radiofrequency/Microwave, UV, VIS, IR, Laser, Ultrasound Energy). In *The Health Physics and Radiological Health Handbook*, 1992 ed. (B. Shleien, ed.), Scinta, Silver Spring, MD, pp. 635–671, and the Non-Ionizing Radiation Glossaries, pp. 693–696, 701–716.

Glaser, Z. R., Observations and RF Field Intensity Measurements at a Commercial FM/TV Transmitter Tower, paper presented at the Annual Convention of the National Association of Broadcasters (NAB), Dallas, TX, March 26–30, 1979.

Glaser, Z. R., Summary of Radiofrequency/Microwave Bioeffects. *Proc. 17th Annual Natl. Conf. Radiation Control*, Conference of Radiation Control Program Directors, Frankfort, KY, Publication No. 85-3, 1985, pp. 283–297.

Glaser, Z. R., Brown, P. F., Allamong, J. M., and Newton, R. C., *Ninth Supplement to the Bibliography of Reported Biological Phenomena ("Effects") and Clinical Manifestations Attributed to Microwave and Radiofrequency Radiation*, National Institute for Occupational Safety and Health, DHEW (NIOSH) Publication No. 78-126, November 1977.

Glaser, Z. R., Cleveland, R. F., Jr., and Kielman, J. K., *Criteria for a Recommended Standard . . . Occupational Exposure to Radiofrequency and Microwave Radiation*, NIOSH Criteria Document, Final Director's Draft, 1979.

Glaser, Z. R., Cleveland, R. F., and Kielman, J. K., Mechanisms of Interaction and Bioeffects of Radiofrequency (RF) and Microwave Radiation. In *Health Implications of New Energy Technologies* (W. N. Rom and V. E. Archer, eds.), Proceedings of the Environmental Health Conference, April 1979, Park City, Utah, Society for Occupational and Environmental Health (SOEH), Washington, DC, 1980, pp. 705–746.

Glaser, Z. R., and Dodge, C. H., Biomedical Aspects of Radio-Frequency and Microwave Radiation: A Review of Selected Soviet, East European, and Western References. In *Biologic Effects of Electromagnetic Waves: Selected Papers of the USNC/URSI Annual Meeting (Boulder, Colorado, Oct. 20–23, 1975)* (C. C. Johnson and M. L. Shore, eds.), HEW Publications (FDA) 77-8010 and 77-8011, December 1976, pp. 2–34.

Glaser, Z. R., and Dodge, C. H., Comments on Occupational Safety and Health Practices in the USSR and Some East European Countries: A Possible Dilemma in Risk Assessment of RF and Microwave Radiation Bioeffects. In *Risk/Benefit Analysis: The Microwave Case* (N. Steneck, ed.), San Francisco Press, San Francisco, 1982, pp. 53–67.

Glaser, Z. R., and Heimer, G. M., Determination and Elimination of Hazardous Micro-wave Fields Aboard Naval Ships, *IEEE Trans. Microwave Theory and Tech.*, MTT-19(2):232–238 (1971).

> Invited paper on the prediction, measurement, and control of the potentially hazardous environment unique to the Navy; special issue on bioeffects of micro-waves. A similar article (invited) by the same authors appeared in *Bioenviron-mental Safety*, 4(1):10–15 (1972).

Hitchcock, R. T., *Radio-Frequency and Microwave Radiation*, 2nd ed., Nonionizing Radiation Guide Series, American Industrial Hygiene Association (AIHA), Fairfax, VA, 1994.

Hitchcock, R. T., McMahan, S., and Miller, G. C., *Extremely Low Frequency (ELF) Electric and Magnetic Fields*, Nonionizing Radiation Guide Series, American Indus-trial Hygiene Association (AIHA), Fairfax, VA, 1995.

Hitchcock, R. T., and Patterson, R. M., *Radio-Frequency and ELF Electromagnetic Energies: A Handbook for Health Professionals*, Van Nostrand Reinhold, New York, 1995.

Manthei, R. C., and Glaser, Z. R., The Sleep Process of Rabbits Exposed to Low Intensity Non-Ionizing Electromagnetic Radiation. I: Development of Methodology (AD #A045-028). In *Biologic Effects of Electromagnetic Waves: Selected Papers of the USNC/URSI Annual Meeting (Boulder, Colorado, Oct. 20–23, 1975)* (C. C. Johnson and M. L. Shore, eds.), HEW Publications (FDA) 77-8010/8011, December 1976, pp. 341–351.

Michaelson, S. M., and Lin, J. C., *Biological Effects and Health Implications of Radio-frequency Radiation*, Plenum Press, New York, 1987.

Moore, J. L., and Glaser, Z. R., *Cumulated Index to the Bibliography of Reported Biological Phenomena ("Effects") and Clinical Manifestations Attributed to Micro-wave and Radiofrequency Radiation*, 1984. (Available from J. Moore & Associates, P.O. Box 5156, Riverside, CA 92517-5156.)

> Development of the index sponsored by Bureau of Radiological Health, Food and Drug Administration.

National Institute for Occupational Safety and Health (NIOSH), National Institute of Environmental Health Sciences (NIEHS), and the U.S. Department of Energy (DoE), *Questions and Answers: EMF in the Workplace*, Joint Publication of NIOSH, NIEHS, and U.S. DoE, September 1996.

Riley, R. M., Glaser, Z. R., and Caira, L., Power Line Concerns—Some Real Estate Perspectives, *Corridor Real Estate J.*, special issue on the environment, 4(39):A-15–A-16 (February 12, 1993).

Slesin, L. (ed.), *Microwave News: A Report on Non-Ionizing Radiation*.

> Published bimonthly, L. Slesin, Publisher, P.O. Box 1799, Grand Central Sta-tion, New York, NY 10163.

Stevens, R. G., Wilson, B. W., and Anderson, L. E. (eds.), *The Melatonin Hypothesis: Breast Cancer and Use of Electric Power*, Battelle Press, Columbus, OH, 1997.

West, D., Glaser, Z., Thomas, A., Alexander, V., Conover, D., Murray, W., Curtis, R., Mallinger, S., Robbins, A., and Bingham, E., Joint NIOSH/OSHA Current Intelligence Bulletin #33, December 1979. (Also published as Radiofrequency (RF) Sealers and Heaters: Potential Health Hazards and Their Prevention, in *American Industrial Hygiene Assoc. J.*, 41(3):A-22–A-38, March 1980.)

Nonionizing Radiation Protection (General)

Conference of Radiation Control Program Directors, Inc., and Bureau of Radiological Health, *Suggested State Regulations for Control of Radiation, Vol. 2, Nonionizing Radiation, Lasers*, U.S. DHHS Pub. FDA 83-8220, Conference of Radiation Control Program Directors, Inc., and Bureau of Radiological Health, Rockville, MD, 1982.

Duchene, A. S., Lakey, J. R. A., and Repacholi, M. H., *The IRPA Guidelines on Protection Against Nonionizing Radiation*, International Radiation Protection Association (IRPA), Kent, UK, 1991.

Kincaid, C., Nonionizing radiation. In *Radiation Safety Handbook*, Bureau of Radiological Health, DHEW Publication (FDA) 76-8005, Rockville, MD, 1975.

Largent, E. J., Olishifaki, J., and Anderson, L. E., Nonionizing Radiation. In *Fundamentals of Industrial Hygiene*, 3rd ed. (B. A. Plog, ed.), National Safety Council, Chicago, IL, 1988, pp. 227–257.

Suess, M. (ed.), *Nonionizing Radiation Protection*, World Health Organization (WHO) Regional Publications, European Series No. 10. Copenhagen, 1982.

World Health Organization (WHO), *Magnetic Fields*, Environmental Health Criteria (EHC) Document No. 69, WHO, Geneva, 1987.

World Health Organization (WHO), *Radiofrequency and Microwaves*, Environmental Health Criteria (EHC) Document No. 16, WHO, Geneva, 1981.

RF Bioeffects/Biohazard Literature and Risk Assessment

National Council on Radiation Protection and Measurements (NCRP), *Biological Effects and Exposure Criteria for Radiofrequency Electromagnetic Fields*, NCRP Report No. 96, NCRP, Bethesda, MD, 1986.

National Council on Radiation Protection and Measurements (NCRP), *Radiofrequency Electromagnetic Fields: Properties, Quantities and Units. Biophysical Interaction, and Measurements*, NCRP Report No. 67, NCRP, Bethesda, MD, March 1981.

Exposure/Emission Standards and Guidelines

American Conference of Governmental Industrial Hygienists, *Documentation of the TLVs, Physical Agents Section*, Publication No. 0205, ACGIH, 1986.

Czerski, P., Radiofrequency Radiation Exposure Limits in Eastern Europe, *J. Microwave Power*, 20(4):233–239 (1985).

Glaser, Z. R., Basis for the NIOSH Radiofrequency/Microwave Radiation Criteria Document. *Nonionizing Radiation: Proc. of a Topical Symposium*, ACGIH, Cincinnati, OH, 1980, pp. 103–116.

Institute of Electrical and Electronics Engineers and Department of the Navy (Co-Secretariat), *Safety Levels with Respect to Human Exposure to Radiofrequency Electromagnetic Fields*, ANSI Standard C95.1-1982, July 1982 (revisions, 1991).

International Radiation Protection Association, International Non-Ionizing Radiation Committee (INIRC), Occupational Exposure Limits for Radiofrequency Electromagnetic Fields, *Health Phys.*, 46(4):975–984 (1984).

Miller, G., Exposure Guidelines for Magnetic Fields, *Am. Ind. Hyg. Assoc. J.*, 48(12):957–968 (1987).

Occupational Safety and Health Administration (OSHA), U.S. Department of Labor,

Standards Relating to Occupational Exposure; 29 Code of Federal Regulations (CFR) Section 1910.97 for RF/Microwave Exposure.

U.S. Food and Drug Administration, *Regulations for the Administration and Enforcement of the Radiation Control for Health and Safety Act of 1968*, Bureau of Radiological Health Report, FDA, U.S. DHHS, Rockville, MD, 1980.

Measurement/Survey Techniques and Instrumentation

Conference of Radiation Control Program Directors, Inc., and the Center for Medical Devices and Radiological Health, FDA, with the assistance of the National Bureau of Standards, *Instrumentation for Nonionizing Radiation Measurement*, HHS Publication FDA 84-8222, Rockville, MD, January 1984.

Phillips, M. L., Industrial Hygiene Investigation of Static Magnetic Fields in Nuclear Magnetic Resonance Facilities, *Appl. Occup. and Environ. Hyg.*, 5(6):353–358 (1990).

Control Techniques and Protective Equipment

Laser Institute of America, *Guide for Selection of Laser Eye Protection*, 2d ed., LIA, Toledo, OH, 1984.

Ruggera, P. S. and Schaubert, D. H., *Concepts and Approaches for Minimizing Excessive Exposure to Electromagnetic Radiation from RF Sealers and Heaters*, Bureau of Radiological Health, FDA, HHS Publ. FDA 82-8192, 1982.

References

1. Glaser, Z. R., Organization and Management of a Nonionizing Radiation Safety Program. In *Handbook of Management of Radiation Protection Programs* (K. L. Miller, ed.), CRC Press, Boca Raton, FL, 1992, pp. 43–52.
2. Schwan, H. P., and Li, K., Hazards Due to Total Body Irradiation by Radar, *Proc. of IRE*, 41:1572–1581 (1956).
3. U.S.A. Standards Institute, *Safety Level of Electromagnetic Radiation with Respect to Personnel*, C95.1, New York, NY, 1966.
4. Cleary, S. F., Biological Effects of Radiofrequency Electromagnetic Fields. In: *Biological Effects and Medical Applications of Electromagnetic Energy* (O. P. Gandhi, ed.), Prentice-Hall, Englewood Cliffs, NJ, 1990, pp. 236–255.

ZORACH R. GLASER

Real-Time Communication

Introduction

When a computer is used to control a physical system, the time scale for the physical system is defined by its dynamics in that the time for the physical system continues to change even if the computer application does not execute due to preemptions, and the like. For such systems, the computer applications must conform to the temporal requirements of the physical system. The term *real-time application* is often used for such applications. Analogous to this, *real-time communication* refers to the communication, which must meet functional and temporal requirements for communications needed by these applications.

Note that the majority of communication networks, such as the Internet, has been designed to deliver what is known as "best-effort" performance. Best effort strives to achieve good average performance. These networks work well for applications for which long delays and high data loss under heavy load conditions are acceptable. These applications are designed so that delays and data loss at arbitrary times during their execution result in no functional harm. They simply try to make progress as fast as they can. In contrast, real-time applications put stringent requirements on data loss, as well as latencies.

Recent technological advances have led to the development of high-performance switches and high-bandwidth links, resulting in *high-speed networking*. However, "real fast is not real time" in that having a very fast network does not assure meeting the temporal requirements of the applications under all conditions. While a lightly loaded high-speed network may be able to meet all the temporal requirements of applications without making any special provisions, it cannot guarantee that it will continue to meet the temporal requirements when the load changes. Since the standard designs of high-speed networks still show a significant variability in performance, real-time guarantees cannot be given to applications. A network capable of supporting real-time communication has to be able to respond to the communication requests with specific temporal requirements.

Traditionally, real-time communication has supported hard/soft real-time applications running in a distributed environment and interacting/controlling a complex physical system such as a power plant, avionics, chemical process plant, automobile, and so on. The current development of high-speed networking enables new application areas for which the communication has to satisfy real-time constraints. Among these applications are *multimedia* applications such as video broadcasts and *process control* applications such as command-and-control systems, automated manufacturing, remote process control, and flight control systems.

When we consider the real-time applications, their communication requirements can be characterized according to their ability to tolerate data loss. Typical process control applications assume that all the data will be delivered in a timely manner with no losses. In contrast, multimedia applications, which often display the results to a human operator, can tolerate some loss of data. Therefore, the requirements these applications place on real-time communications are not only in terms of delays but also of acceptable losses. These combined are usually referred to as *quality of service* (QoS). The parameters of interest for QoS include delay, delay jitter, bandwidth, and loss rate.

Note that process control applications often require a point-to-point communication that is connection oriented. The multimedia applications, on the other hand, require support not only for point-to-point communication, but also for point-to-many or many-to-many communications, leading to multicast and broadcast requirements. It is inevitable that new routing support is required in order to provide QoS requirements while maintaining effective multiparty communications.

This article is dedicated to a discussion of current techniques for the support of real-time communications with special emphasis on multimedia applications.

Background

Consider a packet going from Node S to Node D via Nodes A, B, and C and Links 1, 2, 3, and 4. If the network is carrying no other traffic, the transfer of the packet uses the links and nodes in time order as shown in Fig. 1. The time $t_k - t_0$ is the latency of this transfer. As shown in this figure, the packet processing at each location begins as soon as it can, leading to $t_k - t_0$ as the minimum latency. Achieving this minimum latency requires that the network resources, nodes as well as links, be available and ready to process the packet at the time instants shown in the figure.

As a network typically will handle other traffic also, the links and/or nodes may not be in a position to start processing this packet immediately as they may be processing other packets. In such a case, the packet has to wait, causing

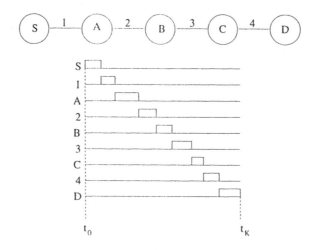

FIG. 1   Minimum latency of a packet transmission.

additional delays. The waiting requires that a buffer be available for the duration of the wait; if that is not the case, the network may drop the packet, resulting in packet loss. Therefore, the real-time communication schemes have to address the problems of resource management and resource reservations in order to assure the QoS.

Network Model

Let us consider an arbitrary topology packet-switched network model with links and switches (or nodes). Links have bounded propagation delay. They can be transmission media or subnetworks with delay bound guarantees, such as asynchronous transfer mode (ATM) and fiber distributed data interface (FDDI) networks. Switches have a number of input and output links. Each output link has a separate buffer of a finite size. A packet arriving to an input link is routed to its output link correctly. Packets at each output link are independently scheduled by a server according to a service discipline. Packets are of variable size.

A source application running on a source node can send a packet to a destination application running on a destination node. Each packet leaving its source node follows a route (or a path), which is a series of (*link*, *switch*) pairs, the last switch being the destination node. The switches between the source and destination nodes are called *intermediate* nodes. A packet transmission is modeled as the traversal of the packet on a series of servers at the output links of the switches along the path.

Source Traffic Models

Typically, there are three different types of traffic in the network: digital continuous media (voice and video) of multimedia applications, periodic sensor/actuator data of process control applications, and the traditional data sources of traditional data applications. These can be categorized based on their generation rate. Some traffic sources generate fixed-size packets at fixed time intervals. These are called *constant bit rate* (CBR) sources. Uncompressed voice and video and data generated at sensors or sent to actuators all fall in this category.

Some continuous media applications apply compression or adaptation techniques to reduce the size of the data to be transmitted. They generate packets at a variable bit rate and are commonly referred to as *variable-bit-rate* (VBR) sources. Specifically, to reduce the size of the voice traffic, no data are generated during the silent periods. This results in *on-off sources*, for which the source alternates between a period in which fixed-size packets are generated at fixed time intervals and an idle period. Size of the video traffic is reduced by compression applied to individual frames. This results in *compressed media sources*, in which variable-size packets are generated at regular intervals.

Traditional data sources include file transfer, interprocess communication in distributed computations, and interactive data (e.g., remote log in). They either involve one long message (as in file transfer) or a number of short messages at widely spaced intervals (as in interprocess communication and interactive data such as rlogin and Telnet). Traditional data sources use best-effort services and do not have stringent QoS requirements.

Informally, traffic sources can also be classified as either *smooth* or *bursty*, depending on the ratio of the peak rate to the average rate of their traffic. A smooth traffic source does not have too much variation in its traffic rate. All CBR sources and some of the VBR sources have smooth traffic. A bursty traffic source, on the other hand, has occasional long bursts of traffic that cause a big variation in its traffic rate. Many of the VBR sources have bursty traffic.

Service Model

The service model should be able to support both real-time and best-effort communication on a packet-switched network.

A real-time communication requires a guaranteed worst-case performance. Network resources required to satisfy the guarantees depend on the traffic volume of the communication. To bound the required resources, traffic is specified through a *traffic characterization* that bounds the volume of data to be sent over a specified period of time. Any application requiring a real-time service needs to supply to the network the traffic characterization and the performance requirements of the communication.

A real-time communication is connection oriented, with the explicit connection establishment phase used as a connection admission control. The network tries to find a route such that there are enough resources over the route to provide the required guarantees for this connection without destroying the guarantees already given to other real-time connections. If such a route is found, then the required resources are reserved and the connection is accepted. Otherwise, the connection is rejected.

Each connection is a contract between the applications that use it and the network. The network promises to give the required performance only if the source satisfies its traffic characterization. A rate-based flow control uses *traffic shaping* at the source and *traffic policing* at the network to guarantee that the source application is not misbehaving.

At each switch, incoming traffic from several connections may be multiplexed for transmission. Multiplexing causes network load fluctuations such that traffic of a connection might get burstier and no longer satisfy its characterization even though it satisfies the characterization at the source. The service discipline has to protect the service guarantees given to one connection from the overloaded traffic of another connection. This is provided by rate-based service disciplines in which each connection is guaranteed a minimum service rate (enough to satisfy its traffic characterization) regardless of the incoming traffic rate of other connections.

Integrating best-effort service into the service model is straightforward. Best-effort communication can be connection oriented or connectionless, and it can choose its own flow control and service discipline independently. Choices are generally a window-based flow control and a first-come, first-served (FCFS) or a round-robin service discipline. Then, each server at an output link of a switch may have two service disciplines at different priority levels. The higher priority service discipline is used to serve real-time traffic. The lower priority service discipline is used to serve best-effort traffic. Since best-effort traffic will be served only when there is no packet to serve from real-time traffic, best-effort traffic cannot hurt the performance guarantees given to real-time connections.

In summary, the components of such a service model include

1. *Connection Specification:* A data structure that is used as an interface between an application and the network. It describes the traffic characterization and the performance requirements (QoS parameters) of the connection.
2. *Connection-Level Control:*
   a. *Routing:* Finds unicast or multicast path(s) that can satisfy the performance requirements.
   b. *Admission Control and Resource Reservation:* Decides whether any of the paths suggested by routing has enough resources to provide the performance guarantees for the given traffic characterization. If it finds such a path, it accepts the connection and reserves the resources. Otherwise, the connection is rejected.
3. *Packet-Level Control:*
   a. *Flow Control:* Protects the guarantees given to connections from misbehaved sources. It forces each connection to stick with its traffic characterization by shaping it at the source and policing it at the network edge.
   b. *Service Discipline:* Schedules incoming packets for transmission. For real-time connections, rate-based service disciplines are used to protect the guarantees given to connections from network load fluctuations and best-effort traffic. They guarantee a minimum service rate to individual connections regardless of the traffic of other connections.

Connection Specification

Quality of Service Performance Parameters

Unlike best-effort communication service, for which the service tries to optimize the average performance, real-time communication service has to guarantee a bound on the worst-case performance of individual connections. Some of the performance parameters (also known as QoS parameters) of interest are bounds on delay, delay jitter, bandwidth, and loss rate.

The most important parameter in a real-time communication is the end-to-end delay bound assigned to individual connections. Successful delivery of a packet depends not only on the intact receipt of the packet, but also on the time it is received. If the time to deliver a packet exceeds its delay bound, then the packet is effectively lost.

Digital continuous media applications such as voice and video need a constant delay between the transmission of a packet at the source and its playback at the receiver. Delay variation in the network delivery necessitates the use of buffers at the receiver. Delay jitter is typically defined as the difference between the maximum and the minimum end-to-end delay. Then, for a lossless playback, the buffer size at the receiver should be the size of the maximum traffic that can be received during a period that is the length of the delay jitter. This is bounded by peak-rate X delay jitter. Therefore, for these applications, providing a tight delay-jitter bound reduces the buffering required at the receiver.

Internal buffering is needed along the path of a connection to minimize packet loss. At a server, the buffer space required for a connection is the maximum amount of traffic that can be received from the connection during a period that equals the longest residence time of a packet of the connection at that server. In all the proposed service disciplines, a tight bound on delay jitter implies a tight bound on the residence times at the intermediate servers. Therefore, controlling the delay jitter reduces the internal buffering required also.

For a real-time connection, the traffic load is bounded by a traffic characterization. This bound also defines the bound on bandwidth guarantee given to the connection.

Packets of a connection are lost in cases of delay-bound violation, delay-jitter bound violation, buffer overflow, and data corruption. Since data corruption is rare in fiber-optic links of high speed, it is generally ignored. The loss-rate bound of a connection is defined as a stochastic bound on the percentage of lost packets of the connection. A *deterministic service* has zero loss rate, and a *statistical service* has a nonzero loss rate.

A deterministic service does not lose any packet even in the worst case, which makes it appropriate for hard real-time, process control applications. A statistical service, on the other hand, can tolerate a certain amount of loss and therefore requires fewer network resources. It is appropriate for multimedia applications for which some loss can be tolerated.

Other performance parameters include bounds on blocking probability and delay distribution. Blocking probability is defined as the probability of a new connection being rejected by the admission control. Delay distribution is the stochastic distribution of delay values for changing loss rates. It can be used to adjust loss-rate-versus-delay guarantees of an already established connection.

Traffic Characterization

Traffic characterization is the term used to specify the data-generation characteristics of the source by identifying the amount of data generated for sending at different time instants. For real-time communications, typically such characterization is specified in terms of bounds on the volumes. Bounding the volume of the traffic limits the amount of network resources necessary to provide the required QoS guarantees. During connection establishment, resources are reserved based on the traffic characterization and QoS requirements. If not protected, a source can misbehave by violating the traffic characterization and sending more traffic than it is allowed. This may result in resource contention and hurts the QoS guarantees of the misbehaving connection. Furthermore, If the service discipline used is not discriminating, then the resource contention may also hurt the guarantees given to other real-time connections. Therefore, a *flow control* mechanism is used that shapes the traffic at the source and polices it at the network to guarantee that it satisfies its characterization.

There are many specification models proposed in the literature for traffic characterization. Among them are (*Xmin,Smax*), (*r,T*) (1,2), (*Xmin,Xave,I, Smax*) (3,4), (σ,ρ) (5), the deterministic bounding interval dependent (D-BIND) (6), and stochastic bounding interval dependent (S-BIND) (7). The (*Xmin, Smax*) and (*r,T*) models are appropriate for smooth traffic sources. They bound the rate of the traffic generated with a peak traffic rate. A connection satisfies the (*Xmin,Smax*) model if the interarrival times between consecutive packets of the connection are always longer than *Xmin* and the largest packet size is bounded by *Smax*. The peak rate of the traffic generated by such a connection is *Smax/Xmin*. In an (*r,T*) model, the time axis is divided into intervals of length *T*, each called a frame. A connection satisfies (*r,T*) if it generates no more than $r \cdot T$ bits of traffic in any of the frames. In this model, *r* is the upper bound on the average rate over the averaging interval *T*. These models are exact in characterizing CBR traffic sources, and they provide a tight bound for smooth traffic. For bursty traffic, however, they overstate the traffic requirements, resulting in low resource utilization.

The (*Xmin,Xave,I,Smax*) and (σ,ρ) models are more appropriate for describing bursty traffic sources. They bound both the average rate and the peak rate of the traffic. A connection satisfies (*Xmin,Xave,I,Smax*) if it satisfies the (*Xmin,Smax*) model and the average interarrival time of consecutive packets is larger than *Xave* during any interval of length *I*. The peak rate and the average rate of traffic are *Smax/Xmin* and *Smax/Xave*, respectively. Alternatively, a connection satisfies (σ,ρ) if, during any interval of length *t*, the number of bits generated by the connection in that interval is less than $\sigma + \rho t$. In the (σ,ρ) model, σ and ρ are the maximum burst size and the long-term average rate of the source traffic, respectively.

A more accurate approximation of bursty traffic sources can be obtained by characterizing the traffic with multiple bounding average rates, each over a different averaging interval. This captures the intuitive property that, as the averaging intervals get longer, a source is bounded by a rate lower than its peak rate and closer to its long-term average rate. The D-BIND model is based on this idea. A connection satisfies a D-BIND model $\{(r_1,T_1),(r_2,T_2),(r_3,T_3), \ldots \}$, where $T_1 < T_2 < T_3 < \ldots$, if over any interval of length T_i, $i = 1, 2, \ldots$, the number of bits generated by the connection is less than $r_i \cdot T_i$.

All traffic models described so far are *deterministic*, that is, there is a deterministic bounding function $b(.)$ such that over any interval of length *t* the number of bits generated during that interval is less than $b(t)$. Alternatively, a traffic model can be *stochastic*, with random variables used to characterize the traffic. In Ref. 7, the S-BIND model is proposed in which a connection satisfies an S-BIND model $\{(\mathbf{R}_1,T_1),(\mathbf{R}_2,T_2),(\mathbf{R}_3,T_3), \ldots \}$, where $T_1 < T_2 < T_3 < \ldots$ and $\mathbf{R}_i$ are random variables if $\mathbf{R}_i$ is stochastically larger than the number of bits generated by the connection over any interval T_i.

Routing

Today, many network applications involve multiple users that have specific QoS
requirements and higher resource requirements. Thus, simple multicast routing
algorithms that often merge multiple unicast paths can no longer support the
needs of current applications. In order to support applications involving multiple users with QoS requirements and higher resource requirements, routing
algorithms must efficiently manage limited network resources to maximize the
rate of successful connection establishments, maximize the probability of successful future connection establishments, and route in real time to a group of
users using multicast techniques when necessary. Straightforward extensions to
the routing algorithms are adequate for point-to-point connections, but new
approaches to routing are required for multicast traffic.

Classification and Design Issues

Multicast routing algorithms fall into one of two categories: shortest path algorithm and minimum Steiner algorithm. Algorithms based on the shortest path
attempt to minimize the cost of each path from the source to a multicast group
member, whereas minimum Steiner algorithms attempt to minimize the total
cost of the multicast tree. However, the minimum Steiner tree is a known
NP-complete problem. Thus, practical solutions are heuristic minimum Steiner
algorithms.

Besides, an algorithm is *dynamic* if it finds the path from the source to one
destination at a time and permits destination nodes to join and leave a multicast
group at any moment. An algorithm is *static* if multicast groups are fixed and
paths from the source to all destinations are computed at the same time.

Multiple-source algorithms allow multiple sources to transmit to the same
multicast group. However, we consider only single-source algorithms since a
single-source algorithm can easily implement a multiple-source algorithm by
having a shared tree rooted at a central node accepting packets from all sources
and then forwarding them to all destinations.

Distributed algorithms need to keep global information about the state of
the network at each node, while centralized algorithms need to know information only about nearby neighbors. Constrained algorithms bound the delay on
the path from the source to each destination by some given delay tolerance Δ.

Reverse-Path Multicasting Algorithm

The reverse-path multicasting algorithm is a technique of transmitting packets
from a single source to a group of recipients (8). It ensures that only the members of the group receive exactly one copy of a packet. This technique, which
evolved from the flooding method, supports the current need of sending packets
to many recipients. However, it does not guarantee the QoS requirements of
applications.

Flooding, the simplest technique for sending packets from a single source to many recipients, delivers multiple copies of packets to all other nodes in the network. Such duplication wastes network resources and, in order to reduce the amount of duplicated packets generated through flooding, reverse-path flooding forwards packets if they arrived via the shortest path from the source to the current node. However, it still faces the problem of delivering multiple copies and not being selective about recipients.

In order to remedy the problem with multiple copies, reverse-path broadcasting modifies reverse-path flooding so that it ensures that exactly one copy of a packet is delivered. Instead of forwarding packets to all outgoing links except the link that the packet just came through, reverse-path broadcasting forwards only to child links. A *child* link is a link that leads to members of the group for which the packet is intended. While it eliminates duplicates, it still has the problem of sending packets to nongroup members.

Truncated reverse-path broadcasting modifies reverse-path broadcasting to ensure that only members of the group will receive packets. It identifies members of the group among all leaves of *child* links. This technique delivers exactly one copy of packets only to members of the group. For this technique to be effective in a large network, however, it needs to reduce the amount of information involved in decision making. Deering et al. suggested the final variation to the simple flooding technique (8).

The very first packet to a multicast group travels according to the truncated reverse-path broadcasting. However, a *non membership report* for the (source, group) pair is sent back if the first packet reaches a router for which all of the child links are leaves and none of them have members of the destination group. Subsequent multicast packets from the same source to the same group do not travel to branches that have no members of the group.

Modified Semi-Constrained Heuristic

Waters presented a heuristic for ATM multicasting routing in (9). This heuristic constructs a broadcast tree that does not violate the maximum end-to-end delay used as the delay constraint is internally generated depending on the network delays. Note that this is not directly related to the application QoS constraints. This delay constraint may be too strict or too lenient as compared to the QoS requirements. This heuristic resembles a semi-constrained minimum spanning tree.

Salama et al. modified the above algorithm in (10). For each node v that is a member of the multicast group, but not yet included in the multicast distribution tree, Waters' heuristic finds a path with the cheapest bounded delay. The difference between two heuristics lies in selection of a path among multiple paths with the cheapest bounded delay. The algorithm presented below shows two approaches.

- Delay bound **dbv** = minimum total delay from s to v;
- Broadcast delay bound **dbB** = maximum dbv, for all $v \in V - \{s\}$;

1. Use an extension to Dijkstra's algorithm to determine dbv, for all $v \in V - \{s\}$ and dbB while constructing two matrices: delay $(j,k) :=$ minimum delay from s_i to node j where k was the last node visited.

$$cost(j,k): = \begin{cases} \text{cost of link } (k,j) & \text{original ATM heuristic} \\ \text{total cost of the minimum} \\ \text{delay path from } s \text{ to } j \text{ where} \\ k \text{ was the last node visited} & \text{modified ATM heuristic} \end{cases}$$

2. If delay $(j,k) >$ dbB, then set delay $(j,k) :=$ cost $(j,k) := \infty$. This results in a directed graph including all edges available to a bounded delay solution.

3. Initially: $T(s) := \{\}$ and $V_T := \{s\}$;
   Repeat{
   Choose a node v with maximum dbv and $v \in V$ and $v \notin V_T$.
   Find the "cheapest" path from s to v.
   For each link (u,v) on that path: $T(s) := T(s) \cup (u,v)$.
   For each node u on that path: $V_T := V_T \cup u$.
   } until $V_T = V$.

4. Prune the broadcast tree beyond the multicast nodes.

Kompella, Pasquale, and Polyzos Heuristics

Kompella, Pasquale, and Polyzos suggested two heuristics, CST_{CD} and CST_C to simultaneously transmit data to multiple destinations (11). These constrained Steiner tree heuristics find a minimum cost tree with a bounded delay for which the delay bound is specified by the application performing the multicast. Both source-based heuristics find such a delay bound minimum cost tree through following three stages. Given a constrained Steiner tree, $G = (V,E)$ (where V is a set of nodes and E is a set of edges) with

s = the source
S = the set of destinations
Δ = the delay tolerance as specified by the application
$C(e)$ = the cost of the edge e
$D(e)$ = the delay of the edge e

a closure graph G' on S is constructed by the constrained cheapest paths between s and v in S, which are the least cost paths from s to v with delay less than Δ. To compute the closure graph G', suggested heuristics determine the constrained cheapest path between all pairs of nodes in the set $S \cup \{s\}$. When Δ is a bounded integer, it is solvable in polynomial time.

Once constrained cheapest paths of all pairs are computed, the heuristics then construct a constrained spanning tree of G' by adding edges to a subtree of the constrained spanning tree until all the destination nodes are covered. CST_{CD} and CST_C differ only by the selection functions they use to determine whether to include some edge adjacent to v, a node already in the tree constructed thus far. The two selection functions used are

$$f_{CD}(v,w) = \begin{cases} \dfrac{C(v,w)}{\Delta - (P(v) + D(v,w))} & \text{if } P(v) + D(v,w) < \Delta \text{ otherwise} \\ \infty \end{cases} \quad (1)$$

and

$$f_C = \begin{cases} C(v,w) & \text{if } P(v) + D(v,w) < \Delta \text{ otherwise} \\ \infty \end{cases} \quad (2)$$

Last, these heuristics expand the edges of the constrained spanning tree into the constrained cheapest paths they represent and remove any loops that may have been created by the expansion. The edge selection functions f_{CD} and f_C give rise to two source-based heuristics: CST_{CD} and CST_C, respectively. CST_{CD} tries to choose the lowest cost edges, but modulates by choosing edges with maximal residual delay as can be shown by its selection function f_{CD}, which uses both cost and delay. It reduces the cost of the tree through path sharing. This heuristic has a tendency to optimize the delay in that it may find paths with delays far lower than Δ at the expense of added cost to the tree. CST_C constructs the cheapest tree possible while ensuring that the delay bound is met. This heuristic tends to minimize the cost of the tree without unduly minimizing the delay. Simulations show better average performance of CST_C than that of CST_{CD}. However, for distributed algorithms for constructing constrained Steiner trees, CST_{CD} performs better than CST_C.

Constrained Adaptive Ordering Heuristic

The constrained adaptive ordering heuristic, developed by Widyono (12), finds paths between a source and a set of destinations while meeting the QoS requirements and minimizing the cost. The cost is computed by the availability of resources and the number of hops. In this heuristic, the cost of an existing path is considered zero. Thus, path sharing is encouraged to minimize the cost.

This heuristic applies the constrained Bellman–Ford algorithm to find independent minimum-cost paths between a source and a set of destinations. But, at each run of the constrained Bellman–Ford algorithm, it selects a path that maximizes the sharing of paths. However, this heuristic must ensure that the minimized cost impact for the destination to be added will not pose a negative impact on the low cost potential of the tree.

Bounded Shortest Multicast Algorithm

The bounded shortest multicast algorithm by Zhu, Parsa, and Garcia-Luna-Aceves is another constrained Steiner tree heuristic (13). It starts by computing a least-delay tree for a given source s and a multicast group G_i. Once the least-delay tree for the source s and the multicast group G_i, $T(s,G_i)$, is constructed, then it iteratively replaces superedges in $T(s,G_i)$ with cheaper paths. A superedge is a path in the tree between two branching nodes or two multicast group members or a branching node and a multicast group member. This algorithm always finds a constrained multicast tree, if one exists, because it is based on the least-delay tree computed by the least-delay algorithm.

The least-delay multicast routing algorithm is an optimal algorithm that is implemented by Dijkstra's shortest paths algorithm. Note that the cost and the delay of an edge is considered equal in Dijkstra's algorithm. It guarantees minimum end-to-end delay from the source to each multicast group member.

Comparison of Multicast Routing Algorithms

Salama et al. give a performance comparison of multicasting algorithms in Ref. 14. In addition to the reverse-path multicasting, two other unconstrained multicasting algorithms were studied in their paper. As expected, the unconstrained algorithms are not suitable for a large network due to their inadequate delay performance. They also found that all three constrained heuristics yield similar performance. Table 1 provides the taxonomy of routing algorithms based on classification and design issues presented above.

Admission Control and Resource Reservation

In order for a network to deliver a quantitatively specified quality of service to a particular connection, it is often necessary for the network to determine if it can support a new connection request without degrading the performance of existing connections and to set aside certain resources, such as a share of bandwidth or a number of buffers, for the new connection. This ability to determine whether to accept a new connection request, to control admission, and to reserve resources to support the new connection request is also a part of the architecture that supports real-time communication requirements. This ability is referred

TABLE 1   Taxonomy of Routing Algorithms

	Constrained/ Unconstrained	Shortest Path/ Minimum Steiner	Distributed/ Centralized	Static/ Dynamic
RPM	Unconstrained	Shortest Path	Distributed	Static
MSC	Semiconstrained	Shortest Path	Centralized	Static
KPP	Constrained	Minimum Steiner	Centralized	Static
CAO	Constrained	Minimum Steiner	Centralized	Static
BSM	Constrained	Minimum Steiner	Centralized	Static

RPM = reverse-path multicasting, MSC = modified semi-constrained, KPP = Kompella, Pasquale, and Polyzos, CAO = constrained adaptive ordering, BSM = bounded shortest multicast.

to as *resource reservation*. Let us consider three recently introduced resource reservation protocols, ST-II, RSVP, and ATM Q.93B, and the admission control used in them.

ST-II

A resource reservation in ST-II is rooted at the source and extends to all receivers via a multicast distribution tree (15). At each intermediate ST agent a connect message with the flow specification and a set of participants generated at the previous ST agent are processed to determine the set of next-hop subnets required to reach all downstream receivers, install multicast forward state, and reserve network-level resources along each next-hop subnet.

On receipt of a connection request, a receiver returns either an accept or a refuse message to the source along with possibly a reduced resource request in case of an accept. The stream source begins data transmission only after receiving replies from each receiver in the multicast distribution tree. When receiving an accept with reduced flow specification, the source either adapts to the lower QoS for the entire stream or rejects the group participation for the specific receiver by sending it a disconnect message.

ST-II supports a dynamic group membership by allowing receivers to be added or removed from the group after the initial stream setup. Unlike the initial stream setup in which the source initiates a connect message to a group of receivers, receivers that wish to join the group after initial stream setup need to contact the source to receive a connect message.

This protocol achieves reliability by sending control messages through reliable hop-by-hop transfers and robustness by sending hello messages to neighbors. In case of failure to receive a response to the hello messages, it reroutes paths to serve the connection continuously. In short, ST-II supports homogeneous reservations over a point-to-multipoint simplex distribution tree, allows a dynamic group membership, and provides reliability and robustness.

RSVP

Similar to ST-II, RSVP is a simplex, yet flexible and scalable protocol (16). It is receiver oriented to accommodate heterogeneous receivers in a multicast group, provides several reservation styles to allow applications to specify resource reservation aggregations for the same multicast groups, and uses a *soft state* to support dynamic membership changes and automatic adaptibility to routing changes.

The design goals of RSVP are

- to provide the ability for heterogeneous receivers to make reservations specifically tailored to their own needs
- to deal gracefully with changes in the multicast group membership
- to allow end users to specify their application needs so that the aggregate

resources reserved for a multicast group can more accurately reflect the resources actually needed by that group
- to enable a channel-changing feature
- to deal gracefully with changes in routes by automatically reestablishing the resource reservation along the new paths as long as adequate resources are available
- to control protocol overhead
- to be relatively independent of the other architectural components

In order to achieve these design goals, the following design principles are used:

- receiver-initiated reservation
- separating reservation from packet filtering
- providing different reservation styles
- maintaining a soft state in the network
- protocol overhead control
- modularity

RSVP establishes a *sink tree* from each receiver to all the sources to forward reservation messages since reservation messages are initiated by each receiver. Multicast routing allows each receiver to join the associated multicast group to receive path messages, initiated by a source application, with flow specifications. The path message serves to distribute the flow specification to receivers and to establish a path state in intermediate nodes. Contrary to ST-II, a source can transmit data without waiting for replies from receivers in RSVP; however, data service guarantees are not enforced. On receipt of each path message, the receiver determines its QoS requirements using the information in that path and any local knowledge and initiates its own reservation request message. Then, it propagates the reservation request along the established path and ends as soon as the reservation "splices" into an existing distribution tree with sufficient resources allocated to meet the requested QoS requirements.

Asynchronous Transfer Mode Q.93B

Q.93B is a setup protocol used in an ATM network to create virtual channels dynamically (17). These dynamic virtual circuits, which remain open until being shut down by either a user request or a network failure, support unidirectional, bidirectional, and unidirectional multicast channels.

A sender initiates a setup message that is acknowledged on receiving a call proceeding message from the network. The setup message contains a flow specification, the address of the receiver, and the ATM adaptation layer (AAL) to be used. The call proceeding message serves to acknowledge the setup and to notify the sender of the network-assigned VCI/VPI (virtual circuit identifier/ virtual path identifier) for the connection.

A receiver initiates a connect message if it decides to accept the call request on receiving the setup message delivered by the network. As for setup messages, connect messages are acknowledged by the network through connect ack messages.

Then, the network notifies the sender that the receiver accepted the requested connection by sending a connect message. Since the data transmission is possible only after the sender receives the connect message, the data transmission on the newly created VCI serves an implicit acknowledgment.

Once a one-way connection is established, add party messages add multiple receivers that receive the same type of service as requested in the original flow specification. However, an individual flow specification could be explicitly specified for two-way connections using setup messages.

Comparison of Admission Control and Resource Reservation Protocols

A simplistic reservation service extends the current Internet Protocol (IP) point-to-point service model with QoS support. Yet, multicast routing in both ST-II and RSVP, enabling technology for supporting multipoint communication, completes the goal of the Integrated Services Packet Network (ISPN) architecture: to provide efficient support for applications requiring QoS support and multipoint communications.

Multicast distribution incorporated into a datagram network improves network resource utilization; however, ST-II and RSVP make different assumptions about the level of multicast support provided by the network. Unlike ST-II, RSVP does not handle the multicast routing and data-forwarding functions and assumes that they are provided by the underlying network.

ST-II supports only a service model of homogeneous reservation over a point-to-multipoint simplex distribution tree, whereas RSVP gracefully handles a service model of heterogeneous reservation over a multipoint-to-multipoint dynamic distribution.

ATM Q.93B meets minimal requirements of resource reservation. This protocol allows applications to specify flow requirements and supports multicasting. Yet, possibly long setup times, inefficient bandwidth allocations, and the lack of support for homogeneous receivers make this protocol less attractive than ST-II and RSVP.

Effects of Reservation on Network Performance

As noted above, ST-II and RSVP take different approaches to reserving resources in order to support QoS requirements of applications. Mitzel et al. observed how reservation protocols change the network behavior and found how application behaviors affect the network performance (18). Resource reservation protocols introduce a new resource contention for the shared network resources.

This resource contention is due to *provisional resources*. Provisional resources are resources reserved but not currently in use. A long end-to-end setup propagation delay, a long admission control delay, a long teardown delay, and circularity in the resource dependencies among independent reservations (i.e., deadlock) all contribute to increased provisional resources. Increased provisional resources lead to throughput degradation and a higher denied access rate. A system in which overall performance and throughput are suffering from these detrimental effects is considered to be *thrashing*. It has been observed that the primary source of provisional resources is long teardown delays.

Real-Time Service Disciplines

As discussed above, service disciplines are the queueing mechanisms at a switch used to schedule incoming packets for transmission. For supporting best-effort service, conventional service disciplines such as FCFS and round-robin queues are sufficient. However, per-connection performance guarantees need more elaborate disciplines. Recently, many new service disciplines have been proposed in the literature. Among them are virtual clock (19,20), weighted fair queueing (WFQ) (21–24), self-clocked fair queueing (SCFQ) (25), delay earliest due date (EDD) (3,4), jitter EDD (26), stop and go (1,27), hierarchical round robin (HRR) (2), and rate-controlled static priority (RCSP) (28–30). In the remainder of the article, each of these disciplines is classified, described, and compared. Similar studies can be found in Refs. 31–36.

Service disciplines can be classified as either *work-conserving* or *non-work-conserving* disciplines. In a work-conserving discipline, a server is never idle when there is a packet to send. In a non-work-conserving discipline, each packet is assigned an eligibility time, a time after which the packet becomes eligible to be serviced. A server is allowed to send only the packets that are eligible, and it is never idle as long as there are eligible packets to send. Among the proposed service disciplines, virtual clock, WFQ, SCFQ, and delay EDD are work-conserving disciplines, whereas jitter EDD, stop and go, HRR, and RCSP are non-work-conserving disciplines.

To provide end-to-end QoS guarantees to individual connections, real-time service disciplines need to supply solutions for per-connection bandwidth and delay allocation in a single switch and traffic distortion handling in a network of switches. The first issue provides local performance guarantees to a connection at a switch under the assumption that input traffic from that connection is characterized. However, multiplexing of connections at switches causes distortion of the traffic pattern, which was shaped at the source. In the worst case, the traffic gets burstier after each multiplexing, and it no longer satisfies its source characterization. The role of the traffic distortion handling is then to *recharacterize* the input traffic either by reconstructing it to its original source characteristics or by deriving the new traffic characterization at the entrance to the switch assuming the worst-case multiplexing. The taxonomy of some of the service disciplines proposed in the literature is given in Fig. 2.

		DELAY/BANDWIDTH ALLOCATION		
		Multilevel Framing	Sorted Priority Queue	Multilevel FIFO Queues
TRAFFIC DISTORTION HANDLING	Characterize Distortion		Delay-EDD Virtual Clock WFQ SCFQ	
	Control Distortion	Stop-and-Go HRR	Jitter-EDD	RCSP

FIG. 2 Taxonomy of service disciplines (EDD = earliest due date; FIFO = first in, first out; HRR = hierarchical round robin; RCSP = rate-controlled static priority; SCFQ = self-clocked fair queueing; WFQ = weighted fair queueing).

Per-Connection Quality of Service Guarantees at a Single Switch

There are three approaches to bandwidth and delay allocation to different connections in a single switch: multilevel framing, sorted priority queue, and multilevel FCFS queue.

Multilevel Framing Strategy

The multilevel framing strategy is used in both stop-and-go and HRR service disciplines. Let us consider one-level framing first. In one-level framing, at each node, the local time axis is divided into fixed intervals T, called frames. Although convenient, the frames of different nodes need not be synchronized. Bandwidth is allocated to each connection as a certain fraction of the frame time, which bounds the average rate of the traffic of a connection, for which the averaging interval is T.

One disadvantage of this scheme is the trade off between the granularity of bandwidth allocation and the delay bound. The minimum granularity of a bandwidth allocation is given by *Minimum Packet Size*/T, which is inversely proportional to T. However, delay bounds for stop and go and HRR presented in the section "Comparison of Service Disciplines" and Table 3 are directly proportional to T. This means the finer the granularity is, the higher the delay bound will be; the lower the delay bound is, the coarser the granularity is.

To overcome the tradeoff problem, multiple frame sizes are introduced. Consider n frame sizes $T_1, T_2, \ldots, T_n$, where each T_{i+1} is a multiple of T_i for every $i = 1, \ldots, n - 1$. At each node, there is a reference point from which the time axis is divided into periods of length T_i, each called a Type i frame, for every $i = 1, 2, \ldots, n$. An example is given in Fig. 3. Frames of Type i are associated with a Service Level i, each offering a different granularity of bandwidth and service priority to its traffic. The shorter the frame size of a service

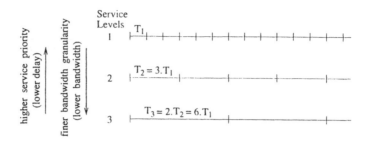

FIG. 3 Three service levels with frames $6T_1 = 2T_2 = T_3$.

level, the higher will be the service priority of its traffic and the coarser will be the granularity of bandwidth allocation. Each connection k is assigned to a service level i and given a bandwidth as a certain fraction of the frame time T_i. As can be seen from Fig. 3, a low-bandwidth, low-delay connection has to be assigned to a high-priority service level to satisfy its delay requirement. However, since the granularity of bandwidth allocation at that level is coarse, the bandwidth allocated for the connection will be higher than the amount needed. This decreases the bandwidth utilization of the network.

Another disadvantage of the framing strategies is their inefficiency in handling the connections with bursty traffic. Since bandwidth is allocated in terms of an average rate over an interval T_i, for bursty traffic the allocation must be done based on the peak rate of the traffic. The high peak-to-average-rate ratio of bursty traffic means low utilization of the transmission capacity unless the unused portion of the allocated bandwidth can be utilized by best-effort connections.

Framing strategies are easy to implement and fast enough to be used at high-speed transmission rates.

Sorted Priority Queue

The sorted priority queue is used to implement dynamic priority schedulers, such as the EDD used in delay EDD and jitter EDD, and others used in virtual clock, WFQ, and SCFQ. It does not have any of the shortcomings of framing strategies. There is no tradeoff between granularity of bandwidth allocation and delay bound, it can handle bursty traffic efficiently, and it can potentially allocate a continuous spectrum of delay bounds to different connections. However, it suffers from a high degree of complexity. Insertion into a sorted priority queue is an $O(\log N)$ problem, which can be unacceptably slow for high-speed networks.

Multilevel First-Come, First-Served Queue

The multilevel FCFS queue is used to implement static priority schedulers. Levels of the FCFS queues correspond to priority levels. Each connection is assigned a priority, and all packets from that connection are inserted into the FCFS queue of that priority level. Multiple connections can be assigned to the same priority level. Packets are scheduled in FCFS order from the highest priority nonempty queue for transmission. This scheme is used in the RCSP discipline.

This scheme is a good compromise between the multilevel framing strategy and the sorted priority queue. It has no tradeoff between bandwidth granularity and delay bound. It can handle the bursty traffic efficiently, and yet its implementation is simple and fast enough to be used in high-speed networks.

The only restriction in using multilevel FCFS queue for static priority schedulers is that the number of different delay bound values that can be provided to the connections are bounded by the number of priority levels in the system. This is because an FCFS queue can only provide one delay bound, and a multilevel FCFS queue can provide as many as its number of levels.

Traffic Distortion Handling

There are two ways to handle the traffic distortion caused by multiplexing at a switch: characterize the distortion or control the distortion.

Characterization of the Distortion

The idea in characterizing the distortion is to let the distortion happen and derive the worst-case traffic characterization at the entrance to a switch from the source traffic characterization and the worst-case distortions along the path from the source to the switch.

In terms of satisfying the performance guarantees, the worst-case distortion is the one making traffic the burstiest. For example, in the (σ,ρ) traffic model if a connection satisfies (σ,ρ) at the source, then its worst-case traffic characterization at the entrance to the ith switch becomes $(\sigma + \Delta\sigma,\rho)$, where $\Delta\sigma = \Sigma_{j=1}^{i-1} \rho \cdot d_j^{max}$ and d_j^{max} is the local delay bound for the connection at the jth switch. For many other traffic models, deriving the traffic characterization inside the network is difficult. But, independent of the model used, the traffic characterizations along the path of a connection have to allocate traffic that is getting burstier and burstier.

This technique is used by all work-conserving service disciplines that are proposed. This is because the alternative technique, controlling the distortion, involves extra holding time for packets, which can only be done in non-work-conserving disciplines.

Controlling the Distortion

Another way to handle the traffic pattern distortion is to control it at every switch by reconstructing the traffic of a connection to its source characterization. All non-work-conserving disciplines proposed in the literature use this technique. They separate the server into two components: a rate controller and a scheduler. The rate controller controls the distortion on each connection using regulators and allocates bandwidth to them. The schedulers, on the other hand, order transmission of packets on different connections and provide per-connection delay bounds. Therefore, this class of service disciplines, called *rate-controlled service disciplines* in Ref. 29, naturally decouples the bandwidth and delay bound allocations.

There are two classes of regulators: *delay-jitter regulators*, in which regulated traffic has exactly the same traffic pattern as the source traffic, and *rate-jitter regulators*, in which the pattern produced by the regulator satisfies the source traffic characterization. Rate-jitter regulators are less restrictive since they do not have to repeat the traffic pattern of the source. The only information they need is the traffic characterization to satisfy. This information is static and can be kept by each switch on the path locally. Delay-jitter regulators, however, need to know the distortions that happened at previous nodes to be able to reconstruct the pattern.

It is shown in Ref. 29 that end-to-end delay does not depend on the holding times of packets at regulators. This implies that local delay guarantees obtained from a node can be extended to end-to-end guarantees by the use of regulators. Also, since regulators prevent the accumulation of burstiness along successive nodes, the buffer requirements along a path are uniform.

Work-Conserving, Real-Time Service Disciplines

Delay Earliest Due Date

Delay EDD was proposed by Ferrari and Verma (3,4). It is a work-conserving discipline in which traffic of a connection is characterized by the ($Xmin$, $Xave,I,Smax$) specification. The main idea is to provide the end-to-end delay bound guarantees for a connection by distributing it to individual nodes as local delay bounds. At each node, a variation of the earliest due date (EDD) is used for scheduling. Each packet is assigned a deadline, which is the time at which it should be sent to satisfy the local delay bound if it was received according to the traffic characterization of its connection. Admission control during connection establishment guarantees that, if the packets are served in the order of increasing deadlines, then every connection satisfies its local delay bounds and hence its end-to-end delay bound. Delay EDD can provide guarantees for bandwidth and delay bounds. It can, further, provide statistical guarantees, for which the loss rate due to missed deadlines can be bounded probabilistically.

As described below, the local delay bound d_k^n for a connection k at a node n is guaranteed by the admission control under the assumption that the traffic of the connection at the entrance to the node satisfies the $Xmin^k$ characterization. However, the actual arrival times might not satisfy this due to misbehaved users and network load fluctuations. Therefore, to avoid missed deadlines in EDD scheduling, deadlines should be assigned to packets after guaranteeing the $Xmin^k$ constraint for their arrivals. This is done by assigning a logical arrival time to each packet such that the logical arrival time for the ith packet of a connection k, LAT_i^k is

$$LAT_0^k = AT_0^k \tag{3}$$

$$LAT_i^k = max(LAT_{i-1}^k + Xmin^k, AT_i^k) \tag{4}$$

where AT_i^k is the actual arrival time of the ith packet of connection k. Since the sequence of logical arrival times $\{LAT_i^k\}_{i=0,1,2,\ldots}$ satisfies the $Xmin^k$ constraint, the scheduler uses them (not the actual arrival times) as the arrival times of packets. When the ith packet of connection k is received, it is stamped with the deadline $LAT_i^k + d_k^n$. The EDD scheduler transmits packets in the order of increasing deadlines.

A packet may be delayed beyond its local delay bound in a node because of two reasons: a temporary saturation of the node's processing or transmission capacity (*node saturation*), and impossible scheduling constraints (*scheduler saturation*). Admission control for connection establishment admits a new connection only if it does not cause node or scheduler saturation in any of the nodes, under the assumption that every connection satisfies its $Xmin$ constraint at every node. If we let K_n denote the set of connections passing through a node n (including the one to be established), and t_k^n denote the maximum service time for a packet of connection k at node n, then the tests performed at each node are as follows:

- *Node saturation test*:

$$\sum_{k \in K_n} \frac{t_k^n}{Xmin^k} < 1 \tag{5}$$

- *Scheduler saturation test*: Finds the minimum local delay bound for a new connection that does not saturate the scheduler even in the worst-case traffic. The worst case happens when a packet from every connection k, $k \in K_n$, is received at a common point in time, called the *critical instant*, and these packets are followed by successive packets from each connection at their maximum rate. The general case is expensive to test, so in Ref. 4 the problem is simplified by assuming that $Xmin^k \geq \Sigma_{j \in K_n} t_j^n, \forall k \in K_n$. Then, the condition to be tested becomes

$$d_k^n \geq \sum_{i \in K_n : d_i^n < d_k^n} t_i^n + t_{max}^n, \quad \forall k \in K_n : d_k^n < \sum_{j \in K_n} t_j^n \tag{6}$$

where t_{max}^n is the maximum service time for a packet in node n.

A new connection is accepted by the admission control if the node saturation test succeeds at each node and the sum of local delay bounds is less than or equal to the end-to-end delay bound requirement.

The properties of delay EDD are as follows:

- For a connection k with N nodes the end-to-end delay bound is $\Sigma_{n=1}^{N} d_k^n$.
- Admission control reserves a bandwidth of $1/Xmin^k$ to each connection k at every node n along its path.
- At each node n of a connection k, the required buffer space is $\dfrac{\Sigma_{i=1}^{n} d_k^i - \Sigma_{i=1}^{n-1} t_k^i}{Xmin^k}$. Note that buffer space requirement increases linearly for each node along the path.

Virtual Clock

The virtual clock service discipline, proposed by Zhang (19,20), tries to emulate the time division multiplexing (TDM) service. It eliminates the interference among connections as in TDM while preserving the flexibility of statistical multiplexing of packet switching. To make a statistical data flow resemble a TDM system, a virtual clock concept is introduced. Packets arriving from a connection are spaced at a constant rate in a virtual timeline such that each packet arrival indicates the corresponding passage of time in the emulated TDM system. It uses a sorted priority queue such that each arriving packet is stamped with the virtual transmission time and packets are served in the order of their increasing stamp values.

Consider the virtual clock service received by a connection k with a constant allocated rate of ρ_k (bytes/second) at a node n. To imitate the transmission ordering of a TDM system, the ith packet of the connection k should be stamped with the following auxiliary virtual clock value, $auxVC_k^i$:

$$auxVC_k^0 = AT_k^0 \tag{7}$$

$$auxVC_k^i = max\,(AT_k^i, auxVC_k^{i-1}) + \frac{S_k^i}{\rho_k}, \quad i > 0 \tag{8}$$

where AT_k^i is the arrival of the ith packet of the connection k and S_k^i is the size of that packet. As shown in Eq. (7), the $auxVC$ value of the first packet of the connection is the arrival time of the packet. In Eq. (8), the term S_k^i/ρ_k is the size of the clock tick for packet i, which would be the time to serve the packet i if a TDM server were used. Emulating a TDM system implies that the $auxVC$ values of consecutive packets are spaced by the corresponding clock ticks, that is, $auxVC_k^i = auxVC_k^{i-1} + S_k^i/\rho_k$. However, this has an undesirable property that, if a connection remains silent for a long time, its $auxVC$ value will get smaller than that of other connections. This implies that connections can increase their priority by staying idle. To overcome this problem, silent periods in a virtual timeline are eliminated by $max(AT_k^i, auxVC_k^{i-1})$ as shown in Eq. (8).

The properties of the virtual clock are as follows:

- Since the virtual clock is a work-conserving emulation of TDM, it provides the property that if a misbehaving connection sends more traffic than it is allowed it will only hurt itself by lowering the priority of its packets. But, it also has the undesirable property that a misbehaving connection will always be punished, even if the excess traffic it introduces does not affect the QoS guarantees of other connections.

- Each connection k is allocated a bandwidth ρ_k, which is used to compute its clock ticks.
- If the source is characterized by (σ, ρ), then it can provide guarantees on end-to-end delay bound. The guarantees are the same as for WFQ, described below in a separate section.

Fluid-Flow Fair Queueing

Fluid-flow fair queueing is a hypothetical service discipline (21) also known as generalized processor sharing (GPS) (22–24); it is based on the idea of perfect fairness in bandwidth allocation to connections competing for the transmission at an output link. In the ideal fairness case, each connection is assigned a weight, and at any time the bandwidth of a link is shared among the competing connections such that each one receives a fraction of the total bandwidth in proportion to its weight. This implies that multiple connections are served simultaneously on a link, and the traffic of each connection is infinitely divisible. However, in reality only one connection can receive service at a time, and service is given in units of packets. Therefore, FFQ is impractical. Weighted fair queueing (WFQ), described in the next section, is an approximation to FFQ that takes into account the noninfinitesimal packet sizes.

Formally, the FFQ discipline can be defined as follows. Assume within each connection the traffic is served according to the FCFS discipline. Further assume there are N connections with their paths passing through a link l where each connection k, $1 \leq k \leq N$, is assigned a weight ϕ_k. Then, FFQ will assign the service rate $r_k(t)$ to a connection k at time t, which is defined as

$$r_k(t) = \begin{cases} \dfrac{\phi_k}{\Sigma_{i \in B(t)} \phi_i} C_l & \text{if } k \in B(t) \\ 0 & \text{if } k \notin B(t) \end{cases} \tag{9}$$

where $B(t)$ is the set of connections backlogged at time t, and C_l is the total bandwidth at link l. As it can be easily seen from Eq. (9), the worst-case bandwidth received by a backlogged connection k at a link l occurs when every connection using that link is backlogged. This gives a bandwidth guarantee of

$$\dfrac{\phi_k}{\Sigma_{i=1}^{N} \phi_i} C_l$$

for each backlogged connection k. It is shown by Parekh and Gallager that, when the source traffic of a connection is constrained by (σ, ρ), the FFQ discipline can also give guarantees on the delay bound (22–24). Their results can be summarized as follows:

- In order to ensure bounded backlogs, the following stability condition has to be satisfied at each link l in the network:

$$\sum_k \rho_k < C_l \qquad (10)$$

where the sum extends over all connections k with paths that pass through a link l. Without the stability condition at a link l, the backlogs at that link grow without any bound, making it impossible to provide delay bound guarantees and lossless communication for connections passing through that link. Therefore, this condition is used for admission control.
- A connection k is called locally stable at a link l if its guaranteed bandwidth is greater than or equal to its average rate, that is,

$$\frac{\phi_k}{\sum_{i=1}^N \phi_i} C_l \geq \rho_k.$$

Then, it was proven in Ref. 23 that, for a connection k locally stable at every link along its path, the end-to-end delay is bounded by σ_k / ρ_k, and the end-to-end backlog is bounded by σ_k, provided the connection is shaped to satisfy the (σ_k, ρ_k) characterization at the source.
- It is also shown in Ref. 23 that, when all connections are shaped by a (σ, ρ) characterization, it is possible to guarantee a bounded delay even for connections that are not locally stable at some links over their paths. The resulting formulas for the end-to-end delay and the end-to-end backlog bounds can be found in Ref. 23.

Weighted Fair Queueing

Weighted fair queueing (WFQ) is a packet-by-packet transmission scheme that closely approximates FFQ in performance (21). It is also known as packet generalized processor sharing (PGPS) (22–24) and as packet-by-packet fair queueing (PFQ) (25). It is a work-conserving discipline that emulates an FFQ server to determine the order of packets for transmission. By convention, it assumes that a packet arrives at a node when its last bit arrives.

Let F_p be the time at which the transmission of packet p finishes in FFQ. Then, a good approximation of FFQ would be a server that serves packets in increasing order of F_p. However, such a work-conserving server does not exist. This is because, when the server is ready to transmit the next packet, there might be some packets backlogged for service but the next packet that completes its transmission under FFQ might not have arrived yet. In this case, a work-conserving server has to pick the packet with the smallest F_p among the ones backlogged for service. This is exactly how WFQ is defined. In WFQ, when a server becomes free to transmit another packet at time t, it picks the first packet that would complete its transmission in the emulated FFQ system if no additional packets were to arrive after time t.

An implementation of WFQ discipline is proposed in Ref. 22. It uses the concept of *virtual time* to keep track of the progress in the emulated FFQ system. Virtual time $v(t)$ is defined as a nondecreasing and piecewise linear function of time. Let an event denote each arrival and departure from FFQ server and t_i as the occurrence time of the ith event. Then, the rate of change of v is fixed during each time interval (t_i, t_{i+1}) and is equal to

$$\frac{\partial v(t + \tau)}{\partial \tau} = \frac{C_l}{\Sigma_{k \in B_i} \phi_k} \quad \forall t \in (t_i, t_{i+1}) \tag{11}$$

where B_i is the fixed set of connections backlogged in (t_i, t_{i+1}), and C_l is the service rate. Thus, v can be interpreted as the marginal service at which the backlogged connections receive service.

For the ith packet of connection k, let a_k^i denote the arrival time of the packet, and let s_k^i and f_k^i denote the service start time and service finish time, respectively, of the packet in the emulated FFQ under the assumption that the set of backlogged connections will remain fixed after a_k^i. Then, it is clear that $s_k^i = max(f_k^{i-1}, a_k^i)$ and

$$f_k^i = max(f_k^{i-1}, a_k^i) + \frac{S_k^i}{\dfrac{\phi_k}{\Sigma_{j \in B(a_k^i)} \phi_j} C_l} \tag{12}$$

It can be easily shown that the order of packets in increasing f_k^i values is the order of packets that we would like to achieve in the WFQ discipline. Since the virtual time function is an increasing function whenever there are packets to serve, the order imposed on packets by the virtual service finish times, that is, $v(f_k^i)$, is the same order we will get using f_k^i. Let $F_k^i = v(f_k^i)$. Then, from Eq. (12)

$$F_k^i = v\left(max(f_k^{i-1}, a_k^i) + \frac{S_k^i}{\dfrac{\phi_k}{\Sigma_{j \in B(a_k^i)} \phi_j} C_l} \right) \tag{13}$$

Since we assume that the set of backlogged connections will not change after a_k^i, Eq. (13) can be simplified to

$$F_k^i = max(F_k^{i-1}, v(a_k^i)) + \frac{S_k^i}{\phi_k} \tag{14}$$

by using the property given in Eq. (11).

Equation (14) is used in implementing a WFQ system. Every time a new packet arrives, its virtual finish time is calculated as given in Eq. (14). Then, every time a server becomes ready to transmit its next packet, it picks the packet with the minimum virtual finish time among the packets backlogged for service.

The properties of a WFQ server are the following:

- Let F_p and $\hat{F}_p$ denote the finish times of packets under FFQ and WFQ, respectively. Then, it is shown in Ref. 22 that

$$\hat{F}_p - F_p \leq \frac{S_{max}}{C_l} \tag{15}$$

It bounds the extra delay that a packet might face in WFQ as compared with FFQ. This can be used to translate the delay bounds in FFQ to corresponding bounds in WFQ. It is also shown in Ref. 22 that the WFQ service can get quite far ahead of the FFQ service, at least as much as $(N - 1)S_{max}$, where N is the number of connections receiving service. Clearly, this bound does not affect the end-to-end delay bound, but according to Refs. 38 and 39, it has an impact on the traffic management algorithms for best-effort traffic. In Refs. 38 and 39, a new service discipline called worst-case fair weighted fair queuing (WF2Q) is proposed that minimizes this difference.
- In WFQ, the end-to-end delay bound is

$$\frac{\sigma_k + NS_{max}}{\rho_k} + \sum_{l=1}^{N} \frac{S_{max}}{C_l} \tag{16}$$

Note that the σ_k/ρ_k part of the delay is the FFQ delay bound. Since WFQ is an approximation of FFQ, at each node we get an additional delay of S_{max}/ρ_k for packetization and S_{max}/C_l for priority inversion.

Self-Clocked Fair Queueing

The fair queueing scheme proposed in WFQ (22,23), which analyzes the performance of a queueing network with a fair queueing service discipline and derives upper bounds on the end-to-end delay is based on a hypothetical fluid-flow reference system to determine the fair order of packet transmissions. To overcome the computational complexity of WFQ, self-clocked fair queueing (SCFQ) introduces a self-contained approach to fair queueing that does not involve a hypothetical queueing system as referenced in defining fairness (25).

As shown earlier, WFQ requires the evaluation of the virtual time $v(t)$ of FFQ (21), the computational complexity of which depends on the frequency of transitions in and out of the set of backlogged sessions. Thus, in a broadband network such as an ATM network, WFQ cannot be accurately implemented due to an infeasible real-time evaluation of $v(t)$. The SCFQ describes a scheme also based on the notion of the system's virtual time, the indicator of progress of work in the system, except that the virtual time is referenced to the actual queueing system itself. It estimates the system's virtual time at any moment t from the service tag of the packet receiving service at t.

The SCFQ operates as follows:

1. Each arriving packet p_k^i is tagged with a service tag $\hat{F}_k^i$ before it is placed in the queue. The packets in the queue are picked up for service in increasing order of the associated service tags.
2. For each session k, the service tags of the arriving packets are iteratively computed as

$$\hat{F}_k^i = \frac{1}{r_k} L_k^i + max(\hat{V}(a_k^i), \hat{F}_k^{i-1}), \tag{17}$$

3. $\hat{V}(t)$, regarded as the system's virtual time at time t, is defined as equal to the service tag of the packet receiving service at that time. $\hat{V}(t) \equiv \hat{F}_l^j, \hat{s}_l^j < t < \hat{d}_l^j$, where $\hat{s}_l^j$ and $\hat{d}_l^j$, respectively, denote the times packet p_k^i starts and finishes service.
4. Once a busy period is over, the algorithm is reinitialized by setting to zero the virtual time $\hat{V}(t)$, and the packet counts i for each session k.

FIFO+

All of the service disciplines presented in this paper, with the exception of FIFO+ (first in first out), aim to provide guarantees on worst-case performance of individual connections. Naturally, they are best suited for intolerant and rigid applications such as process control applications. The FIFO+ (40), on the other hand, aims to provide *predictive service*, which tries to minimize the actual *post facto* delay bounds. Therefore, FIFO+ should be the choice for tolerant and adaptive applications such as multimedia applications. It is claimed in Ref. 40 that FFQ-based algorithms, with their emphasis on isolation, are not well suited for these applications.

Under FFQ-based algorithms, a burst originating from a source has a minimal effect on the delays seen by the packets originating from other sources. However, the burst will induce jitter on packets coming from the source. FIFO+ argues that this isolation, while necessary for providing guaranteed service, is counterproductive for predicted service. The *post facto* jitter is smaller when the delays are shared as in FIFO rather than when they are isolated as in FFQ-based algorithms.

The queueing presented in FIFO+ is a combination of the effective sharing scheme of the FIFO scheme and the complete isolation scheme of FFQ. By organizing the traffic with similar service requirements into classes, FIFO+ makes the queueing decision in two steps: it isolates classes and shares within each class. FIFO+, qualitatively very similar to the *least slack* scheduling algorithm, induces FIFO-style sharing (equal jitter for all sources in the aggregate class) across all the hops along the path to minimize jitter.

For each hop, FIFO+ measures the average delay seen by packets in each aggregate class at that switch and then computes the difference between its particular delay and the class average. This difference, which accumulates the total offset for this packet from the average for its class, allows each switch to compute when the packet was expected to arrive. Then, the switch schedules the packet by the order of the expected arrival times.

Non-Work-Conserving Real-Time Service Disciplines

Stop and Go

Stop-and-go discipline was proposed by Golestani (1,27). It is a non-work-conserving discipline based on multilevel framing strategy. Each connection is assigned to a service level i of frame length T_i, and its traffic is shaped at the source and preserved at every switch as (r,T_i)-*smooth*. The main idea is to synchronize the frames of incoming and outgoing traffic at every switch and to transmit the incoming traffic in the frame next to the one in which it is received. This scheme can provide guarantees for delay, delay jitter, and bandwidth. First, details of the discipline are given in terms of one-level framing, and then it is extended for multilevel framing.

At every node in the network, time is divided into frames of length T. Frames of a node are used to regulate the traffic at the outgoing links; they are called *departing frames* of that node. Departing frames of different nodes need not be synchronized. The receiving end of a link l sees the departing frames of the sending end with a constant shift π_l, which is the propagation delay over the link l and the processing time of a frame at the receiving end. These frames are called *arriving frames* from a link l.

In stop-and-go scheduling, any packet received during an arriving frame f from link l will be transmitted in the next departing frame that starts after the end of frame f (Fig. 4). The scheduling does not specify the order of transmission for packets inside a frame, so any queueing mechanism, such as a simple FCFS, could be used for that purpose. The consequences are the following:

- It is a non-work-conserving discipline, where (r_k,T)-*smoothness* of a connection k will be preserved throughout its path if it is (r_k,T)-*smooth* at the source.
- Let θ_l be the constant delay introduced at the receiving node to synchronize the arriving frames from link l and departing frames. As can be seen in Fig.

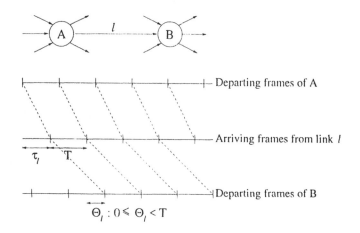

FIG. 4   Timing properties of stop-and-go at a single hop.

4, the start times of the corresponding departing frames on consecutive nodes over a link l are separated by $\pi_l + T + \theta_l$. Furthermore, since scheduling of packets inside a frame can relocate them anywhere in the frame, there is a variable delay J, where $-T < J < T$. Then, end-to-end delay for a packet of a connection k can be given as

$$\sum_l (\pi_l + T + \theta_l) + J, \tag{18}$$

where the sum extends to all the links over the path of connection k. The formula can be simplified by considering the fact that $0 \leq \theta_l < T$. If π denotes the end-to-end propagation and processing delay and N denotes the number of hops of the connection, then the formula becomes

$$\pi + \alpha NT + J, \qquad 1 \leq \alpha < 2, -T < J < T \tag{19}$$

- As shown in Fig. 4, the start of the receipt of an arriving frame and the end of the transmission of the corresponding departing frame are separated by $2T + \theta_l$. Since a packet can be relocated anywhere inside a frame, the residence time of a packet at a node will be bounded by this value. Therefore, at a node, a buffer space of $(2T + \theta_l)r_k$ is required for a connection k received from a link l.
- At each node, traffic from different connections, each coming from an arriving frame of an incoming link, is multiplexed on a single departing frame. An admission control for connection establishment has to guarantee that departing frames can allocate this traffic. If the set of links over the path of a new connection is denoted P and the set of connections passing through a link l is denoted K_l, then the admission control for one-level stop and go would be simply

$$(\forall l \in P) \sum_{k \in K_l} r_k \leq C_l \tag{20}$$

where C_l is the total bandwidth on link l.

In stop and go with n frame sizes, $T_1 < T_2 < \ldots < T_n$, and the scheduling of packets from a connection of level i will be as follows:

- As in stop and go with one-level framing, each such packet is transmitted in the next departing Type i frame that starts after the end of the arriving Type i frame in which it is received. The ordering of such packets inside a single Type i frame is not specified.
- Any such packet has nonpreemptive priority over the packets from a connection of level $j > i$, that is, the levels with larger frame sizes.

Scheduling at a level i is the same as scheduling at one-level framing in terms of timing properties. Therefore, formulas for delay, delay jitter, and buffer space will remain the same except that each T is now replaced with T_i.

The admission control in multilevel framing has to guarantee that the departing frames from a level i, $1 \leq i \leq n$, can allocate the incoming traffic from higher or equal-priority levels, that is, levels $j \leq i$. Since the scheduling is non-preemptive, in the worst case they also have to be able to allocate a packet from a lower priority level. If the set of links over the path of a new connection is denoted P, and the set of connections at a level j passing through a link l is denoted K_l^j, then the admission control for n-level stop and go would be

$$\sum_{j=1}^{i} \sum_{k \in K_l^j} r_k + \frac{Smax}{T_i} \leq C_l, \qquad \forall l \in P, \qquad i = 1, 2, \ldots, n-1 \qquad (21)$$

where C_l is the total bandwidth on link l and $Smax$ is the largest packet size in the network.

Hierarchical Round Robin

The hierarchical round robin (HRR) discipline was proposed by Kalmanek, Kanakia, and Keshav (2). It uses a multilevel framing strategy in which time is divided in units of slots, and each level is given a fixed set of slots in each of its frames. This allows each service level to offer a share of the total bandwidth at a particular granularity. Each connection is allocated a number of slots at a selected level. At every frame of a service level, connections of the level will be given round-robin service over the fixed set of slots reserved for the level. The HRR can guarantee a fixed bandwidth and a bounded delay for each connection.

The HRR is designed to be used in networks with fixed-size cells, such as ATM. At each node, time is divided into fixed-size intervals, each called a slot, in which the length of a slot is the transmission time of a cell at the output link. Slots are used as the granularity of time.

In HRR, service levels provide a menu of granularities for bandwidth allocation. Assume there are n service levels, and each service level i, $i = 1, 2, \ldots, n$, has frames of length T_i slots. Then, granularity of bandwidth at a level i will be $(1/T_i)$ of the total bandwidth capacity B. Bandwidth share given to each granularity at a service level i depends on how many slots of T_i slots are allocated to that service level in each Type i frame. If we let n_i be the number of slots in a Type i frame not used by higher priority service levels and b_i be the number of slots of these n_i slots that will be given to lower priority service levels, then there will be $n_i - b_i$ unites from granularity B/T_i for a total bandwidth share of $(n_i - b_i)/T_i$. An example is given in Fig. 5. Note that to get n_i slots from higher service levels to service level i, the frame size T_i should be $T_{i-1}(n_i/b_{i-1})$ since each Type $i - 1$ frame provides b_{i-1} slots to lower priority levels. Given the base case $T_1 = n_1$, this recursive formula generalizes to $T_i = n_1(n_2/b_1)(n_3/b_2) \ldots (n_i/b_{i-1})$. As can be seen in Fig. 5, not only the number of slots given to each service level, but also their positions inside a frame are fixed. For each service level i, the first $n_i - b_i$ slots not used by higher priority levels are reserved for the current level and the remaining b_i slots are left for lower levels.

In HRR, each connection k will be assigned to a service level i and will be

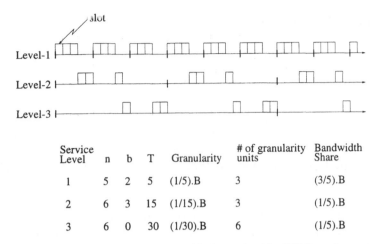

Service Level	n	b	T	Granularity	# of granularity units	Bandwidth Share
1	5	2	5	$(1/5).B$	3	$(3/5).B$
2	6	3	15	$(1/15).B$	3	$(1/5).B$
3	6	0	30	$(1/30).B$	6	$(1/5).B$

FIG. 5 An example of hierarchical round-robin (HRR) service.

given a service quantity of α_k slots in every Type i frame based on bandwidth and delay bound requirements of the connection. Connections at a service level i will compete for the fixed $n_i - b_i$ slots in every Type i frame in a round-robin fashion. At each round, each connection gets service for a number of slots that is the minimum of two values: the number of cells currently queued for that connection and the number of slots left in the current frame. If during a slot reserved for level i traffic, all connections from level i that have something to send have already used up their quantas for the current Type i frame, then the server will stay idle for that slot unless it is used for best-effort traffic.

The properties of HRR scheduling are as follows:

- It is non-work-conserving, and it guarantees that each connection k at a level i will be given exactly α_k slots in each Type i frame, whether it is used or not. This implies a constant bandwidth allocation of $(\alpha_k/T_i) \cdot B$, where B is the bandwidth of the output link.
- Jitter in HRR is defined as the maximum number of cells transmitted by a node in a short interval called the jitter-averaging interval. It captures the burst length received by a node. In the worst case, a connection k at level i can be served for α_k cells at the end of one Type i frame, followed by another α_k cell at the beginning of the next Type i frame. Therefore, the jitter bound is $2 \cdot \alpha_k$ cells during one frame time at that level. This also bounds the size of the buffer needed.
- In the worst case, a cell from a connection k at level i arrives at a node right after the quota for connection k gets used in the current Type i frame. This cell might not get serviced until the end of the next Type i frame, so the delay bound at a node is bounded by $2T_i$. For a path of length N, the end-to-end delay bound will be $2NT_i$. Minimum delay for a cell, on the other hand, can be as low as the sum of propagation and processing delays over its path, so there is no tight bound for delay jitter.

Rate-Controlled Static Priority

Rate-controlled static priority (RCSP) was proposed by Zhang and Ferrari (28,29). It is a non-work-conserving discipline that uses the ($Xmin,Xave,I,$ $Smax$) specification for characterizing the traffic of connections. The RCSP server has two components: a rate controller that reconstructs the traffic of every connection to its original characterization and a static priority scheduler that serves packets from different connections based on their priorities. During its establishment, each connection is assigned a priority at each node, which will later be inherited by the packets of that connection for scheduling. The main idea is that rate controllers characterize the traffic of every connection to ($Xmin,$ $Xave,I,Smax$) at the entrance to every scheduler, and static priority schedulers provide local guarantees on delay bounds under this characterization. These properties allow the local guarantees to be extended to end-to-end guarantees on the delay and delay-jitter bound.

An RCSP server operates as shown in Fig. 6. The rate controller is implemented by traffic regulators for each connection, and the static priority scheduler is implemented by a multilevel FCFS queue. A packet arriving at a node is first held at the regulator of its connection until it becomes eligible to be serviced. Once eligible, a packet will be inserted into the FCFS queue of its priority level. At any time, the scheduler will serve the first packet in the highest priority non-empty queue.

There are two regulators that are proposed for use in RCSP: the *DJ regulator*, a delay-jitter regulator reconstructing the original traffic pattern, and the *RJ regulator*, a rate-jitter regulator constructing a pattern that satisfies ($Xmin,$ $Xave,I,Smax$) characterization. In this text, RCSP-DJ and RCSP-RJ denote the RCSP disciplines using the DJ regulator and RJ regulator, respectively.

In the DJ regulator, the traffic pattern is fully reconstructed at every node by forcing the same constant delay (the worst-case delay) to happen to every packet of a connection between the schedulers of two consecutive nodes. This

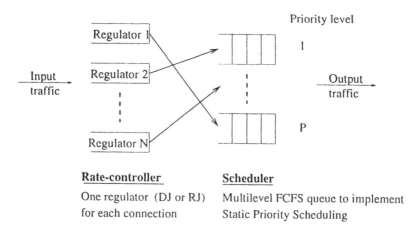

FIG. 6   Rate-controlled static priority (DJ = delay jitter; FCFS = first come, first served; RJ = rate jitter).

worst-case delay is the maximum queueing delay at the first scheduler and the maximum link propagation delay between nodes. The former is guaranteed by static priority scheduling and the latter is assumed by the network model. For the packets of a connection, let d_n^{max} represent the maximum queueing delay at node n and π_n^{max} represent the maximum propagation delay between nodes $n - 1$ and n. Then, the eligibility time for the ith packet of the connection at node n, ET_n^i, is defined as

$$ET_0^i = AT_0^i \tag{22}$$

$$ET_n^i = ET_{n-1}^i + d_{n-1}^{max} + \pi_n^{max}, \qquad n > 0 \tag{23}$$

where AT_n^i is the arrival time of the ith packet at node n.

The RJ regulator assigns the eligibility time of a packet based on the eligibility times of the earlier packets from the same connection, such that the original traffic characterization is still satisfied. Clearly, it depends on the traffic model being used. For an $(Xmin, Xave, I, Smax)$ traffic for which $Xmin \le Xave < I$ holds, the eligibility times will be

$$ET_n^i = -I, \qquad i < 0 \tag{24}$$

$$ET_n^1 = AT_n^1 \tag{25}$$

$$ET_n^i = max(ET_n^{i-1} + Xmin, ET_n^{i - \lfloor \frac{I}{Xave} \rfloor + 1} + I, AT_n^i), \qquad i > 1 \tag{26}$$

The properties of RCSP are as follows. A static priority scheduler provides one delay bound for every priority level. Connections assigned to a priority level have the same local delay bound guarantees, which are the delay bound provided by that priority level. The longest delay for a packet of priority level p, d_p^{max}, occurs when a lower priority packet with the maximum size is being transmitted at the time the packet arrives at the scheduler and is followed immediately by the longest possible transmission of packets with higher or equal priority. If C_q denotes the connections at priority level q, then d_p^{max} will be the solution of

$$\sum_{q=1}^{p} \sum_{k \in C_q} \lceil \frac{d_p^{max}}{Xmin^k} \rceil \cdot Smax^k + Smax^{max} \le d_p^{max} C_l \tag{27}$$

where $Smax^{max}$ is the largest packet that can be received by the node and C_l is the link speed of the output link. This was proved in Ref. 28. Here, the delay bound calculation is based solely on the maximum rate $Xmin$ of the connections, not on their average rate $Xave$. A tighter delay bound calculation is given in Ref. 30 that utilizes the $Xave$ values.

Also, it was shown in Ref. 29 that, for a connection passing through N nodes,

1. End-to-end delay is bounded by $\sum_{n=1}^{N} d_n^{max} + \sum_{n=2}^{N} \pi_n^{max}$ if DJ or RJ regulators are used.

2. Delay jitter is bounded by d_n^{max} if a DJ regulator is used.
3. Maximum residence time at a node n is the sum of the maximum holding time at the regulator $d_{n-1}^{max} + \pi_n^{max} - \pi_n^{min}$ and the maximum delay at the scheduler d_n^{max}. For $(Xmin, Xave, I, Smax)$ traffic, this means a buffer space of $\left(\dfrac{d_{n-1}^{max} + \pi_n^{max} - \pi_n^{min} + d_n^{max}}{Xmin} \right)$ $Smax$ is needed at node n to prevent any packet loss.

Jitter Earliest Due Date

Jitter EDD was proposed by Verma, Zhang, and Ferrari (26). It is a non-work-conserving discipline extended from delay EDD to provide delay-jitter guarantees. The regulator at a node is used to smooth out the local delay variation experienced by a packet at the previous node. After a packet has been served at a node, it is stamped with the difference between its deadline and actual finish time, called the *slack time*. The regulator at the next node holds the packet for this period before it is made eligible. Since packets of a connection obtain a constant delay between the eligibility times at consecutive nodes, they can be provided an end-to-end delay-jitter bound.

Operation of jitter EDD servers for the packets of a connection k is depicted in Fig. 7. Jitter EDD assumes a constant propagation delay over the links and a bounded service delay at nodes. Let d_n^{max} denote the local delay bound for the packets of connection k at the nth node and π_{n+1} denote the propagation delay of the link between the nth and $(n+1)$th nodes. When a packet i of connection k becomes eligible at node n at time t, its deadline is set to $t + d_n^{max}$ and it is inserted into the EDD queue. As in delay EDD, admission control guarantees that the packet will be served by its deadline. When it is actually served at time t', there may be a slack time of $(t + d_n^{max}) - t'$. This slack time is stamped in

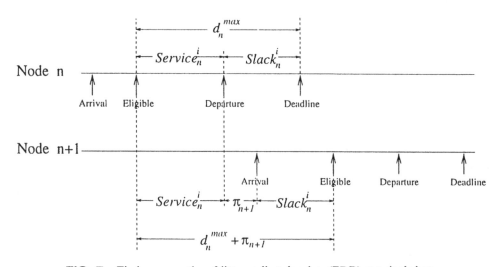

FIG. 7   Timing properties of jitter earliest due date (EDD) at a single hop.

the header of the packet, and when the packet is received by the next node at time $t' + \pi_{n+1}$, it is held for the slack time. As shown in the figure, the eligibility times of all the packets of connection k at consecutive nodes n and $n + 1$ are separated by $d_n^{max} + \pi_{n+1}$. This results in the following properties:

- End-to-end delay bound for connection k is $\Sigma_{n=1}^{N} d_n^{max} + \Sigma_{n=2}^{N} \pi_n$ if there are N nodes. Assuming that the destination node does not have a regulator, the only delay variation happens at the last node, so the delay jitter is bounded by d_N^{max}.
- Since packets of a connection are provided a constant delay between the schedulers of two consecutive nodes, the source traffic pattern will be preserved throughout the network. If the source traffic obeys an (*Xmin*, *Xave,I,Smax*) characterization, then so does the incoming traffic at the schedulers. However, if the source traffic is misbehaving, then logical arrival times (as in delay EDD) can be used at the first node to regulate the traffic to its characterization. Once regulated at the first node, the following nodes will preserve the characterization.
- The maximum residence time at a node n is $d_{n-1}^{max} + \pi_n + d_n^{max}$. Therefore, for an (*Xmin,Xave,I,Smax*) characterization, the buffer requirement of the connection is $\lceil \dfrac{d_{n-1}^{max} + \pi_n + d_n^{max}}{Xmin} \rceil Smax$.

Comparison of Service Disciplines

A comparative assessment of the service disciplines discussed in this section is presented in Tables 2 and 3.

Real-Time Communication in Local-Area Multiple-Access Networks

Hard real-time applications such as embedded command-and-control systems, automated manufacturing, scientific visualization, and industrial process control are generally distributed in nature. They cover a geographical span of a building or a group of buildings or a single ship or an aircraft, making local-area multiple-access networks sufficient for their communication needs. In these applications, deadline constraints on task execution induce deadline constraints on their communications. Each packet is assumed to have a deadline for its transmission, and a packet delivered after its deadline is considered lost. This section discusses hard real-time communication in local-area multiple-access networks. Excellent surveys in this area can be found in Refs. 41 and 42. Also, schedulability analysis for several access protocols are derived in Ref. 43.

In a multiple-access network, nodes communicate via a single shared channel. At any time, only a single node can access the channel to send a packet to

TABLE 2   Service Guarantees and Buffer Requirements in Work-Conserving Disciplines

	Traffic Constraint	End-to-End Delay Bound	End-to-End Delay-Jitter Bound	Buffer Space at hth Switch
Delay EDD	$b_k(\cdot)$	$\sum_{n=1}^{N} d_k^n$	$\sum_{n=1}^{N} d_k^n$	$b_k\left(\sum_{n=1}^{h} d_k^n\right)$
FFQ	(σ_k, ρ_k)	$\dfrac{\sigma_k}{\rho_k}$	$\dfrac{\sigma_k}{\rho_k}$	σ_k
Virtual clock	(σ_k, ρ_k)	$\dfrac{\sigma_k + NS_{max}}{\rho_k} + \sum_{l=1}^{N} \dfrac{S_{max}}{C_l}$	$\dfrac{\sigma_k + NS_{max}}{\rho_k}$	$\rho_k + hS_{max}$
WFQ	(σ_k, ρ_k)	$\dfrac{\sigma_k + NS_{max}}{\rho_k} + \sum_{l=1}^{N} \dfrac{S_{max}}{C_l}$	$\dfrac{\sigma_k + NS_{max}}{\rho_k}$	$\sigma_k + hS_{max}$
SCFQ	(σ_k, ρ_k)	$\dfrac{\sigma_k + NS_{max}}{\rho_k} + \sum_{l=1}^{N} K_l \dfrac{S_{max}}{C_l}$	$\dfrac{\sigma_k + NS_{max}}{\rho_k} + \sum_{l=1}^{N} (K_l - 1) \dfrac{S_{max}}{C_l}$	$\sigma_k + hS_{max}$

TABLE 3 Service Guarantees and Buffer Requirements in Non-Work-Conserving Disciplines

	Traffic Constraint	End-to-End Delay Bound	End-to-End Delay-Jitter Bound	Buffer Space at hth Switch
Stop and go	(r_k, T_k)	$NT_k + \sum_{l=1}^{N} \theta_l + T_k + \pi$	$2T_k$	$r_k(2T_k + \theta_h)$
HRR	(r_k, T_k)	$2NT_k$	$2NT_k$	$2r_k T_k$
Jitter EDD	$b_k(\cdot)$	$\sum_{n=1}^{N} d_n^{max} + \sum_{n=2}^{N} \pi_n$	d_N^{max}	$b_k(d_{h-1}^{max} + \pi_h + d_h^{max})$
RCSP-DJ	$b_k(\cdot)$	$\sum_{n=1}^{N} d_n^{max} + \sum_{n=2}^{N} \pi_n^{max}$	d_N^{max}	$b_k(d_{h-1}^{max} + \pi_h^{max} - \pi_h^{min} + d_h^{max})$
RCSP-RJ	$b_k(\cdot)$	$\sum_{n=1}^{N} d_n^{max} + \sum_{n=2}^{N} \pi_n^{max}$	$\sum_{n=1}^{N} d_n^{max} + \sum_{n=2}^{N} \pi_n^{max}$	$b_k(d_{h-1}^{max} + \pi_h^{max} - \pi_h^{min} + d_h^{max})$

another node or nodes. A media access control (MAC) protocol is used to arbitrate access to the network. It schedules the packet transmission over the channel. It consists of two processes: an *access arbitration process* that determines *when* a node can use the channel and a *transmission control process* that determines *how long* a node can continue to use the channel. Hard real-time communication is achieved by a real-time MAC protocol.

Real-time network scheduling in local-area multiple-access networks is very similar to real-time centralized processor scheduling. Both have a single server, a shared channel in network scheduling, and a processor in processor scheduling. In network scheduling, the scheduling entity is a packet with a constant transmission time and a deadline for its transmission. Correspondingly, in processor scheduling the scheduling entity is a task with a bounded execution time and a deadline for its execution. However, there are fundamental differences between the characteristics of the processor and network scheduling. Network scheduling is distributed in nature, and it requires substantial overhead to obtain the global state information at every node. Therefore, often accuracy of the global information is traded off for a reduced overhead. Because of this nature, many of the centralized scheduling algorithms perform poorly if used in network scheduling, and the optimality results in centralized scheduling do not always carry over to network scheduling.

Most of the real-time MAC protocols proposed in the literature follow one of the following three scheduling techniques:

1. *Priority Scheduling*. Each message is assigned a priority and the MAC protocol tries to ensure that at any time the channel is serving the highest priority message in the network.
2. *Bounded Access Scheduling*. Nodes are arranged in a logical ring in which they take turns accessing the channel. Real-time guarantees are obtained by bounding the access time of every node at each turn. When a node gets its turn, it tries to transmit as many packets as it can during its bounded access time. The order of packets for transmission at a node is determined by a local scheduling algorithm.
3. *Time-Based Scheduling*. Each packet transmission is explicitly reserved a time interval on the time axis that guarantees exclusive access to the shared channel. When a node needs to schedule a packet, it reserves a time interval and broadcasts that information to all other nodes.

Network Model

The two most common models of multiple-access networks are *ring networks* and *bus networks*. In ring networks (44), nodes are physically arranged in a ring. A special bit pattern called a *token* circulates among the nodes and is used to control access to the ring. A node wishing to send a packet has to seize the token first and possess it during the transmission of the packet.

Bus networks have a single cable that is connected to all nodes using OR logic. If several nodes transmit a bit simultaneously, then the value read back from the bus is the logical OR of the values transmitted.

The most commonly used protocol for access control in bus networks is called the Carrier Sense Multiple Access with Collision Detection (CSMA/CD) Protocol. In bus networks, when two or more nodes send a packet simultaneously, they overlap in time and the resulting signal is garbled. This event is called a *collision*. In CSMA/CD networks, it is assumed that nodes can sense the channel. Any node can determine whether the channel is idle with no packet on it or is busy transmitting a packet. The node can also detect a collision when it occurs. A node wishing to send a packet senses the channel first and transmits the packet only if the channel is idle. But, this does not guarantee exclusive access to the channel. Several nodes might sense the channel as idle and start transmitting simultaneously, resulting in a collision. In the worst case, two nodes can sense the channel as idle and start transmitting a packet at different times separated by the maximum propagation delay. After a node starts transmitting a packet, it checks for a collision and stops transmitting if there is one. This check and the start of the transmission are separated by a time equal to the maximum propagation delay of the channel. Clearly, large propagation delays increase the probability of collisions.

Traffic Model

The local-area networks are primarily used for real-time communications in process control applications. As a consequence, a delay-bound-based traffic model is considered appropriate for this environment.

Real-time traffic in the network is composed of messages, high-level entities sent from a source application to a number of destination applications. Each message is divided into packets, and each packet is individually transmitted over the channel without any preemption allowed. However, packets of different messages can be interleaved, allowing preemption of message transmission.

Each message is given an application layer end-to-end delay bound. It is the maximum delay allowed for the message to be transmitted from the source application to the destination application. The total delay is composed of the processing of the protocol layers in both the source node and the destination node, the transmission delay of the packets comprising the message, and the propagation delay over the channel. It is assumed that, for a message, all these delay components are constant and known in advance except the processing time of the MAC protocol at the sending node. Note that the MAC protocol at the sending node decides when the packets of a message can access the channel and has a variable delay depending on the MAC protocol used and the traffic loads at the network nodes. Therefore, the application layer end-to-end delay bound of a message is converted to an MAC layer delay bound at the sending node.

Messages can be categorized as *periodic messages* and *aperiodic messages*. Periodic messages are composed of a set of message streams $\{S_1, S_2, \ldots, S_n\}$. Each stream S_i represents a sequence of messages with an interarrival time or period T_i, a transmission time C_i, and a transmission deadline D_i. The utilization of the periodic message set is defined as $\sum_{i=1}^{n} (C_i/T_i)$. It is the proportion of the time needed for the transmission of periodic messages. Sensory data and inter-task communication between periodic tasks are examples of periodic messages. Since the timing properties of all periodic messages are known in advance, they can be given hard real-time guarantees.

Aperiodic messages, on the other hand, are not generally known before their arrival. Each such message has an arrival time, a transmission time, and a transmission deadline. When they arrive, the network tries to meet their deadlines, but no hard guarantees can be given. Therefore, they are generally treated as soft real-time communication requests.

Priority Scheduling

In priority scheduling, each message is assigned a priority, and the network tries to ensure that at any given time the channel transmits the packets of the highest priority message. Each packet of a message inherits the priority of the message. The network operates in alternating processes of access arbitration and transmission control. The access arbitration process uses a global priority arbitration protocol to find the highest priority pending packet in the network. Then, the transmission control process serves a single packet with a priority that is higher than or equal to that priority.

Priority scheduling algorithms can be classified as *preemptive* and *nonpreemptive*. A preemptive scheduling algorithm in multiaccess networks requires the preemption of the transmission of a low-priority message when a high-priority message arrives. Since transmission of individual packets cannot be preempted, the preemption can only be approximated by dividing messages into small packets. There is a tradeoff in the packet size by which a smaller packet size increases the preemptability but decreases the throughput by increasing the header-to-packet-size ratio. A nonpreemptive scheduling algorithm, on the other hand, can be easily implemented by transmitting each message in a single packet.

Priority scheduling algorithms can also be classified as being *static* or *dynamic*. In static priority scheduling algorithms, the priorities are assigned for each message stream at a prerun time, whereas in dynamic scheduling algorithms the priorities are assigned for an individual message at a run time when they arrive.

There are two priority scheduling algorithms discussed in this section: the rate monotonic algorithm, a preemptive static scheduling algorithm for scheduling periodic messages, and the minimum laxity first algorithm, a preemptive dynamic algorithm for scheduling aperiodic messages.

Global Priority Arbitration Protocols

The global priority arbitration protocols try to find the priority of the highest priority packet currently pending in the network. There are several protocols in the literature proposed for priority arbitration; some of them are designed specifically for certain types of networks. Let us consider some of the protocols:

Token-Passing Protocols Token-passing protocols are designed for ring networks in which the token is used to implement global priority arbitration. The token contains a priority field in which the arbitration protocol tries to keep the priority of the highest priority packet currently pending in the network. A node can seize the token only if it has a packet to transmit with a priority that is greater than or equal to the priority field of the token.

One such arbitration protocol proposed is the Institute of Electrical and Electronics Engineers (IEEE) 802.5 Protocol (45). It has two alternating modes for the token: a reservation mode and free mode. In the reservation mode, the token is seized by a node for a single-packet transmission. The node transmits its highest priority packet appended with the token in the reservation mode. During transmission of the token, any node will update the priority field with the priority of its highest priority pending packet, if the pending packet's priority is higher than that of the priority field. When the token in the reservation mode returns to the sending node, it is put into the free mode and released. Then, the first node that has a packet with a priority higher than or equal to the priority field seizes the token and puts it back into the reservation mode. Clearly, this arbitration protocol is not exact, meaning it might not always choose the highest priority packet. In the reservation mode, if the highest priority packet arrives at a node after the token in the reservation mode visits the node, then the token will not have the highest priority in the system, and therefore the next packet transmitted might not be the highest priority packet.

Countdown Protocol. A countdown protocol is designed for bus networks. In this protocol, time is divided into alternating periods of priority arbitration and packet transmission. During priority arbitration, time is further divided into equal-length slots, with the length of each slot equal to the end-to-end propagation delay. Over these slots, each node transmits the priority of its highest priority pending packet, one bit at a slot, starting from the most significant bit. Since simultaneous transmission follows the OR logic, a node that broadcasts 0 and reads 1 in a slot realizes that there is at least one node with a higher priority pending packet and drops out of contention. Notice that this scheme might not elect a single node in cases of priority ties. To prevent this, each node is given a unique identification (ID) that is appended to the priorities. Then, in the packet transmission period, the node elected at the arbitration period transmits its highest priority packet. This is not an exact protocol in implementing arbitration because if the highest priority packet arrives at a node after the arbitration period starts, it will not be included in the arbitration process.

Virtual Time Protocols. Virtual time protocols are designed for use in CSMA/CD networks (46). A node uses the state of the channel to reason about the priorities of pending packets residing at other nodes. Let us assume that priorities are represented by real numbers, and they increase with decreasing values. Then, the global priority arbitration is performed as follows.

When the channel becomes idle (signifying the end of a packet transmission), each node with a packet to send waits for an interval of time proportional to the priority of its highest priority pending packet. There is a fixed proportionality constant used to determine the waiting times. At the expiration of its waiting time, the node checks the status of the channel. If it is busy, meaning some other node had a higher priority packet to send and is sending that packet, then the node waits for another idle period. Otherwise, the node starts transmission of its highest priority pending packet. However, collision will occur if another node with a packet of the same priority also starts transmission. This problem is resolved probabilistically. Nodes involved in a collision either retransmit the packet with a probability p or wait for another interval of time proportional to their priority.

Because of the propagation delay in the network, nodes cannot instantaneously detect when another node starts transmission. In the worst case, a collision may occur between two transmissions that have their start times separated by the maximum propagation delay. This bounds the difference between the priorities of two packets with transmissions that may collide. The bound is the maximum propagation delay divided by the proportionality constant. Therefore, a small proportionality constant may increase the number of collisions and the number of priority inversions in the system. A large constant, on the other hand, increases the idle intervals by increasing the waiting periods of nodes. If not chosen carefully, a large or a small constant will decrease the utilization of the network. Note that the arbitration is not exact since collisions may favor a lower priority packet.

Window-Based Protocols. Nodes in the network agree on a common window (*low,high*), which is the range of priorities; *high* is the highest and *low* is the lowest priority (41). A packet is regarded as being in the window if its priority is in this range. Priority arbitration starts after every packet transmission with a window (*low,high*), with *high* the highest priority that a packet can get and *low* determined by the protocol. Then, the protocol checks the number of nodes that have a packet in the window. If there is only one such node, then that node will be allowed to send its highest priority packet. In cases for which there are more than one node or no nodes with a packet in the window, then the window sizes will be decreased and increased, respectively, by changing the value of *low* until exactly one node remains with a packet in the window.

This protocol can be implemented in both CSMA/CD networks and ring networks. In CSMA/CD networks, each node stores a local copy of the window. Whenever the channel becomes idle after the transmission of a packet, each node with a packet in the current window starts transmission. The channel state is then used to determine the number of such nodes. If the channel remains idle, meaning there are no nodes with a packet in the window, then the window size will be increased. If a collision occurs, meaning there is more than one node with a packet in the window, then the window size will be decreased. Otherwise, if the channel is busy with only one packet, then that packet will be transmitted.

In a ring network, the window bounds are stored only in the token. The token also has a counter field keeping track of the number of nodes with a packet in the window. After every packet transmission, the sender will circulate the token around the ring. Every node with a packet in the window will increment the counter field and, when the token is received back at the sender node,

It will have the number of such nodes. If the value of the counter is zero or more than one, then the window size will be adjusted accordingly. If the counter is one, then the node with a packet in the window will be allowed to send its highest priority packet.

The Rate Monotonic Algorithm

The rate monotonic algorithm is a preemptive static scheduling algorithm for scheduling a set of periodic tasks on a uniprocessor (47). Each task is given a priority inversely proportional to its period. Tasks with a smaller period are assigned higher priorities.

If we assume that the deadline of each task is the end of the period in which it arrived, then the rate monotonic algorithm is the *optimal* preemptive static scheduling algorithm. Optimality here means that if a set of such tasks is able to be scheduled by a preemptive static scheduling algorithm, then it is also able to be scheduled by the rate monotonic algorithm. It also has a nice property that its worst-case-achievable utilization can be bounded in terms of number of tasks in the system. It was shown in Ref. 47 that any set of n periodic tasks are able to be scheduled as long as their total utilization is bounded by $n(2^n - 1)$. For a large n, this bound approaches to $\ln 2 \approx 0.69$.

The rate monotonic algorithm can be used to schedule periodic message transmissions in a multiaccess network when the deadline of each message is assumed to be the end of the period in which it arrived. Messages with a lower period are assigned a higher priority. In the literature, the rate monotonic implementations that are proposed use a token-passing protocol or a countdown protocol. In Refs. 48–50, the IEEE 802.5 Protocol is used for its implementation.

Note that rate monotonic algorithm implementation cannot be exact. There are several reasons for that. First, it is a preemptive algorithm that can only be approximated by dividing messages into small packets and transmitting each packet nonpreemptively. The second problem is inherited from priority arbitration protocols used to implement the rate monotonic algorithm. These protocols are not exact in implementing arbitration, meaning they cannot always find the highest priority packet in the network.

Clearly, the approximated rate monotonic algorithm has a smaller worst-case-achievable utilization than the exact one. The value of this utilization is still an open area of research, but it was shown in Ref. 49 that this value must be greater than or equal to $0.69 - B_{max}/T_{min}$, where B_{max} is the maximum blocking time caused by a priority inversion and T_{min} is the smallest period in the system.

The Minimum Laxity First Algorithm

The minimum laxity first (MLF) algorithm is the optimal dynamic scheduling algorithm for scheduling a set of aperiodic tasks (51). Therefore, it is adopted for use in multiaccess networks to provide soft real-time guarantees to aperiodic messages, with the objective to minimize the number of messages that miss their deadlines. *Laxity* of a message is defined as the amount of time remaining before transmission of the message must begin if the message is to meet its deadline. At any time, the MLF policy tries to serve the packets of the message that has the minimum non-negative laxity. Note that a packet with a negative laxity cannot make its deadline.

The implementation of MLF policy can use any of the global priority arbitration protocols discussed above. Particularly, the use of token-passing protocols and countdown protocols is examined in Refs. 52 and 53, the virtual time protocols in Refs. 54 and 55, and the window-based protocols in Refs. 41 and 56–60. In the window-based protocols, the window that is used is a time interval (*low*,*up*), for which *low* maintains the current time in the network that is the highest priority a packet can receive and *up* maintains an upper bound of the interval that is the lowest priority for a packet to be in the window.

Implementation of MLF policy cannot be exact. It suffers from the same problems the rate monotonic implementation does. In an MLF implementation, an additional problem might occur as follows. A message with a small non-negative laxity at the time of its priority check in the arbitration process might have a negative laxity by the end of the arbitration process, meaning that the message cannot make its deadline. If priority arbitration chooses such a message, then the arbitration process has to be repeated. High overheads in the priority arbitration process might result in repeated unsuccessful arbitration processes and might cause a substantial amount of message losses.

Bounded Access Scheduling

Timed Token Protocol

The Timed Token Protocol is a token-passing protocol in ring networks in which hard real-time guarantees are given to a set of periodic messages by bounding the amount of time that each node may hold the token (61). It has been incorporated into many network standards, including the fiber distributed data interface (FDDI) (62), the IEEE 802.4 token bus standard (63), and the Survivable Adaptable Fiber Optic Embedded Network (SAFENET) (64,65).

At network initialization time, the expected token rotation time around the ring is specified by a parameter called the target token rotation time (TTRT). The FDDI standard requires the TTRT to be no more than half of the minimum message deadline. The TTRT is distributed among individual nodes according to the timing characteristics of the periodic messages originating from each node. Each node i is assigned a synchronous bandwidth H_i, which is the maximum amount of time the node is allowed to hold the token in each turn to transmit periodic messages. If the token arrives at a node earlier than expected, then the extra time can be used to send aperiodic messages with no real-time guarantees.

Guaranteeing the deadlines of periodic messages depends on an appropriate distribution of the token rotation time TTRT into synchronous bandwidths H_i allocated to individual nodes i. Synchronous bandwidth allocation schemes were studied in Refs. 66 and 67. Let us assume, without loss of generality, that each node i has a single message stream with a period T_i, transmission time C_i, and transmission deadline D_i. Two synchronous bandwidth allocation schemes proposed in Refs. 66 and 67 are described below.

1. *Normalized proportional allocation scheme.* Note that if the propagation delay is denoted by τ, then only $TTRT - \tau$ time units can be used for sending periodic messages in one rotation. In this scheme, this time is distributed among the nodes according to the ratio of the utilization of individual nodes to the total utilization:

$$H_i = \frac{C_i/T_i}{\sum_{j=1}^{n} C_j/T_j} (TTRT - \tau) \tag{28}$$

2. *Local allocation scheme.* At each node i a message of transmission time C_i arrives with an interarrival time of T_i. It is known that the token visits the node i at least $\lfloor T_i/TTRT \rfloor - 1$ times during the time T_i. In this scheme, the transmission time C_i of each message is divided by the worst-case number of visits to the node i during a single message period:

$$H_i = \frac{C_i}{\lfloor T_i/TTRT \rfloor - 1} \tag{29}$$

Both schemes described above have a worst-case-achievable utilization of $(1 - \tau/TTRT)/3$ under the assumption that the deadline of each message is the end of the period in which it arrived. An advantage of the local allocation scheme over the normalized proportional allocation scheme is that, since the local allocation scheme uses only the characteristics of the stream S_i and global parameters TTRT and τ to calculate H_i, changing the characteristics of a stream S_i changes only the synchronous bandwidth H_i but not the synchronous bandwidths of the other nodes in the network.

Conservative Estimation Method

The conservative estimation method provides hard real-time guarantees to aperiodic messages when a bounded access time protocol is used for packet scheduling. The basic idea is as follows. When a new message arrives at a node, the node calculates the worst-case time by which the transmission of the message will be completed. If this time is less than the deadline of the message, then the message is given the guarantee.

Generally, in bounded access time networks, the worst-case transmission completion time is computed as follows. Assume a message M arrives at node n. The total access time T node n needs to complete the transmission of the message M is the sum of the transmission times of the previously guaranteed messages that will be sent from node n plus the transmission time of the message

M. If we assume that the access time of node *n* is bounded by *t* in each turn, then the node *n* will need T/t turns to complete the transmission of message *M*. Finally, we assume that every node in the network will always have messages to send, and they will always use their access time bound in full when they get their turns. Notice that this method is too pessimistic in granting guarantees when the network is lightly loaded.

In Ref. 68, a conservative estimation method is used with timed token protocols, and in Ref. 69 it is used with a CSMA/CD window protocol that bounds the access time of each node.

Time-Based Scheduling

One example of time-based scheduling is the dynamic reservation method.

Dynamic Reservation Method

The dynamic reservation method provides hard real-time guarantees to aperiodic messages by explicitly scheduling the channel access time of each message such that each message meets its deadline. Each node maintains a calendar structure in which the access time reservations of guaranteed messages are kept. When a message arrives at a node, the node attempts to reserve a future time interval by broadcasting a control message to all other nodes informing them about the reservation. Then, each node receiving the control message inserts the reserved interval in its calendar structure so that no other node will try to reserve that interval in the future. The control message itself has a deadline that corresponds to the scheduling deadline of the message it is attempting to guarantee. A message will not be guaranteed if its corresponding control message cannot be sent by the scheduling deadline or there is no free interval before the transmission deadline long enough to transmit the whole message.

The dynamic reservation method has been adopted for use with both a CSMA/CD window protocol and a token passing protocol (68). They perform very well under low network load conditions, but as the load increases the extra overhead of control messages results in poorer performance.

References

1. Golestani, S. J., Congestion-Free Transmission of Real-Time Traffic in Packet Networks, *Proc. IEEE INFOCOM '90*, 527–542 (June 1990).
2. Kalmanek, C., Kanakia, H., and Keshav, S., Rate Controlled Servers for Very High Speed Networks, IEEE Global Communications Conference, San Diego, CA, December 1990.

3. Ferrari, D., and Verma, D., *A Scheme for Real Time Channel Establishment in Wide-Area Networks, Technical Report TR-89-036*, International Computer Science Institute, Berkeley, CA, May 1989.

4. Ferrari, D., and Verma, D., A Scheme for Real-Time Channel Establishment in Wide-Area Networks, *IEEE J. Sel. Areas Commun.*, 8(3):368–379 (April 1990).

5. Cruz, R., A Calculus for Network Delay, Part I: Network Elements in Isolation, *IEEE Trans. Info. Theory*, 37(1):114–121 (1991).

6. Knightly, E., and Zhang, H., Traffic Characterization and Switch Utilization Using Deterministic Bounding Interval Dependent Traffic Models, *Proc. IEEE IN-FOCOM '95* (April 1995).

7. Zhang, H., and Knightly, E., Providing End-to-End Statistical Performance Guarantees with Interval Dependent Stochastic Models, *ACM SIGMETRICS '94*, 211–220 (May 1994).

8. Deering, S., and Cheriton, D., Multicast Routing in Datagram Networks and Extended LANs, *ACM Trans. Comp. Sys.*, 8:85–110 (1990).

9. Waters, A., A New Heuristic for ATM Multicast Routing, *Proc. 2d IFIP Workshop on Performance Modeling and Evaluation of ATM Networks*, Vol 8 (1994).

10. Salama, H., Reeves, D., Viniotis, I., and Sheu, T., *Comparison of Multicast Routing Algorithms for High-Speed Networks*, IBM Technical Report IBM-TR29.1930, IBM, Research Triangle Park, NC, 1994.

11. Kompella, V., Pasquale, J., and Polyzos, G., Multicast Routing for Multimedia Communication, *IEEE/ACM Trans. Networking*, 1:286–292 (1993).

12. Widyono, R., *The Design and Evaluation of Routing Algorithms for Real-Time Channel*, Technical Report TR-94-024, Tenet Group, Department of EECS, University of California at Berkeley, 1994.

13. Zhu, Q., Parsa, M., and Garcia-Luna-Aceves, J., A Source-Based Algorithm for Near-Optimum Delay-Constrained Multicasting, *Proc. IEEE INFOCOM '95* (1995).

14. Salama, H., Reeves, D., Viniotis, I., and Sheu, T., Evaluation of Multicast Routing Algorithms for Real-Time Communication on High-Speed Networks, *Proc. 5th IFIP Conference on High Performance Networks, HPN 94*, (1995).

15. Mitzel, D., Estrin, D., Shenker, S., and Zhang, L., An Architectural Comparison of ST-II and RSVP, *Proc. IEEE INFOCOM '94* (1994).

16. Zhang, L., Deering, S., Estrin, D., Shenker, S., and Zappala, D., RSVP: A New Resource ReSerVation Protocol, *IEEE Network Mag.* (September 1993).

17. Partridge, C., *Gigabit Networking*, Addison-Wesley, Reading, MA, 1994.

18. Mitzel, D., Estrin, D., Shenker, S., and Zhang, L., A Study of Reservation Dynamics in Integrated Services Packet Networks, *Proc. IEEE INFOCOM '96* (1996).

19. Zhang, L., VirtualClock: A New Traffic Control Algorithm for Packet Switching Networks, *Proc. ACM SIGCOMM '90*, 19–29 (1990).

20. Zhang, L., A New Architecture for Packet Switched Network Protocols, Ph.D. dissertation, Massachusetts Institute of Technology, Cambridge, MA, July 1989.

21. Demers, A., Keshav, S., and Shenker, S., Analysis and Simulation of a Fair Queueing Algorithm, *Proc. ACM SIGCOMM '89*, 1–12 (September 1989).

22. Parekh, A., and Gallager, R., A Generalized Processor Sharing Approach to Flow Control in Integrated Services Networks—the Single Node Case, *Proc. IEEE INFOCOM '92*, 915–924 (1992).

23. Parekh, A., and Gallager, R., A Generalized Processor Sharing Approach to Flow Control in Integrated Services Networks—The Multiple Node Case, *Proc. IEEE INFOCOM '93*, 521–530 (March 1993).

24. Parekh, A., A Generalized Processor Sharing Approach to Flow Control in Inte-

grated Services Networks, Ph.D. dissertation, Massachusetts Institute of Technology, Cambridge, MA, February 1992.

25. Golestani, S. J., A Self-Clocked Fair Queueing Scheme for Broadband Applications, *Proc. IEEE INFOCOM '94*, 636–646 (1994).

26. Verma, D., Zhang, H., and Ferrari, D., Guaranteeing Delay Jitter Bounds in Packet Switching Networks, *Proceedings of TRICOM '91*, Chapel Hill, NC (April 1991).

27. Golestani, S. J., A Stop-and-Go Queueing Framework for Congestion Management, *SIGCOMM '90 Symposium, Communications Architecture and Protocols*, 8–18 (September 1990).

28. Zhang, H., and Ferrari, D., Rate-Control Static-Priority Queueing, *Proc. IEEE INFOCOM '93*, 1:227–236 (1993).

29. Zhang, H., and Ferrari, D., Rate-Controlled Service Disciplines, *J. High Speed Networks*, 3(4):389–412 (1994).

30. Zhang, H., Service Disciplines for Integrated Services Packet-Switching Networks, Ph.D. dissertation, University of California at Berkeley, April 1993.

31. Zhang, H., Service Disciplines for Guaranteed Performance Service in Packet-Switching, *Proc. IEEE*, 85:10 (October 1995).

32. Zhang, H., and Keshav, S., Comparison of Rate-Based Service Disciplines, *Proc. ACM SIGCOMM '91*, 113–121 (September 1991).

33. Aras, C., Kurose, J., Reeves, D., and Schulzrinne, H., Real-Time Communication in Packet-Switched Networks, *Proc. IEEE*, 82(1):122–138 (January 1994).

34. Kurose, J., Open Issues and Challenges in Providing Quality of Service Guarantees in High Speed Networks, *ACM Computer Commun. Rev.*, 23(1):6–15 (January 1993).

35. Zhang, L., A Comparison of Traffic Control Algorithms for High-Speed Networks, *2d Annual Workshop on Very High Speed Networks, Proceeding Supplement*, Greenbelt, MD, March 1991.

36. Zhang, H., Providing End-to-End Performance Guarantees Using Non-Work-Conserving Disciplines, *Computer Commun.: Special Issue on System Support for Multimedia Computing*, 18(10) (October 1995).

37. Cruz, R., A Calculus for Network Delay, Part II: Network Analysis, *IEEE Trans. Info. Theory*, 37(1):121–141 (1991).

38. Bennett, J., and Zhang, H., WF2Q: Worst-Case Fair Weighted Fair Queueing, *Proc. IEEE INFOCOM '96* (March 1996).

39. Bennett, J., and Zhang, H., Why WFQ Is Not Good Enough for Integrated Services Networks, *Proc. NOSSDAV '96* (April 1996).

40. Clark, D., Shenker, S., and Zhang, L., Supporting Real-Time Applications in an Integrated Services Packet Network: Architecture and Mechanism, *ACM SIGCOMM Symp. Commun. Architectures and Protocols*, 14–26 (1992).

41. Kurose, J. F., Schwartz, and Yemini, Y., Multiple-Access Protocols and Time Constrained Communication, *Computing Surveys*, 16(1):43–70 (1984).

42. Malcolm, N., and Zhao, W., Hard Real-Time Communication in Multiple-Access Networks, *J. Real-Time Sys.*, 8:35–77 (1995).

43. Tindell, K., Burns, A., and Wellings, A. J., Analysis of Hard Real-time Communications, *J. Real-Time Sys.*, 9:147–171 (1995).

44. Liu, M. T., Distributed Loop Networks, *Advances in Computers*, 17:163–221 (1978).

45. Institute of Electrical and Electronics Engineers, *Token Ring Access Method and Physical Layer Specifications*, IEEE Standard 802.5-1989.

46. Molle, M. L., and Kleinrock, L., Virtual Time CSMA: Why Two Clocks Are Better Than One, *IEEE Trans. Commun.*, COM-33(9):919–933 (1985).

47. Liu, C. L., and Layland, J. W., Scheduling Algorithms for Multiprogramming in a

Hard Real-Time Comparative Study of Three Token Ring Protocols for Real Time Communication, *Proc. 11th IEEE Int. Conf. Distributed Computing Sys.*, 308–317 (1991).

48. Strosnider, J. K., and Marchok, T. E., Responsive, Deterministic IEEE 802.5 Token Ring Scheduling, *J. Real-Time Sys.*, 1(2):133–158 (1989).

49. Strosnider, J. K., Marchok, T. E., and Lehoczky, J., Advanced Real-Time Scheduling Using the IEEE 802.5 Token Ring, *Proc. IEEE Real-Time Sys. Symp.*, 42–52 (1988).

50. Strosnider, J. K., Highly Responsive Real-Time Token Rings, Ph.D. dissertation, Carnegie Mellon University, Pittsburgh, PA, 1988.

51. Panwar, S. S., Towsley, D., and Wolf, J. K., Optimal Scheduling Policies for a Class of Queues with Customer Deadlines to the Beginning of Service, *J. Assoc. Computing Machinery*, 35(4):832–844 (1988).

52. Shin, K. G., and Hou, C. J., Analysis of Three Contention Protocols in Distributed Real-Time Systems, *Proc. IEEE Real-Time Sys. Symp.*, 136–145 (1990).

53. Yao, L. J., and Zhao, W., Performance of an Extended IEEE 802.5 Protocol in Hard Real-Time Systems, *Proc. IEEE INFOCOM '91*, 469–478 (1991).

54. Zhao, W., and Ramamritham, K., A Virtual Time CSMA Protocol for Hard Real-Time Communications, *Proc. IEEE Real-Time Sys. Symp.*, 120–127 (1986).

55. Zhao, W., and Ramamritham, K., Virtual Time CSMA Protocol for Hard Real-Time Communications, *IEEE Trans. Software Eng.*, SE-13(8):938–952 (1987).

56. Kurose, J. F., Schwartz, and Yemini, Y., Controlling Window Protocols for Time-Constrained Communication in Multiple Access Networks, *IEEE Trans. Commun.*, 36(1):41–49 (1988).

57. Lim, C. C., Yao, L. J., and Zhao, W., A Comparative Study of Three Token Ring Protocols for Real-Time Communication, *Proc. 11th IEEE Int. Conf. Distributed Computing Sys.*, 308–317 (1991).

58. Zhao, W., Stankovic, A., and Ramamritham, K., A Multi-Access Window Protocol for Time Constrained Communications, *Proc. 8th IEEE Int. Conf. Distributed Computing Sys.*, 384–392 (1988).

59. Zhao, W., Stankovic, A., and Ramamritham, K., A Window Protocol for Transmission of Time Constrained Messages, *IEEE Trans. Computers*, 39(9):1186–1203 (1990).

60. Znati, T., Deadline Driven Window Protocol for Transmission of Real-Time Traffic, *Proc. 10th IEEE Int. Conf. Computers and Commun.*, 667–673 (1991).

61. Grow, R. M., A Timed Token Protocol for Local Area Networks, *Proc. Electro/82, Token Access Protocols*, paper 17/3 (1982).

62. American National Standards Institute, *FDDI Media Access Control (MAC)*, ANSI Standard X3T9.5/88-139, Rev. 4.0, 1990.

63. American National Standards Institute/Institute of Electrical and Electronics Engineers, *Token-Passing Bus Access Method and Physical Layer Specifications*, ANSI/IEEE Standard 802.4-1985, 1985.

64. Green, D. T., and Marlow, D. T., SAFENET — A LAN for Navy Mission Critical Systems, *Proc. IEEE Conf. Local Computer Networks*, 340–346 (1989).

65. Paige, J. L., SAFENET — A Navy Approach to Computer Networking, *Proc. IEEE Conf. Local Computer Networks*, 268–273 (1990).

66. Agrawal, G., Chen, B., Zhao, W., and Davari, S., Guaranteeing Synchronous Message Deadlines in High Speed Token Ring Networks with Timed Token Protocol, *Proc. 12th IEEE Int. Conf. Distributed Computing Sys.*, 468–475 (1992).

67. Agrawal, G., Chen, B., and Zhao, W., Local Synchronous Capacity Allocation Schemes for Guaranteeing Message Deadlines with the Timed Token Protocol, *Proc. IEEE INFOCOM '93*, 186–193 (1993).

68. Malcolm, N., Zhao, W., and Barter, C., Guarantee Protocols for Communication in Distributed Real-Time Systems, *Proc. IEEE INFOCOM '90*, 1078–1086 (1990).
69. Arvind, K., Protocols for Distributed Real-Time Systems, Ph.D. dissertation, University of Massachusetts at Amherst, 1991.

ASHOK K. AGRAWALA
ARDAS CILINGIROGLU
SUNG LEE

Regulation of Cable Television
(see The 1992 Cable Television Act)

Regulation of Telecommunications at the State Level in the United States

Why State Regulation Is Important

The United States stands virtually alone in having subnational governments play a significant role in regulating telecommunications. Canada, for example, recently ended its provincial regulation of telecommunications. The 50 American states plus the District of Columbia have the authority to regulate telecommunications rates, services, and investments *within* their jurisdiction. Interstate activities are under the regulatory purview of the Federal Communications Commission (FCC) and, ultimately, Congress.

While this sounds like a neat division of labor, in fact many political and legal battles have been fought over where and how to draw the jurisdictional boundaries between interstate and intrastate telecommunications. This problem became even more intense in 1996 following the passage of the Federal Telecommunications Act (FTA) and will continue in the future as new technologies further blur jurisdictional boundaries.

The independent existence of state telecommunications regulation comes from explicit language in the 1934 Communications Act, which avoided using the "Shreveport" legal precedent from transportation regulation that gave the federal government more potential regulatory authority over intrastate transportation. For over 60 years this ensured an important state role. In February 1996, however, Congress passed and the President signed the first major telecommunications legislation in 60 years. The FTA set the general guidelines for expanded competition in the telecommunications industry. In terms of federal-state relationships, it established national policy goals and limited somewhat the role of the states, even in their intrastate activities, in achieving these policy goals. For example, the FTA established dates by which states must achieve certain policy outcomes. The FTA also left the states significant discretion in terms of how to carry out new policies encouraging competition. Subsequent decisions by the FCC, however, in the summer of 1996 narrowed further the range of state policy discretion, causing the National Association of Regulatory Utility Commissioners (NARUC) to appeal the FCC's decisions. Thus, while state regulation continues to hold firm legal status, its scope has been reduced somewhat.

Prior to the passage of the FTA, as the Clinton administration advanced its goal of a National Information Infrastructure (NII) leading to an "information superhighway," issues of state regulation increased in importance as state regulators retained many crucial decisions that shape the extent and timing of development of such an infrastructure. The local-exchange companies (LECs), even though facing greater competition for their most lucrative customers and eventually for all customers, continue to provide most telecommunications users with the "access ramps" or "local roads" onto the information superhighway. These are the most expensive parts of the system to build and to maintain and remain likely to be the least desirable for new competitors because of their ubiquity and expense. And, they are largely under the jurisdiction of state regulators.

Ironically, at the same time the information superhighway was being discussed, new wireless technologies forced analysts to question the assumption that wireline companies would dominate the future expansion of telecommunications markets. While no one can be certain about exactly how the future of telecommunications will evolve except to be sure that it will be different from the past and present industry, it seems likely that the state regulatory role in wireless technologies will be less than that over wireline operations. In 1993, Congress passed rules on spectrum allocation that reduced substantially the role of the 19 states, including New York and California, that regulated cellular telephone and other wireless service rates. This Congressional decision to minimize the state role in setting cellular and mobile telephone services rates partially foreshadowed Congressional decisions concerning the 1996 FTA.

While state regulation was somewhat important prior to the AT&T divestiture in 1984, it was not politically salient and it was not in the media spotlight. With AT&T providing all of America's long-distance service and 80% of all local services, the choices that state regulators made focused mainly on the appropriate rate of return (ROR) to the investments that AT&T and other telecommunications companies made. This ROR regulatory process also established local residential and business rates, as well as long-distance rates within the states. Until the 1980s, since improving technology had reduced costs for decades, particularly for long-distance telecommunications, the state ROR process often involved the relatively easy political task of apportioning rate cuts for different classes of consumers and different services.

After the AT&T divestiture was implemented in 1984, state regulation was placed into the spotlight. Seven new regional holding companies (RHCs), spun off from AT&T and that owned the LECs in the states, had been created in the divestiture process. The future of these firms was closely tied to the type of state regulation they would face. The states then addressed a much broader range of possible policies in the deregulated environment, including changing rate structures to better match economic efficiency and reduce cross-subsidies and allowing competition in various segments of the industry, such as deciding which long-distance carriers could provide what kind of service within their states. With potential competition as the justification for divestiture, many analysts believed that competition would cause pressure for rates to become aligned closer to costs. Largely, this would mean pressure on state regulators to do that which they like the least — raise local rates for residential consumers.

In most states, telecommunications regulators are appointed officials who serve on a multimember commission (usually with three, five, or seven commissioners), usually called a public utility commission (PUC), public service commission, utility board, or a similar name. In about three-quarters of the American states, such regulators are appointed by the state governor, usually for staggered terms of four years or more. Such appointments typically have limits on how many Republicans and how many Democrats can serve, and the regulators must be approved by the state senate. In a few states, such as Florida and Ohio, a bipartisan citizen group develops a list of qualified candidates from which the governor must choose. In the other one quarter of the states, commis sion members are elected, by the state legislators in South Carolina and Virginia, and directly by the voters at large in the other states.

The idea behind the appointed commissioners is that they will be somewhat insulated politically to make difficult decisions and tradeoffs. Election of commissioners, on the other hand, provides a more direct accountability to voters. Research has focused on the different decisions made by appointed versus elected commissioners, but a clear and consistent pattern has yet to be found (1).

Advantages and Disadvantages of State-Level Regulation

Theoretical Advantages

With all of the technological and market changes in telecommunications, some have questioned whether state regulation still makes sense, especially as no other major countries have such regulation. But, theoretically, state regulation can provide some benefits. First, state regulation takes place "closer" to the citizens and businesses of a state. Thus, state regulators know more about their own economy and society and can tailor their policies appropriately according to these local needs. Needs, preferences, and tastes for telecommunications services may vary across states, related to such issues as subsidizing poor and rural people onto networks, the desire to have many competitive alternatives, and the tradeoff of current prices versus future network services that is implicitly reflected in the arcane accounting details of depreciation rates. As Andrew Varley, former chair of the Iowa State Utility Board, argues:

> It is essential to allow state laboratories to experiment and see what works. . . . If you live in Washington, D.C., you can reach with a local call roughly the same number of people as the entire population of the states of Iowa and North Dakota. . . . These people (FCC) should totally understand our problems before prescribing solutions. (2)

In parallel, state regulation should also lead to more public participation in decisions and more responsiveness by politicians and their appointees to local interests, which are important democratic values (3). Traditionally, however, scholars have found business interests to be more dominant on the state level than on the federal level because of less intense scrutiny of regulation by the media and consumer groups at the state level and states' needs for economic development from mobile businesses (4). At least in the larger states, however, this situation has been changing. Beginning in the 1980s, consumer groups and the media realized that not all important regulatory activity takes place in Washington, D.C., and earlier state procedural and institutional reforms reduced the likelihood of complete business dominance (5).

Another important advantage is that states can act as experimental laboratories from which analysts, other states, and the federal government can learn. In the area of economic development, many analysts in the late 1980s argued that states have provided important policy laboratories (see, e.g., Ref. 6). This type of policy response fosters a kind of incrementalism in American policy that illustrates the value of federalism. According to Stanford Levin, former Illinois Commerce Commission regulator: "State diversity is helpful to learn more about the elements of uncertainty" (2). Issue networks created through the NARUC, the National Governors Association, and regional groups such as the Council of Great Lakes Governors encourage such learning. According to Eliot Maxwell, former executive director of external affairs for Pacific Telesis: "State regulation might converge to become more homogeneous in ten years, but the differences today might help choose among policy approaches" (2).

In some cases, state policies have guided later federal policies. A few states, particularly New Jersey, New York, and Massachusetts, acted before the federal government, for example, on the increasingly important issue of network interconnection agreements. The FCC has learned from those state policy experiments and developed its own interconnection policy with an attentive eye on these state successes and failures.

As a related subset to the policy experimentation issue, some states are attempting to use their telecommunications regulatory innovations to stimulate economic development. Many analysts and firms argue that there is a critical link between economic development and state telecommunications regulatory policies. Nebraska found this argument convincing enough to deregulate radically in the hope that it could enhance its status as the "1-800" number capital of the United States (7). The benefits and costs of interstate economic development competition generally are the subject of widespread debate; some feel that competition promotes policy innovation that "expands the pie," while others suggest that it is usually a "zero-sum game" that creates a prisoner's dilemma for states, ultimately leading to taxpayer subsidies of private corporations. To the extent that the first view is accurate, varying state regulatory models can illustrate the importance of the telecommunications infrastructure in stimulating economic growth (8).

Theoretical Disadvantages

While these theoretical benefits may justify ongoing state regulation, in theory state regulation also may increase costs to our society. Telecommunications networks require standards so that all subscribers can communicate with each other, regardless of where they live and what local preferences are on various policy issues. Americans need an interconnected national network rather than 51 (including the District of Columbia, which has its own regulators) separate state networks. This need may become even more critical as newer and more complicated communications services become feasible on the information superhighway. Superior telecommunications is a vital infrastructure for international competition, and state regulatory fragmentation could hinder this development (9). As Noll noted before the 1996 FTA:

> The existing jurisdictional separation arises from the grossly outdated Communications Act of 1934, written when almost all of the revenues in the telecommunications

industry came from local telephone service, and before such innovations as television, satellites, microwave, and computers. The notion that there is a meaningful technical and economic distinction between federal and state services was always a fiction, but it has become increasingly so. This fiction is likely to become ever more costly as the federal government withdraws from regulation, thereby increasing the capability of the states to impose mutually inconsistent requirements on what amounts to a national network providing national information services using equipment manufactured for a national market. (10)

The Integrated Systems Digital Network, or ISDN, is a good example, as are many new services provided over the Internet, of the need for standardization across local networks now regulated by different state commissions. If every state takes a different approach, problems could arise. As in the computer industry, however, component compatibility can come about in many ways — voluntary, enforced, or mixed models are possible. While single, centrally planned technological standards are often touted as desirable, there may be tradeoffs between enforced uniform standard setting and further technological innovation. The federal government could choose the "wrong" standard, for example, as many have argued they have done with high-definition television (HDTV). A mixed model with minimum standards but room for innovation, as practiced on today's rapidly growing Internet, may be appropriate.

To this end, Noll outlined four options for jurisdictional standard-setting disputes in telecommunications:

1) full federal control; 2) state control; 3) a consortium akin to the old Ozark plan . . . to develop consensus among all participants; or 4) minimum standards for elements important in the federal jurisdiction set by federal officials, with states deciding as they wish within those federally-imposed strictures. (11)

Noll predicted that the fourth option was most likely to emerge as a compromise. Based on the 1996 FTA and subsequent FCC interpretations, it appears that Noll's prediction was fairly accurate.

Issues of multiple jurisdictions and standards setting are not simply scholarly debates; comparative advantage is involved for many segments of the industry. Some firms may gain more than others from single national standards. According to Northern Telecom's Patrick Keenan: "Many of the potential conflicts between federal and state regulators are in fact driven by enhanced service providers (ESPs) who complain they do not want to work with 51 separate state jurisdictions" (11). In another example, "Daniel Kelley of MCI said interexchange carriers, not just ESPs, have strong fears of too much variation across state jurisdictions" (11).

Related to the standards issues, positive network externalities do not end at state boundaries, they can flow from one state to another, as with new information services. Therefore, a second advantage of one national regulatory policy is the positive network externalities to all subscribers from expanded subscribership in other states. Any network becomes more valuable to all of its existing subscribers when more subscribers can be reached; positive externalities were the implicit economic justification for the national goal of "universal service." If some states develop policies that lose subscribers or that act to retard the expansion of new network services, customers in other states will be harmed by the smaller network size. Some believe these externalities are the most important consideration opposing the advantages of state decision making (12).

So far, there is no conclusive evidence on whether or not states are retarding new service expansion. According to a summary of a 1988 expert panel discussion:

> It appeared from the discussion that while there was great concern about possible *future* impediments to the development of enhanced services that state regulation might erect, in practice the ESPs are so new that there has not been time to see whether states actually will create significant obstructions. (11)

While many new and potential information services are developing rapidly that could render this statement obsolete, it remains essentially accurate in 1997.

A third normative argument against state regulation is that there may be economies of scale in government's production of regulatory policy; 51 jurisdictions may waste substantial resources, and 1 federal regulator might be cheaper and more effective than 51. The U.S. Department of Commerce's National Telecommunications and Information Administration estimated that direct telecommunications regulatory costs were about $1 billion in 1985, less than 1% of industry revenues, but that indirect costs of regulation were much higher. Quantitative evidence from the 50 states points to some economies of scale in developing analytically informed policy about rate structure (11). In particular, small state public utility commission staffs may not have the capacity to regulate thoroughly, as well as to oversee elements of this growing industry. Some states have already recognized this by cooperating on a regional basis to deal with RHC-level telecommunications issues.

In considering these theoretical advantages and disadvantages of state regulation, there are some important questions that have not yet been answered completely. These include:

Do preferences for telecommunications policies really vary greatly by state?
Does being closer to constituents really help improve policy, or, instead, does it prevent changes in rates that would enhance economic efficiency?
Are state regulators inherently and inevitably preoccupied with local residential rates to the exclusion of all other important factors?
Can states with small regulatory staffs really do an adequate job regulating?
Are the states really innovating or just being dragged along by exogenous changes in technology and markets?
Do states imitate successful policy experiments in other states?

History of State Regulation

American telecommunications history includes periods of competition and monopoly, with regulated monopoly as the most dominant form in the 20th century. After the expiration of its initial patent monopoly, before 1900, AT&T faced competition all over the country. AT&T consolidated its holdings and agreed to accept regulation as a tradeoff for its near-monopoly status. As the federal government began to consider regulating AT&T, AT&T used state regulation to gain their regulated monopoly, to avoid municipal regulation or ownership, and to maintain a continuing cross-subsidy that was beneficial to the growth of their business and profits (13,14).

Thus, state regulation of telecommunications predates federal regulation as the early growth of networks was largely localized. Some municipalities even regulated the industry prior to the states. The first state regulatory commissions were established in Wisconsin and New York in 1907, utilizing PUCs that were already dealing with electricity regulation (13). State telecommunications regulation spread rapidly, with 42 states plus the District of Columbia regulating by 1920 (14). While the Interstate Commerce Commission (ICC) regulated AT&T somewhat after 1914, the federal role was strengthened in the 1934 Communications Act with the establishment of the FCC. Unfortunately for later policy disputes, this act did not adequately clarify the jurisdictional issues between the federal government and the states, in part because at that time only 10% of AT&T's revenues came from the interstate jurisdiction.

The most contentious issue of debate at that time was the rate structure. The 1931 U.S. Supreme Court case *Smith versus Illinois Bell* argued that, since toll calls required the existence of local networks, toll call prices should include a contribution toward paying these local costs. Today, most economists argue that such logic runs counter to economic efficiency. Still, this method was adopted, and the cross-subsidy from toll calls to residential service grew steadily after World War II. It was particularly tempting for regulators to expand this practice as technology steadily drove down the costs of toll calls.

This policy of subsidizing residential end users from toll call revenue aided the federal goal of universal service, and, with the regulated natural monopoly protected from "cream-skimming" competition, it continued with few changes until competitors began to chip away at the edges of the Bell monopoly in the 1960s. With the development of microwave communications and subsequent FCC decisions expanding long-distance competition in the 1970s, the previously stable environment began to unravel. Inflation and slower technological progress meant that local service costs no longer fell as rapidly as in the past. Technological innovation continued to drive toll costs down, however, so regulators, particularly Congress and state regulators, pushed to increase the subsidy from toll prices to local rates to retain the political advantage of falling local rates. With the emergence of MCI as a competitor and with the court-mandated divestiture in 1982, however, this subsidy system faced severe constraints. Federal regulators and the state PUCs both faced difficult decisions and a more complex relationship.

For most of its history, then, state telecommunications regulation was not a highly salient area of political interest. The most important issue was the ROR for the local carriers, which affected the prices users paid for services. But, the method of finding an appropriate ROR was a complex, technical one to which few observers, other than those with high direct stakes, paid much attention. Thus, it was not an intensely political issue in the way that it has become today. Another important reason for the lack of political conflict was that rates, especially for long-distance service, were generally falling as technology improved. These lower costs were translated into lower prices but also into higher cross-subsidies to reduce the price of local service.

Divestiture and State Regulation

After divestiture, much more attention was focused on state regulation. The new competitive environment put substantial pressure on all regulators to raise local rates, and state regulators would be forced to make most of the critical decisions. Huge new companies had been created, and much of their viability and success would depend on state regulation. Deregulation and competitive markets were the ideas that motivated the divestiture; this raised the question of how soon state regulation would adapt to accommodate this new model.

After the divestiture in 1984, states varied greatly in the extent to which they followed the main federal policy of changing rate structures by reducing long-distance rates and increasing end-user charges and opening markets to competitive entry (7). While some states made major regulatory policy changes, many took only small, gradual steps. Thus, the states did in fact pursue a variety of policy models.

Why did the states develop different policies? These policy choices are partly the result of different political environments and structures in each state and partly the result of choices made by legislators and regulators themselves (15). While consumer preferences or tastes may vary across states, the preferences and decision calculi of political officials and their staffs probably vary more substantially and importantly across the states.

The politics of rate setting are controversial because the exact extent of the cross-subsidy is not easily resolved either by economic theory or by available empirical data. Still, most economists call for increased end-user charges and reduced toll rates to price telecommunications more efficiently (16). Such rate structure changes discourage large-business users of telecommunications from bypassing the switched network (which tends to reduce the source of the subsidy) and reducing their telecommunications bills substantially. However, these changes tend to increase total telephone bills for a majority of residential consumers, at least in the short run. The possibility of a majority of consumers facing higher bills obviously causes political resistance and made state regulators, to whom local increases are most easily traced, less likely to change rate structures than their federal counterparts. Noll characterized state goals to keep residential prices low as leading to "the possibility of turning the BOCs [Bell Operating Companies] into the railroads of the 21st Century: sluggish, inefficient firms that are moribund because of the constraints imposed by regulation" (17).

These issues are also contentious because of the competitive questions of which businesses should be allowed to participate in which markets. Rather than directly involving residential consumers, this is more of an interest group battle in which some firms want regulatory protection and others want to be allowed into as many markets as possible. There are tradeoffs involved. Economies of scope, for example, may be best exploited by allowing the regulated LECs to participate in a wide range of new and enhanced information services, such as database access or "video on demand." But, the LECs still control the "bottleneck" local transport and switching functions that nearly all carriers must use to connect to ultimate users. Thus, the local companies may be able to choke off competitors and cross-subsidize services to their own advantage (which is why AT&T was divested and Judge Greene continued to monitor the LECs closely for these antitrust issues for over a decade). The LECs argued for more than 10 years that they should be allowed into new markets, while less-regulated competitors began to nibble into the margins of their formerly captive markets, particularly in the most profitable segments of the local market (see, e.g., Ref. 18). New competitive access providers (CAPs) began to bypass the local telephone providers, and, in the 1990s, cable television firms, with plant and equipment that already passed by most American households, began to seek to provide telephone service.

Prior to the 1980s, federal and state goals were largely similar (19). After 1980, the FCC pushed hard for competition. To advance deregulation and to push the states along further, both before and after divestiture, the FCC successfully preempted the authority of state PUCs in a number of rulings (20). This trend was abruptly reversed in 1986, however, with the U.S. Supreme Court *Louisiana* ruling in favor of the states on the issue of intrastate depreciation rates. A few other subsequent cases tended to support the states' rights to maintain regulatory control over intrastate concerns. Congressional action in 1993, however, that preempted state regulation of cellular and wireless service rates set the tone for further Congressional preemption of state authority, which is much more likely to withstand court scrutiny than FCC preemption. As noted above, the 1996 FTA established deadlines for state action and national policy goals that could not be frustrated by state policies.

Postdivestiture State Policies

In general, after divestiture local rates did increase substantially; interstate long-distance toll rates, as (de)regulated by the FCC, fell by over 40%; and intrastate long-distance rates also fell, but by a lesser amount. The interstate drop is largely a result of reversing part of the cross-subsidy by increasing flat end-user charges and reducing carrier access charges. Thus, federal policy has moved in the direction favored by economists since divestiture. With some states holding back the pace of change, however, access charges remain about 50% of the total cost of long-distance calls and still represent a substantial cross-subsidy.

Although it oversimplifies the variation among the states, to some extent state regulatory actions after divestiture can be categorized into three overlapping phases. Sharon Nelson, former president of the NARUC, coined phrases for the subsequent state phases, which she dubbed "the 3 R's": reaction, re-trenchment, and restructuring.

The first phase was reaction, which occurred from 1982 when the AT&T breakup was announced until about 1986. The states needed to learn quickly how to regulate new companies in a new environment. Most of all, they did not want to create major problems, rate shocks, or disruptions of telephone service. There was considerable uncertainty and a gradual, careful approach seemed most appropriate. Only in a few states did legislators and regulators move quickly to respond to the new environment. But, nearly all states did need to respond over the next few years as many had regulatory statutes that reflected the old monopoly environment and did not allow PUCs even to consider compe-tition and other new policy decisions. Thus, state legislators, in addition to PUCs, often became involved in these policy choices.

After the reaction period, which was associated with large local rate in-creases, came the retrenchment phase, from about 1986 through 1990. Rate pressure relaxed somewhat, particularly as the 1986 Tax Reform Act reduced most LEC corporate taxes and as general price inflation was reduced. State regulators were concerned about the local increases that had been implemented in the reaction phase and used this period to retrench as much as possible, if only to gain some time and consumer goodwill. Several states started to establish local rate freezes or moratoria at this time.

As the pressure on local rates receded and after legislators and regulators saw that increases in local rates, combined with lower long-distance rates and more telecommunications options, did not cause consumers to revolt and to dislodge politicians and regulators, the third phase of restructuring began in the 1990s. At this time, legislators and regulators began to map a plan for the future that would allow competition to exist, at least temporarily, alongside continued regulation of near-monopoly elements of state telecommunications. Restructur-ing includes new policy directions like long-term price caps on local residential service, incentive plans to allow LECs pricing flexibility for more competitive services, incentives to cut costs to earn greater profits (rather than an ROR cap), and other programs that fundamentally changed the way in which telecommuni-cations firms were regulated at the state level. Econometric study has shown that states with very high or low earnings, with high-quality commissions, with limited commission resources, with elected commissioners, and with Democratic governors were more likely to adopt incentive regulation plans (21).

Even before the restructuring phase, some states proved to be innovative. Before divestiture, California had developed a lifeline program for low-income subscribers. After divestiture, Illinois utilized an aggressive, competitive, eco-nomics-oriented approach to deregulation. Nebraska approved radical legisla-tive telecommunications deregulation focused on achieving economic develop-ment goals. Vermont adopted a "social contract" approach designed to freeze local rates but allow pricing flexibility for the LECs in other services. On the other hand, states like Idaho tried to stick to a traditional approach to regula-tion.

In the restructuring phase of the 1990s, states tried many new approaches. New York allowed expanded local competition and encouraged interconnection agreements. New Jersey gave regulatory incentives to New Jersey Bell to have the state completely wired with fiber optics by 2010, before any other state. Kansas and Wisconsin provided examples of rate moratoria. Oregon and Rhode Island pursued price cap regulation, along with 8 other states. By 1994, 19 states, including Florida and Texas, had adopted earnings sharing plans designed to promote incentives for LECs to cut costs. In all phases, but particularly in the restructuring era, states acted somewhat as a policy laboratory, with real policy variation. Recent FCC decisions that start implementing the 1996 FTA explicitly recognize the innovative policy roles of some states.

Cole argues that although states have acted as laboratories, the analysis of success and failures has not been systematic (22). Only Mueller has examined a state telecommunications policy experience critically and objectively, that of Nebraska, and it focuses only on a single state over a relatively short period of time (23). Megdal points out that these cannot be considered pure or untouched experiments; they involve many changing variables, and specific causal links are hard to pinpoint (24). A few econometric studies have found that, controlling for other factors, innovative regulatory plans, particularly price caps and incentive regulation, have encouraged LECs to reduce intrastate toll prices and to invest in more modern telecommunications infrastructures (25–28). While some interested stakeholders, as well as some scholars, definitely will be analyzing this cross-state data to argue that certain policies promote the public interest, it is less clear which government agencies have the resources and incentives to do such needed analysis in a time of federal and state budget cutbacks.

Thus, after 10 years of new policies following divestiture, most states had opened their intra-local-access-and-transport-area (intra-LATA) market to competition. A handful had also opened up local services to competition and were developing interconnection agreements to allow new competitors to access a broader market on fair terms. Based on the FTA, all states must soon address interconnection and local competition issues under federal guidelines and time constraints. Many states had already moved to reduce the cross-subsidy to local service, but almost none had completely reduced it. Most states had developed new regulatory paradigms for the LECs, often with price caps on local services, with pricing flexibility for other services, and with an ability to earn a higher rate of return by cutting costs.

In addition to state regulation, state telecommunications policies include other closely related elements. One is taxation. Most states tax telecommunications services, some states more than others, and this has created a problem in high-tax states like New York (29). Like other utilities, telephone services often acted as "hidden tax collectors." As with cross-subsidies, policy makers could easily get away with such practices in a monopoly environment. As competition developed and as telecommunications costs became crucial to state economic development efforts, the tax component became a more explicit element of public policy. New York, for example, cut its telecommunications taxes in the late 1980s as the taxes represented about 18% of the telecommunications costs of large-business users, compared to only 6% in neighboring New Jersey.

The tax issues lead to the next point. As its importance to infrastructure and economic development became more clear, state telecommunications policy was influenced increasingly by other actors in addition to the PUC regulators, including legislators, governors, and economic development agencies. The Advisory Commission on Intergovernmental Relations considered this important enough to call it "Phase Two" of state telecommunications policy after divestiture (30). Teske and Bhattacharya argue that this is a positive development that will tend to increase innovation in the state regulatory environment (31). New York State's Telecommunications Exchange, a collection of public and private actors interested in influencing telecommunications policy, may be at the leading edge of state efforts to bring a wider array of concerns into state policy making.

The idea gradually emerged in the 1980s that advanced telecommunications technology was critical to developing state economies and jobs. As the service economy expanded and more workers needed to be linked by networks, this slowly led to a dissolution of the need for concentrated economic activities. Back offices of large service operations, 1-800 operations, and home shopping all require good quality, affordable telecommunications. Some states saw advantages in their economic development strategies, with Nebraska, already with a good infrastructure because of the long-time presence of the Strategic Air Command, moving strongly in this direction. Other states felt that they had to follow, and by the late 1980s and early 1990s, more than half of the states had commissioned task forces or consulting studies to determine how best to exploit this potential in their state.

An important related issue was the telecommunications facilities and usage of the largest user in most states, the state government itself. When universities, hospitals, and other facilities are included in addition to the standard state executive branch agencies, government is a huge user of telecommunications services. State policy makers in the late 1980s began to explore the possibility of using assured demand by state government to stimulate expansion of the development of improved technologies. Maine was a leader in this movement, which often focused on bringing advanced technologies to more-rural areas before they otherwise might have received them due to a relative lack of aggregate private-sector demand. New York State coordinated a considerable amount of its internal telecommunications activities through its cross-agency Forum for Information Management, which led to the development of the innovative, award-winning Center for Technology in Government (32).

Thus, while regulation by the PUCs is still the most important element of telecommunications policy at the state level, other actors and other influences have been playing stronger roles over time. These are likely to continue to grow in importance as more American workers become reliant on telecommunications at the office or as telecommuters.

Politics and the Continuing Existence of State Regulation

The theoretical debates about the appropriate levels at which Americans should regulate telecommunications may not be as important in regulatory politics as explanations focusing on why we do regulate at the state level. In reality, states do not only make regulatory policies because of arguments that they should, but because specific strong interests favor their involvement and policies (33). Positive explanations of state regulation must consider which interests are favored by state regulators and how they might be harmed if the system were changed, as well as analysis of bureaucratic incentives.

Residential consumers largely have been protected by state regulation and seem to want it to remain. For most of the last 15 years, the FCC was committed to raising their local rates (while cutting their toll rates) far more than were most state regulators. Consumers may feel that slower (than federal) state deregulation moderates change in an era marked by a transition away from the protection and stability of "Ma Bell."

Of course, residential consumers (who are also voters) were not directly active participants in many of the most important regulatory decisions in telecommunications. Historically, they were not clamoring for state regulation around 1907, nor did they favor AT&T being dismantled in 1982. Opinion polls show that residential consumers largely do not favor faster deregulation of telecommunications services. A December 1989 survey found that more consumers felt divestiture was a bad idea rather than a good idea (34). This contrasts with an extremely favorable public opinion about telecommunications services prior to divestiture (35).

Some other interest groups favor continuing state regulation. The LECs are highly strategic actors that prefer state regulation when it maintains healthy profits by protecting and nurturing them and by keeping competitors out, or at least placing competition on a "level playing field" (36). The specific preference of a local company depends largely on their reading of the politicians and bureaucrats in the state and their comparative advantage in different markets (7). Some local companies, such as Ameritech in the midwestern states and Rochester Telephone in New York, have tried to maximize opportunities by bargaining across federal and state rule makers, hoping to be allowed into prohibited markets in exchange for allowing competitors into their domains.

The U.S. Congress generally seems to have favored state regulation as it may have taken some political heat off them on this contentious issue. Consequently, Congress has not moved to preempt state regulation as it did for airlines, railroads, trucking, and savings and loan regulation. The 1980s were not the first time that Congress and state regulators worked together; they blocked the FCC from reducing the subsidy from toll prices to local prices in 1970. According to Noll,

> Initially the FCC's criterion was to deregulate whenever competition could be relied upon to control AT&T's market power, but by the mid-1980s the primary criterion has apparently become political feasibility to deregulate as fast as possible without causing a pro-regulatory backlash in Congress. (17)

The oversight relationship between Congress and the FCC at the federal level and state legislatures and state regulators is interesting and appears opposite in some important ways. In the 1980s, the FCC pushed deregulation and rate structure changes, while Congress resisted, sending signals without actually passing significant legislation (37). While state regulators show great variety, as a whole they have deregulated far less than the FCC, and in several cases state legislatures have passed laws to push regulators to increase the pace of deregulation.

Three factors help explain why some state legislatures passed procompetitive, deregulatory legislation when Congress did not pass any significant telecommunications legislation and leaned in the opposite direction. First, less than half of the state legislatures actually passed these deregulatory bills. Second, a majority of the bills were passed in states served by U.S. West, with economic development clout and aggressive political pressure that was a strong factor state legislators were more likely to be concerned about than regulators (7). Third, in a few states laws were required simply to update public utility enabling legislation that was 75 years old and that did not anticipate deregulation. Thus, on closer inspection, and exempting the U.S. West states, state legislatures have not been strongly out of step with state regulators or with Congress.

In considering the incentives and preferences of critical actors, regulators themselves must be part of the picture. Regulators and their staffs may want to keep their jobs and bias their decisions in favor of continued state regulation (38). Indeed, in contrast to deregulation at the federal level for airlines (the Civil Aeronautics Board is out of business), trucking, and railroads (the ICC staff has declined by more than 70%), state telecommunications regulatory bureaucracies generally have grown, with the increased need for analysis and monitoring, since the AT&T divestiture. In these other industries, federal bureaucrats helped to start deregulating themselves out of jobs (39), which has not happened in state telecommunications regulation. This explanation does not focus on a large number of bureaucrats; the typical state telecommunications regulatory staff is smaller than 20 people, with only a handful of such staff in smaller states. However, these bureaucrats hold power more from control over information delivered to decision makers than from their numbers, and a bias in favor of continued regulation could hardly be considered surprising.

With so many powerful advocates—Congress, many LECs, residential consumers, and state regulators—it is not entirely surprising that state regulation remains in place. There are some opponents, however. As noted above, some LECs that are not regulated favorably by their states are opponents. Many interexchange carriers (IXCs) and ESPs dislike the possibilities of regulatory balkanization and the increased costs of monitoring, lobbying, and complying. Some large-business users of telecommunications also prefer a single set of federal regulations and favor the overall pricing and competitive policies of the FCC. Lobbying in 50 states and dealing with 50 different sets of rules is expensive and cumbersome for all of these groups, especially if they feel that state regulators and legislators are essentially captured by the LECs on issues on which they are opposed.

Powerful interests take their stands on both sides of this debate, but state regulation is also aided by its historical entrenchment. Regulation is often harder to remove than to initiate. The choices made in 1996 were only a little more than one decade away from the AT&T divestiture. State regulation also remains important in telecommunications because more than half of all revenue is still generated at the intrastate level.

Since important interests want it to remain, there are few signs that Congress will completely preempt the state role. The tradition, range, and power of state regulation suggest that it will not soon wither away in telecommunications. As Noll notes: "I believe that state regulation probably will have a role, at least for the next decade or two, but not because I believe that a major state role is either inevitable or desirable" (10). And, as Doug Jones, director of the National Regulatory Research Institute, notes: "The decline and fall of state regulation has been prematurely implied or predicted a number of times by practitioners as well as academics" (40).

Likely Future of State Regulation

Until and unless Congress preempts them entirely, state regulators have a critical role to play in the future development of American telecommunications. In their regulation of LECs, they have regulatory authority over a majority of all new telecommunications investments and thus a good portion of the future American infrastructure. The role of state regulators has been circumscribed somewhat by the 1996 FTA and subsequent FCC decisions. As one analysis described the legislation, it

> ushers in a brave new world of telecommunications competition, sweeping away a decades-old tradition of state regulators acting as gatekeepers to the telecommunications market. . . . In contrast to the jurisdictional dichotomy reflected in existing Section 2(b) of the 1934 Act, Section 253(a) applies to purely intrastate services as well as interstate services. (41, p. 10)

But the states do retain an important policy role in establishing competition, interconnection, and rate-setting decisions.

Inevitably, though, it appears that all government regulation of telecommunications will be reduced over time, and the relative share of that regulation performed at the state level will also likely be reduced. New technologies are making state boundaries far less relevant than they used to be. It will be increasingly difficult to categorize telecommunications traffic as intrastate as diverse networks and new technologies make these geographic boundaries ever more irrelevant (42). As the telecommunications networks of the future become more complex, modularized, and interconnected, drawing the precise regulatory boundaries will not only become more difficult, it may become a very costly and inefficient exercise. Thus, Eli Noam argues that the near-term future for state regulation is regulating a "shrinking share of a shrinking pie" (43).

Some prominent critics believe that state regulation should be reduced. Noll argues that "in the long run the telecommunications system might better serve society's objectives if, as in broadcasting, state regulation played no role at all" (10). In part, critics make these arguments because the states have not completely resolved and are still dealing with the issues that they faced immediately after divestiture. Local rates are probably still below their costs in many states, leaving more competitive services to pay disproportionately high rates. As local competitors gain ground, this subsidy continues to be unsustainable even if the subsidy is smaller than it was a decade ago. This creates incentives for state regulators to drag their feet, explicitly or through time lags in regulatory policy, on competition and interconnection issues. The FTA, however, reduces the ability of states to drag their feet on these issues.

It would not be unprecedented in America for state regulation to recede or even disappear over time. After being the initiators of regulation in this country, states no longer play an active role in the regulation of railroads as Congress preempted them in 1980. Congress preempted state trucking regulation in 1994. The FTA and FCC implementation decisions have already reduced the state role considerably by forcing action by certain deadlines and establishing default policies if states do not act. Although recent court decisions favored the states, they were based on the 1934 Communications Act, and Congress has been much clearer about jurisdictional roles in the 1996 FTA.

As Congress did finally pass its first major telecommunications regulatory action in 60 years in 1996, it is also timely to ask whether the American reliance on state regulation is at odds with experience in other parts of the world. For example, while the Western European nations will maintain substantial control over their networks in the coming years, they are coordinating a growing portion of their telecommunications activities through the European Community. At least for now, the United States appears to be alone in emphasizing a strong role for the states, even if that role is a little smaller in 1996 than it was in 1995.

Conclusions

The 1996 FTA reflects an emerging consensus at the national level about what American telecommunications policy should look like, a necessary condition for preemption of the states in favor of a single national policy. Most officials seem to support a government policy with competition on a "level playing field," with open architecture and interconnected networks, with prices that reflect economic costs, with antitrustlike regulatory safeguards, and with explicit programs to maintain universal access. The FCC is engaged actively in the process of developing specific regulations to implement these goals. As we have seen, however, the implementation and timing of introduction of each of these policy elements will affect stakeholders differently and are thus controversial and subject to being blocked politically. Thus, states are likely to adopt somewhat different approaches to these issues, but under a structure and timetable given to them by the federal government.

State regulation paints a picture of contrast, with rapidly changing technologies and markets regulated by a fairly stable regulatory structure. That structure served well in the past, with little need for political attention for decades and as a moderating force since the AT&T divestiture, but the structure may require further revisions in the future. The states have been making changes in their regulatory structure over the past decade, but these changes generally have been quite gradual. Thus, state regulation will influence the development of information superhighways over the next several years. But, over time, if competition works as well as many policy makers expect, it may slowly fade in importance and perhaps eventually fade away completely.

State regulation has served an important purpose in a large and diverse nation. In the face of the breathtaking technological changes that lie ahead, it may lose its purpose and role. State regulation will not end before the turn of the century or perhaps for years, or even decades, beyond that, but it almost certainly will lose much of its role over a longer span of time. How that transition is completed, how states cooperate with national objectives and with one another, and how states handle the important policy choices they still control will greatly affect the success of the American telecommunications industry and, indeed, of the American economy in global competition.

One recent observer is optimistic about continuing state regulation: "Most state governments will succeed, and a few will excel, in meeting these policy challenges. At the risk of being polemical, states and territories will manage telecommunications policy issues at least as well as the federal government" (44, p. 127).

References

1. Teske, P., The State of State Regulation. In *Handbook of Regulation and Administrative Law*, Marcel Dekker, New York, 1994, pp. 117–137.
2. Teske, P., *State Telecommunications Regulation: Assessing Issues and Options in the Midst of Changing Circumstances. Report of an Aspen Institute Program in Communications and Society Conference*, Aspen, Colorado, 1987.
3. Gormley, W., Comments—Alternative Perspectives on Intergovernmental Relations. In *American Regulatory Federalism and Telecommunications Infrastructure* (P. Teske, ed.), Lawrence Erlbaum Associates, Hillsdale, NJ, 1995, pp. 105–12.
4. Schattschneider, E., *The Semi-Sovereign People: A Realist's View of Democracy in America*, Holt, Rinehart and Winston, New York, 1960.
5. Scholz, J., State Regulatory Reform, *Policy Studies Review*, 1:347–360 (July 1982).
6. Osborne, D., *Laboratories of Democracy*, Harvard Business School Press, Cambridge, MA, 1987.
7. Teske, P., *After Divestiture: The Political Economy of State Telecommunications Regulation*, SUNY Press, Albany, NY, 1990.
8. Wilson, R., and Teske, P., Telecommunications and Economic Development: The State and Local Role, *Economic Development Quarterly* 4:158–174 (October 1990).
9. Harris, R., Telecommunications Policy in Japan: Lessons for the U.S. Paper presented to the Rutgers University Conference on Public Utility Economics and Regulation, Monterey, CA, 1988.
10. Noll, R., *Additional Comments on Statement of Goals and Strategies for State Telecommunications Regulation*. Report of an Aspen Institute Program in Communications and Society Conference, Wye, MD, 1989.
11. Entman, R., *State Telecommunications Regulation: Developing Consensus and Illuminating Conflict*. Report of an Aspen Institute Program in Communications and Society Conference, Wye, MD, 1988.

12. Haring, J., and Levitz, K., The Law and Economics of Federalism in Telecommunications, *Federal Communications Law J.*, 41:261–330 (Fall 1989).
13. Gabel, D., Federalism: An Historical Perspective. In *American Regulatory Federalism and Telecommunications Infrastructure* (P. Teske, ed.), Lawrence Erlbaum Associates, Hillsdale, NJ, 1995, pp. 19–34.
14. Cohen, J., *The Politics of Telecommunications Regulation: The States and the Divestiture of AT&T*, M. E. Sharpe, New York, 1993.
15. Teske, P., Rent-Seeking in the Deregulatory Environment: State Telecommunications, *Public Choice*, 68:235–243 (January 1991).
16. Wenders, J., *The Economics of Telecommunications*, Ballinger, Cambridge, MA, 1987.
17. Noll, R., State Regulatory Responses to Competition and Divestiture in the Telecommunications Industry. In *Antitrust and Regulation* (R. Grieson, ed.), Lexington Books, Lexington, MA, 1986, pp. 165–200.
18. Teske, P., and Gebosky, J., Local Telecommunications Competitors: Strategy and Policy, *Telecom. Policy*, 15:429–436 (October 1991).
19. Noam, E., *Federal and State Roles in Telecommunications: The Effects of Deregulation*, Research Working Paper at the Columbia University Institute for Tele-Information, 1983.
20. Maher, W., *Legal Aspects of State and Federal Regulatory Jurisdiction Over the Telephone Industry: A Survey*, Center for Information Policy Research Monograph, Harvard University, Cambridge, MA, 1985.
21. Donald, S., and Sappington, D., Explaining the Choice Among Regulatory Plans in the U.S. Telecommunications Industry. Paper presented at the AEI Conference: Telecommunications Summit: Competition and Strategic Alliances, Washington, DC, 1994.
22. Cole, B., State Policy Laboratories. In *American Regulatory Federalism and Telecommunications Infrastructure* (P. Teske, ed.), Lawrence Erlbaum Associates, Hillsdale, NJ, 1995, pp. 35–46.
23. Mueller, M., *Telephone Companies in Paradise: A Case Study in Telecommunications Deregulation*, Transaction Books, New Brunswick, NJ, 1993.
24. Megdal, S., The Benefits of State Regulation. In *American Regulatory Federalism and Telecommunications Infrastructure* (P. Teske, ed.), Lawrence Erlbaum Associates, Hillsdale, NJ, 1995, pp. 85–94.
25. Mathois, A., and Rogers, R., The Impact of Alternative Forms of State Regulation of AT&T on Direct Dial Long Distance Telephone Rates, *RAND J. Econ.*, 20: 437–453 (Autumn 1989).
26. Mathois, A., and Rogers, R., The Impact and Politics of Entry Regulation on Intrastate Telephone Rates, *J. Regulatory Econ.*, 2:53–68 (March 1990).
27. Kahn, B., Price Caps Versus Cost of Service Regulation: Evidence from Intrastate Toll Pricing. Paper presented to the Rutgers University Conference on Public Utility Economics and Regulation, Monterey, CA, 1988.
28. Greenstein, S., McMaster, S., and Spiller, P., The Effect of Incentive Regulation on Local Exchange Companies' Deployment of Digital Infrastructure. Paper presented to the AEI Conference: Telecommunications Summit: Competition and Strategic Alliances, Washington, DC, 1994.
29. Teske, P., State Telecommunications Policy in the 1980s, *Policy Studies Rev.*, 11: 118–125 (Spring 1992).
30. McCray, S., *Intergovernmental Regulation of Telecommunications*, Advisory Commission on Intergovernmental Relations, Washington, DC, 1990.
31. Teske, P., and Bhattacharya, M., State Government Actors Beyond the Regulators. In *American Regulatory Federalism and Telecommunications Infrastructure* (P. Teske, ed.), Lawrence Erlbaum Associates, Hillsdale, NJ, 1995, pp. 47–65.

32. Benton Foundation and Center for Policy Alternatives, *State and Local Strategies for Connecting Communities: A Snapshot of the 50 States*, Benton Foundation and Center for Policy Alternatives, Washington, DC, 1996, p. 79.

33. Noam, E., and Wenders, J., Economic Theories of Regulation in Telecommunications. In *The Froehlich/Kent Encyclopedia of Telecommunications*, Vol. 6 (F. E. Froehlich and A. Kent, eds.), Marcel Dekker, New York, 1994, pp. 417–426.

34. The AT&T Divestiture After 5 Years, *Washington Post*, December 29, 1989, p. A1.

35. Coll, S., *The Deal of the Century: The Breakup of AT&T*, Atheneum Books, New York, 1986.

36. Campbell, H., The Politics of Requesting: Strategic Behavior and Public Utility Regulation, *J. Policy Analysis and Management*, 15:3 (Summer 1996).

37. Ferejohn, J., and Shipan, C., Congress and Telecommunications Policymaking. In *New Directions in Telecommunications Policy* (P. Newberg, ed.), Duke University Press, Durham, NC, 1989.

38. Niskanen, W., *Bureaucracy and Representative Government,* Aldine, Chicago, 1971.

39. Derthick, M., and Quirk, P., *The Politics of Deregulation,* Brookings Institution, Washington, DC, 1985.

40. Jones, D., Comment—Institutional Issues. In *American Regulatory Federalism and Telecommunications Infrastructure* (P. Teske, ed.), Lawrence Erlbaum Associates, Hillsdale, NJ, 1995, pp. 99–104.

41. Emeritz, R., Tobias, J., Berthot, K., Dolan, K., and Eisenstadt, M. (eds.), *The Telecommunications Act of 1996: Law and Legislative History*, Pike and Fischer, Bethesda, MD, 1996, p. 10.

42. Pierce, A., Computer Inquiry I, II, and III—Computers and Communications: Convergence, Conflict or Policy Chaos? In *The Froehlich/Kent Encyclopedia of Telecommunications*, Vol. 4 (F. E. Froehlich and A. Kent, eds.), Marcel Dekker, New York, 1994, pp. 219–329.

43. Noam, E., The Federal-State Friction Built Into the 1934 Act and Options for Reform. In *American Regulatory Federalism and Telecommunications Infrastructure* (P. Teske, ed.), Lawrence Erlbaum Associates, Hillsdale, NJ, 1995, pp. 113–124.

44. Bonnett, T., *Telewars in the States: Telecommunications Issues in a New Era of Competition*, Council of Governors' Policy Advisors, Washington, DC, 1996, p. 127.

PAUL E. TESKE

Regulation of Telecommunications by the U.S. Federal Government—A Guide to the 1996 Telecommunications Act

Introduction

On February 8, 1996, the tortoise of federal law finally caught up with the hare of communications technology. After more than a half century of broadcast regulation under the Communications Act of 1934,[1] a dozen years tinkering with cable television,[2] and more than a decade of telephone supervision by Judge Greene,[3] Congress passed, and President Clinton signed, the Telecommunications Act of 1996.[4]

This law represents a vision of a telecommunications marketplace in which the flexibility and innovation of competition replace the heavy hand of regulation. It is based on the premise that technological changes will permit a flourishing of telecommunications carriers engaged in head-to-head competition, resulting in a multitude of communications carriers and programmers being made available to the American consumer.

Ironically, it is the convergence of technology that is to lead to a diverse telecommunications marketplace.[5] If the same physical plant can offer local and long-distance telephone service, provide cable television programming, and carry voice, data, and video signals, then competing systems offering the whole package, as well as selected subparts, can replace localized monopolies.

There is no guarantee, however, that true competition will flourish, and it is certainly possible that unregulated fiefdoms will soon dot the electronic landscape. The 1996 act is an experiment as, one would have to admit, all telecommunications regulation is an experiment.[6]

Whether or not the devil is in the details, the future of telecommunications regulation can only be appreciated with, at least, a preliminary understanding of the specific details of the 1996 act.[7] This article represents an effort to provide a guided tour through the major provisions of the 1996 act.

The next section describes the new regime for telecommunications in general and telephone in particular. The third section describes the newest cable regulation, while the fourth section details the changes in broadcast regulation. The fifth section describes the major attempt to regulate content, rather than conduit, in the 1996 act.

From Telephone to Telecommunications

The goal of Congress was to create a legislative change as dramatic as the evolution of the old-fashioned telephone, carrying voices over distant wires into telecommunications, the transmission of "information," including data and video, as well as aural communications.[8] Accordingly, Congress decided that the monopolistic local telephone company must be forced to share its market, while at the same be permitted into both the free-wheeling competitive world of long-distance service and the potentially competitive video market as well. In the words of the Federal Communications Commission (FCC): "In the old regulatory regime, government encouraged monopolies. In the new regulatory regime, we and the states remove the outdated barriers that protect monopolies from competition and affirmatively promote efficient competition using tools forged by Congress."[9]

The most difficult piece in the deregulatory puzzle is how to create competition for local telephone service. Telephone (in fact all telecommunications) service is generally divided between local and long-distance service. Current technology has created a peculiar reality by which it is far easier to carry information thousands of miles across the country than the last mile into a recipient's business or home. It is the market for that last mile, so to speak, that must be competitive for the 1996 act to achieve its far-reaching goals.

The geographical dividing line for a local telephone region is termed a *local access transport area*, or "LATA."[10] A company that offers long-distance service is said to be offering *inter-LATA service,* meaning the communications are between points not within the same local area.[11]

Following the breakup of AT&T in 1982, long-distance, or inter-LATA, telephone service became a highly competitive market with both large and small players.[12] The local telephone market was initially divided among seven Bell Operating Companies (BOCs),[13] known colloquially as "Baby Bells." But these babies were not only large, they continued to enjoy virtual monopolistic control over their area's local telephone service. There were numerous much smaller local telephone companies that had monopolies over rural or much smaller geographic areas.

The expense of duplicating the local phone company's infrastructure, and the necessity of interconnecting with its plant, also made it obvious that competition for the local market would be impossible without the active assistance of the very companies with whom competition was sought. The question for lawmakers was how local telephone service, long considered a "natural monopoly,"[14] could be opened for competition.

Part II of the Communications Act, "Development of Competitive Markets," creates the blueprint for what Congress hoped would become the future of telecommunications.[15] Section 251 details the substantive framework necessary for achieving competition in telecommunications, and Section 252 describes the procedural mechanism for implementing that framework.[16]

The Duties of Competitors

Congress divided telecommunication carriers into four classifications and varied the degree of regulation with each category. The broadest group is the general telecommunications carrier, then there is the subgroup local-exchange carriers (LECs), which is further subdivided into "incumbent" local-exchange carriers, and, finally, the most detailed regulatory provisions are for the BOCs.[17] The duties of each are described below.

A *telecommunications carrier* is defined as any entity offering, for a fee to the public, to transmit information without changing the content of that which is transmitted.[18] The primary duty imposed on all telecommunications carriers is interconnection. In other words, all telecommunications carriers must connect directly or indirectly with other carriers.[19] In addition, carriers are prohibited from designing their networks to thwart the ability of other carriers to interconnect with them.[20]

Far more detailed requirements are imposed on the LECs, which provide either telephone exchange service or service access.[21] This is the area in which the creation of competition was seen as most important if there were to be true telecommunications competition, yet was recognized as the most difficult goal to achieve. Despite the initial situation of general monopoly LEC status, the 1996 act described five obligations to be shouldered by all future LECs, both dominant and challenging.

The first of these obligations involves resale.[22] The LECs are barred from either prohibiting or imposing discriminatory or unreasonable conditions on the resale of telecommunications services.

Next, all LECs must provide "number portability."[23] This will permit users to switch from one telecommunications carrier to another without having to change their existing telecommunications numbers.[24] The act recognizes the potential difficulty in implementing this mandate, so it provides that number portability will be required only "to the extent technically feasible," and in accordance with FCC requirements.[25] In addition, the act states that the FCC must ensure that the costs of establishing number portability are "borne by all telecommunications carriers on a competitively neutral basis."[26]

The third requirement for all LECs is that they provide dialing parity.[27] The term *dialing parity* is defined as the ability of customers to dial the same number of digits to use any available telecommunications provider.[28]

Next, all LECs must provide their competitors with access to their poles, conduits, and other rights-of-way.[29] Congress first protected the rights of cable operators to use telephone poles in the Pole Attachment Act of 1978.[30] Under the 1996 act, access to poles will be even easier. In addition to requiring LECs to share their poles, the act requires utilities,[31] such as gas and electric companies, to provide access on a nondiscriminatory basis to cable operators and other telecommunications carriers.[32]

The final obligation imposed on all LECs is that they establish "reciprocal compensation arrangements."[33] Such arrangements provide that a network in which a call originates compensates the network in which that call terminates. For simplicity and efficiency, the act permits arrangements such as the so-called bill-and-keep arrangement by which two networks agree to waive their rights to recover from one another under this section.[34]

Congress was well aware that existing LECs would have an enormous potential advantage over potential competitors in the local market. Accordingly, several additional restrictions were placed on incumbent LECs, either those providing telephone exchange service on the enactment date of the act (February 8, 1996) or those newcomers determined by the FCC to have obtained a market position comparable to that of an incumbent.[35]

In addition to the requirements imposed on all LECs, incumbent LECs must provide interconnection for other telecommunications carriers at "any technically feasible point" in the incumbent's network.[36] This means that interconnection must be provided for all competitors that wish to provide local telephone exchange service and exchange access. There is no similar requirement that an incumbent LEC permit interconnection by a cable television operator or other sort of information service except to the extent that they are providers of telecommunications service. All interconnection under this provision must be at least of the same quality as that available for either the incumbent LEC itself or its affiliates, and it must be made available at reasonable and nondiscriminatory rates and terms.

Incumbent LECs are also required to make available to competing telecommunications carriers "unbundled access" to "network elements."[37] A *network element* is defined to include not only the physical equipment used to provide telecommunications service, but also significant functions, systems, and information that are made available by, or are used in the transmission of, telecommunications service.[38] These would include local loops and subloops, switching, and signaling functions. *Unbundled access* means the availability of access to distinct parts of the incumbent's network at an appropriately lower cost than access to all of the elements of the network. Thus, a competitor can purchase only those network components and functions that it needs to offer service. That competitor is then free to combine these unbundled elements in the manner it deems best for providing service.

In order to make interconnection and unbundled access economically feasible, incumbent LECs are required to permit physical co-location of their competitors' equipment.[39] In other words, incumbent LECs must allow other telecommunications carriers to place their equipment at the site of the incumbent's own switching center. Again, rates charged for using these premises must be reasonable and nondiscriminatory.

A different type of competition is made possible by the requirement that the incumbent LEC sell to other carriers, at "wholesale" rates, the same telecommunications service it provides to retail customers.[40] The availability of wholesale pricing will enable the other carriers to offer for sale the same service to the incumbent's customers.[41]

The most contentious part of this issue is the determination of wholesale rates. If the wholesale rates are too close to retail rates, it will discourage competition in the local market.[42] On the other hand, if wholesale rates are too low, it may discourage the construction of facilities-based competition.[43]

The 1996 act states that wholesale rates are to be calculated by subtracting from retail subscriber rates any "costs that will be avoided by the local exchange carrier."[44] Arithmetically, wholesale rates equal retail rates minus costs avoided. Even though the 1996 act says that each "state commission" should determine the wholesale rates,[45] the FCC has adopted a "minimum set of criteria for [the] avoided cost studies" that will be conducted by the states.[46]

Most notably, the FCC ruled that the "avoided costs" are not to be limited to only the costs that an incumbent LEC will actually avoid by selling wholesale rather than directly to subscribers.[47] If that were the case, the incumbent would have an incentive to keep its expenditures high so that its competitors would pay a higher resale price. The FCC instead ruled that avoided costs means all of the costs "that an incumbent LEC would no longer incur if it were to cease retail operations and instead provide all of its services through resellers,"[48] whether these savings were actually realized or not.

The FCC also decreed a default wholesale rate, one that is to be used by state commissions that either have not yet conducted an "avoided retail cost study" or choose not to undertake such a study.[49] In either case, the FCC decreed that interim wholesale rates must be set at between 17% and 25% below the incumbent LECs retail rates.[50]

The Special Case of Bell Operating Companies

The strongest monopolists in the telecommunications universe are the BOCs, who spun off from AT&T and carved up the local telephone market. If a competitive telecommunications system is to evolve, the BOCs must face direct competition for local service. The BOCs, meanwhile, have long been eager to enter the long-distance market.

In a sense, the most important piece of legislative strategy in the 1996 act was the provision that the BOCs are to be given permission to offer long-distance service to their local customers only upon fulfilling a "competitive checklist" that ensures or permits competition for local service. Because of the lack of a similar danger of unfair competition, BOCs are free to offer long-distance service to those not within their local service areas immediately.[51] For a BOC to be given FCC permission to offer long-distance service to its own local clientele, though, there must be either (1) an agreement with an existing competitor for that BOC's local service or (2) if no competitor has come forward, a statement indicating that the BOC is ready to provide access and interconnection for potential competitors. Before the FCC will consider a BOC's request to provide long-distance service to its local customers, the state commission with jurisdiction over that locality must give its approval to the agreement statement.

The Competitive Checklist

The competitive checklist for BOCs wanting to offer long-distance service includes many of the provisions required under the provisions governing general incumbent LECs.[52] The act, however, adds several requirements that are specific to telephone service. To gain permission to enter the long-distance market, a BOC must offer access and interconnection to others that wish to compete for the local market. *Access and interconnection* are defined to include all of the following:

From the incumbent LEC checklist

1. Interconnection for other carriers offering intra-LATA service at any technically feasible point in the BOC's network
2. Unbundled access to network elements

From the general LEC checklist

3. Access to BOC poles, conduits and other rights-of-way
4. Number portability
5. Dialing parity
6. Reciprocal compensation arrangements
7. Availability for resale

Specific for BOCs checklist

8. Local loop transmission (from the central office to the customer's premises)
9. Local transport
10. Local switching
11. Nondiscriminatory access to emergency numbers (911 and E911), directory assistance, and operator call-completion services
12. White page directory listing for competitors' customers
13. Nondiscriminatory access to telephone numbers
14. Nondiscriminatory access to databases and signaling necessary for call routing and completion

Facilities-Based Competitors for the Local Market

The above checklist encompasses the minimum requirements for agreements with competitors for the local telephone market. In order to ensure that a powerful, independent competitor exists for local service prior to a BOC's entry into the long-distance market, the act requires that any competitor be facilities based.[53] The term *facilities based* means that the competitor is providing local service either exclusively or predominantly over its own facilities. It excludes a competitor merely reselling the BOC's telephone exchange service.[54] The prime example given by Congress of an effective "two-wire" policy is the provision of local telephone service through the facilities of a local cable television system.[55]

Moreover, to ensure that local competition is in place, the competitor must be operational, not merely in the planning stage.[56] This will also make it easier to ensure that the full checklist is in place.

The Absence of Local Competitors

It is, of course, possible that no one will want to take on a particular BOC on its home court. The drafters of the 1996 act did not want to deny a BOC under those circumstances all opportunity to offer long-distance service to its local customers. Accordingly, the act provides that a BOC that has not received a request for interconnection may still apply for permission to provide such long-distance service. Instead of an agreement, the BOC must file a statement of the terms and conditions under which it is ready and willing to offer the components of the competitive checklist.

It was also foreseeable that, if a signed agreement was a necessary prerequisite for BOC entry into the long-distance market, some player in that market might try to delay BOC entry by engaging in bad faith negotiation or otherwise acting improperly. To prevent such subterfuge, the statement described above will also suffice if the only providers to request access have failed to negotiate in good faith.[57]

Separate Affiliates

Even with the competitive checklist in place, Congress feared that a BOC could use its local power to leverage an unfair advantage over competitive markets. As an additional safeguard, the 1996 act requires that a BOC create a separate affiliate if it wants to offer certain services.

First, a BOC must use an affiliate to offer its local customers long-distance service.[58] These services include all long-distance telephone, telecommunications, or information services (other than "incidental" services)[59] and services that had been authorized prior to the 1996 act.[60]

Second, a separate affiliate is needed for BOCs engaged in manufacturing activities.[61] The term *manufacturing* includes all the activities that were previously covered by that term in the AT&T Consent Decree.[62] A BOC cannot discriminate in favor of its own affiliate in the procurement of manufacturing equipment.[63]

The separate affiliate must operate independently from its BOC parent.[64] It must keep separate books and records and must have separate officers, directors, and employees. Further, all transactions with the BOC must be "on an arm's length basis."[65]

With the hope that true competition against BOCs will flourish quickly, the 1996 act provides for a sunsetting of the separate affiliate requirements. Three years after a BOC is authorized to offer long-distance service, it will no longer need to either offer long-distance telecommunications service or conduct manufacturing activities through a separate affiliate.[66] Using a different starting point, the act similarly states that four years after the date of the 1996 act (February 8, 2000), a separate affiliate will not be needed for the provision of long-distance information services.[67] The timing of either of the sunset provisions can be extended by the FCC if the risk of anticompetitive abuse remains.[68]

Obtaining an Interconnection Agreement

Most of the issues surrounding interconnection will be resolved, to a large extent, on a case-by-case basis. The final rates, terms, and conditions governing a particular interconnection must reflect the legitimate needs of all parties. Thus, competitors must reach agreement on a host of sensitive areas. There are obvious problems involved in reaching a mutually beneficial agreement between business adversaries, especially when one party, in effect, holds all the cards.[69]

The 1996 act creates a multilayered scheme by which interconnection agreements are reached either through negotiation or binding arbitration and then are subject to review by the local state commission.

It will be most cost efficient if parties will resolve their differences through voluntary negotiations.[70] Both sides have the duty to negotiate "in good faith."[71] This provision requires more cooperation than the analogous requirement under federal labor law. While unions and management are required to "confer in good faith," that duty is specifically restricted so as not to "compel either party to agree to a proposal or require the making of a concession."[72] There is no similar restriction in the 1996 act. Although the act is not more specific, the very purpose of the act, to "accelerate rapidly private sector deployment of advanced telecommunications and information technologies . . . by opening all telecommunications markets to competition"[73] implies a more cooperative mindset. While reaching an agreement is not required, it would violate the 1996 act if a party "negotiates without serious intent to contract."[74]

If voluntary agreement is not reached, either side can petition the state commission to conduct binding arbitration.[75] In setting rates for interconnection or access to unbundled network elements, the state commission is not to use the traditional rate-of-return formula. Instead, the rates are to be based on actual cost, including a reasonable profit, and must be nondiscriminatory.[76] Similarly, rates charged for reciprocal compensation (for termination of calls originating on a competitor's network) must either reflect the actual costs associated with the termination of calls or be supplanted by a bill-and-keep arrangement.[77]

The 1996 act requires that all interconnection agreements, whether obtained through voluntary negotiations or binding arbitration, must be submitted to the state commission for approval.[78] It is not entirely clear why an agreement that was created through arbitration by a state commission should need to be submitted to the same commission for approval, but so be it.[79] Negotiated agreements must be approved unless the agreement is found to either discriminate against a carrier not party to the agreement or be otherwise against the public interest.[80] Arbitrated agreements are to be approved unless they conflict with the provisions of Section 251 or 252.[81]

There are only two permissible ways to attack the decisions, action, or inaction of a state commission in this area. First, an aggrieved party can petition the FCC to preempt the state commission and take over the proceedings.[82] Second, a complaint can be filed in federal court to determine whether an agreement complies with the act.[83] State courts are denied jurisdiction in this matter.

States are further limited by Section 253, which preempts any local law or regulation that creates a barrier to entry into the telecommunications market.[84] The local government is not entirely out of the regulatory picture, though. The drafters of the 1996 act intended that state and local governments would retain their ability to "manage the public rights-of-way" in a nondiscriminatory, competitively neutral way and charge "fair and reasonable" fees for use of those rights-of-way.[85]

This savings clause is likely to be the source of much litigation. Not only is the act silent as to what makes a fee "reasonable," there is also the question as to whether cable operators, who already pay a franchise fee,[86] can be required to pay a second time for the same wire just because it is carrying telecommunications information as well as video programming. Moreover, local regulators will contend that the power to manage rights-of-way encompasses the power to condition use by telecommunications carriers to a variety of regulatory obligations, while those carriers will argue that management of rights-of-way is limited to safety-type concerns and does not include the manner in which the carrier provides service.

While neither the 1996 act nor the conference report define "manage," the restricted interpretation is probably more consistent with the intent of Congress. First, the section is entitled, "Removal of Barriers to Entry."[87] In addition, the act specifically permits local governments to protect other important interests, such as the rights of consumers and the promotion of universal service.[88] Such specific permission would not have been needed had the authority to manage rights-of-way been all-inclusive. The FCC is charged with preempting any local government that oversteps its management powers and creates an impermissible barrier to telecommunications service.[89]

Universal Service

The concept of universal service is one of the most important links between the old regulatory scheme and the 1996 act. From the very beginnings of the FCC, universal service has been at the center of telecommunications policy.[90] The 1934 Communications Act, for example, mandated that the FCC regulate electronic communications "so as to make available, so far as possible, to all people of the United States a rapid, efficient, Nation-wide and world-wide wire and radio communication service with adequate facilities at reasonable charges."[91]

The 1996 act represents an attempt to fulfill the commitment to universal service while adapting to ongoing changes in both technology and the marketplace.[92] The 1996 act includes within its list of universal service principles that rates be "just, reasonable and affordable"; that access to "advanced telecommunications and information services" be provided to consumers living in rural, low-income, or high-cost areas; and that all providers of telecommunications services make an "equitable and non-discriminatory contribution" to universal service.[93]

There are, of course, two preliminary questions for any universal service policy: which services must be provided, and how universal must their provision be?

Most important, the 1996 act recognizes that the services that, as a matter of national policy, should be available to all Americans can no longer be a static concept. Thus, the act decrees that universal service "is an evolving level of telecommunications service."[94] The FCC's definition of universal service is to be established "periodically" based on which services have become essential to education, health, or safety; deployed and subscribed to by a "substantial majority" of residential customers; and otherwise consistent with the public interest.[95]

The actual mechanism for ensuring universal service is to be decided by the FCC. All telecommunications carriers that provide interstate service must contribute to this mechanism on an equitable and nondiscriminatory basis.[96] These contributions will then go to "eligible telecommunications carriers"[97] — that is, those carriers that offer and advertise the components of universal service throughout a designated service area.[98] Both the scope of the designated area and a carrier's status as "eligible" are determined by each state's commission.[99] In nonrural areas, the state commissions must designate more than one carrier as eligible if multiple carriers request the designation and meet the statutory requirements.[100]

In addition to receiving monetary contributions, an eligible carrier also has the right to demand that incumbent LECs share their infrastructure in order to receive the benefits of the incumbent's economics of scale and scope.[101] The incumbent must make infrastructure, technology, information, and facilities available at a "just and reasonable rate" for the purpose of providing universal service.[102]

Eligible carriers may only use universal service support for the provision, maintenance, and upgrading of facilities and services related to the provision of universal service.[103] This is consistent with the general requirement that carriers cannot use noncompetitive services to subsidize services subject to competition.[104] For interstate services, the FCC must establish whatever cost-allocation rules and accounting safeguards are necessary to ensure that universal services bear no more than a reasonable share of the costs of the facilities providing all services.[105] The states have the same responsibility for intrastate services.[106]

Cable Television and Video Programming

Like the BOCs, cable operators enjoyed a virtual monopoly in their service areas prior to the 1996 act. True, there was the legal right under the 1992 Cable Television Act for competitors to obtain franchises to lay a second cable in an area.[107] Also, competition of a limited kind was being offered by the newer video distribution systems, such as direct broadcast satellites (DBSs) and multichannel, multipoint distribution services (MMDSs). Nonetheless, cable television still had many of the earmarks, and much of the power, of a traditional monopoly.[108]

Cable television had faced a dizzying array of changing regulatory environments in the 12 years prior to the 1996 act. In 1984, cable operators received a large amount of freedom, for the first time, from rate and other forms of regulation.[109] With consumers angry about skyrocketing rates, and broadcasters anxious to ensure that their programming would be carried on cable systems, Congress overrode a presidential veto and tightened the regulatory reins in 1992.[110]

The 1996 act represents a different balancing act. Cable television is to be partly deregulated so it can compete in the broader telecommunications market. Full deregulation, however, is avoided because of the likelihood of its continuing (at least for the immediate future) market dominance in the local video arena. Meanwhile, provisions are made to lift once and for all the legal barriers

to both telephone company provision of cable and other video programming
and cable entry into the local telephone market.

Effective Competition

The provisions involving the definition of effective competition are important
for many reasons. Effective competition is a label desired by cable operators
seeking rate deregulation, as well as the ultimate goal of the drafters of the 1996
act.

Under the 1992 Cable Act, the term *effective competition* was limited and
rarely found. These standards continue after the 1996 act, but an additional
means to determine effective competition was added.

Under the 1992 Cable Act, cable systems with very low penetration rates,
under 30% of franchise-area households subscribing, are deemed to face effective competition.[111] The vast majority of cable operators, those with higher
penetration, need to prove that they face direct head-to-head competition from
a multichannel programmer (such as another cable operator, MMDS system, or
DBS operator).[112] To qualify though, that competitor needs to offer service to
at least half of the cable operator's franchise area and provide service to at least
15% of the area's households.[113] The last way the 1992 Cable Act provided for
a declaration of effective competition was if the local government itself offered
a comparable service to at least half of the franchise area's households.[114]

The 1996 act adds a fourth way for a cable operator to claim effective
competition and thus obtain rate deregulation. If a telephone company (an LEC
in the act's parlance) offers comparable video programming, either directly
or through an affiliate, the cable operator will be deemed to face effective
competition.[115] Included in this definition are telephone companies that offer
video service through any means, including MMDS, as well as a wire, other than
direct-to-home satellite, in an unaffiliated cable operator's service area. Unlike
the 1992 Cable Act's provisions regarding other private multichannel competitors, the telephone company need not serve any particular number of video
subscribers. Instead, the telephone company must simply "offer" service, that
is, be physically able to provide video programming service with only a minimal
additional investment.[116]

Rate Deregulation

The battle over the extent to which cable rates should be regulated has always
been the emotional highlight of debates over cable regulation in general.[117] The
1996 act provides greater deregulation immediately, with the promise of far
greater pricing freedom for cable operators in the years to come.

The 1996 act continues the distinction, created in the 1992 Cable Act, between basic service and cable programming service.[118] The basic tier of cable
programming contains broadcast channels and channels that offer public, educational, and governmental (PEG) access programming. Cable programming
service includes all other tiers of cable programming excepting programming
offered on a per channel or per program (such as pay-per-view) basis.

The basic tier is subject to local regulation, but that local regulation must follow strict FCC guidelines.[119] The FCC's guidelines are extraordinarily complicated, even for the telecommunications field, with the commission's implementation order totaling more than 500 pages.[120] Ironically, all the legal and mathematical factors represent nothing more than an attempt to define which basic rates are reasonable.[121]

Cable operators facing effective competition are not subject to regulation of the basic tier. In addition, small cable operators that only offered basic service at the end of 1994 are similarly free of basic service rate regulation.[122]

For cable operators, the most important change brought by the 1996 act is that the other tiers of cable programming, those providing cable programming service, will be free from rate regulation after March 31, 1999.[123] In the belief that there will be true competition for delivering "cable-type" programming to consumers after April 1, 1999, governmental rate regulation of cable programming service will be no more. Such rate regulation is, of course, already ended for those operators facing effective competition, but this deregulation will cover the entire cable industry.

Until March 31, 1999, for cable operators not facing effective competition, rates for these tiers, as for the basic tier, are required to be reasonable. The primary difference is that the setting of these rates is done by the FCC, not the local franchising authorities.[124]

While the rate relief reflects a belief in the impending arrival of competition, the 1996 act did not relieve cable operators of many other rules that were designed to protect the video marketplace. For example, the 1992 Cable Act contained strong program access requirements.[125] Programmers vertically integrated with cable operators must sell their programs to competing distribution services at reasonable and nondiscriminatory prices.[126] The requirement was not lessened by the 1996 act.[127] Similarly, the must-carry rules, which require cable operators to carry local broadcast channels, was substantially unchanged by the 1996 act.[128]

Ownership of Cable Systems and Other Video Providers

Other than rates, the other area in which the 1996 act promises to create major changes in the cable industry involves the loosening of the rules involving cable ownership. Cable television operators are free to enter into the larger telecommunications market but may face stiff competition on their home turf from local telephone companies.

Telephone Company Provision of Video Programming

Ending a lengthy legal battle, the 1996 act eliminated the ban on cable telephone cross ownership. Previously, cable television operators and local telephone companies were barred from entering the other's field in the same location.[129] A string of lower federal courts had struck down the cross-ownership ban as violative of the First Amendment, and the Supreme Court had agreed to hear the issue.[130] Before the Court could rule however, the 1996 act's removal of the ban was enacted,[131] and the Court dismissed the challenge as moot.[132]

Under the 1996 act, there are four ways a local phone company can offer video programming in its local area. A telephone company can provide video either as a pure common carrier or a traditional cable operator.[133] As a video common carrier, the telephone company is treated as a classic common carrier, subject to the common carrier provisions of Title II of the Communications Act of 1934. As a traditional cable operator, the telephone company would be subject to all of the requirements of the 1984 and 1992 Cable Acts.

A third possibility for video distribution is through radio-based communication.[134] If a telephone company provides a wireless, radio-based, multichannel video programming distribution service, it will not be subject to cable act restrictions.

The most innovative part of the 1996 act in this area was the creation of a fourth way for telephone company delivery of video programming: the "open video system."[135] An open video system is a hybrid, of sorts. It permits some programming control for the telephone company, but reserves other channels for use by nonaffiliated programmers. The drafters of the 1996 act hoped that the open video system would become the predominant model for telephone entry into the video marketplace.[136]

Specifically, the local telephone company can only select the programming for one-third of the open video system's channel capacity if demand for channels exceeds the system's supply. The other two-thirds of the system must be made available to nonaffiliated program providers. The 1996 act, however, places no upper limit on the number of channels a telephone company or its affiliate can program. This provides an incentive for the creation of open video systems with large channel capacity.

As with so much of the 1996 act, regulation of open video systems contains significant antidiscrimination provisions. Most broadly, the operators of an open video system may not discriminate in regard to carriage of programming, and rates and other conditions of carriage must be just, reasonable, and non-discriminatory.[137] Moreover, operators of an open video system may not discriminate in favor of their own programming or that of their affiliates with regard to information presented to subscribers.[138] Thus, in its advertising or its provision of technical means of program selection, the open video systems operator cannot favor its own programming over that offered by nonaffiliated entities.

In keeping with its hybrid nature, an open video system is not to be treated, aside from the above requirements, as a common carrier[139] and faces only a limited amount of cable-type regulation. There is no need for a cable franchise, and rate regulation, leased access, cable equipment, or consumer service rules do not apply.[140] Other cable act provisions designed to increase the variety of programming choices, such as PEG access, must-carry rules, and program access rules, remain applicable to open video systems.[141]

Cable Provision of Telephone Service

The 1996 act clears away much of the regulatory underbrush that kept cable operators from providing local telephone service. Most basically, the act preempts much of the state and local regulations that governed the provision of noncable service by cable operators.

First, franchising authorities are barred from imposing any limit on the provision of telephone or telecommunications service by a cable operator.[142] Second, franchising authorities are barred from requiring that cable operators obtain a franchise prior to offering telephone or telecommunications service.[143] Finally, the franchising authority may not use revenue from a cable operator's telephone or telecommunications service to calculate the franchise fee owed by the operator.[144]

Mergers Between Cable Operators and Local Telephone Companies

The major restriction on the competitive free-for-all for video programming is the continued restriction on mergers and buyouts between cable companies and local telephone companies within their respective service areas. This is in keeping with the two-wire dream of direct head-to-head competition between cable and local telephone companies.[145]

The 1996 act contains parallel prohibitions: a local telephone company cannot acquire more than a 10% financial interest in a cable operator providing service in the telephone company's service area, and a cable operator cannot acquire more than a 10% financial interest in a local telephone company providing service in the cable operator's franchise area.[146] Not only are direct mergers prohibited, but joint ventures between cable operators and telephone companies in the same market are also proscribed by the 1996 act.[147]

The joint venture ban is limited, though, to the provision of video programming and telecommunications services. A joint venture for other purposes, such as constructing the physical facilities for providing the programming and services, would be permitted.[148] Similarly, a local telephone operator can use a cable system's subscriber drops, the last link between the cable operator's network and the individual subscriber.[149] This use requires the approval of both the cable operator, as to rates and conditions, and the FCC to ensure that this sharing is of limited scope and duration.[150]

The FCC was also given authority to issue waivers permitting cable/telephone company combinations.[151] Prior to issuing such a waiver, the FCC must determine that the cable or telephone company faces economic distress. the cable system or telephone facilities would not be economically viable, or the public interest clearly outweighs the anticompetitive effects of the combination.[152] In addition, the local franchising authority must approve such a waiver before it becomes effective.[153]

Other Cable Ownership Issues

While the drafters of the 1996 act maintained numerous provisions to limit co-ownership of cable and telephone company systems, a deregulatory mindset pervaded other cable ownership issues. In an effort to strengthen cable as a player in the new competitive marketplace, many previous restrictions on cable sale and ownership were lifted.

Under the former cable act provisions, "trafficking" in cable systems was limited. Cable operators were barred from selling a cable system for three years after acquisition or initial construction.[154] That three-year holding period has now been eliminated.[155] As under the old law, franchising authorities are given 120 days to decide whether to approve a request for transfer of system ownership, with the transfer treated as granted if no decision is rendered within that time.[156]

Similarly, restrictions on cable operator co-ownership of other forms of electronic communication were also eased. First, the FCC was instructed to eliminate its restriction on cable operator ownership of a broadcast network.[157] To prevent anticompetitive abuse by such co-ownership, the FCC was also instructed to ensure carriage, channel positioning, and nondiscriminatory treatment of nonaffiliated broadcasters by the cable/network combination.[158]

The 1996 act is not quite so bold with the issue of cable cross ownership of broadcast stations in the same market. While the 1996 act removes the statutory ban on such combinations,[159] the FCC is left to determine the ultimate question as to their permissibility. In fact, the drafters of the 1996 act specified that they did not intend, by their statutory repeal, to indicate one way or the other whether the FCC should change its existing cross-ownership ban.[160]

Finally, the ban on cable ownership of either co-located SMATV (satellite master antenna television) systems or co-located MMDS systems is eased. Cable ownership of either of these two video delivery systems will now be permitted in any area a cable operator is subject to effective competition.[161]

Broadcasters in the New Telecommunications Marketplace

The lowest tech players in the telecommunications marketplace revolution, the broadcasters, received significant regulatory relief from the 1996 act. In addition to the changes in the rules governing cable ownership of stations and networks,[162] broadcasters also benefited from the deregulatory tradewinds. Ironically, though, most of the changes affecting broadcasters actually serve to limit intramedia competition.

For example, in deciding whether to renew a broadcast license, the FCC is now barred from considering the proposal of any alternate potential broadcaster.[163] Instead, the FCC must only consider whether the broadcaster has committed "serious" violations of FCC rules and has served the public interest.[164]

Not only is renewal easier,[165] the terms of the license have been increased. Instead of a five-year license for television and seven-year license for radio, all broadcasters will now enjoy an eight-year license period.[166]

Many of the limits on multiple ownership of broadcast licenses have been eased or eliminated. On a national level, the limit on the number of AM (amplitude modulation) or FM (frequency modulation) radio stations that can be controlled by one entity was eliminated.[167] The national limit on the number of television stations was also removed,[168] and now the only remaining national ceiling is that one entity cannot own television stations that, together, reach more than 35% of the nation's television households.[169]

The rules governing multiple ownership within a particular local market have been relaxed, though not eliminated. For radio, a complicated matrix was created, with the number of permissible co-owned stations dependent on both the number of available commercial radio stations and whether the stations are concentrated in the same "service," either the AM or FM band (see table below).[170]

Number of Commercial Stations in a Market	Maximum Number of All Stations in that Market	Maximum Number of Same-Service Stations
45 or more	8	5
30–44	7	4
15–29	6	4
14 or fewer	5	3

Congress was not willing to make a final decision as to local ownership limits for television stations. Instead, the 1996 act directs the FCC to conduct a rulemaking proceeding to determine whether to retain, modify, or eliminate the current ban on owning more than one station in a market (the so-called duopoly rule).[171] While not expressing any position on this issue directly, Congress did state in the Conference Report that if the duopoly rules were revised, VHF-VHF (very high frequency–very high frequency) combinations should only be allowed in "compelling circumstances."[172]

Congress also liberalized the rules as to television/radio co-ownership. Prior to the act, the FCC permitted common ownership of a radio and a television station in the same market only if 30 broadcast owners were operating in that market and the market was in any of the top 25 largest.[173] The 1996 act extends that policy to include any of the top 50 markets.[174]

The 1996 act also provides a little more flexibility than in the past for broadcast television networks to combine. Previously, all television networks were barred from common ownership.[175] The "Big 4" networks (ABC, CBS, NBC, and Fox) are still barred from merging with either each other or the fledgling WB and UPN networks. Mergers are permitted, though, between either the WB and UPN networks or between any existing network and some new network formed after the act.[176]

Perhaps the strangest part of the 1996 act involves advanced television (ATV) services, the allocation of the spectrum that will be used for digital, high-definition television.[177] The entity awarded this portion of the spectrum stands to enjoy an enormous windfall, in large measure due to the variety of services that will be able to be transmitted simultaneously by the license holder.

The early winners in this legislative slugfest were the existing broadcasters. The 1996 act states that if ATV licenses are issued, they are to be awarded only to those already licensed broadcasters.[178]

But even that requirement in the act did not end the dispute. Shortly before

Congress completed its work on the act, political opposition to this "governmental giveaway" began to build. In order to prevent the issue from scuttling the rest of the act and to avoid a painful rewriting process, Congressional leaders secured a promise from the FCC that the commission would not issue ATV licenses until Congress held further hearings and had the opportunity to revise the plan laid out in the 1996 act.[179] The FCC finally decided to award the spectrum to broadcasters in April, 1997.[180]

Direct Regulation of Content of Telecommunications

There is an almost Holmesian feel to much of the 1996 act, an underlying conviction that the creation of multiple carriers of information will benefit society by multiplying the number of voices in the marketplace of ideas.[181] Or, as Judge Learned Hand saw it, "right conclusions are more likely to be gathered out of a multitude of tongues, than through any kind of authoritative selection."[182] The 1996 act, accordingly, focuses largely on creating competition between carriers of information and trusts the market to determine the content that is carried.

All freedoms, though, are capable of being abused.[183] In particular, not only was the marketplace for violent and sexually oriented programming getting larger, but Congress feared that too many children were wandering through its stalls. Accordingly, in contradistinction to the content neutrality of the rest of the 1996 act, several provisions dealing directly with such programming were included under the appellation "Communications Decency Act of 1996."[184]

Broadcast Violence and Indecency

For broadcast television, Congress's primary interest was to create a technological barrier to objectionable programming but leave control over that barrier in the hands of individual parents. The easier part of that task was the mandating of the technological barrier. First, all video programming that has been rated must be transmitted with its rating.[185] Second, all television sets sold in the United States will contain a so-called V chip that will block out all programming rated unsuitable for children due to its sexual or violent content.[186] The FCC is charged with choosing the starting date for this requirement as long as that date is after February 8, 1998, two years after enactment of the 1996 act.[187]

The tricky part of the plan, however, is to design a rating system that does not violate the First Amendment. Any system in which the government is choosing what programs to bar is, of course, fraught with constitutional peril.[188] The 1996 act contains the threat that the FCC will create its own ratings guidelines.[189] The only way to avoid a governmentally created rating system is if the broadcasters, along with cable television operators, follow the model of the film industry[190] and do the rating themselves.[191]

There are several possible side effects from the V-chip proposal that may prevent it from achieving its goal. First, since the 1996 act requires that the ratings identify "programming that contains sexual, violent, or other indecent material about which parents should be *informed*,"[192] it is possible that much prime time television programming will be "identified." If that happens, parents will be forced to choose between having almost all broadcasting blocked in the evenings or letting all of the material, appropriate and inappropriate, into their homes.

Second, the FCC's current ban on indecent broadcasting was upheld by the Court because "broadcasting is uniquely accessible to children" since offensive broadcasts could not "be withheld from the young without restricting the expression at its source."[193] Once the V-chip is in place, though, such selective withholding of offensive material will be possible, and the prime rationale for distinguishing broadcast indecency from that in bookstores and movie theaters will be gone. The unintended result of the V-chip proposal, then, would be that broadcasters would show far *more* sexually explicit, indecent, and violent programming, contending that children were now to be protected by technology.[194]

Cable Indecency

Cable television, in a sense, may present a picture of what the future of broadcasting looks like. Lockboxes and similar devices to block out particular, unwanted programming have not only been available since 1984, cable operators have been required to offer them to subscribers.[195] Accordingly, courts have uniformly found bans on cable indecency to be unconstitutional.[196]

Nonetheless (or perhaps, predictably), complaints concerning the explicitness of cable programming have continued. Both the 1992 Cable Act and the 1996 Telecommunications Act contain numerous provisions attempting to deal with the issue, but much of the legislative plan became entangled in constitutional challenges.

The 1996 act, for example, required the scrambling of any "sexually explicit adult programming" or indecent programming on a channel "primarily dedicated to sexually-oriented programming."[197] Prior to scrambling the signal, the program could not be shown during any time of day "when a significant number of children are likely to view it."[198] This provision has been found to be a constitutional means of protecting children from indecent programming.[199]

A more modest, content-neutral scrambling requirement imposed by the 1996 act has not been challenged. This section requires the scrambling, at no cost, of any channel at the request of a cable subscriber.[200] An earlier Senate version of this provision had included the additional standard that the programming to scrambled be, in the judgment of the subscriber, "unsuitable for children," but this was dropped from the final bill.[201]

Another area of interest to legislators has been programming offered on leased and public access channels. These channels are programmed by those not affiliated with the cable operator, and traditionally have been carried free from operator censorship.[202] Both the 1992 Cable Act and the 1996 Telecommunications Act contain provisions permitting cable operators to refuse to carry offensive programming on these channels.[203] In a split decision, of sorts, the Supreme Court ruled that the 1992 provisions on leased access were constitutional, but that the public access provision violated the First Amendment.[204] The primary differences between the two provisions, according to a plurality of the Court, was that public access was imposed and regulated by franchising authorities and not the federal government, and that there was no record of a nationwide problem with "patently offensive" public access programming.[205] Granting cable operators the power to bar public access programming would "greatly increase the risk that certain categories of programming (say borderline offensive programs) will not appear."[206]

Although the 1996 act's provisions were not at issue in this case, it seems likely that only the leased access provisions will be enforceable. One court has suggested that the 1996 act provision permitting operators to bar "obscene" public access programming would be constitutional since such programming is unprotected by the First Amendment.[207] The same reasons that led to the 1992 public access provisions being held unconstitutional—the role of local franchising authorities, the lack of a record of national problems with public access, and the risk that borderline public access programming will be barred—apply with equal force to the 1996 public access provisions.

One area of potential difficulty arises because the 1992 act imposed liability on cable operators for carrying obscene public and leased access programming.[208] It is, however, unconstitutional to impose liability for programming that one is mandated to carry.[209] Unless the cable operator is relieved of all liability for public access programming, control of such programming, with the attendant risk that certain programming "will not appear," will need to be returned to the operator.

Internet Indecency

The 1996 act does not deal in great detail with the Internet.[210] Generally, the philosophy seems to be that the highly populated world of networks, web pages, and on-line services was sufficiently competitive without federal intervention. The major area in which Congress did attempt to regulate computer services involved the presentation of indecent material that could be accessed by minors.

In addition to several noncontroversial provisions,[211] the 1996 act criminalized the use of interactive computer services to display "patently offensive" sexually explicit material so that it was "available" to minors.[212] This provision was held unconstitutional by two different three-judge courts.[213] Each court found that there was no practical way, under current technology, for most providers of on-line information to control who receives their communication. As Judge Sloviter wrote: "[I]t is either technologically impossible or economically prohibitive for many of the plaintiffs to comply with the [Act] without seriously impeding their posting of online material which adults have a constitutional right to access."[214]

A perhaps more successful approach will be the encouragement of nongovernmental players to police the Internet. The 1996 act provides protection that it terms "good Samaritan" blocking of certain programming.[215] This section states that those who run interactive computer services may not "be held liable" if they voluntarily restrict access to material they consider, in good faith, to be "obscene, lewd, lascivious, filthy, excessively violent, harassing or otherwise objectionable."[216]

The stated purpose of this provision is "to overrule"[217] the decision of the New York trial court in *Stratton-Oakmont versus Prodigy*.[218] The issue in that case was whether Prodigy, an on-line computer service that operated numerous "forums" for subscribers to use to share information, should be viewed as a "publisher" responsible for defamatory comments in its forums or a "conduit" with minimal liability. Because Prodigy had declared itself to be family oriented and used both software and personnel to police the forums for inappropriate language and topics, the court held that Prodigy would bear legal responsibility, just like a traditional newspaper publisher.[219] This theory would force on-line service providers to choose between foregoing all control of their service or engaging in the task of reviewing, and censoring, thousands of postings daily.[220]

The 1996 act attempts to give service providers a middle ground. They will not be held responsible for content they do not produce simply "because they have restricted access to objectionable material."[221] Thus, a service provider need not adopt a totally hands-off policy to escape liability for bulletin board comments.[222]

While this should give some comfort to service providers that want to offer a family service, the protection given by the 1996 act is not complete. If a service provider bars a message for a reason other than those listed in the act, it presumably would be treated as a publisher of *all* the messages it posts.

A Prodigy spokesperson had stated, "What we do with our bulletin boards is identical to the policy taken by most newspapers on letters to the editor. No obscenity, no slander, no libel, no commercialism."[223] It is not at all clear that this policy is covered by the good Samaritan provision. Under the doctrine of *ejusdem generis*, the phrase "otherwise objectionable" probably will not be interpreted so broadly as to cover anything to which a service provider like Prodigy might object.[224] Ironically, then, an on-line service provider that removes some defamatory material may well end up responsible for any defamatory material that remains, while a similar provider that permits all the material to be posted would escape liability.

Conclusion

The 1996 Telecommunications Act has forever transformed the regulatory landscape.[225] The act itself is complex, simultaneously detailed and incomplete. The thousands of pages of FCC rulemaking only increase the difficulty of comprehending the enormous changes brought about by the act.

Nonetheless, there are basic themes that permeate the act. The act contemplates the creation of competition across the full telecommunications field, even in areas such as local telephone service and cable television service, which had previously been monopoly controlled.[226] The main combatants in this new marketplace will tend to be even larger companies than those currently dominating the scene. One can well "envision a future of titanic telecommunications and titanic telecommunicators, a competitive field dominated by highly capitalized, deep-pocket giants."[227]

The hope is that this new marketplace will create not only the advantages of competition, but the unforeseeable benefits that result from a new synergetic relationship between previously separated businesses and technologies: "The opening of all telecommunications markets to all providers will blur traditional industry distinctions and bring new packages of services, lower prices and increased innovation to American consumers."[228]

There are numerous dangers, however, that will have to be averted in order for the act to be successful. The first is that existing monopolies, such as the BOCs or cable operators, will leverage their current power either to gain an unfair advantage in a competitive market or to retain their advantage in the local arena.[229]

The second danger is that the cure to the first is worse than the disease. The primary strategy for creating new competition is that the act permits, indeed encourages, smaller players "to combine, collude, and combat" the entrenched monopolies.[230] Accordingly, there is a lessening of intramedia competition (such as the ability for one entity to control more broadcast stations) in the hope of creating intermedia competition. In addition, certain cross-media combinations (such as between cable operators and broadcast networks or between long-distance and local telephone providers) are permitted in the hope of improving the chances of "intermodal" competition.[231]

If these new combinations do not compete with one another, then the act may have only permitted the creation of large, deregulated monopolists (or oligopolists). The FCC, local regulators, and Congress must watch carefully the unfolding of the new telecommunications field so that we may see these large entities truly battling each other for the hearts and wallets of consumers.[232]

Even if there is such competition, the FCC will have one more critical task: the need to ensure that there is a place for the smaller player. Be it in reselling of local phone service or a programmer seeking one channel on a cable or open video system, there must always be some way for new entry into the telecommunications field.

Finally, amid all the wiring and rewiring, merging and affiliating, one thought should be kept in mind. At the end of the day, what will be most important for the American citizen is not the quantity of fiber optics, coaxial cable, or microwave antennas that line our streets, but the quality of the information that enters our businesses and homes.

Glossary

ADVANCED TELEVISION (ATV) SERVICES. Digital, high-definition television.
AFFILIATE. A person or corporation that, directly or indirectly, controls or possesses an equity interest of more than 10% in another.

BASIC TIER. The least expensive tier of cable programming containing broadcast channels and channels offering public, educational, and government (PEG) access programming.

BELL OPERATING COMPANIES (BOCs). The local telephone companies spun off from AT&T.

BILL-AND-KEEP ARRANGEMENTS. Agreements in which two networks agree to waive their rights to recover from one another compensation for calls terminating in one another's network. *See* Reciprocal compensation arrangements.

CABLE PROGRAMMING SERVICE. All tiers of cable programming, other than basic, except that programming offered on a per channel or per programming (such as pay-per-view) basis.

CABLE TELEVISION SYSTEM. A facility that uses the public rights-of-way to provide video programming within a franchise area via a set of closed transmission paths.

COMPETITIVE CHECKLIST. The list of access and interconnection services the BOCs must provide before they are permitted to offer long-distance service.

DIALING PARITY. When customers can dial the same number of digits to use any available telecommunications provider.

DIRECT BROADCAST SATELLITE (DBS). A communications service in which signals are transmitted by space stations for direct reception by the general public.

DUOPOLY RULE. Federal ban on owning more than one broadcast television station in a single market.

EFFECTIVE COMPETITION. Status of cable television system that permits immediate total rate deregulation. Effective competition occurs for cable systems (1) with penetration rates under 30% of franchise area households subscribing, (2) that face direct head-to-head competition from a multi-channel programmer (such as another cable operator, MMDS system, or DBS operator) that offers service to at least half of the cable operator's franchise area and provides service to at least 15% of the area's households, (3) in areas where the local government itself offered a comparable service to at least half of the franchise area's households, or (4) that face competition from a local-exchange carrier that offers comparable video programming.

ELIGIBLE TELECOMMUNICATIONS CARRIERS. Those carriers designated by a state commission as responsible for providing universal service throughout a designated service area.

FACILITIES-BASED COMPETITOR. A competitor of a local-exchange company that is providing local telephone service either exclusively or predominantly over its own facilities.

FRANCHISING AUTHORITIES. A governmental entity empowered by law to grant a cable franchise.

INCUMBENT LOCAL EXCHANGE COMPANIES. LECs that provide telephone exchange service on the date the Telecommunications Act was enacted, February 8, 1996, or those newcomers determined by the FCC to have obtained a comparable market position.

INTERLATA SERVICE. Long-distance telephone service, communications between points not within the same local area.

LOCAL ACCESS AND TRANSPORT AREA (LATA). A contiguous geographic area that delimits the scope of local telephone service.

LOCAL-EXCHANGE CARRIERS (LECs). Provide local telephone service, either telephone exchange service or service access.

MULTICHANNEL MULTIPOINT DISTRIBUTION SERVICE (MMDS). A one-way communications service, carried on microwave frequencies, capable of providing multiple channels of programming.

NETWORK ELEMENT. The physical equipment used to provide telecommunications service, as well as significant functions, systems, and information that are made available by, or are used in, the transmission of telecommunications service (such as subscriber numbers, databases, signaling systems, and information sufficient for billing and collection or other features used in the transmission, routing, or provision of a telecommunications service).

NUMBER PORTABILITY. The ability for consumers to switch from one telecommunications carrier to another without having to change their existing telecommunications numbers.

OPEN VIDEO SYSTEM. A model for telephone entry into the video marketplace in which, when the demand for video channels exceeds the system's supply, the local telephone company selects the programming for a maximum of one-third of the system's channel capacity and the rest of the channels are made available to nonaffiliated program providers.

RECIPROCAL COMPENSATION ARRANGEMENTS. Agreements that provide that a network in which a call originates compensates the network in which that call terminates.

SATELLITE MASTER ANTENNA TELEVISION (SMATV). So-called private cable television because programming is distributed without crossing public rights-of-way; SMATV typically serves individual or connecting multiple-unit dwellings.

STATE COMMISSION. A state commission with jurisdiction over a public utility company.

TELECOMMUNICATIONS. The transmission of information of the user's choosing without change in the form or content of the information as sent and received.

TELECOMMUNICATIONS CARRIER. Any entity offering, for a fee to the public, to transmit information without changing the content of that which is transmitted.

TELEPHONE EXCHANGE SERVICE. The provision to telephone subscribers of intercommunicating capability within a service area.

UNBUNDLED ACCESS. The availability for purchasing distinct parts of an incumbent phone company's network at an appropriately lower cost than that for access to all of the elements of the network.

UNIVERSAL SERVICE. The minimum telecommunications services that should be available to all consumers, including those living in rural, low-income, or high-cost areas. The FCC is charged with periodically reviewing its definition of universal service to include services that have become essential to education, health, or safety; deployed and subscribed to by a substantial majority of residential customers; or otherwise consistent with the public interest.

V-CHIP. A device that will enable parents to block out all programming rated unsuitable for children due to its sexual or violent content.

Notes

1. Pub. L. No. 73-416, 48 Stat. 1064 (1934) (codified at 47 U.S.C. §§151 et seq.).
2. See, e.g., *Cable Communications Policy Act of 1984*, Pub. L. No. 98-549, 98 Stat. 2779 (codified at 47 U.S.C. §§521–559); *Cable Television Consumer Protection and Competition Act of 1992*, Pub. L. No. 102-385, 106 Stat. 1460 (some sections codified throughout 47 U.S.C.; other sections uncodified). In this article, "the cable act" refers to cable law at the time the Telecommunications Act of 1996 was passed. "1984 Cable Act" or "1992 Cable Act" refer to particular sections from those laws.
3. Judge Harold Greene oversaw the breakup of AT&T, beginning with approval of an antitrust settlement in *United States versus AT&T*, 552 F.Supp. 131 (D.D.C. 1982), aff'd sub nom *Maryland versus United States*, 460 U.S. 1001 (1983) and numerous decisions that followed.
4. *Telecommunications Act of 1996*, Pub. L. No. 104-104, 110 Stat. 56 (1996). Unfortunately for lawyers and judges trying to discuss this act, there is no uniform way to refer to all the sections. Some are codified, some are not. For ease of reference, I use the following convention: When a provision is codified, it is referred to by its U.S.C. citation; uncodified sections are referred to by their section number within the act itself, for example, "1996 Act, Section 301 (i)." The most important legislative history for the 1996 act is the report that was issued by the House-Senate Conference, which reported the final version of the bill, *Joint Explanatory Statement of the Committee of Conference Report*, 104-458, H.R. Conf. Rep. No. 458, 104th Cong., 2d Sess. 142 Cong. Rec. H1078 (January 31, 1996) (hereinafter *Conference Report*).
5. One definition of *convergence* is "the combination of both new and existing media — e.g., broadcasting, cable, fiber optics, satellites — into one integrated system for delivery of video, voice, and data." Michael H. Botein, Antitrust Issues in the Telecommunications and Software Industries, *Sw. U. L. Rev.*, 25:569 (1996).
6. Cf. *Abrams versus United States*, 250 U.S. 616, 630 (1919) (Holmes, J., dissenting) (describing the theory of free speech by stating, "It is an experiment, as all life is an experiment").
7. Those struggling to attain this understanding may well feel new sympathy for the sentiments of Henry David Thoreau: "Our life is frittered away by detail. . . . Simplify, simplify." Henry David Thoreau, *Walden*.
8. See 47 U.S.C. §153 (r) (48).
9. Implementation of the Local Competition Provisions in the Telecommunications Act of 1996, Part II, 61 Fed. Reg. 45476, 45478 (1996) (*Implementation Order, Part II*).
10. 47 U.S.C. §153 (43). A LATA is a contiguous region encompassing no more than one metropolitan statistical area. The area can be greater if permitted under the AT&T Consent Decree or by the FCC. Id.

11. 47 U.S.C. §133 (42).
12. The three largest companies (AT&T, U.S. Sprint, MCI) are joined by hundreds of smaller long-distance companies. See, generally, Debra Kay Thomas Graves, The Consumer Protection Myth in Long-Distance Telephone Regulation: Remedies for the "Caveat Dialer" Attitude, *Tex. Tech. L. Rev.*, 27:383, 391 (1996).
13. 47 U.S.C. §153 (35).
14. "The incumbent LECs have economies of density, connectivity, and scale; traditionally, these have been viewed as creating a natural monopoly"; *Implementation Order, Part II*, 61 Fed. Reg. at 45481. See, generally, Lawrence Sullivan, Elusive Goals Under the Telecommunications Act, *Sw. U. L. Rev.*, 25:487, 494–507 (1996).
15. 1996 Act, Section 101.
16. 47 U.S.C. §§251 and 252.
17. 47 U.S.C. §§251 and 271.
18. 47 U.S.C. §153 (49), (50), and (51).
19. 47 U.S.C. §251 (a) (1).
20. 47 U.S.C. §251 (a) (2).
21. 47 U.S.C. §153 (44).
22. 47 U.S.C. §251 (b) (1).
23. 47 U.S.C. §251 (b) (2).
24. 47 U S.C. §153 (46).
25. 47 U.S.C. §251 (2). For the FCC's regulations, see 47 C.F.R. §§52.21–52.31.
26. 47 U.S.C. §251 (e) (2).
27. 47 U.S.C. §251 (3).
28. 47 U.S.C. §153 (39). See, generally, 47 C.F.R. §§51.205–51.215.
29. 47 U.S.C. §251 (b) (4).
30. 47 U.S.C. §224.
31. 47 U.S.C. §224 (a) (1).
32. 47 U.S.C. §224 (f) (1). Such access may be denied if there is insufficient pole capacity or problems with safety or reliability. 47 U.S.C. §224 (f) (2). By 1998, the FCC must devise a means, to be phased in over a five-year period, for resolving disputes over the rates charged for this pole attachment. 47 U.S.C. §224 (e). For now, cable operators using the poles of others can continue to rely on the rate formula used prior to the 1996 act. 47 U.S.C. §224 (d) (3). The new FCC regulations, when they do become effective, will apply to all telecommunications carriers except cable television systems providing only cable service. Id.
33. 47 U.S.C. §251 (b) (5).
34. 47 U.S.C. §252 (d) (2) (B) (i).
35. 47 U.S.C. §251 (h). As the FCC stated, "The rules we adopt . . . will benefit consumers by making some of the strongest aspects of local exchange carrier incumbency—the local dialing, telephone numbers, operator services, directory assistance, and directory listing—available to all competitors on an equal basis." *Implementation of the Local Competition Provisions of the Telecommunications Act of 1996, Part III*, 61 Fed. Reg. 47284, 47284 (1996).
36. 47 U.S.C. §251 (c) (2). According to the FCC, the term *interconnection* in this section "refers only to the physical linking of two networks for the mutual exchange of traffic." *Implementation Order, Part II*, 61 Fed. Reg. at 45500. See 47 C.F.R. §51.305 for the FCC's regulations on interconnection.
37. 47 U.S.C. §251 (c) (3). See 47 C.F.R. §§51.307–51.321 for the relevant FCC regulations.
38. 47 U.S.C. §153 (45).
39. 47 U.S.C. §251 (c) (6). If physical co-location is either technically impractical or

impossible due to space limitations, the act permits a state regulatory commission to authorize "virtual co-location" instead. Id. See, generally, 47 C.F.R. §51.323.

40. 47 U.S.C. §251 (c) (4) (A).

41. In keeping with this particular provision's goal of creating direct competition, only one limitation is placed on the resale of telecommunications service: Where a service has not been made universally available by the incumbent, but has only been available to a particular category of subscribers, a state commission can prohibit the resale of that service to a different category of subscribers. 47 U.S.C. §251 (c) (4) (B).

42. See, generally, *Implementation Order, Part II*, 61 Fed. Reg. at 45563–45567. The FCC stated that resale is an "important entry strategy for many new entrants, especially in the short term when they are building their own facilities," and also for "small businesses that may lack capital to compete in the local exchange market . . . by building their own networks." Id. 61 Fed. Reg. at 45564.

43. Id., 61 Fed. Reg. at 45565.

44. 47 U.S.C. §252 (d) (3).

45. Id.

46. *Implementation Order, Part II*, 61 Fed. Reg. at 45565.

47. Id.

48. Id. See 47 C.F.R. §51.609. This avoided cost includes both the direct costs of providing retail service and a pro rata share of indirect costs, meaning costs like general corporate operating expenses that are shared between retail and wholesale operations. 47 C.F.R. §51.609 (c).

49. *Implementation Order, Part II*, 61 Fed. Reg. at 45565.

50. 47 C.F.R. §51.611 (b). Although the FCC's rules on "wholesale rates" were stayed by the court in *Iowa Utility Board versus FCC*, 109 F.3d 418 (1996), many states are still using the FCC's criteria as guidelines.

51. 47 U.S.C. §271 (b) (2). Some services, such as 800 services that terminate inside a BOCs local area and permit subscribers to choose their long-distance carriers, are considered in-region services subject to the competitive checklist requirement. 47 U.S.C. §271 (j).

52. See Notes 36–44.

53. 47 U.S.C. §271 (c) (1) (A). According to the Conference Report, this requirement is to ensure "that an unaffiliated competing provider is in the market." *Conference Report*, at 148.

54. *Conference Report*, at 148.

55. Id.

56. Id.

57. 47 U.S.C. §271 (c) (1) (B). The statement will also suffice if the only providers to request access have failed to comply, in a timely fashion, with the implementation schedule of an interconnection agreement. Id.

58. 47 U.S.C. §272 (a) (2) (B).

59. 47 U.S.C. §272 (a) (2) (B) (i). *Incidental services* are defined to include such services as alarm monitoring services, audio and video programming (and the capability of interaction for subscriber selection of such programming), commercial mobile services, and signaling information used in connection with telephone exchange services. 47 U.S.C. §271 (g).

60. 47 U.S.C. §272 (a) (2) (B) (iii).

61. 47 U.S.C. §272 (a) (2) (A).

62. 47 U.S.C. §273 (h). This section states that the definition of "manufacturing" is to be the same as in the AT&T Consent Decree.

63. 47 U.S.C. §273 (e) (1).

64. 47 U.S.C. §272 (b).
65. 47 U.S.C. §272 (b) (5).
66. 47 U S.C. §272 (f) (1).
67. 47 U.S.C. §272 (f) (2).
68. 47 U.S.C. §272 (f) (1) and (2).
69. "As distinct from bilateral commercial negotiation, the new entrant comes to the table with little or nothing the incumbent LEC needs or wants." *Implementation, Part II*, 61 Fed. Reg. at 45481. As one commentator noted, "I would compare it to going to the Department of Motor Vehicles and trying to negotiate more favorable terms there. It is very hard to negotiate with somebody who has 100% of the market, and a very strong desire to keep that situation in place." See Implications of the New Telecommunications Legislation, Fordham, *Media and Ent. L. J.*, 6: 517, 534 (1996) (statement of J. Richard Devlin, Executive Vice President, Sprint Corporation).
70. 47 U.S.C. §252 (a) (1).
71. Sec. 251 (c) (1). Although this requirement is contained in the subsection entitled "Additional Obligations of Incumbent Local Exchange Carriers," the section explicitly places the duty to negotiate in good faith on both the incumbent and "[t]he requesting telecommunications carrier." Id. See also 47 C.F.R. §51.301 (Duty to Negotiate).
72. 29 U.S.C. §158 (d).
73. *Conference Report*, at 113.
74. R. Summers, "Good Faith" in General Contract Law and the Sales Provisions of the Uniform Commercial Code, *Va. L. Rev.*, 54:195, 221 (1968). See, generally, E. Allan Farnsworth, *Contracts*, 1982, pp. 187–190, Little Brown, Boston, MA.
75. 47 U.S.C. §252 (b) (1). The demand for arbitration must be filed within a relatively short window, between the 135th and 160th day (inclusive) after the incumbent has received a request for negotiation. Id. The state commission must reach its arbitration decision within nine months from the date of that request for negotiation. 47 U.S.C. §252 (b) (4).
76. 47 U.S.C. §252 (d) (1).
77. 47 U.S.C. §252 (d) (2). The bill-and-keep arrangement provides that each network will waive it recovery rights in exchange for the other network's agreement to do the same. See text at Note 34.
78. 47 U.S.C. §252 (e). If the state commission does not act with 90 days of submission of a negotiated agreement, or 30 days of an arbitrated agreement, the agreement will be deemed approved. 47 U.S.C. §252 (e) (4).
79. While the arbitration procedure, designed to reach a mediated agreement, serves a different purpose from the approval procedure, which is designed to ensure that the substance of the agreement squares with the procompetitive sections of the act, it would have no doubt been simpler to require the state commission, as arbitrator, to ensure that the final agreement is consistent with the act.
80. 47 U.S.C. §252 (e) (2) (A).
81. 47 U.S.C. §252 (e) (2) (B).
82. Sec. 252 (e) (5).
83. Sec. 252 (e) (6).
84. Sec. 253 (a). State and local governments are also prohibited from regulating direct broadcast satellite (DBS) service, 47 U.S.C. §303 (v), and the FCC must regulate to prevent local zoning or other regulation that impairs a viewer's ability to receive DBS or multichannel, multipoint distribution service (MMDS). 1996 Act, Section 207. The new federal regulations bar restrictions not only by zoning and building laws, but by private covenants and homeowner's associations, unless

narrowly tailored to protect interests in safety or historic preservation. 47 C.F.R. §1.4000.

85. 47 U.S.C. §253 (c). See also *Conference Report*, at 180 (stating that even though franchising authorities cannot regulate cable television operators in their provision of telecommunications services, local governments retain their authority over rights-if-way and can charge reasonable fees for their use).

86. See 47 U.S.C. §542 (b).

87. 47 U.S.C. §253.

88. 47 U.S.C. §253 (b).

89. 47 U.S.C. §253 (d).

90. Though, as the FCC correctly points out, the costs of universal service are sometimes hidden from view: "The current universal service system is a patchwork quilt of implicit and explicit subsidies." *Implementation Order, Part II*, 61 Fed. Reg. at 45480.

91. 47 U.S.C. §151.

92. See, generally, *Conference Report*, at 130–134.

93. 47 U.S.C. §254 (b).

94. 47 U.S.C. §254 (c).

95. 47 U.S.C. §254 (c), (a), and (d). The FCC's action is preceded by the recommendation of a Federal-State Joint Board. 47 U.S.C. §254 (a) (1).

96. 47 U.S.C. §254 (d).

97. 47 U.S.C. §254 (e).

98. 47 U.S.C. §214 (e) (1) (A) and (B).

99. 47 U.S.C. §214 (e) (2) and (5).

100. 47 U.S.C. §214 (e) (2). For rural areas, the decision whether to designate more than one carrier as "eligible" is left to the discretion of the state commission. Id.

101. 47 U.S.C. §259 (a) and (b) (4). If an eligible carrier enjoys its own economics of scale and scope, there is no right under this provision to share infrastructure. 47 U.S.C. §259 (d).

102. 47 U.S.C. §259 (b) (4).

103. 47 U.S.C. §254 (e).

104. 47 U.S.C. §254 (k).

105. Id.

106. Id.

107. 47 U.S.C. §541 (a) (1). See also *Preferred Communications, Inc., versus City of Los Angeles*, 754 F.2d 1396 (9th Cir. 1985) (holding exclusive cable franchises unconstitutional), aff'd on other grounds, 476 U.S. 488 (1986).

108. "Limited non-cable competition exists today from several non-cable technologies, such as DBS, MMDS, and SMATV. Since the total penetration of such alternative providers today accounts for less than 10% of the country's multichannel video offerings, however, the competition is a long way from being effective in most areas of the country." See Botein, Note 5, pp. 596–97.

109. See 1984 Cable Act. This was not, by any means, a law that deregulated cable. See, generally, Michael I. Meyerson, The 1984 Cable Act: A Balancing Act on the Coaxial Wires, *Ga. L. Rev.*, 19:543 (1985).

110. See 1994 act.

111. 47 U.S.C. §543 (1) (1) (A).

112. 47 U.S.C. §543 (1) (1) (C).

113. 47 U.S.C. §543 (1) (1) (B). See also *Time Warner Entertainment Co., L.P., versus FCC*, 56 F.3d 151 (D.C.Cir. 1995).

114. 47 U.S.C. §543 (1) (1) (C).

115. 47 U.S.C. §543 (1) (1) (D). "Comparable" programming means at least 12 chan-

nels of programming, at least some of which are television broadcast channels. *Conference Report*, at 170.

116. 47 C.F.R. §76.905 (e). In addition, potential subscribers must be reasonably aware that the telephone company's service is available. Id.

117. The 1992 Cable Act, in fact, begins with the finding that cable rates were rising three times faster than the rate of inflation since the 1984 act. See Section 2 (a) (1), 1992 Cable Act.

118. See 47 U.S.C. §543 (b) (7) (A) and (1) (2).

119. 47 U.S.C. §543.

120. *Rate Regulation, Report and Order*, MM Docket No. 92-266, 8 F.C.C.R. 5631 (1993).

121. 47 U.S.C. §543 (b) (1).

122. 47 U.S.C. §543 (m). A "small" cable operator means one not affiliated with either any company serving more than 1% of the nation's subscribers or any business with gross annual revenue greater than $250 million. 47 U.S.C. §543 (m) (2).

123. 47 U.S.C. §543 (c) (4).

124. 47 U.S.C. §543 (c). It will be harder to file a complaint concerning rate increases under the 1996 act. In a change from the 1992 Cable Act, only complaints from franchising authorities, rather than individual subscribers, will trigger FCC investigation. Id.

125. 47 U.S.C. §548. The constitutionality of this requirement was upheld in *Daniels Cablevision, Inc., versus United States*, 835 F.Supp. 1 (D.C.Cir. 1993).

126. 47 U.S.C. §548 (c).

127. In fact, the provision was extended to telephone companies providing video programming. See text accompanying Note 141 infra.

128. 47 U.S.C. §§534 and 535. The only changes made to the must-carry rules were minor. The FCC was permitted to use a variety of measures to determine the market for broadcasters, §534 (h) (1) (C), and the FCC was forced to rule on petitions to modify a television market within 120 days of a request, 47 U.S.C. §534 (h) (1) (C) (iv).

129. 47 U.S.C. §533 (b), now repealed. There were a few small exceptions, such as for cross ownership in rural areas. See 47 C.F.R. §63.56 (a), now repealed.

130. See, e.g., *Chesapeake and Potomac Telephone Co. versus United States*, 830 F. Supp. 909 (E.D.Va. 1993), aff'd 42 F.3d 181 (4th Cir. 1994), vacated 116 S.Ct. 1036 (1996); *U.S. West, Inc.,* versus United States, 855 F.Supp. 1184 (W.D. Wash. 1994), aff'd 48 F.3d 1092, (9th Cir. 1994), vacated 116 S.Ct. 1037 (1996); *Pacific Telesis Group versus United States*, 48 F.3d 1106 (9th Cir. 1994), vacated sub nom; *U.S. West, Inc., versus United States*, 116 S.Ct. 1037 (1996).

131. 1996 act, Section 302(b) (1).

132. See, e.g., *Chesapeake and Potomac Telephone Co. versus United States*, 116 S.Ct. 1036 (1996), vacating 42 F.3d 181 (4th Cir. 1994); *U.S. West, Inc., versus United States*, 116 S.Ct. 1037 (1996) vacating, 48 F.3d 1092 (9th Cir. 1994); and *Pacific Telesis Group versus United States*, 48 F.3d 1106 (9th Cir. 1994).

133. 47 U.S.C. §651 (a) (2) and (3).

134. 47 U.S.C. §651 (a) (1).

135. 47 U.S.C. §651 (a) (3). The FCC's regulations implementing this provision are found at 47 C.F.R. §76.5100 et. seq. The open video system concept replaces the former FCC attempt to create a telco-cable hybrid, the video-dialtone regulations. 1996 act, Section 302 (b) (3).

136. *Conference Report*, at 187 ("[T]he conferees hope that this approach will encourage common carriers to deploy open video systems and introduce vigorous competition in entertainment and information markets").

137. 47 U.S.C. §653 (b) (1) (a).
138. 47 U.S.C. §653 (b) (1) (e) (i).
139. 47 U.S.C. §653 (c) (3).
140. 47 U.S.C. §653 (c) (1) (C).
141. 47 U.S.C. §653 (c) (1) (B) and (C) (stating, *inter alia*, that Cable Act Sections 611 [47 U.S.C. §531], 614 [47 U.S.C. §534], and 628 [47 U.S.C. §548] will apply to open video systems).
142. 47 U.S.C. §541 (b) (3) (B). See also 47 U.S.C. §253 (a) (preempting state and local regulation having the effect of prohibiting any entity from providing telecommunications service), discussed in text at Note 84 supra. Local governments are also prohibited from requiring that cable operators offer telecommunications services, except for PEG and leased access channels and institutional networks. 47 U.S.C. §541 (b) (3) (D).
143. 47 U.S.C. §541 (b) (3) (A) (i).
144. 47 U.S.C. §542 (b). Under the Cable Act, franchise fees are capped at 5% of a cable operator's gross revenue. Id.
145. There is much uncertainty as to how soon it will be economically practicable for either the cable or telephone company to use one wire to carry both cable service and telephone service: "[I]t is physically impossible to send a telephone conversation over a contemporary unswitched cable system, or to push a full-motion video signal through a switched but low-capacity telco." Botein, Note 5, p. 594. As Professor Botein has noted, although both cable systems and LECs send electronic signals through wires, "the resemblance between the two technologies just about ends there; for the foreseeable future, the two distribution systems are about as similar as an electric utility and a gas pipeline" (p. 569).
146. 47 U.S.C. §652 (a) and (b).
147. 47 U.S.C. §652 (c).
148. *Conference Report*, at 174.
149. *Conference Report*, at 174.
150. There were some narrow exceptions made to the ban on cable/telco merger. Telephone companies can combine with co-located cable operators in rural areas. 47 U.S.C. §652 (d) (1). Under this provision, the combined entity must serve a location with fewer than 35,000 inhabitants outside an urbanized area and must serve no more than 10% of the households in the telephone company's service area.

   There were also some provisions that were written to apply to only a tiny number of situations. For example, under one exception, merger is permitted if the cable system either (1) not owned by one of the 50 largest cable operators, is outside the top 100 television markets, and serves no more than 17,000 subscribers, with at least 8000 urban and 6000 nonurban; or (2) serves fewer than 20,000 subscribers, of whom at most 12,000 live in urban areas, and is combining with a small telephone company, one with less than $100 million in annual revenue. 47 U.S.C. §652 (d) (4) and (5).

   A special exception was also carved out for some of the extremely few areas where a competitive cable market existed prior to the 1996 act. Under this provision, in all but the top 25 largest television markets, a telephone company will be able to merge with a local cable operator as long as (1) it is not the largest cable operator in the area; (2) the acquired cable system is not owned by one of the 50 largest multiple system operators (MSOs); (3) the area's larger cable system is owned by one of the 10 largest MSOs; and (4) the acquired cable system must have obtained a franchise covering the same area as the largest system as of May 1, 1995. 47 U.S.C. §652 (d) (3).

151. 47 U.S.C. §652 (d) (6).
152. 47 U.S.C. §652 (d) (6) (A).
153. 47 U.S.C. §652 (d) (6) (A).
154. 47 U.S.C. §537 (a), now repealed.
155. 1996 act, Section 301 (i).
156. 47 U.S.C. §537.
157. 1996 act, Section 202 (f) (1) (eliminating 47 C.F.R. §76.501).
158. 1996 act, Section 202 (f) (2).
159. 1996 act, Section 202 (i) (eliminating 47 U.S.C. §533 (a) (1)).
160. *Conference Report*, at 164 (discussing FCC review of 47 C.F.R. §76.501).
161. 47 U.S.C §533 (a) (3) For a discussion of effective competition, see text accompanying Notes 111–116 supra.
162. See text accompanying Notes 157–160 supra.
163. 47 U.S.C. §309 (k) (4). In fact, this subsection is entitled, "Competitor Consideration Prohibited." Id.
164. 47 U.S.C. §309 (k) (1).
165. In practice, though, broadcast renewal has always been a virtual sure thing. See, e.g., *Monroe Communications Corp. versus FCC*, 900 F.2d 351, 359 (D.C.Cir. 1990) (Silberman, J., concurring) ("Quite obviously the FCC shrinks from the prospect of taking the license away from the incumbent, but . . . it is hard to see how the FCC can justify the weight it places on incumbency in [the renewal] case").
166. 47 U.S.C. §307 (c) (1).
167. 1996 act, Section 202 (a).
168. 1996 act, Section 202 (c) (1) (A).
169. 1996 act, Section 202 (c) (1) (B). This is an increase from the previous national cap of 25%. See 47 C.F.R. §73.3555 (now modified).
170. 1996 act, Section 202 (b).
171. 1996 act, Section 202 (c) (2) (referring to current FCC duopoly rule at 47 C.F.R. §73.3555).
172. *Conference Report*, at 163.
173. 47 C.F.R. §73.3555 (now revised).
174. 1996 act, Section 202 (d).
175. 47 C.F.R. §73.658 (g), now revised.
176. 1996 act, Section 202 (e). See 47 C.F.R. §73.658 (g) (as revised).
177. 47 U.S.C. §335 (g).
178. 47 U.S.C. §335 (a) (1).
179. See, e.g., Jonathan D. Blake and Ellen P. Goodman, Second Byte: Congressional Excursion into Digital TV, *Communications Lawyer*, 14:3–5 (Summer 1996).
180. See Mass Media Action: Commission Adopts Rules for Digital Television Service, 1997 FCC LEXIS 1733 (April 3, 1997).
181. See, e.g., *Abrams versus United States*, 250 U.S. 616, 630 (1919) (Holmes, J., dissenting) ("[T]he best test of truth is the power of the thought to get itself accepted in the competition of the market.").
182. *United States versus Associated Press*, 52 F.Supp. 362 (S.D.N.Y. 1943), aff'd 326 U.S. 1 (1944).
183. "Some degree of abuse is inseparable from the proper use of every thing; and in no instance is this more true than that of the press." *New York Times Co. versus Sullivan*, 376 U.S. 254, 271 (1964), quoting James Madison in 4 Elliot's Debates on the Federal Constitution (1876), p. 571.
184. 1996 act, Section 506.
185. 47 U.S.C. §303 (w) (2).

186. 47 U.S.C §303 (x). Actually, this only applies to television sets with a diagonal screen of 13 inches or greater. Id.

187. 1996 act, Section 551 (e) (2).

188. See, e.g., *Freeman versus Maryland*, 380 U.S. 51 (1965); *Interstate Circuit versus Dallas*, 390 U.S. 676 (1968).

189. 47 U.S.C. §303 (w) (1) and 1996 act, Section 551 (e) (1). These guidelines, which are to be promulgated after an advisory board issues recommendations, are not intended to be formal "requirements" that broadcasters must use. *Conference Report* at 195. However, if any rating system is used, the rating must be transmitted with the programming. 47 U.S.C. §303 (w) (2).

190. See, generally, Douglas Ayer, Roy Bates, and Peter Herman, Self-Censorship in the Movie Industry: An Historical Perspective of Law and Social Change, *Wisc. L. Rev.*, 791 (1970).

191. 1996 act, Section 551 (e) (1). The broadcast industry, in part due to their desire to please the same congressional powers that would be distributing the lucrative spectrum for advanced television (see text accompanying Notes 177–180) established "voluntary" rules for rating programming.

192. 1996 act, Section 551 (e) (1) (italics added).

193. *FCC versus Pacifica*, 438 U.S. 726, 748 (1978).

194. This is what occurred in the dial-a-porn case, in which a ban on telephone indecency was struck down due to the availability of "less restrictive means" for protecting children. *Sable Communications versus FCC*, 492 U.S. 115, 129 (1989).

195. 47 U.S.C. §544 (d) (2).

196. See, e.g., *Cruz versus Ferre*, 755 F.2d 1415, 1420–21 (11th Cir. 1985); *Community Television, Inc., versus Wilkinson*, 611 F.Supp. 1099 (D.Utah 1985), aff'd sub nom; *Wilkinson versus Jones*, 800 F.2d 989 (10th Cir. 1986), aff'd without opinion 480 U.S. 926 (1987). Obscenity, which is not constitutionally protected, can be banned from cable television, and the 1996 act raised the penalty for transmitting obscene cable programming from $10,000 to $100,000. 47 U.S.C. §559.

197. 47 U.S.C. §641 (a). *Scramble* is defined to mean rearranging the content of the signal sent into the home so that the programming cannot be seen or heard in an understandable manner. 47 U.S.C. §641 (c).

198. 47 U.S.C. §641 (b). The FCC ruled that unscrambled adult programming could only be shown between the hours of 10 P.M. and 6 A.M. In the Matter of Implementation of Section 505, 11 FCC Rcd. 5386 (1996).

199. *Playboy Entertainment Group, Inc., versus United States*, 945 F. Supp. 772 (D.Del. 1996), aff'd 117 S.Ct 1309 (1997).

200. 47 U.S.C. §640 (a).

201. *Conference Report*, at 192.

202. 47 U.S.C. §§531 and 532.

203. The 1992 Cable Act permitted operators to refuse to carry access programming that depicted "sexual or excretory activities or organs in a patently offensive manner," Section 10 (a) and 10 (c). The 1996 Telecommunications Act permitted operators to refuse to carry access programming "which contains obscenity, indecency or nudity." 47 U.S.C. §§531 (e) and 532 (c) (2).

204. *Denver Area Educational Tele-Communications Consortium, Inc., versus FCC*, 116 S.Ct. 2374 (1996). The Court also struck down a 1992 provision requiring cable operators to segregate "patently offensive" leased access programming on a single channel and only permit subscriber access on written request. 1992 Cable Act, Section 10 (b).

205. Denver Educational, 1165 S.Ct. at 2394–97. This part of Justice Breyer's opinion was joined by only Justices Stevens and Souter.

206. Id., 116 S.Ct. at 2394.

207. *Time Warner Entertainment Co. versus FCC*, 93 F.3d 957, 981 n.7 (D.C.Cir. 1996).

208. 47 U.S.C. §558. This provision was upheld in *Time Warner Entertainment Co. versus FCC*, 93 F.3d 957 (D.C.Cir. 1996).

209. See *Farmers Educational and Cooperative Union versus WDAY, Inc.*, 360 U.S. 525, 535 (1959).

210. For a general description of the Internet, see *Shea versus Reno*, 930 F. Supp. 916, 925 (S.D.N.Y. 1996) (describing the Internet as "a collection of more than 50,000 networks linking some 9 million host computers in 90 countries.").

211. For example, the 1996 act makes it a crime to use telecommunications devices to induce a minor to engage in any illegal sexual act, 18 U.S.C. §2422 (b), or to annoy or harass another person either with obscene and indecent communication or by repeated telephone calls. 47 U.S.C. §223 (a) (1) (B), (D), and (E). The act also clarifies that it is a felony to use a computer to transmit obscene material. 18 U.S.C. §1462. This last amendment probably does not change preexisting obscenity law, which was generally interpreted to reach that result. See *United States versus Thomas*, 74 F.3d 701, 704–05 (6th Cir. 1995) (affirming obscenity convictions for the operation of a computer bulletin board).

212. 47 U.S.C. §223 (d).

213. *Shea versus Reno*, 930 F. Supp. 916 (S.D.N.Y. 1996); *ACLU versus Reno*, 929 F. Supp. 824 (E.D.Pa. 1996). As this volume was going to press, the Supreme Court affirmed the lower courts and ruled this provision unconstitutional. *Reno versus ACLU*, 1997 U.S. LEXIS 4037 (June 26, 1997).

214. *ACLU versus Reno*, 929 F. Supp. at 853.

215. 47 U.S.C. §230 (c).

216. 47 U.S.C. §230 (c) (2). The 1996 act also protects those who provide connections to the Internet or networks they do not control, and are not responsible for on-line content. 47 U.S.C. §223 (e). This protection is reserved for "entities that simply offer general access to the Internet and other online content." *Conference Report*, at 190.

217. *Conference Report*, at 194.

218. 1995 N.Y.Misc. LEXIS 229 (N.Y.S.Ct. 1995).

219. Id.

220. See, generally, Michael I. Meyerson, Authors, Editors, and Uncommon Carriers: Identifying the "Speaker" Within the New Media, *Notre Dame L. Rev.*, 71:79, 116–124 (1995).

221. *Conference Report*, at 194.

222. Cf. *Cubby, Inc., versus CompuServe, Inc.*, 776 F. Supp. 135, 140 (S.D.N.Y. 1991) (holding that CompuServe was not responsible because it had "little or no editorial control" over the content of the postings).

223. Felicity Barringer, Electronic Bulletin Boards Need Editing. No They Don't, *New York Times*, March 11, 1990, Section 4, p. 4.

224. This doctrine states that "where general words follow an enumeration . . . by words of a particular and specific meaning, such general words are not to be construed to their widest extent, but are to be held as applying only to persons or things of the same general kind or class as those specifically mentioned." *Black's Law Dictionary*, 1979, p. 464, West Publishing, St. Paul, MN.

225. "From this point forward, telecommunications law starts with the Telecommunications Act of 1996." Jim Chen, Antitrust Issues in the Telecommunications and Software Industries: Titanic Telecommunications, *Sw. U. L. Rev.*, 25:535, 537 (1996).

226. See text accompanying Notes 13–21 and 107–108 supra.

227. Chen, Note 225, p. 545. See also p. 551 (stating that in the local telephone market, "we should expect only one type of entrant: big").

228. *Implementation Order, Part II*, 61 Fed. Reg. at 45480.

229. See, e.g., Sullivan, Note 15, p. 522 (stating "There is no question that the risk of cross-subsidy is highest when the regulated monopolies enter an adjacent market with high joint costs"). But, see Chen, Note 225, p. 552 (stating local telephone service is "a contestable, albeit imperfectly competitive market.").

230. Chen, Note 225, p. 558.

231. Chen, Note 225, p. 551.

232. See, e.g., Botein, Note 5, p. 597 (stating that "both the telco and cable industries may end up with three or four dominant players—which in turn may merge or form strategic alliances with each other"); Chen, Note 225, p. 557 (referring to "the possibility of procompetitive combinations").

MICHAEL I. MEYERSON

Review of Electromagnetic Compatibility in Telecommunications

Introduction

Our society relies on the ability to establish and maintain extensive, reliable communications. In general, the requirement for use of the electromagnetic spectrum for communication, navigation, and radar systems has been rapidly increasing. Our military strategy is based on the rapid deployment of dynamic forces supported by an extensive command, control, communications, and intelligence (C^3I) network to provide the information required for battle management. In the civilian sector, our communication requirements have increased drastically as a result of the mobility of our society and our dependence on computers. The cellular telephone has significantly increased the capacity of our mobile communications, and fixed point-to-point microwave and satellite communications systems provide an extensive data-transmission network for computer systems.

One of the most important considerations in the design, installation, and operation of a communications electronic system is that of achieving and maintaining electromagnetic compatibility (EMC) between the system and the other communications electronic equipment in the immediate vicinity. EMC is the ability of equipment or systems to function as designed without degradation or malfunction in an intended operational electromagnetic environment. The equipment or system should not adversely affect the operation of, or be adversely affected by, any other equipment or system.

In order to succeed in achieving EMC and to permit efficient use of the frequency spectrum, it is essential that engineers, technicians, and users responsible for the planning, design, development, installation, and/or operation of communications electronic equipment employ suitable analysis techniques. These techniques permit them to identify, localize, and define EMI problem areas prior to, rather than after, expenditures of time, effort, and dollars. More timely and economical corrective measures may then be taken.

The primary purpose of this article is to provide analysis techniques and tools that may be used in planning, designing, installing, and operating communication equipment or systems that are free from electromagnetic interference (EMI) problems. Careful application of these techniques at appropriate stages in the system's life-cycle will ensure EMC without either the wasteful expense of overengineering or the uncertainties of underengineering.

The Communication System Electromagnetic Interference Problem

In a typical communication situation, the receiver must be able to "pick up" its intended signal, which is probably relatively weak, while operating in the pres-

Courtesy of EMF-EMI Control, 6193 Finchingfield Road, Gainesville, VA 22065; http://www.emf-emi.com.

ence of a number of relatively strong, potentially interfering, signals that result from other communications electronic systems operating in close proximity. Conversely, the transmitter must be able to transmit a relatively strong signal without causing interference to sensitive receivers operating in the immediate vicinity.

The basic EMC requirement is to plan, specify, and design systems, equipment, and devices that can be installed in their operational environments without creating or being susceptible to interference. In order to help satisfy this requirement, careful consideration must be given to a number of factors that influence EMC. In particular, it is necessary to consider major sources of EMI, modes of coupling, and points or conditions of susceptibility. The EMC technologist should be familiar with the basic tools (including analysis, measurement, control, suppression, specifications, and standards) that are used to achieve EMC.

This article presents a methodology for the EMC design of telecommunications systems and describes analysis techniques that may be used to identify and define potential EMI problems. The techniques are specifically oriented toward EMI signals generated by potentially interfering transmitters and propagated and received via antennas and that cause EMI in receivers associated with telecommunications systems.

Major Electromagnetic Interference Interactions Between Transmitters and Receivers

In the planning and design of communication system, it is important to recognize that there are several different means by which EMI may occur; for each situation, the appropriate types of EMI must be considered. The important types of EMI, which are shown in Fig. 1, may be considered as one of three basic categories: (1) co-channel, (2) adjacent signal, or (3) out of band. These categories are defined as follows:

- *Co-channel EMI* refers to interference resulting from signals that exist within the narrowest passband of the receiver. For superheterodyne receivers (the type used for many applications) the frequency of co-channel interference must be such that the interference is translated to the intermediate frequency (IF) passband in the same manner as the desired signal. This requires that the frequency of co-channel interfering signals equals the tuned radio frequency plus or minus one-half the narrowest IF bandwidth. Although the receiver is most sensitive to this type of interference, it is usually easily controlled by avoiding co-channel assignments within a relatively large control zone over which this type of interference may occur.
- *Adjacent signal EMI* refers to potentially interfering signals that exist within or near the receiver radio frequency (RF) passband but after conversion fall outside of the IF passband. The most significant adjacent signal EMI effects

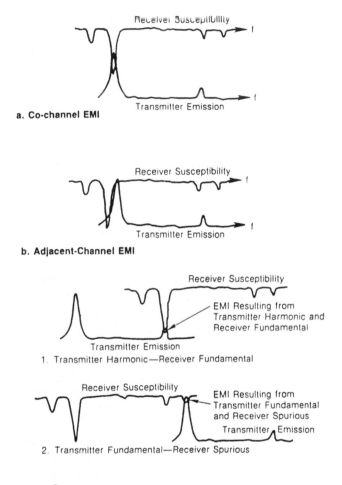

FIG. 1   Types of transmitter-receiver electromagnetic interference (EMI).

result from intermodulation and transmitter noise. Although Fig. 1 illustrates a situation in which a transmitter and a receiver are tuned to adjacent channels, the adjacent signal EMI region may extend over a considerable range of frequencies on each side of the tuned frequency. For example, for a typical ultra-high frequency (UHF) communication transceiver having 25-kilohertz (kHz) channel spacing, the adjacent signal EMI region may include 400 channels (i.e., 10 megahertz [MHz]) on each side of the desired channel. Although the adjacent signal EMI region includes a relatively wide range of frequencies, the receiver is not particularly sensitive to these signals. As a result, adjacent signal EMI is usually limited to co-site situations involving transceivers located within 1 or 2 miles of each other.

- *Out-of-band EMI* refers to signals having frequency components that are significantly outside of the widest receiver passband. The most significant out-of-band EMI effects result from transmitter harmonics interfering with receiver fundamentals or transmitter fundamentals interfering with receiver spurious responses. EMI between transmitter harmonics and receiver spurious responses is also possible but extremely unlikely. Because of the power levels involved, out-of-band EMI is usually restricted to co-site situations.

Basic Electromagnetic Compatibility Analysis

In order to determine whether an EMI problem exists between a potentially interfering transmitter and a receiver, it is necessary to consider the susceptibility of the receiver to both the design and spurious outputs (individually and collectively) of the potentially interfering transmitters. The factors that must be included in the analysis for each transmitter output (or group of transmitter outputs) include the transmitter power P_T, the transmitting antenna gain in the direction of the receiver G_{TR}, the propagation loss between the transmitter and receiver L, the receiver antenna gain in the direction of the transmitter G_{RT}, and the amount of power required to produce interference in the receiver P_R in the presence of the desired signal.

Factors that must be considered in interference analysis include both (1) the design (intentional) and operational performance characteristics of equipment and (2) the nondesign (unintentional) and nonoperational characteristics. This article discusses equipment characteristics that are used in EMC design and analysis and describes methods and techniques that are employed for representing these equipment characteristics in the form of general mathematical models for analysis. The necessity for considering parameters such as transmitter spurious output emissions, receiver spurious responses, antenna side-and-back lobe radiation, and unintentional propagation paths introduces complications since it is now necessary to obtain information on equipment nondesign characteristics. Unlike equipment design characteristics (which are usually well defined and may be readily obtained from equipment specifications), equipment spurious characteristics are not usually identified or described in equipment specifications. Therefore, it is difficult to obtain information on the spurious characteristics of specific equipment.

For situations in which specific detailed data are not available, several different sources of input information may be used to derive "default models." This article presents equipment EMC default models that have been derived from statistical summaries of measured equipment characteristics and from MIL-STD-461 limits.

The procedure used for each transmitter output emission can be demonstrated by considering the interference situation that exists between a particular output of one of a number of potentially interfering transmitters and a specimen receiver. For the case of a particular transmitter output (which may be either a fundamental or a spurious emission), the power available at the receiver is given by

$$P_A(f,t,d,p) = P_T(f,t) + C_{TR}(f,t,d,p) \qquad (1)$$

where

$P_A(f,t,d,p)$ = power available at the receiver (in dBm) as a function of frequency f, time t, distance separation d, and direction p, of both transmitter and receiver and their antennas

$P_T(f,t)$ = transmitter power (in dBm)

$C_{TR}(f,t,d,p)$ = transmission coupling between transmitter and receiver (in decibels [dB])

In problems involving interference coupled from a transmitting antenna to a receiving antenna, the transmission coupling function is represented by

$$C_{TR}(f,t,d,p) = G_{TR}(f,t,d,p) - L(f,t,d,p) + G_{RT}(f,t,d,p) \qquad (2)$$

where

$G_{TR}(f,t,d,p)$ = the transmitting source antenna gain in the direction of the receiver (in dB)

$L(f,t,d,p)$ = the propagation loss function (in dB)

$G_{RT}(f,t,d,p)$ = the receiving antenna gain in the direction of the transmitter (in dB)

By comparing the power available at the receiver input terminals to the power required at the input to produce interference in the receiver at the frequency in question $P_R(f,t)$, it is possible to determine the interference situation for the particular transmitter output being considered. The requirement for EMC is that the power available at the receiver be less than the power required to produce interference in the receiver. Thus, the condition for electromagnetic compatibility is

$$P_A(f,t,d,p) < P_R(f,t) \qquad (3)$$

On the other hand, if the power available at the receiver input terminals is equal to or greater than the power required to produce interference in the receiver, an electromagnetic interference problem may exist, that is,

$$P_A(f,t,d,p) \geq P_R(f,t) \qquad (4)$$

When $P_A = P_R$, EMC is marginal and an EMI problem may or may not exist.

An indication of the magnitude of a potential interference problem may be obtained by considering the difference between the power available and the susceptibility threshold. This difference is termed *interference margin* (IM) and provides a measure of the total contribution to interference, that is,

$$IM(f,t,d,p) = P_A(f,t,d,p) - P_R(f,t) \qquad (5)$$

The interference margin is defined such that there is a potential interference problem if the margin is positive and there is little to no chance of interference if the interference margin is negative.

The expression $IM(f,t,d,p)$ in Eq. (5) can be considered to represent an equivalent on-tune interference-to-noise ratio (I/N) at the receiver input terminals. If the expressions for $P_A(f,t,d,p)$ and $P_R(f,t)$ are expanded, Eq. (5) becomes

$$IM(f,t,d,p) = I/N = P_T(f_E) + G_{TR}(f_E,t,d,p) - L(f_E,t,d,p)$$
$$+ G_{RT}(f_E,t,d,p) - P_R(f_R) + CF(B_T,B_R,\Delta f) \qquad (6)$$

where

$P_T(f_E)$ = power transmitted in dBm at emission frequency f_E

$G_{TR}(f_E,t,d,p)$ = transmitter antenna gain in dB at emission frequency f_E in the direction of the receiver

$L(f_E,t,d,p)$ = propagation loss in dB at emission frequency f_E between transmitter and receiver

$G_{RT}(f_E,t,d,p)$ = receiver antenna gain in dB at emission frequency f_E in direction of transmitter

$P_R(f_R)$ = receiver susceptibility threshold in dBm at response frequency f_R

$CF(B_T,B_R,\Delta f)$ = factor in dB that accounts for transmitter and receiver bandwidths B_T and B_R, respectively, and the frequency separation Δf between transmitter emission and receiver response

The final term in Eq. (6), $CF(B_T,B_R,\Delta f)$, takes into account the relative bandwidths, transmitter modulation envelope, receiver selectivity curve, and the frequency separation, if any, between the transmitter output and the receiver response. The procedure used for determining $CF(B_T,B_R,\Delta f)$ is illustrated by considering the various possibilities that may exist between particular output response pairs (see Fig. 2).

First, if the output and response occur at the same center frequency (i.e., $\Delta f = 0$), there are two basic co-channel possibilities that may be considered:

1. Receiver bandwidth is either equal to or larger than the transmitter bandwidth ($B_R \geq B_T$). For this case, all the power associated with the transmitter output is received, and no correction is necessary, that is, $CF(B_T,B_R,\Delta f) = 0$.

2. Receiver bandwidth is less than the transmitter bandwidth ($B_R < B_T$). For this case, only a portion of the power associated with the emission output is

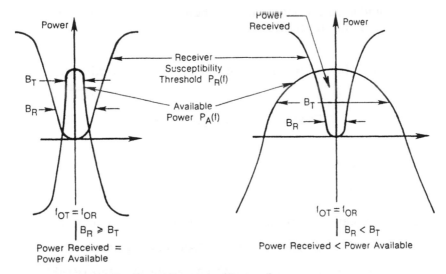

a. On-Tune Case (Co-channel Frequency Alignment)

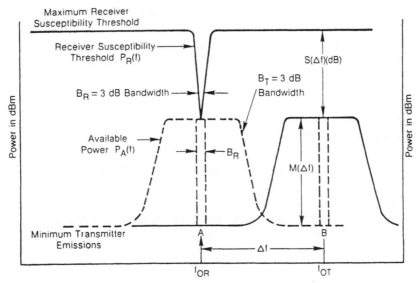

b. Off-Tune Case (Spurious Frequency Alignment)

FIG. 2   Frequency bandwidth analysis.

received, and it is necessary to apply a bandwidth correction CF to account for the bandwidth differences.

This correction for $\Delta f = 0$ is dependent on the bandwidth ratios and is of the form

$$CF(\Delta f = 0) = K \log_{10}(B_R/B_T) \text{ dB} \tag{7}$$

where

B_R = receiver 3-dB bandwidth in hertz (Hz)
B_T = transmitter 3-dB bandwidth in Hz

K = a constant for a particular emission-response combination

  = 0 for $B_R \geq B_T$ and co-channel frequency alignment

  = 10 for noiselike signals for which root mean square (RMS) levels apply and $B_R < B_T$

  = 20 for pulse signals for which peak levels apply and $B_R < B_T$ $\quad$ (8)

As the transmitter and receiver center frequencies are separated, the transmitter power can enter the receiver by either of two other possible means (see Fig. 1):

1. The transmitter emission modulation sidebands can enter the receiver at the main response frequency. For this case, the correction factor is

$$CF_R(\Delta f) = [K \log_{10}(B_R/B_T) + M(\Delta f)] \text{ dB} \tag{9}$$

where
  $M(\Delta f)$ = modulation sideband level in dB above transmitter power at frequency separation Δf
  K = definition from Eq. (8)

2. The power at the transmitter main output frequency can enter the receiver off-tune response. For this case, the correction factor is

$$CF_T(\Delta f) = -S(\Delta f) \text{ dB} \tag{10}$$

where $S(\Delta f)$ is the receiver selectivity in dB above receiver fundamental susceptibility at frequency separation Δf.

The final bandwidth correction factor that must be applied to the interference margin due to nonalignment of the transmitter output and receiver response is either $CF_R(\Delta f)$ or $CF_T(\Delta f)$, whichever is the larger of the two.

The prediction equations presented above are applicable to various types of interference problems. In most cases, the major difficulty is determining the parameters in the equations. Although this may appear to be a relatively simple undertaking when transmitting and receiving equipment are involved, it is not. This occurs because each transmitter produces a number of undesired spurious emissions and each receiver has a number of spurious responses and information is not usually available on spurious characteristics. Furthermore, it is necessary to consider radiation in unintended directions via unintended propagation paths. Interactions between transmitters and receivers having totally different operational functions, purposes, and technical characteristics also must be determined. Hence, for the simple case of an EMI prediction involving a single transmitter and receiver pair, information must be obtained for each transmitter output and receiver response, and the basic prediction equation must be applied for each output-response combination.

The following sections describe transmitter, receiver, antenna, and propagation characteristics that may be used for EMC analysis.

Transmitter Emission Characteristics

The primary function of a transmitter is to generate radio-frequency power containing direct or latent intelligence within a specified frequency band. In addition to the desired power, transmitters produce numerous unintentional emissions at spurious frequencies (see Fig. 3). A *spurious emission* is any radiated output that is not required for transmitting the desired information. The desired and/or undesired radio-frequency power generated by transmitters may produce EMI in receivers or other equipment. Therefore, in evaluating EMC, it

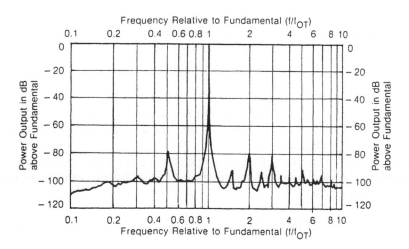

FIG. 3 Transmitter output spectrum resulting from composite of broadband noise and discrete emissions.

is necessary to consider all transmitter emissions as potential sources of interference.

Fundamental Emissions

In order to consider the transmitter fundamental output in EMC analysis, it is necessary to define the transmitter operating frequency, the fundamental power output, the bandwidth associated with the fundamental emission, and the modulation envelope in the vicinity of the fundamental emission.

The operating frequency is obtained from frequency assignment data or operational information or is defined as part of the statement of the problem. The transmitter fundamental power output and bandwidth are nominal data that should be available from the manufacturer's specifications on the transmitter. The modulation envelope describes the relative power in the sidebands around the carrier frequency and may be represented as described in the following paragraphs.

The transmitter fundamental output is not actually confined to a single frequency; it is distributed over a range of frequencies around the fundamental. The characteristics of the power distribution in the vicinity of the fundamental are determined primarily by the baseband modulation characteristics of the transmitter. The resulting spectral components are termed *modulation sidebands*. The power distribution in the modulation sidebands is represented by a modulation envelope function. In general, the modulation envelopes are described by specifying bandwidths or frequency ranges and functional relationships that describe the variation of power with frequency, $M(\Delta f)$. The modulation envelope model is

$$M(\Delta f) = M(\Delta f_i) + M_i \log_{10} \frac{\Delta f}{\Delta f_i} \qquad (11)$$

where

Δf = separation from reference frequency
Δf_i = initial frequency of applicable region
M_i = slope of modulation envelope for applicable region (dB/decade)

An example of the resulting functional relationship is shown in Fig. 4. The parameters required to specify the modulation envelope are the bandwidths of applicable regions of constant slope and the rate at which the envelope falls off over the frequency region of interest.

Table 1 summarizes modulation envelope parameter values for some of the more commonly used types of modulation. The off-tune transmitter emission level is given by

$$P_T(f_{OT} \pm \Delta f) \text{ dBm/channel} = P_T(f_{OT}) \text{ dBm} + M(\Delta f) \text{ dB} \qquad (12)$$

For adjacent signal frequencies that are sufficiently removed from the transmitter tuned frequency, the major source of interference may result from the broad-

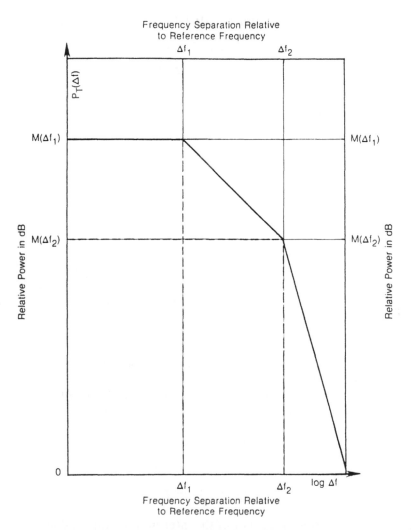

FIG. 4   Modulation envelope representation.

band noise generated by the transmitter. This transmitter noise may be considered to be included in the modulation envelope and may be represented as a noise floor that extends over a large portion of the frequency spectrum.

Transmitter Intermodulation

Intermodulation is the process by which two or more undesired signals mix in a nonlinearity to produce additional undesired signals at frequencies that are the sum or difference of the input frequencies or their harmonics. In general, intermodulation may occur in both transmitters and receivers. To determine which type of intermodulation predominates for a given EMI situation, it is necessary to calculate the equivalent interference level that results from both

TABLE 1   Summary of Transmitter Modulation Envelope Parameters

Modulation	i	Δf_i	Constants for Modulation Envelope Model	
			$M_T(\Delta f_i)$ (dB above fundamental)	M_i (dB/decade)*
AM communication	0	$0.1\ B_T$	0	0
	1	$0.5\ B_T$	0	133
	2	B_T	-40	60
	3	$10\ \ B_T$	-100	0
FM communication	0	$0.1\ B_T$	0	0
	1	$0.5\ B_T$	0	333
	2	B_T	-100	0
Pulse	0	$1/10\ \tau$	0	0
	1	$1/\pi(\tau + \Delta\tau)$	0	20
	2	$1/\pi\Delta\tau$	$-20\log_{10}\left(1 + \dfrac{\tau}{\Delta\tau}\right)$	40

*The slope M in dB/decade is negative on the upper side of the carrier and positive on the lower side of the carrier.

transmitter and receiver intermodulation and consider the case that results in the largest potential interference.

In general, the most serious problems result from third-order intermodulation and will result from mixing products that are given by

$$f_{IM} = 2f_1 - f_2 \tag{13}$$

or

$$f_{IM} = 2f_2 - f_1 \tag{14}$$

where f_{IM} is the resulting frequency of the intermodulation product.

The transmitter third order intermodulation problem is illustrated in Fig. 5. Referring to the figure, it is seen that intermodulation will occur in both of the two transmitters. The predominant transmitter intermodulation situation depends on the geometry and the power levels and frequencies of the two transmitters. In general, it will be necessary to consider both transmitter intermodulation situations to determine which one produces the largest signal at the receiver.

For cases for which the frequency separation f between the transmitters is less than or equal to 1% of the transmitter frequency, the equivalent transmitter intermodulation power P_E may be approximated by Eq. (15).

$$P_E(\text{dBm}) = P_1(\text{dBm}) - 10\ \text{dB} \tag{15}$$

where P_1 (dBm) is the interfering power available at the transmitter where the intermodulation occurs.

For cases for which the frequency separation is greater than 1%, P_E may be approximated by Eq. (16).

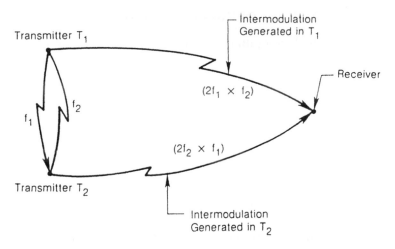

FIG. 5 Transmitter intermodulation.

$$P_E(\text{dBm}) = P_1(\text{dBm}) - 10\,\text{dB} - 30\log_{10}\Delta f(\%) \tag{16}$$

It should be noted that P_E is the intermodulation signal level at the transmitter where the intermodulation occurs. To determine the level at a receiver, it is necessary to include the effects of propagation loss.

Harmonic Emission Levels

Referring to Fig. 3, it is readily observed that transmitter emissions are present at frequencies harmonically related to the transmitter fundamental frequency. For the example illustrated in Fig. 3, there are other outputs (of lesser amplitude) present at frequencies that are harmonics of the master oscillator frequency. However, because of their reduced amplitude, these master oscillator harmonics do not usually create EMI problems. The frequencies of harmonics of the fundamental output are given by

$$f_{NT} = Nf_{OT} \tag{17}$$

where
f_{NT} = frequency of Nth harmonic of transmitter
N = integer associated with harmonic
f_{OT} = operating frequency of transmitter

The general math model used to describe transmitter harmonic emission levels is

$$P_T(f_{NT})\,\text{dBm} = P_T(f_{OT})\,\text{dBm} + [(A\log_{10}N) + B] \tag{18}$$

TABLE 2   Values for B Based on MIL-STD-461

Transmitter Power (dBm)	B (dB Above Transmitter Power)
20	−38
50	−80
70	−100
100	−118

where
  A = slope of harmonic levels in dB/decade
  B = intercept in dB relative to fundamental emission

If data on transmitter harmonic emission outputs are available from spectrum signature measurements or other information sources, they should be used to determine specific harmonic output models. Conversely, in many instances specific data are not available. Thus it is necessary to employ other techniques for determining specific models to be used in EMC analysis.

Harmonic Emission Levels Based on MIL-STD-461

One source of information regarding transmitter spurious output levels is the specification or standards associated with the particular communications electronics equipment. Transmitter specifications impose a limit on spurious outputs, and, for certain types of problems, it may be desirable to use these levels in performing an EMC analysis. If this approach is used, the resulting transmitter harmonic amplitude models would be obtained by setting A to zero and B to the specification limit. Thus, for example, if transmitter harmonic amplitude models were based on MIL-STD-461, the constants for the model would be $A = 0$ and B as indicated in Table 2.

Statistical Summary Harmonic Amplitude

In order to provide transmitter harmonic amplitude models that may be used in the absence of specific measured data, statistical summary models have been derived from available spectrum signature data. The results obtained by summarizing data for approximately 100 different transmitter nomenclatures are presented in Table 3. The specific values of A and B that correspond to the har-

TABLE 3   Harmonic Average Emission Levels

Harmonic	2	3	4	5	6	7	8	9	10
Average emission level (dB above fundamental)	−51	−64	−72	−79	−85	−90	−94	−97	−100

monic emission levels in Table 3 are -70 dB/decade and -30 dB, respectively. The resulting math models for the harmonic emission level is

$$P_T(f_{NT}) \text{ dBm} = P_T(f_{OT}) \text{ dBm} - 70 \log_{10} N - 30 \qquad (19)$$

Receiver Susceptibility Characteristics

Receivers are designed to respond to certain types of electromagnetic signals within a predetermined frequency band(s). However, receivers also respond to undesired signals having various modulation and frequency characteristics. Thus, it is necessary to treat a receiver as potentially susceptible to all transmitter emissions.

There are a number of interference effects that an undesired signal can produce in a receiver. In order to represent receiver composite susceptibility it is necessary to consider these effects and to determine which effect(s) dominate within a given range of frequencies. Figure 6 is a functional diagram useful in discussing various receiver EMI effects. A superheterodyne receiver generally employs RF stages that provide frequency selectivity and/or amplification and one or more mixers that translate the RF signal to intermediate frequencies (IFs). It also contains IF stages that provide further frequency selectivity and amplification, a detector that recovers the modulation, and postdetection stages that process the signal and drive one or more output displays. Since tuned-radio-frequency (TRF) and crystal-video receivers do not use the superheterodyne principle, they do not contain mixers and IF amplifiers. In specifying receiver susceptibility, it is necessary to consider the effects of an interfering signal on each of these stages. The resulting susceptibility function (Fig. 7) represents a composite of the most significant effects.

Co-Channel Interference

Co-channel interfering signals are amplified, processed, and detected in the same manner as the desired signal. Thus the receiver is particularly vulnerable

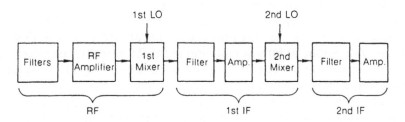

FIG. 6 Superheterodyne receiver (IF = intermediate frequency; LO = local oscillator; RF = radio frequency).

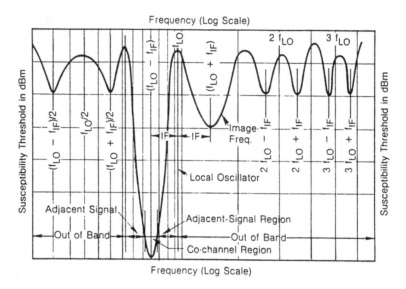

FIG. 7   Receiver susceptibility characteristics.

to these emissions. Co-channel EMI may either desensitize the receiver or override or mask the desired signal. It may also combine with the desired signal to cause serious distortion in the detected output or cause the automatic frequency control circuitry to retune to the frequency of the interference, if this is applicable.

For co-channel signals, the receiver susceptibility threshold may be represented by the receiver (or environment) noise (i.e., signals below the noise can be considered to be noninterfering). The receiver noise level is directly related to the receiver sensitivity, which may be obtained from nominal data on the receiver.

Receiver Adjacent Signal Interference

Adjacent signal interference can produce any one of several effects in a receiver. The interference may be translated through the receiver together with the desired signal; both appear at the input to an IF stage. In this case, the IF selectivity and the adjacent signal emission spectrum will both influence the relative level of the interfering signal appearing at the input to the detector. Alternately, one or more interfering emissions may produce nonlinear effects such as desensitization, cross-modulation, or intermodulation in the RF amplifier or mixer. Desensitization is a reduction in the receiver gain to the desired signal as a result of an interfering emission-producing automatic gain control (AGC) action or causing one or more stages of the receiver to operate nonlinearly due to saturation. Cross-modulation is the transfer of the modulation from an undesired emission to the desired signal as a result of the former causing one or more stages of the receiver to operate nonlinearly. Intermodulation is the generation of undesired signals from the nonlinear combination of two or more input signals that produce frequencies existing at the sum or difference of the input frequencies or their harmonics.

Although desensitization and cross-modulation effects can occur in receivers, recent improvements in receiver design have significantly reduced EMI problems due to these effects. In many cases, transmitter noise and/or transmitter or receiver intermodulation are the limiting factors in adjacent signal operation. Because intermodulation is often the most serious receiver nonlinear adjacent signal effect, the models for this effect only are discussed in this article.

Receiver Selectivity

The receiver selectivity determines the amount of attenuation or rejection provided to off-tuned signals by the receiver. In general, the receiver susceptibility threshold for off-tuned signals is increased by the receiver selectivity for the frequency separation in question. The mathematical model used to represent receiver IF selectivity $S(\Delta f)$ is expressed as a piecewise linear function of the logarithm of frequency separation Δf.

$$S(\Delta f) = S(\Delta f_i) + S_i \log_{10}(\Delta f/\Delta f_i) \qquad \text{for } \Delta f_i < \Delta f < \Delta f_{i+1} \qquad (20)$$

where

S_i = slope of selectivity curve for applicable region
Δf_i = initial frequency of applicable region
$\Delta f = |f - f_{OR}|$

The above model can be used by specifying the frequency deviations associated with the 3-dB and 60-dB selectivity levels. The resulting selectivity model is shown in Fig. 8. Notice that a maximum value of 100 dB is assumed for receiver selectivity. This implies that any emission source greater than 100 dB above receiver sensitivity may penetrate and become a source of EMI.

One good indicator for the selectivity characteristics is given by the shape factor: the ratio of the 60-dB bandwidth to the 3-dB bandwidth. Applying the shape factor (SF) concept to Eq. (20), the IF selectivity relation yields

$$S(\Delta f) \, \text{dB} = 60 \, \frac{\log_{10}(\Delta f/\Delta f_1)}{\log_{10} SF} \, \text{dB} \qquad (21)$$

When $\Delta f/\Delta f_1$ is chosen to equal the shape factor, Eq. (21) yields 60 dB.

A typical value for the receiver shape factor is 4. When this value is substituted into Eq. (21), the selectivity model parameters can be determined. The resulting values are summarized in Table 4. The receiver susceptibility to narrowband off-tune signals is given by

$$P_R(f_{OR} \pm \Delta f) \, \text{dBm} = P_R(f_{OR}) \, \text{dBm} + S(\Delta f) \, \text{dB} \qquad (22)$$

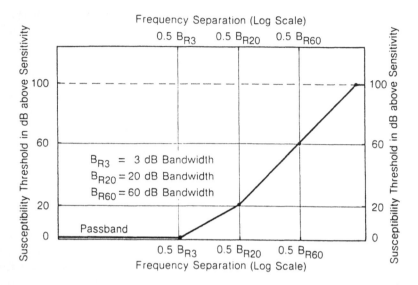

FIG. 8 Receiver selectivity model.

Receiver Intermodulation

In order for an intermodulation product to cause interference, it must be transformed to a frequency within or near the IF passband for detection to occur. The method considered here is intermodulation in the RF amplifiers and first mixer, which results in an intermodulation frequency at or near the receiver tuned frequency f_{OR}.

Signals that are capable of producing intermodulation interference in a receiver must statisfy the following relationship:

$$mf_1 \pm nf_2 - f_{OR} \leq B_R/2 \tag{23}$$

TABLE 4 Summary of Receiver Selectivity Parameters

	Constants for IF Selectivity Model		
i	Δf_i	$S(\Delta f_i)$ (dB)	S_i^* (dB/decade)
0	$0.1\ B_R$	0	0
1	$0.5\ B_R$	0	100
2	$5\ B_R$	100	0

*The slope S_i in dB/decade is positive on the upper side and negative on the lower side of the receiver tuned frequency.

where

f_1, f_2 = frequencies of two interfering emissions
f_{OR}　= receiver tuned frequency
B_R　= IF bandwidth in which intermodulation products are significant
m, n = integers

The only signals that are potentially serious sources of intermodulation are those that are in the vicinity of the receiver frequency and produce intermodulation products that fall within the receiver operating or immediately adjacent channels.

The following equations present the frequency criteria that two interfering signals must meet to satisfy these constraints.

$$f_N \pm f_F - f_{OR} \leq B_R/2 \text{ (second order)}$$

$$2f_N - f_F - f_{OR} \leq B_R/2 \text{ (third order)}$$

$$3f_N - 2f_F - f_{OR} \leq B_R/2 \text{ (fifth order)}$$

$$4f_N - 3f_F - f_{OR} \leq B_R/2 \text{ (seventh order)}$$

where

f_{OR} = receiver RF tuned frequency
f_N = frequency of interfering emission nearest to f_{OR}
f_F = frequency of interfering emission farthest from f_{OR}

Equation (23) may be normalized to the receiver fundamental frequency and solved to show the relationship between two culprit signals that will produce an intermodulation product at the receiver fundamental frequency.

$$m \frac{f_1}{f_{OR}} \pm n \frac{f_2}{f_{OR}} = 1 \tag{24}$$

Figure 9, obtained from Eq. (24), shows the resulting chart for second- and third-order intermodulation products. Intermodulation signal combinations falling on or near one of the lines are capable of generating an intermodulation product in the vicinity of the receiver tuned frequency.

The area on the chart marked "Region of Major Significance" is particularly important because of the proximity of the signals to the receiver tuned frequency. Signals within this region will in general experience less RF selectivity (rejection) than will signals outside the region. Thus they are more likely to produce significant intermodulation products. The extent of this region is in general a function of RF selectivity, but the area indicated is representative of typical receivers.

In order to evaluate the impact of receiver intermodulation, it is convenient to express the effect in terms of an equivalent interference margin corresponding to the margin resulting from two interfering signals that produce an intermodulation product falling within the receiver overall 3-dB passband. If the intermodulation product is off tuned from the receiver tuned frequency, the IF selectivity

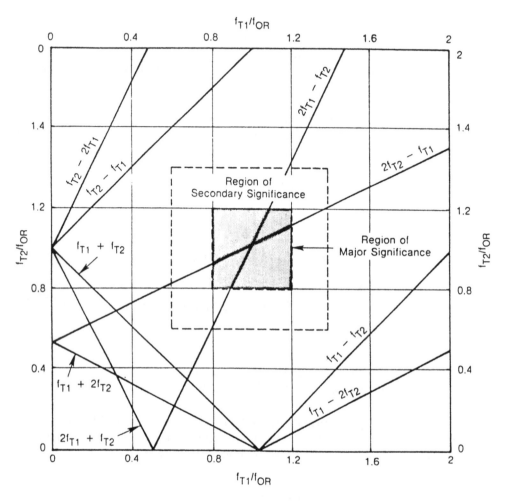

FIG. 9 Second- and third-order intermodulation chart.

can be applied to determine the resulting off-tune interference margin. If the input signals producing the intermodulation do not produce hard saturation in the receiver front end and the desired signal and the resulting intermodulation do not exceed the receiver automatic gain control (AGC) threshold, the equivalent interference margin (IM) resulting from intermodulation is

$$IM \text{ (dB)} = mP_N + nP_F + IMF - P_R(f_{OR}) \qquad (25)$$

where

P_N, P_F = power in dBm at receiver input resulting from interfering signals at frequencies f_N and f_F, respectively

m, n = constants associated with intermodulation order (m corresponds to the harmonic of the near signal and n corresponds to the harmonic of the far signal that are mixing to produce the intermodulation product)

IMF = intermodulation factor that depends on receiver non-linearity and RF selectivity

From an EMI standpoint, third-order intermodulation is usually the most serious offender. For this case, the equivalent interference margin is

$$IM\ (\text{dB}) = 2P_N + P_F + IMF - P_R(f_{OR}) \tag{26}$$

In order to use Eq. (26) in an EMC analysis, it is necessary to determine the value of the intermodulation factor (IMF). If measured data on receiver intermodulation characteristics are available, these data may be used to evaluate IMF. If measured data are not available, IMF may be evaluated from MIL-STD-461 limits to provide a default model, given below.

Intermodulation Models from MIL-STD-461

Intermodulation measurements made in accordance with MIL-STD-461B are performed in a manner such that the two interfering signals are equal in amplitude and the resulting intermodulation product produces a standard response in the receiver. The MIL-STD-461 limits (CS03) for conducted susceptibility to intermodulation interference specifies that no intermodulation responses shall be observed when the interfering signals are 66 dB above the on-tune level required to produce a response. The resulting MIL-STD-461 default model for third-order intermodulation interference is

$$IM\ (\text{dB}) = 2P_N + P_F - 3P_R(f_{OR}) - 198 \tag{27}$$

Receiver Spurious Responses

Strong out-of-band interference may produce spurious responses in a receiver. The superheterodyne receiver is most susceptible to those out-of-band signals that mix with local oscillator harmonics to produce a signal at the IF. Spurious responses in such a receiver usually occur at specific frequencies, and other out-of-band frequencies are attenuated by the receiver IF selectivity.

For a tuned radio-frequency or crystal-video receiver, the receiver will be susceptible to those out-of-band interfering signals that are not adequately rejected by the RF selectivity.

There are several means by which an out-of-band emission can be translated to one of the pass band frequencies of a superheterodyne receiver. The most significant of these occurs in the first mixer stage. Here the desired signal is heterodyned with the local oscillator (LO) to translate the incoming signal to the intermediate frequency. In addition to desired signals, interfering emissions at many different frequencies are capable of being heterodyned with the LO or other signals and translated to the receiver IF. The amplitude of responses produced in this manner is directly proportional to the strength of the original signals. Because the level of the LO is typically on the order of 120 dB greater than desired and interfering signals present at the input to the first mixer stage, heterodyne products that involve the LO are much larger in amplitude than those heterodyne products that do not involve the LO. Thus, superheterodyne receivers are most susceptible to out-of-band signals that heterodyne with the LO to produce a product in or near the IF passband.

In this article, the term *spurious response* when applied to superheterodyne receivers refers specifically to those undesired responses that result from the mixing of an LO and an undesired emission. Those input interfering frequencies capable of appearing at the IF as a result of mixing with the LO are known as *spurious response frequencies*. The amount of power necessary to cause interference at any particular spurious response frequency is a function of receiver susceptibility to the response.

The frequencies for which spurious responses will occur are given by the following expression:

$$f_{SR} = \frac{pf_{LO} \pm f_{IF}}{q} \tag{28}$$

where

f_{SR} = spurious response frequencies
p, q = integers associated with the local oscillator and interference signal
f_{LO} = local oscillator frequency
f_{IF} = intermediate frequency

Figure 7 illustrates receiver spurious response susceptibility. In general, the most significant responses are those for which q is equal to 1, and higher values of q do not need to be considered for most EMI situations.

The general math model that is used to describe receiver spurious response susceptibility for a given value of q is

$$P_R(f_{SR}) = P_R(f_{OR}) + I\log_{10} p + J \tag{29}$$

where

I = slope of spurious response susceptibility in dB/decade
J = intercept in dB relative to fundamental susceptibility

Receiver Response Models Based on MIL-STD-461

If specific information on the spurious response characteristics of a receiver are not available, it may be desirable to use default models that are based on the spurious response limits specified in CS04 of MIL-STD-461. These limits may be used to solve for I and J of Eq. (29) for the various regions of interest; the resulting default models are provided in Table 5.

Statistical Summary of Receiver Spurious Response Levels

When specific receiver measured data are not available, one alternative for obtaining an out-of-band susceptibility model is to derive statistical summaries

TABLE 5 MIL-STD-461 Models for Spurious Response

For interfering signals within the receiver 80 = dB bandwidth
(i.e., $f_o - BW/2 < f_{SR} < f_o + BW/2$)

$$P(f_{SR}) \text{ dBm} = P_R(f_{OR}) \text{ dBm} + \left(\frac{160}{BW} \right)[f_{SR} - f_o]$$

For interfering signals outside the receiver 80-dB bandwidth but within the overall tuning range of the receiver
(i.e., $f_L \le f_{SR} \le f_o - BW/2$ or $f_o + BW/2 \le f_{SR} \le f_H$)

$$P_R(f_{SR}) \text{ dBm} = P_R(f_{OR}) \text{ dBm} + 80 \text{ dB}$$

For interfering signals outside the tuning range of the receiver
(for $f_{SR} < f_L$ or $f_{SR} > f_H$)

$$P_R(f_{SR}) = 0 \text{ dBm}$$

where
$\quad f_o \quad$ = receiver tuned frequency
$\quad BW$ = receiver 80-dB bandwidth
$\quad P_R \quad$ = receiver sensitivity
$\quad f_L \quad$ = lowest tuned frequency of receiver
$\quad f_H \quad$ = highest tuned frequency of receiver

from data on receivers. Statistical summary models have been evaluated from available spectrum signature data, and the specific values for I and J are 35 dB/decade and 75 dB, respectively. The corresponding spurious response model is

$$PR(f_{SR}) = P_R(f_{OR}) + 35 \log_{10} p + 75 \tag{30}$$

Table 6 presents the average spurious response susceptibility levels obtained from measured data.

Antenna Radiation Characteristics

Antennas are designed to radiate and/or receive signals over a specific solid angle and within a specified frequency range. For land mobile or broadcast applications, the antenna is usually designed to radiate or receive uniformly over all sectors surrounding the antenna. Other systems such as fixed point-to-point communications, radar, and certain telemetry systems are designed to confine the functional radiated or received signals to certain limited sectors. In practice, however, it is not possible to accomplish perfect discrimination with antennas in either the spatial or frequency domains. Thus, antennas that are intended to restrict radiation to specific regions also radiate into or receive signals from

TABLE 6   Summary of Spurious Response Average Susceptibility ($q = 1$)

Local Oscillator Harmonic (p)	1 (Image)	2	3	4	5	6	7	8	9	10
Average susceptibility level (dB above fundamental sensitivity)	75	82	92	96	99	102	105	107	108	110

other unintentional regions. In addition, undesired signals at nondesign frequencies are inadvertently radiated or received by antennas, and the spatial characteristics of an antenna for spurious frequencies are significantly different from characteristics at the design frequency.

Design Frequency and Polarization

Figure 10 illustrates the design frequency and polarization radiation characteristics for a number of typical antenna types. Referring to Fig. 10, the radiation characteristics for each antenna may be considered to consist of intentional and unintentional radiation regions.

Typical gains in the intentional radiation region (for the design frequency and polarization) are given in the column on the right in Fig. 10. Antennas are often categorized according to these gains G as follows:

- Low gain: $G < 10$ dB
- Medium gain: 10 dB $\leq G \leq 25$ dB
- High gain: $G > 25$ dB

For the unintentional radiation region, typical mean gain levels relative to an isotrope would typically be

- -3 dB for low-gain antennas
- -10 for medium- and high-gain antennas

Gain levels at a specific orientation may exhibit large variations from the levels given above.

Polarization Dependence

If an antenna is linearly polarized, there will be a significant difference between antenna gain, in the intentional radiation region, for vertical and horizontal polarizations. This effect will be most pronounced at the design frequency, and the gain will be maximum for the predominant mode of polarization. In general,

Type Antennas	Pattern		Gain in dB/Isotrope
	Horizontal	Vertical	
Quarter-Wave Vertical Monopole			3
Half-Wave Horizontal Dipole			3
Vertical Loop			3
Long Wire			6-10
Colinear Array			6-10
Broadside Array			6-10
End-Fire Array			6-10
Discone			6-10
Slot		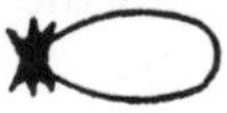	6-10
Helix (Omnidirectional Mode)			6-10
Yagi			10-15
Broadside Curtain			10-15
End-Fire Curtain			10-15
Rhombic			15-25
Horn			15-25
Corner Reflector			15-20
Log Periodic			10-15
Helix Axial Mode			10-15
Aperture or Array			25-60

FIG. 10   Radiation characteristics for typical antennas.

the discrimination afforded by using antennas that are orthogonally polarized will be on the order of 16 dB to 20 dB; this provides one means of reducing the probability of interference between different users (e.g., land mobile applications typically use vertical polarization, whereas television broadcast utilizes horizontal polarization).

Nondesign Frequencies

For nondesign frequencies, the antenna gain in the intentional radiation region would typically be reduced by the following:

- 13 dB for high-gain antennas
- 10 dB for medium-gain antennas
- 0 dB for low-gain antennas

The antenna gain at specific nondesign frequencies may exhibit variations from the values specified above.

The overall characteristics of the unintentional radiation region are not significantly affected by frequency.

Propagation Effects

In discussing concepts regarding propagation, it is helpful to begin with a discussion of free-space propagation between lossless isotropic antennas. Once the principles governing propagation under these conditions are understood, it is easier to follow the concept of propagation between either omnidirectional or directional antennas in the presence of earth and reflecting and scattering objects such as buildings, trees, and so on.

Because many EMI situations involve transmitters and receivers that are co-located or located in close proximity, free-space propagation conditions are often assumed for the purpose of performing an EMC analysis. If a transmitted signal is radiated from an isotropic antenna in free space, the signal spreads uniformly in all directions. Thus, at a distance d from the source, the power density is

$$P_D = \frac{P_T}{4\pi d^2} \tag{31}$$

where
$\quad P_D$ = power density (i.e., power per unit area)
$\quad P_T$ = transmitted power
$\quad d\;$ = distance from antenna to observation point

The power available at the terminals of a lossless receiving antenna having an area A_R and a gain G is

$$P_R = P_D A_R = P_D \frac{G\lambda^2}{4\pi}$$

$$= \frac{P_T\lambda^2}{(4\pi d)^2} \qquad \text{for } G = 1 \text{ (isotropic)} \qquad (32)$$

where λ is the wavelength in the same units as d.

The above relation can be expressed in terms of frequency in MHz (f_{MHz}) and distance in statute miles (d_{mi}) by substituting for λ:

$$\lambda \text{ (feet)} = \frac{984}{f_{MHz}} \qquad (33)$$

or

$$\lambda \text{ (miles)} = \frac{984}{5280 f_{MHz}} = 0.186/f_{MHz}$$

so that

$$P_T/P_R = \frac{(4\pi)^2 (5280)^2}{(984)^2} f^2_{MHz}\, d^2_{mi}$$

$$= 4560\, f^2_{MHz}\, d^2_{mi} \qquad (34)$$

Therefore, free-space attenuation in dB between lossless isotropic antennas for far-field conditions is

$$L(f,d) = 10 \log_{10} P_T/P_R$$

$$= 37 + 20 \log_{10} f_{MHz} + 20 \log_{10} d_{mi}$$

$$L(f,d) = 32 + 20 \log_{10} f_{MHz} + 20 \log_{10} d_{km} \qquad (35)$$

Sample Electromagnetic Compatibility Analysis

There are many EMC analysis problems for which only a few transmitter-receiver pairs need to be considered, and the prediction is either performed manually or with the aid of a small computer program that may be run on a personal computer or a time-share terminal. This section presents a step-by-step process for performing a manual EMC analysis by use of a special form. Although the particular form presented in this section was designed for analyzing AM and FM analog voice communication systems such as those used for land mobile applications, similar forms may be used for other types of communication systems.

Transmitter Noise

Consider the case (Example 1) of a land mobile receiver operating at 150 MHz. Determine whether EMI will result if a land mobile transmitter, operating at 150.1 MHz, is installed 400 ft (120 m) from the receiver. The pertinent transmitter and receiver characteristics are

Adjacent Signal Interference*	
Transmitter Noise	
1. Transmitter Power, P_T (dBm/Channel)	50
2. Noise Constant	56
3. $20 \log \Delta f_{TR}$ (kHz)	40
4. Noise per Channel dBm/Channel) (1) − (2) − (3)	−46
5. Transmitter Antenna Gain, G_{TR} (dB)	3
6. Effective Radiated Noise Power (dBm/Channel); (4) + (5)	−43
7. Propagation Constant	32
8. $20 \log d_{TR}$ (km)	−18
9. $20 \log f_R$ (MHz)	44
10. Propagation Loss, L (dB): (7) + (8) + (9)	58
11. Receiver Antenna Gain, G_{RT} (dB)	3
12. Noise Power Available, P_A (dBm); (6) − (10) + (11)	−98
13. Receiver Sensitivity Level (dBm)	−107
14. Allowable Degradation of Receiver Sensitivity (dB)	0
15. Receiver Susceptibility Level, P_R (dBm); (13) + (14)	−107
16. Interference Margin (dB); (12) − (15)	9
Third Order Intermodulation Frequency Check	
• Select Receiver fo Analysis	
17. Receiver/Frequency, f_R (MHz)	
• Select Cosite Transmitter, T_1, with Frequency Nearest to f_R	
18. Transmitter Frequency, f_{T_1} (MHz)	
19. Frequency Separation ΔF_{TR} (MHz); (18) − (17)	
20. Frequency, f_{T_2}, for Intermodulation; (18) + (19)	
21. Channel Width, (MHz)	
22. Band for Intermodulation; (20) ± (21)	(−) | (+)

• Check Other Cosite Transmitters for Frequency within Band Specified by (22). If one is found, continue with analysis. If none, eliminate selected transmitter from consideration and repeat process with another transmitter.

Interference Margin < 0.10 dB, EMI Highly Improbable
10 dB < Interference Margin < 10 dB, EMI Marginal
Interference Margin > 10 dB, EMI Probable.

*Applies to co-site transmitters and receivers with frequency separations (Δf)

EXAMPLE 1   Transmitter noise.

Transmitter power P_T = 50 dBm
Transmitter antenna gain G_T = 3 dB
Receiver antenna gain G_R = 3 dB
Receiver sensitivity = −107 dBm
Allowable degradation = 0 dB

This is clearly a co-site adjacent signal situation, and the primary cause of potential interference would be transmitter noise. The completed short form is provided for this example. The results indicate that a +8-dB interference margin will be obtained, and a marginal interference situation exists.

Intermodulation

Consider the case (Example 2) of a land mobile receiver operating at 450 MHz in the vicinity (40 ft or 12 m) of a land mobile transmitter at 451 MHz. Determine whether an intermodulation problem will result if a second transmitter operating at 452 MHz is located 100 ft or 30.5 m from the receiver on a site that is 80 ft or 24.5 m from the first transmitter. The pertinent transmitter and receiver characteristics are

Transmitter power (T_1 and T_2) = 50 dBm
Transmitter antenna gain (G_{T1} and G_{T1}) = 3 dB
Receiver antenna gain G_R = 3 dB
Receiver sensitivity = −107 dBm
Allowable degradation = 0 dB
Channel width = 50 kHz

This situation could potentially result in either transmitter or receiver third-order intermodulation. In order to determine whether third-order intermodulation is possible, it is necessary to first perform the frequency check indicated on the short form. This has been performed on the accompanying form, and the results indicate that an intermodulation problem may occur.

Next, it is necessary to calculate the interference margin resulting from both receiver and transmitter intermodulation situations to determine the corresponding interference potential. These calculations, which are straightforward, have been performed on the accompanying form. The calculations indicate that receiver intermodulation results in a +30-dB interference margin, and transmitter intermodulation results in a +57-dB interference margin. For this situation, transmitter intermodulation will predominate, and EMI is probable.

Out-of-Band Electromagnetic Interference

Consider that an industrial user desires to operate a land mobile base receiver at 158.1 MHz. The receiving antenna will be located on top of a building, and a

Adjacent Signal Interference*	
Transmitter Noise	
1. Transmitter Power, P_T (dBm/Channel)	
2. Noise Constant	56
3. 20 log Δf_{TR} (kHz)	
4. Noise per Channel dBm/Channel (1) − (2) − (3)	
5. Transmitter Antenna Gain, G_{TR} (dB)	
6. Effective Radiated Noise Power (dBm/Channel); (4) + (5)	
7. Propagation Constant	32
8. 20 log d_{TR} (km)	
9. 20 log f_R (MHz)	
10. Propagation Loss, L (dB): (7) + (8) + (9)	
11. Receiver Antenna Gain, G_{RT} (dB)	
12. Noise Power Available, P_A (dBm); (6) − (10) + (11)	
13. Receiver Sensitivity Level (dBm)	
14. Allowable Degradation of Receiver Sensitivity (dB)	
15. Receiver Susceptibility Level, P_R (dBm); (13) + (14)	
16. Interference Margin (dB); (12) − (15)	

Third Order Intermodulation

Frequency Check

• Select Receiver fo Analysis	
17. Receiver/Frequency, f_R (MHz)	450
• Select Cosite Transmitter, T_1, with Frequency Nearest to f_R	
18. Transmitter Frequency, f_{T_1} (MHz)	451
19. Frequency Separation ΔF_{TR} (MHz); (18) − (17)	+ 1
20. Frequency, f_{T_2}, for Intermodulation; (18) + (19)	452
21. Channel Width, (MHz)	.050
22. Band for Intermodulation; (20) ± (21) (−) 451.95 (+) 452.05	

• Check Other Cosite Transmitters for Frequency within Band Specified by (22). If one is found, continue with analysis. If none, eliminate selected transmitter from consideration and repeat process with another transmitter.

Interference Margin < 0.10 dB, EMI Highly Improbable
10 dB < Interference Margin < 10 dB, EMI Marginal
Interference Margin > 10 dB, EMI Probable.

*Applies to cosite transmitters and receivers with frequency separations (Δf)

EXAMPLE 2   Intermodulation.

Adjacent Signal Interference*	T_1	T_2
Receiver Intermodulation		
23. Transmitter Power, P_T (dBm)	50	50
24. Transmitter Antenna Gain, G_{TR} (dB)	3	3
25. Effective Radiated Power (dBm) (23) + (24)	53	53
26. Propagation Constant	32	32
27. 20 log d_{TR} (km)	−38	−30
28. 20 log f_T (MHz)	53	53
29. Propagation Loss (dB); (26) + (27) + (28)	47	55
30. Receiver Antenna Gain, (dB)	3	3
31. Power Available at Receiver, (dBm); (25) − (29) + (30)	9	1
32. Multiply T_1 Power Available, Line (31), by Two	18	
33. T_2 Power Available, Line (31)		1
34. Intermodulation Constant		−93
35. Frequency Separation, Δ f% [(19) − (17)] × 100		0.22
36. 60 log Δ f% or 0		0
37. Equivalent Intermodulation Power (dBm); (32) + (33) + (34) − (36)		−74
38. Receiver Susceptibility Level, P_R (dBm)		−107
39. Interference Margin, (dB); (37) − (38)		+33
Transmitter Intermodulation		
40. Power of T_2 (dBm)		50
41. T_2 Antenna Gain (dB)		3
42. T_2 Effective Radiated Power (dBm), (40) + (41)		53
43. Propagation Constant		32
44. 20 log $d_{T_1T_2}$ (km)		−32
45. 20 log f_{T_2} (MHz)		53
46. Propagation Loss L (dB), (43) + (44) + (45)		53
47. T_1 Antenna Gain (dB)		3
48. T_2 Signal at T_1 (dBm); (42) − (46) + (47)		3
49. Intermodulation Constant		10
50. 30 log (Δ f% , (line 35), or 0; Whichever Is Larger		0
51. Intermodulation Power at T_1 (dBm); (48) − (49) + (50)		−7
52. T_1 Antenna Gain (dB)		3
53. Intermodulation ERP (dBm); (51) + (52)		−4
54. Propagation Constant (dB)		32
55. 20 log d_{T_1R} (km)		−38
56. 20 log f_R (MHz)		53
57. Intermodulation Propagation Loss (dB); (54) + (55) + (56)		47
58. Receiver Antenna Gain (dB)		3
59. Intermodulation Power at Receiver (dBm); (53) − (57) + (58)		−48
60. Receiver Susceptibility Level (dBm)		−107
61. Interference Margin (dB)		59
Interference Margin < .10 dB, EMI Highly Improbable − 10 dB < Interference Margin < 10 dB, EMI Marginal Interference Margin > 10 dB, EMI Probable.		

EXAMPLE 2 Continued.

survey of the immediate vicinity reveals that there is a public safety transmitter operating at 39.525 MHz and a land transportation transmitter operating at 452.9 MHz. The separations between the industrial receiver and the public safety and land transportation transmitters are 100 ft or 30.5 m and 20 ft or 6 m, respectively. Determine whether an EMI problem exists if the system characteristics are as follows:

Industrial receiver
    Frequency = 158.1 MHz
    Intermediate frequency = 10.7 MHz
    Local oscillator = 147.4 MHz
    Fundamental sensitivity = −107 dBm
    Antenna gain = 3 dB
Public safety transmitter [Example 3(a)]
    Frequency = 39.525 MHz
    Power output = 100 watts
    Antenna gain = 0 dB
Land transportation transmitter [Example 3(b)]
    Frequency = 452.9 MHz
    Power output = 50 watts
    Antenna gain = 6 dB

These two potential interference situations are clearly examples of out-of-band EMI. The most probable causes of interference for these situations would be a harmonic of the public safety transmitter interfering with the industrial receiver fundamental and a spurious response of the industrial receiver being interfered with by the fundamental of the land transportation transmitter. The calculations have been performed on the accompanying forms. The results indicate that both of these transmitters pose a potential EMI problem to the receiver.

Computer Electromagnetic Compatibility Analysis

The previous section presented forms that may be used to perform a manual EMC analysis. All of the operations indicated on the forms may be easily programmed on a computer or calculator to assist the system designer in performing an EMC analysis. If one is to be involved in the planning and design of a large system (e.g., a statewide public safety system), it is recommended that a computer or calculator be used to assist in the many calculations that will be required to perform an EMI analysis. Also, it is suggested that a computer database be established and maintained concerning all other users in the area.

Out of Band Interference*	
Transmitter Harmonic to Receiver Fundamental; $f_R > f_{Tq}$	
1. Receiver Frequency, f_R (MHz)	158.1†
2. Transmitter Frequency, f_T (MHz)	39.525†
3. (1) ÷ (2) and Round Off to Nearest Integer, N	4
4. Transmitter Harmonic Frequency, Nf_T (MHz); (3) × (2)	158.1
5. Frequency Separation, $\mid$ (4) − (1) $\mid$, (MHz)	0
6. Receiver Bandwidth	0.015†
• If (5) > (6) No Harmonic Interference	
If (5) < (6) Continue	
7. Transmitter Power, P_T (dBm)	50
8. Harmonic Correction, (dB); from Table 3	−72
9. Harmonic Power (dBm); (7) + (8)	−22
10. Propagation Constant	32
11. 20 log d_{TR} (km)	−30
12. 20 log f_R (MHz)	44
13. Propagation Loss, L, (dB) (10) + (11) + (12)	46
14. Receiver Antenna Gain, G_R (dB)	3
15. Power Available at Receiver (dBm); (9) − (13) + (14)	−65
16. Receiver Susceptibility Level, P_R (dBm)	−107
17. Interference Margin, (dB); (15) − (16)	+42
Transmitter Fundamental to Receiver Spurious: $f_T > f_R$	
18. (2) ÷ (1) and Round Off to Nearest Integer, P	
19. Local Oscillator Frequency, f_{LO} (MHz)	
20. Intermediate Frequency, f_{IF} (MHz)	
21. $\mid Pf_{LO} \pm f_{IF} - f_T\mid$;(18) × (19) ± (20) − (2)$\mid$	
If (21+) or (21−) > (6) No Spurious Interface	
If (21+) or (21−) < (6) Continue	
22. Transmitter Power, P_T(dBm)	
23. Transmitter Antenna Gain, G_T (dB)	
24. Propagation Constant	32
25. 20 log d_{TR} (km)	
26. 20 log f_T (MHz)	
27. Propagation Loss, L (dB); (24) + (25) + (26)	
28. Power Available at Receiver, (dBm); (22) + (23) − (27)	
29. Receiver Fundamental Susceptibility, P_R (dBm)	
30. Spurious Correction, from Table 6	
31. Spurious Susceptibility, (dBm); (29) + (30)	
32. Interference Margin, (dB); (28) − (31)	

Interference Margin < −10 dB, EMI Highly Improbable
−10 dB < Interference Margin < 10 dB, EMI Marginal
Interference Margin > 10 dB, EMI Probable.

*Applies to cosite transmitters and receivers with frequency
separations (Δf) greater than 10% of operating frequency.
†These entries are also required for transmitter fundamental to receiver
spurious.

(a)

EXAMPLE 3   Electromagnetic interference: *a*, from a public safety transmitter; *b*, from a land transportation transmitter.

Out of Band Interference*	
Transmitter Harmonic to Receiver Fundamental; $f_R > f_{Tq}$	
1. Receiver Frequency, f_R (MHz)	158.1†
2. Transmitter Frequency, f_T (MHz)	452.9†
3. (1) ÷ (2) and Round Off to Nearest Integer, N	
4. Transmitter Harmonic Frequency, Nf_T (MHz); (3) × (2)	
5. Frequency Separation, | (4) − (1) |, (MHz)	
6. Receiver Bandwidth	0.015†
• If (5) > (6) No Harmonic Interference If (5) < (6) Continue	
7. Transmitter Power, P_T (dBm)	
8. Harmonic Correction, (dB); from Table 3	
9. Harmonic Power (dBm); (7) + (8)	
10. Propagation Constant	32
11. 20 log d_{TR} (km)	
12. 20 log f_R (MHz)	
13. Propagation Loss, L, (dB) (10) + (11) + (12)	
14. Receiver Antenna Gain, G_R (dB)	
15. Power Available at Receiver (dBm); (9) − (13) + (14)	
16. Receiver Susceptibility Level, P_R (dBm)	
17. Interference Margin, (dB); (15) − (16)	
Transmitter Fundamental to Receiver Spurious: $f_T > f_R$	
18. (2) ÷ (1) and Round Off to Nearest Integer, P	3
19. Local Oscillator Frequency, f_{LO} (MHz)	147.4
20. Intermediate Frequency, f_{IF} (MHz)	10.7
21. $\|Pf_{LO} \pm f_{IF} - f_T\|$; (18) × (19) ± (20) − (2) (+) 0	(−) 21.4
If (21 +) or (21 −) > (6) No Spurious Interface If (21 +) or (21 −) < (6) Continue	
22. Transmitter Power, P_T (dBm)	47
23. Transmitter Antenna Gain, G_T (dB)	6
24. Propagation Constant	32
25. 20 log d_{TR} (km)	− 44
26. 20 log f_T (MHz)	53
27. Propagation Loss, L (dB); (24) + (25) + (26)	41
28. Power Available at Receiver, (dBm); (22) + (23) − (27)	− 12
29. Receiver Fundamental Susceptibility, P_R (dBm)	− 107
30. Spurious Correction, from Table 6	92
31. Spurious Susceptibility, (dBm); (29) + (30)	− 15
32. Interference Margin, (dB); (28) − (31)	+ 3

Interference Margin < − 10 dB, EMI Highly Improbable
− 10 dB < Interference Margin < 10 dB, EMI Marginal
Interference Margin > 10 dB, EMI Probable.

*Applies to cosite transmitters and receivers with frequency separations (Δf) greater than 10% of operating frequency.
†These entries are also required for transmitter fundamental to receiver spurious.

(b)

EXAMPLE 3 Continued.

Acknowledgment: The author would like to express his appreciation to Janet Agee for her assistance in the preparation of this article.
The author assumes sole responsibility for the material presented in this article.

Glossary

ELECTROMAGNETIC COMPATIBILITY (EMC). The ability of equipment or a system to operate in its normal electromagnetic environment without degradation and without adversely affecting the operation of other equipment or systems.

ELECTROMAGNETIC INTERFERENCE (EMI). The inability of equipment or a system to operate in its normal electromagnetic environment without degradation or without adversely affecting the operation of other equipment or systems.

HARMONICS. Spurious signals that occur at integer multiples of a fundamental frequency.

INTERMEDIATE FREQUENCY (IF). A frequency that results from the mixing of the intended received signal and the local oscillator.

INTERMODULATION. The process involving the mixing of two or more interfering signals in a nonlinear device to produce signals at new frequencies.

SPURIOUS RESPONSE. The response of a receiver to a signal other than the intended operational signal.

Bibliography

Duff, W. G., *Electromagnetic Compatibility in Telecommunications*, Vol. 7, A Handbook Series on EMI and EMC, EMF-EMI Control, Gainesville, VA, 1988.

Duff, W. G., *Fundamentals of Electromagnetic Compatibility*, Vol. 1, A Handbook Series on EMI and EMC, EMF-EMI Control, Gainesville, VA, 1988.

Duff, W. G., *A Handbook on Mobile Communications*, Don White Consultants, Gainesville, VA, 1976.

Duff, W. G., and White, D. R. J., *Electromagnetic Interference and Compatibility*, Vol. 5, EMI Prediction and Analysis Techniques, Don White Consultants, Gainesville, VA, 1972.

WILLIAM G. DUFF

Review of Management Modeling for Telecommunications

Introduction

The globally accepted basis for the management of telecommunications networks is ITU-T (International Telecommunication Union Telecommunications Standardization Sector) Recommendation M.3010 on the Telecommunications Management Network (TMN) (1). This standard describes both the entities that make up a management system for a telecommunications network and the interfaces between them. These entities are the operations systems (OSs), mediation devices (MDs), and network elements (NEs), and they are related by management interfaces as shown in Fig. 1. There are also interfaces between OSs (X interfaces) and between OSs and workstations (F interfaces).

Network elements (e.g., exchanges) are the entities that provide the basic telecommunications functions of the telecommunications network. A true network element must have a standard TMN management interface, known as a Q interface. Old exchanges may not be true TMN network elements because they may not have Q interfaces; however, these may be managed by a TMN through another TMN entity, a Q adapter (QA).

A Q adapter allows a non-TMN network element with a proprietary or nonstandard management interface, an M interface, to be managed over a TMN Q interface. Q adapter functions can be physically implemented either in a discrete unit as an independent Q adapter or in a mediation device. A mediation device can concentrate management traffic, but it may also have additional functions, such as the filtering of alarms, that may not be performed by older equipment.

Layering and Functional Architectures

The modeling of telecommunications networks used as the basis of communication over the TMN management interfaces is derived from the general functional architecture of transport networks. This architecture is described in ITU-T Recommendation G.805, which describes telecommunications networks in terms of layers of transport functions and reference points between them (2). Each layer is classified according to its functions. Between adjacent layers there is a client/server relationship. Layers may also be partitioned according to geographic regions or management domains; this is described in more detail below in the discussion of technology-independent network modeling.

On each layer of the functional architecture there are trails that may be divided into a series of connections (Fig. 2). Each connection may be the client of a trail on an adjacent server layer which lies below the client layer. Certain

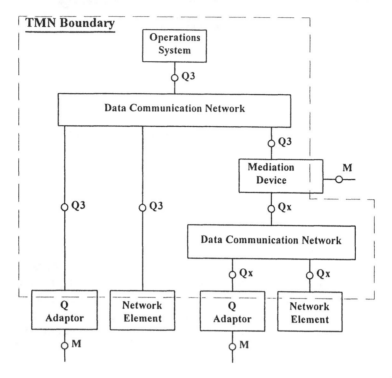

FIG. 1 Telecommunications Management Network (TMN) elements and interfaces (Q3, Qx = standard; M = proprietary).

functions are performed within the functional architecture, notably the trail termination functions at the end points of trails and the interlayer adaptation functions between the layers. Many of the basic classes of managed objects used in the modeling of telecommunications networks represent various parts of the functional architecture.

Often, a higher architectural layer corresponds to a lower degree of multiplexing since the layer that performs the higher level of multiplexing serves the layers with traffic that is multiplexed onto it. However, this is not always true. For example, the trails of an HDSL (high-speed digital subscriber loop) transmission layer may serve a connection of a 2048- or 1544-kilobit-per-second (kb/s) trail, but the trails of the HDSL transmission layer operate at a lower rate than those of the client layer.

At the termination of a trail that acts as the source of the information flow, functions such as framing and encoding may be performed. Complementary functions, such as decoding and detection of framing, may be performed at a sink termination. The adaptation functions between different architectural layers are performed at the terminations of connections. The point between the trail termination functionality and the adaptation functionality to the serving layer is called the access point, and the point at which the termination of one connection is joined to the termination of a trail or to the termination of another connection is known as a connection point.

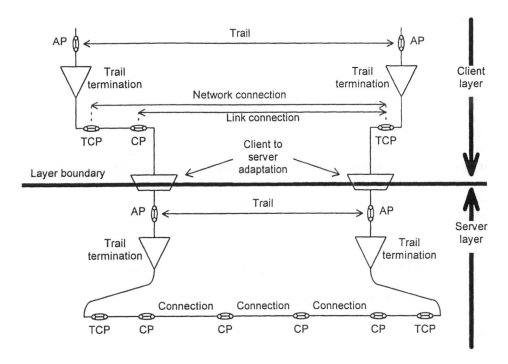

FIG. 2 Functions and points of a functional architecture (AP = access point; CP = connection point; TCP = termination connection point).

At the higher architectural layers, the trails are typically composed of streams of discrete units such as bits, octets, or cells. These units can be processed in two different ways at the terminations of the trails. The discrete units can be assembled or disassembled or, alternatively, formatting within the units themselves may be generated or detected. The architectural layers at which the assembly and disassembly are performed are the path layers; these are the clients of the section layers at which formatting is processed. The clients of the path layers are the circuit layers, but these may again consist of formatting layers that support further assembly and disassembly, so the layering can be recursive (Fig. 3).

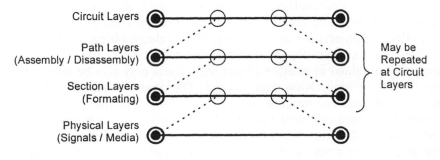

FIG. 3 Layers of a transport network.

The Definition of Managed Objects

Managed object classes are formally defined in terms of a number of packages, some mandatory and some conditional. These packages consist of the definitions of properties (attributes, actions, notifications, and behavioral descriptions) and any associated parameters with data fields specified using ASN.1 (Abstract Syntax Notation One) definitions, which are similar to type definitions such as "Integer" or "Character" in a programming language. The ASN.1 definitions allow characteristics to be encoded as octets that can be manipulated over a management interface. New classes are specified by creating subclasses from more general classes by adding new packages or refining existing packages. When a new class is defined by subclassing from an already existing class, the derived class is known as the subclass, and the existing class is known as the superclass. A subclass possesses all the packages of its superclass plus some additional packages. Conditional packages of the superclass can become mandatory classes in the subclass.

The attributes are the observable fields of an object; these take two forms. Data attributes hold the data characteristic of an object, while pointer attributes identify associated objects. Certain attributes are read only while others can be written to. Attributes that are read only can still be changed due to other events within the managed system; this may cause the managed system to send a change notification to the operations system. Notifications are messages that can be sent autonomously by an object, for instance, to raise an alarm or to indicate a change of state. Behavioral descriptions often correspond to both the comments statements in a programming language and the algorithms formally defined in the programming language.

Actions are the specific functions that an object can be requested to perform; however, objects may be controlled either by invoking actions or by changing attributes. Both approaches can be used to initiate a function; it can be a matter of taste as to which is used. Often, it may be better to initiate a function by changing an attribute if the associated data needs to be stored by the object, and it may be better to use an action to initiate a function if no associated data needs to be stored in the object. There are also cases for which it is better to use an action to modify a read-only attribute because this allows the changes to the attribute to be constrained.

Objects can be created either directly through a CMIP (Common Management Information Protocol) creation message (3) or indirectly through an action on another object. The creation of objects through actions on other objects ensures that the objects are not created unless a related object exists.

All of the managed objects in a management model, apart from the root object, are contained in another object. An object may also be related to additional objects by association relationships. The object classes and the relationships between them are summarized in what is known as the entity-relationship diagram for the management model. The containment of objects is similar to the structure of directories and subdirectories used for computer files. The full name of an object consists of the name of the root object, followed in turn by the names of all the relevant containing objects down the containment tree until

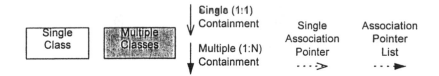

FIG. 4   Conventions for classes of managed objects and relationships between them.

the final object is reached. The conventions used here for classes of managed objects and the relationships between them are given in Fig. 4.

The Generic Modeling of Network Elements

The generic management model for telecommunications networks is defined in ITU-T Recommendation M.3100 (4). This covers both the logical resources associated with the functional architecture and the physical resources such as cards and shelves. The object classes it defines are applicable across different technologies, architectures, and services and are typically subclassed for particular applications, such as SDH (synchronous digital hierarchy) and ATM (asynchronous transfer mode). Classes are defined for networks, network elements, connections, trails, connection termination points, trail termination points, and cross-connections. Generic classes are also defined that may be subclassed to model equipment and software.

Physical equipment is represented by instances of equipment object classes; these objects may send notifications to indicate equipment faults. Equipment objects and the objects representing logical components of the functional architecture are related by pointer attributes. Other functions such as testing and reporting, filtering and logging of events, and the associated generic object classes are defined in the ITU-T X.700 series of recommendations.

The adaptation functions of the functional architecture and the side of the connection point closest to them can be represented by the connection termination point (CTP) classes of managed objects (Fig. 5). The trail termination functions and the side of the connection point associated with them can be represented by the trail termination point (TTP) classes. If there is no flexibility within an architectural layer, then the adaptation functions to the serving layer that are linked at the connection point with the trail termination functions of the client layer can also be included in the definition of a TTP class without the need to specify a CTP class. If there is a high level of flexibility, then the connection termination point may be represented by the cross-connection class of managed object.

Instances of the TTP classes are contained in the root object that represents the network element. Instances of the TTP classes, in turn, contain instances of the CTP classes that represent the adaptation to the higher layer of the functional architecture served by the trails associated with the TTP classes.

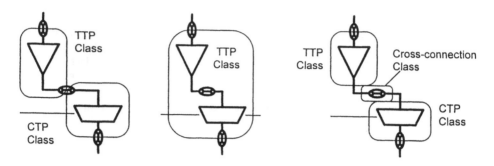

FIG. 5   Mapping the functional architecture to managed objects (TTP = trail termination point; CTP = connection termination point).

A TTP instance that represents the source of a trail has a connectivity pointer attribute that indicates the object representing the destination of its information flow. This is referred to as a downstream pointer because it points down the stream of the information flow. A TTP instance that represents the sink of a trail has an upstream connectivity pointer attribute that points upstream to the object representing the source of the information. If TTPs represent trail terminations at the boundary of the network element, then their connectivity pointers are null since the model of a network element has no information about the structure of a network.

The CTPs served by, and hence contained in, a source TTP must also be sources since they represent the origins of the connections served by the source TTP. Source CTPs must have upstream connectivity pointer attributes, unlike source TTPs, since these indicate the objects that represent the origin of the information flows (Fig. 6). The situation for sink termination points is reversed from that for source termination points.

Termination points that can be flexibly assigned through cross-connections have a read-only crossConnectionObjectPointer. If such a termination point is connected, then this pointer points to an instance of the class crossConnection, which represents a single active cross-connection. If the termination point is not

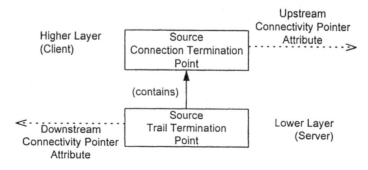

FIG. 6   Termination points and connectivity pointers.

connected, then the crossConnectionObjectPointer points to an instance of the class fabric, which represents the associated cross-connection matrix and contains the instances of the class crossConnection. The connectivity pointer attribute of a CTP may point to either a TTP object or to an instance of crossConnection or fabric, but not both since it cannot simultaneously identify both a cross-connection and a TTP as the source or sink of its information flow.

The TTPs always have a read-only data attribute that indicates whether an instance is enabled or disabled. They may also have a read-write data attribute that indicates whether their administrative state, which is controlled by the operations system, is locked, unlocked, or shutting down. Both TTPs and CTPs may emit various notifications to indicate the detection of faults or the occurrence of other events. The CTPs, unlike TTPs, do not have the data attribute for the administrative state unless this has been introduced by technology-specific subclassing, since this is determined from the containing TTP object. However, CTPs may have a read-only data attribute that indicates their channel number.

It is not appropriate for an object model to have objects that represent the adaptation functions and trail termination functions for each layer of the complete functional architecture because in many cases adjacent layers can be consolidated and treated as a single layer with an enhanced functionality. For example, if a continuous framed transmission is broadcast from the common OLT (optical line termination) to all remote ONUs (optical network units) on an optical access network, then it may be simpler to treat all of the formatting for framing and error detection as a single layer that has functionality equivalent to both the regenerator section layer and the multiplex section layer of an SDH system.

Synchronous Digital Hierarchy Transmission Systems

One of the first places TMN was applied was to SDH transmission systems. Three fundamental layers were identified for the SDH architecture (Fig. 7):

1. The transmission layer, which is subdivided into a physical media layer, a regenerator section layer, and a multiplex section layer.
2. The path layer, which consists of the SDH payloads and is subdivided into high- and low-order path layers.
3. The circuit layer, which is served by the SDH-specific transport (transmission and path) layers.

The functional architecture for SDH follows the identification of these layers and the associated clarification of the adaptation and termination functions. From the lowest level upward, these are (Fig. 8)

Multiplex Section Termination (MST) — handles the multiplex section overhead (MOH). Since the corresponding regenerator section termination (RST) handles the regenerator section overhead (ROH), together they handle the sec-

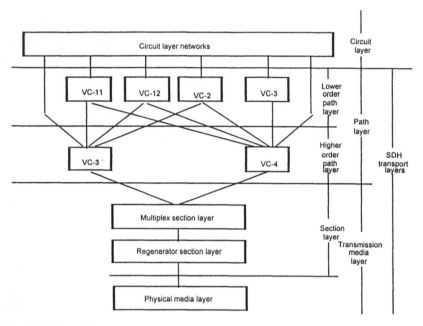

FIG. 7   Architectural layers for synchronous digital hierarchy (SDH) networks (VC = virtual container).

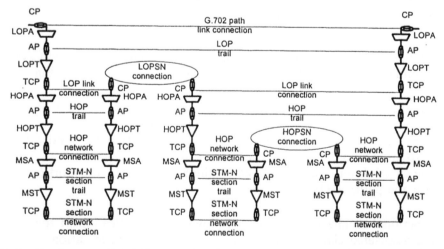

FIG. 8   Synchronous digital hierarchy functional architecture (section layer and above) (HOP = higher order path [e.g., VC-4]; HOPA = higher order path adaptation; HOPT = higher order path termination; HOPSN = higher order path subnetwork; LOP = lower order path [e.g., VC-12]; LOPA = lower order path adaptation; LOPT = lower order path termination; LOPSN = lower order path subnetwork; MSA = multiplex section adaptation; MST = multiplex section termination).

tion overhead (SOH) and the STM (synchronous transfer mode) signal that is transmitted. This signal is finally adapted to the optical or electrical physical medium.

Multiplex section adaptation (MSA)—handles the pointer for the higher order path layers and handles the associated SDH adaptation units (AUs) and adaptation unit group (AUG).

Higher order path termination (HOPT)—provides the path overhead (POH) for the tributary unit group (TUG) to form the VC-4 (virtual container 4) container.

Higher order path adaptation—defines the pointer and hence defines the tributary units (TUs) and handles the multiplexing of the TUs and their groups (TUGs) into the high order virtual container (VC).

Lower order path termination (LOPT)—provides the path overhead (POH) to the 2048 kb/s or 1544 kb/s based lower order path payloads, forming the VC-12 or VC-11.

Lower order path adaptation—maps the 2048 kb/s or 1544 kb/s connections from the client circuit layer into the appropriate lower order path payloads.

The classes of managed objects correspond to parts of the functional architecture for SDH transport from the SDH transmission fragment; these classes can be grouped into subfragments according to the layers of the functional architecture (Fig. 9). The root class for this model is SDH network element (sdhNE), which is a subclass of the generic managedElement class and represents managed elements.

In addition to the transmission fragment, there are other SDH fragments that allow the modeling of SDH physical equipment and SDH protection switching.

Transmission Media Layer Subfragments

The entity-relationship diagrams for the classes of the subfragments for the transmission media layers are shown in Fig. 10.

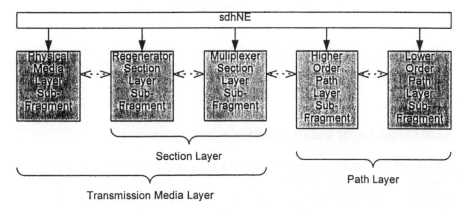

FIG. 9 The synchronous digital hierarchy transmission fragment (sdhNE = synchronous digital hierarchy network element).

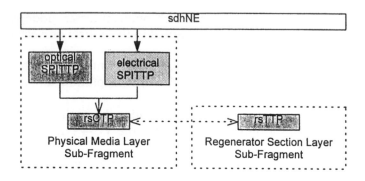

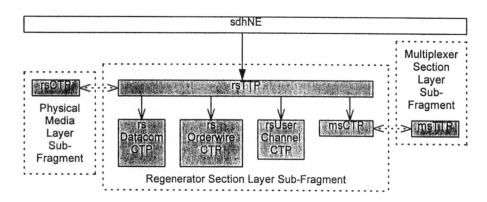

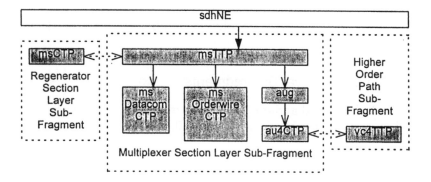

FIG. 10 Transmission media layer subfragments (aug = adaptation unit group; ms = multiplexer section; rs = regenerator section; SPITTP = synchronous physical interface trail termination point).

Physical Media Layer Subfragment

The classes of the media layer subfragment are the electrical synchronous physical interface trail termination point (electricalSPITTP), the optical synchronous physical interface trail termination point (opticalSPITTP), and the regenerator section connection termination point (rsCTP).

The electrical synchronous physical interface trail termination point (electricalSPITTP) classes represent the end points of the trails of the physical media layer when electrical transmission is used. Trail sink functions include the transformation of the incoming signal into an internal logic level, recovering a bit clock from the incoming signal, and the detection of a loss of signal (LOS) condition. The source function consists of the generation of a synchronous transport module level N (STM-N) signal from the internal bit streams. These classes are contained in the class that represents the sdhNE.

The optical synchronous physical interface trail termination point (optical-SPITTP) classes represent the end points of the trails of the physical media layer when optical transmission is used. The sink and source functions include those specified for electrical transmission but, in addition, the source includes detection of the failure of the laser transmitter. These classes are contained in the class that represents the sdhNE.

The regenerator section connection termination point (rsCTP) classes represent the intermediate points of trails at the regenerator section layer of the functional architecture. They are contained in the corresponding physical interface TTP classes, and their connectivity pointers identify either the end points of the regenerator section trails, if the trails are terminated, or other intermediate points of the trails if a regenerating repeater is being modeled.

Regenerator Section Layer Subfragment Classes

The classes of the regenerator section layer subfragment are the regenerator section trail termination point (rsTTP), the regenerator section data communications connection termination point (rsDatacomCTP), the regenerator section orderwire connection termination point (rsOrderwireCTP), the regenerator section user channel connection termination point (rsOrderwireCTP), and the multiplex section connection termination point (rsCTP).

The regenerator section trail termination point (rsTTP) classes represent functions such as the insertion and extraction of the regenerator section overhead, frame detection, and the scrambling and descrambling of the bit stream. They are contained in the class that represents the sdhNE.

The regenerator section data communications connection termination point (rsDatacomCTP) classes represent the generation and termination of the D0 to D3 bytes of the SDH section overhead, which are used to carry information between regenerator sections. They are contained in the corresponding regenerator section TTP classes.

The regenerator section orderwire connection termination point (rsOrderwireCTP) classes represent the generation and termination of the E1 byte of the section overhead that is allocated to the SDH orderwire function for the regenerator section. They are contained in the corresponding regenerator section TTP classes.

The regenerator section user channel connection termination point (rsOrderwireCTP) classes represent the generation and termination of the F1 byte of the section overhead that is made available to the user. They are contained in the corresponding regenerator section TTP classes.

The multiplex section connection termination point (rsCTP) classes represent the intermediate points of trails at the multiplex section layer of the functional architecture. They are contained in the corresponding regenerator section TTP classes, and their connectivity pointers identify either the end points of the multiplex section trails, if the trails are terminated, or other intermediate points of the trails if the trails are not terminated.

Multiplex Section Layer Subfragment Classes

The classes of the multiplex section layer subfragment are the multiplex section trail termination point (msTTP), the multiplex section data communications connection termination point (mssDatacomCTP), the multiplex section orderwire connection termination point (rsOrderwireCTP), the adaptation unit group (AUG), and the adaptation unit 4 connection termination point (au4CTP).

The multiplex section trail termination point (msTTP) classes represent the source and sink functions of SDH multiplexer sections. These functions include the insertion and extraction of the multiplex section overhead and the detection of faults, which include excessive bit error rates, far-end receive failure (FERF), and adaptation unit path errors. They also include the corresponding return of a FERF signal or emission of an AIS (alarm indication signal). These classes are contained in the class that represents the sdhNE.

The multiplex section data communications connection termination point (mssDatacomCTP) classes represent the generation and termination of the D4 to D12 bytes of the SDH section overhead, which are used to carry information between multiplex sections. They are contained in the corresponding multiplex section TTP classes.

The multiplex section orderwire connection termination point (rsOrderwireCTP) classes represent the generation and termination of the E2 byte of the section overhead that is allocated to the SDH orderwire function for the multiplex section. They are contained in the corresponding multiplex section TTP classes.

The adaptation unit group (AUG) classes represent points at which the adaptation unit group pointer is processed. They are contained in the corresponding multiplex section TTP classes.

The adaptation unit 4 connection termination point (au4CTP) classes represent the points at which the SDH adaptation unit connections are processed and the loss of alignment pointer is detected. They are contained in the corresponding AUG classes and their connectivity pointers identify either the end points of the VC-4 trails, if the trails are terminated, or other intermediate points of the trails if the trails are cross-connected.

Path Layer Subfragments

The entity-relationship diagram for the classes of the subfragments for the path layers are shown in Fig. 11.

The virtual container trail termination point (vc4TTP, vc3TTP, vc2TTP, vc12TTP, vc11TTP) classes represent the source and sink functions of trails of

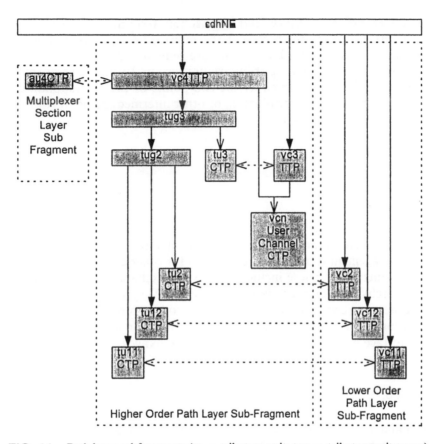

FIG. 11   Path layer subfragments (tu = tributary unit; tug = tributary unit group).

the SDH path layers, in particular, the processing of the path overhead. They are contained in the class that represents the sdhNE.

The tributary unit group 3 (tug3) classes represent groups of tributary units. They are contained in the corresponding VC-4 TTP classes (vc4TTP).

The tributary unit group 2 (tug2) classes represent the point at which the VC-3 or VC-4 containers are assembled or disassembled and the tributary unit pointer is processed, taking account of the phase of the lower order path trails relative to the higher order path overheads. They are contained in the corresponding tributary unit group 3 (tug3) classes.

The virtual container N user channel connection termination point (vcn-UserChannelCTP) classes represent the generation and termination of the F2 byte of the path overhead that is made available to the user. They are contained in the corresponding high order path layer TTP classes (vc4TTP or vc3TTP).

The tributary unit 3 connection termination point (tu3CTP) classes represent the intermediate points of VC-3 trails at which tributary units for VC-3s are generated or terminated. They are contained in the corresponding tributary unit group 3 (tug3) classes and their connectivity pointers identify either the end points of the VC-3 trails, if the trails are terminated, or other intermediate points of the trails if the trails are cross-connected.

The lower order tributary unit connection termination point (tu2CTP, tu12CTP, tu11CTP) classes represent the intermediate points of lower order trails at which lower order tributary units are generated or terminated. They are contained in the corresponding tributary unit group 2 (tug2) classes, and their connectivity pointers identify either the end points of the corresponding lower order trails, if the trails are terminated, or other intermediate points of the trails if the trails are cross-connected.

Other Models

Following the development of the management model for SDH transmission systems, management models were developed for other elements of the telecommunications network. These include models for exchanges, for plesiochronous digital hierarchy (PDH) transmission, and for ATM systems. In addition, it has now become more common for management models to be developed to support other telecommunications interfaces, for example, for management related to the V5 interface between access networks and exchanges.

Exchanges

In the management model for narrowband customer administration for on-demand switches, the central reference from which the services and resources associated with a customer or group of customers can be found is represented by an instance of the class customerProfile (Fig. 12). These instances are contained in the object that represents the host exchange or central office (typically, a subclass of managedElement). Each instance of customerProfile has a list of pointers to instances of subclasses of directoryNumber that represent the directory numbers associated with a customer or group of customers and a list of pointers to instances of subclasses of accessPortProfile. Instances of accessPort-Profile represent a reference point to various port-related resources.

The different type of directory numbers (E.164 and X.121) are represented by different subclasses of directoryNumber. The subclasses inherit a list of pointers to the related instances of customizedResource, which represent how access ports and channels are used to support customized services. They also inherit a list of pointers to the related instances of accessPortProfile and a single pointer back to the related instance of customerProfile.

The instances of subclasses of accessPortProfile have a list of pointers to the related instances of accessPort that represent terminations of the trails between the exchange and the customer premise equipment. This list of pointers is structured to enable signaling for ports to be nonassociated (i.e., carried on a different port). Instances of accessPortProfile also have a list of pointers to the related instances of customizedResource and a list of pointers back to the related instances of customerProfile. They also have an attribute to indicate which

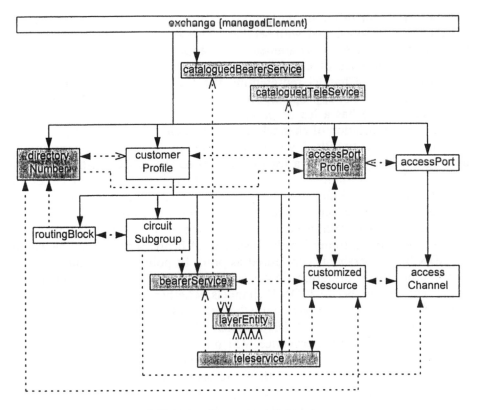

FIG. 12 Customer administration.

circuits are sensitive and require special clearance before being put out of service, for example, if circuits are used by emergency services.

Instances of accessPort have a single pointer back to the associated instance of accessPortProfile and an attribute that indicates the equipment used to terminate the customer's line. They contain instances of accessChannel that represent intermediate points on the trails served by the trail terminating at an instance of accessPort.

Every instance of customizedResource is contained in the appropriate instance of customerProfile and indicates the number of related ISDN (Integrated Services Digital Network) B channels. Each instance has lists of pointers to instances of accessPortProfile, to instances of accessChannel indicating the nature of the customization, and to instances of directoryNumber. Each instance also has a list of pointers to related instances of subclasses of bearerService, which represent common aspects of ISDN customized bearer services at OSI (Open System Interconnection) layers 1 to 3, and a list to instances of teleservice, which represent communications services with a communications protocol at OSI layers 4 to 7.

Instances of accessChannel indicate their alarm status and may also indicate the equipment used to support the channel. They also have a list of pointers to the related instances of customizedResource.

Channels from different ports may be grouped together and ordered so that they can be associated with services. This is represented by instances of circuitSubgroup; these are contained in the appropriate instance of customer-Profile. Each instance of circuitSubgroup has an ordered list of pointers to instances of accessChannel representing the ISDN B channels and a list of pointers to instances of appropriate service-related classes such as bearerService. Each instance also has a list of pointers to instances of routingBlock, which represent the association of channel subgroups with directory numbers.

Instances of routingBlock are also contained in the appropriate instances of customerProfile. They have a list of pointers to the related instances of sub-classes of directoryNumber and a list of pointers to instances of circuitSub-group, which also includes how digits from the circuit subgroup are modified. Instances also have actions that control these modifications.

Instances of bearerService and teleservice are again contained in the appropriate instance of customerProfile. Instances of bearerService describe whether the ISDN bearer service is configured as point to point, multipoint, or broadcast; whether the information flow is established on demand, is permanent, or is reserved; the number of ISDN B channels; whether the flow is unidirectional or bidirectional and, if bidirectional, whether it is symmetrical or asymmetrical; and the type (B or H) of the ISDN channel. There is also a single pointer to an instance of cataloguedBearerService, which represents noncustomizable bearer services, and a list of pointers to related instances of customizedResource. For appropriate on-demand services, there are also single pointers to instances of subclasses of layerEntity, contained in the appropriate instance of customerPro-file, that represent the customizable characteristics of layers 2 and 3 of the OSI stack.

Instances of teleservice have single pointers to the related instances of bearer-Service and catalogedTeleservice, which represents the noncustomizable aspects of OSI layers 4 to 7. They also have a list of pointers to the related instances of customizedResource. Instances of teleservice and catalogedTeleservice may also have pointers to objects representing the characteristics of layers 4 to 7 of the OSI stack.

In addition, there are also classes that represent customizable and noncus-tomizable user facilities and supplementary services and to support provisioning when detailed knowledge is absent.

Plesiochronous Digital Hierarchy Modeling

The PDH management model was developed by the European Telecommunica-tion Standardization Institute (ETSI) to represent the functionality of both European PDH equipment and that of European PDH ports on SDH equip-ment. The object classes are subclasses of the equivalent TTP and CTP classes defined in the global standards for SDH equipment. The main difference from the SDH modeling is that in the PDH modeling the multiplexing functions and the datastream formatting functions are combined in the layers of the PDH functional architecture. This is because PDH multiplexing is recursively nested in layers at the various PDH rates with formatting added at each layer, unlike

SDH multiplexing, which has a single level of formatting. The PDH basic model consists of four generic object classes:

1. European PDH trail termination point sink, source, and bidirectional (ePDHTTP). This set of object classes represents the termination points of PDH trails. Source objects occur at the origin of a trail and sink objects occur at the destination of the trail. Sink objects are able to indicate the detection of a loss of frame (LOF).
2. European PDH connection termination point sink, source, and bidirectional (ePDHCTP). This set of object classes represents the termination points of PDH connections, including adaptation to trails of the serving PDH layer. Sink objects are able to indicate the detection of an alarm indication signal (AIS).
3. European PDH AIS trail termination point sink, source, and bidirectional (ePDHATTP). This object class represents the combined functionality of both TTPs and CTPs. It is useful for the modeling of simple PDH multiplexers for which the modeling of a cross-connection should not be permitted. Sink objects are able to indicate the detection of an alarm indication signal (AIS).
4. Sink, source, and bidirectional European PDH TTPs for interworking transport of SDH virtual containers (VCs) and ATM cells (ePDHIntTTP). This object class represents TTPs that allow the PDH to transport SDH VCs or ATM cells.

Specific object classes are subclassed from these generic classes and labeled according to the plesiochronous hierarchy. The label *0* following the letter *e* denotes 64 kb/s, 1 denotes 2 megabits per second (Mb/s), and so on. For example, a 140-Mb/s trail termination point is represented by an instance of e4TTP. The PDH model also includes a plesiochronous physical interface trail termination point class (pPITTP) that is used to represent the functionality of TTPs at the physical interface layer. In particular, this class represents the conversion between transmitted signals and logic levels, the recovery of timing information, and the detection of loss of signal (LOS).

Asynchronous Transfer Mode

The management model for ATM was derived from the ATM functional architecture (Fig. 13). The lowest layers of the functional architecture are the physical layers that support the ATM transmission, for example, within SDH VC-4 containers. Cell headers are handled at the transmission convergence layer; this supports the virtual path and the virtual channel layers in turn. The ATM adaptation layer and the application layers sit on top of the virtual channel layer.

The ATM virtual path connections (VPCs) are the trails of the virtual path (VP) level of the functional architecture. The end points of these trails are represented by vpTTPBidirectional objects and intermediate points; their func-

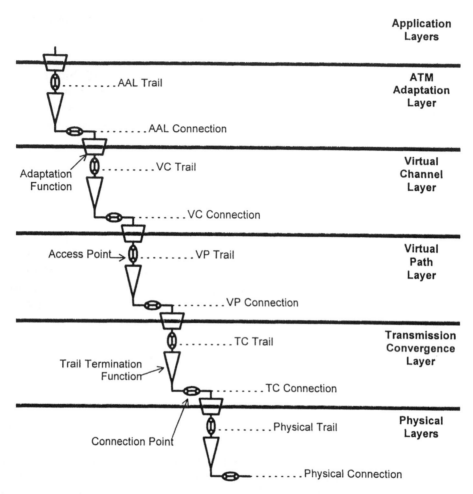

FIG. 13   Asynchronous transfer mode (ATM) functional architecture (AAL = ATM adaptation layer; TC = transmission convergence; VC = virtual channel; VP = virtual path).

tionalities are represented by vpCTPBidirectional objects (Fig. 14). The adaptation between the end points of the VPCs and the supported links at the VC layer of the functional architecture at the intermediate points of the virtual channel connections (VCCs) is represented by vcCTPBidirectional objects. The end points of the VCCs are represented by vcTTPBidirectional objects.

The VP layer of the ATM functional architecture is supported by the ATM transmission convergence (TC) layer. The end points of this layer, at which ATM cell headers are assembled and header error rates are monitored, are represented by tcAdaptorTTPBidirectional objects. Characteristics of the ATM interface are represented by associating a user–network interface (UNI) object with the end point of the transmission convergence layer if the interface is a UNI or, for a network–network interface (NNI), an inter- or intra-NNI object (interNNI or intraNNI). Characteristics of supported VPs or of supported VCs

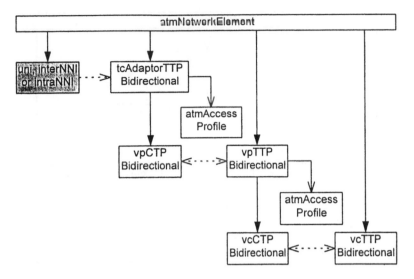

FIG. 14   Modeling of the asynchronous transfer mode layers.

are represented by atmAccessProfile objects that are contained in the object representing the end point of the supporting layer (i.e., the tcAdaptorTTPBidirectional object or the vpTTPBidirectional object).

A VP cross-connection is modeled by creating flexible associations between the intermediate points of the VP trails. Each flexible association is represented by an atmCrossConnection object associated with the adjacent vpCTPBidirectional objects that represent the intermediate points and are served by the end points of the transmission convergence trails associated with the ATM interfaces (Figs. 15 and 16). All of the individual atmCrossConnection objects are contained in an atmFabric object that represents the entire cross-connection matrix.

Fault and performance management aspects are represented by objects contained in the objects representing the ATM interfaces or the end points or intermediate points of ATM trails. The generation and monitoring of OAM

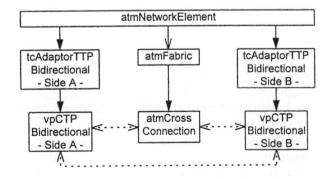

FIG. 15   Virtual path cross-connection modeling.

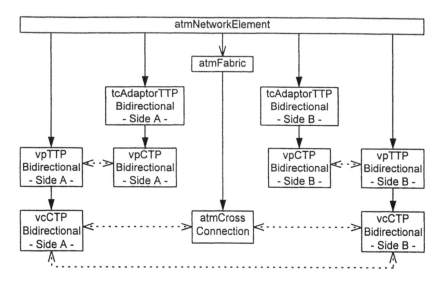

FIG. 16   Virtual circuit cross-connection modeling.

(operations, administration, and maintenance) cells for F4 and F5 OAM flows are managed using performanceMonitor and continuityMonitor objects contained in the relevant objects representing the end points or intermediate points of the trails. Statistical information currently being gathered is represented by currentData objects, and statistical information previously gathered is represented by historyData objects.

V5 Modeling

Access networks are the increasingly sophisticated transmission and multiplexing systems used to link customers to the core of the telecommunications network, often using optical fiber (5). The V5 interface is the interface between an access network and its host exchange for narrowband services. The object modeling of a V5 interface has two forms because a V5 interface has both an exchange side and an access network side. The exchange side is consistent with the object model for customer administration for the services and resources associated with customers.

V5 Modeling in the Host Exchange

Within the host exchange, the termination of the trail for a customer's access to services is represented by the object class accessPort. This is subclassed for "real" ports, which are directly connected to the exchange, and for virtual ports, which are the images of the ports in an access network and are connected via a V5 interface. There are virtual ports for analog PSTN (public switched telephone network) accesses, basic and primary rate ISDN accesses, and for leased-line accesses. The bearer channels on these ports are represented by instances of virtualAccessChannel.

For V5.1 interfaces, the virtualAccessChannel objects are associated with particular v5Timeslot objects (Fig. 17), which represent the bearer channels of the V5.1 interface. This association is not modeled for the V5.2 interface because it is controlled by the V5.2 bearer channel connection (BCC) protocol and is not visible to the management system. In both cases, however, the D channels at the ISDN ports have to be associated with V5 communications paths and channels; these are associated in turn with the v5Timeslot objects.

The core functionality of a V5 interface is represented by an instance of the object class v5Interface, which indicates the associated virtual access ports and also the 2.048-Mb/s links of the interface, which in turn are represented by instances of v5Ttp. Setting the administrative state attribute of the v5Ttp to locked or shutting down initiates the V5 link blocking procedure for the represented link.

Each V5 interface object also contains instances of the protocol-specific communications path object classes that represent the particular paths for the signaling protocols. In the case of the ISDN signaling protocols, the communications path objects identify the particular virtual ports that use the represented paths. All of the communications path objects, except that for the protection protocol, identify the commChannel objects that represent the logical 64-kb/s signaling channels. These in turn are associated with the v5Timeslot objects that represent the physical 64-kb/s channels of the V5 interface.

The protection switching of a V5.2 interface is modeled using the protCommPath and v5ProtectionUnit object classes. Two instances of v5ProtectionGroup are used to represent the two different types of protection. Each instance of

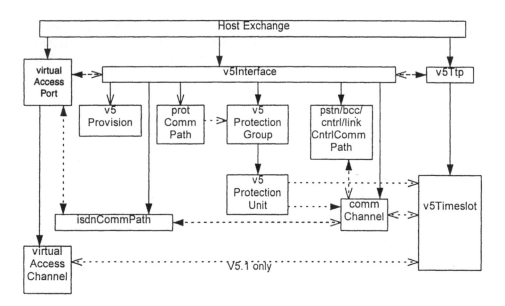

FIG. 17 Model for V5 configuration in the host exchange.

v5Timeslot that represents a physical communications channel is associated with a v5ProtectionUnit object indicating if the physical communications channel is active or on standby. By association with the commChannel objects, the v5ProtectionUnit objects identify the physical time slot in use at any time for the protected logical communications channels. The protCommPath object is associated with the physical time slots identified for use with V5 protection group 1.

If the V5 provisioning messages are used to coordinate reprovisioning across the V5 interface, then the V5 messages may be relayed between the V5 interface and the Q3 interface using the v5Provision object. The v5Provision object is also contained in the instance of v5Interface.

The measurement of traffic statistics on V5 bearer channels and communications channels is modeled using currentData objects (Fig. 18). The measured information may be collected using a simpleScanner object and results may be recorded using scanReportRecord objects kept in a log object.

V5 Modeling in the Access Network

Much of the access network management modeling associated with V5 interfaces is equivalent to the modeling already described for the host exchange. The differences in the management modeling are due to the real user ports and channels being in the access network and only virtual images of these ports existing in the host exchange. The virtualAccessPort and virtualAccessChannel objects in the exchange model are replaced by userPortTtp objects and userPort-BearerChannelCtp objects in the access network model. The userPortTtp objects represent the end points of the port layer trails, and the userPortBearer-ChannelCtp objects represent the intermediate points of the 64-kb/s trails served by the port layer trails (Fig. 19).

The testing of the user ports also has to be performed in the access network since these ports are no longer situated in the exchange. Tests are initiated by sending a message to the testActionPerformer object in the access network (Fig.

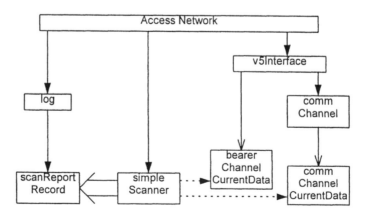

FIG. 18   Modeling traffic measurement.

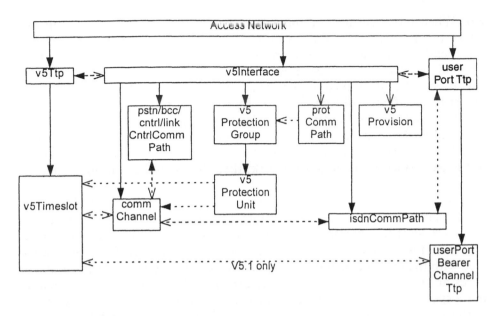

FIG. 19 Model for V5 configuration in the access network.

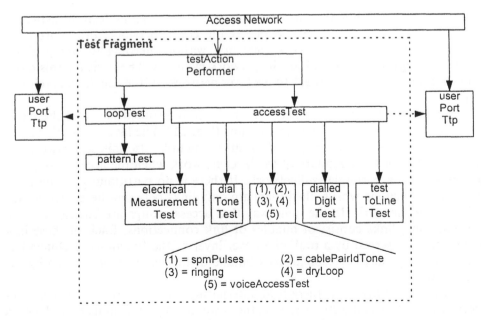

FIG. 20 Model for user port testing.

20). When tests can give an immediate result and there is no requirement to interrupt or to stop the test, uncontrolled tests may be used, and no further specification is required other than the clarification of the test action and the responding notification. Tests that last for a significant period of time or that may need to be interrupted or stopped are delegated by the testActionPerformer object to specific test objects that exist for the duration of each test.

Delegated loopback testing is modeled by a loopTest object that allows loops to be applied at various locations and is associated with the user ports being tested. If the pattern generation and comparison associated with the loopback test is performed in the access network, then this is modeled using a contained patternTest object.

Loopback testing is more appropriate to ISDN line testing since the PSTN is not designed to support loopback testing. Often, a series of tests other than the loopback have to be applied to clarify the nature of a fault, especially for the PSTN. A test session of this sort is represented by an accessTest object associated with the ports being tested. An accessTest object reserves the ports and any necessary resources required for the test session. The specific tests applied during a specific test session are then represented by temporary objects contained in the accessTest object. Some of these specific test objects model the measurement of electrical parameters and the testing of the dialed digits and the dial tone. Others allow a variety of other specialized tests and test conditions to be applied to the line or to the line circuit.

Partitioning and Network Level Modeling

In addition to modeling the various elements of a telecommunications network, it is also necessary to be able to model the entire network, or a part of it containing one or more network elements, without needing to worry about the technology-specific details of the network elements. The basis for this modeling is a more generic approach to the layer networks of the functional architecture.

The extremities of a layer network are its trail terminations, and these can be bundled together to form access groups (Fig. 21). The layer network can be partitioned into subnetworks joined by links, and links also join subnetworks to access groups at the boundary of the layer network when the access groups do not belong to a particular subnetwork. If there is no partitioning of the layer network into subnetworks, then there are only links between the access groups of the layer network. In the same way that access groups are bundles of trail terminations, links consist of bundles of link connections. Each of these link connections is served by a trail on a lower layer of the functional architecture, and a link typically corresponds to a bundle of link connections served by the same trail.

Within a subnetwork, there is an alternating sequence of subnetwork connections and link connections joining the connection points on its boundary via connection points internal to the subnetwork. The connection points on the

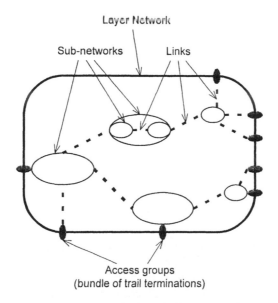

FIG. 21   Partitioning of a layer network.

boundary of the subnetwork are always joined to internal connection points by subnetwork connections. If a subnetwork has no internal structure, then all of its connection points are on its boundary, and the appropriate connection points are joined by its subnetwork connections. A subnetwork with internal structure has internal connection points joined to its external connection points by subnetwork connections, and a complete connection between connection points on its boundaries consists of an alternating sequence of subnetwork connections and link connections.

Each subnetwork connection can in turn be decomposed into a sequence of subnetwork connections and link connections at a more detailed level of partitioning (Fig. 22). The connection points of the decomposed subnetwork

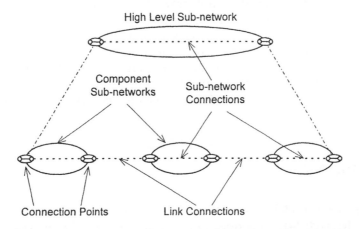

FIG. 22   Decomposition of a high-level subnetwork into component subnetworks.

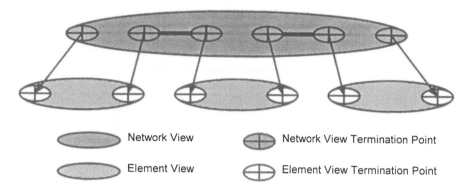

⬭ Network View	⊕ Network View Termination Point
⬭ Element View	⊕ Element View Termination Point

FIG. 23   Relationship between network and element views.

connection correspond to the external connection points of the subnetwork, which provides the more detailed partitioning.

Decomposing a network connection in this way shows that it also consists of an alternating sequence of subnetwork connections and link connections, each joining a connection point. The outermost connection points are the termination connection points for the trail terminations on the boundary of the layer network. The decomposition of a network connection may differ from that of a subnetwork connection because it may start or end with a link connection.

The Flatland principle is the key to the flexibility and technology independence of network-level modeling. In the story of Flatland, the various shapes existing in a two-dimensional world had no immediate knowledge of the three-dimensional world, although three-dimensional beings could see the two-dimensional world and the internal structure of the objects there. In a similar way, network-level modeling allows visibility of and access to the element-level modeling, which has no knowledge of the network level or its interconnections. This link from the network level to the element level can be accomplished by a pointer attribute by which an abstract generic network object can identify the corresponding object of a technology- or application-specific element view (Fig. 23).

Network-level modeling does not prevent proprietary features from being added since these can be included as additional objects contained in the objects of the generic network-level view. This avoids the need to use proprietary subclasses of the network classes, so the generic network-level model remains unaltered. Proprietary details can also be added at the element level in the same way, but the advantage of avoiding subclassing at the element level is smaller because element-level modeling is neither generic nor independent of technology.

Wider Aspects

The management models described so far have been based on ITU-T Recommendation M.3100 (4). This does not take account of the alternative Simple

Network Management Protocol (SNMP) approach developed for the manage ment of the Internet, and it does not take account of the open distributed processing (ODP) approach developed so that certain management specifications can be made futureproof.

Simple Network Management Protocol and the Internet

The SNMP was developed as a fast track approach for the management of the hubs, bridges, and routers of the Internet. The planned longer term goal was to use a CMIP approach to manipulate managed objects in a management model similar to those already described. However, this longer term goal has been dropped because there is now a very large existing base of SNMP-managed devices, and there is no obvious evolution path toward CMIP.

There are two key differences between SNMP and CMIP. The first is the obvious difference that, since they are different protocols, they support different functions and have different features. The second difference is the more subtle difference that they lead to different structures in the management information.

The CMIP management models are constructed from managed objects that have various relationships to each other. The SNMP management information has a tree structure with leaves on the tree ordered sequentially, which allows all of the information to be read using SNMP Get-next commands. Certain ordered sequences correspond to sequential entries in a table, and larger sequences of tables and individual items form groups, which are analogous to chapters in a book. Other sequences are intentionally reserved for vendor-specific extensions. Here are four of the commonly used groups (there are many more):

1. System Group. This is a list that gives general information about the managed device.
2. Interfaces Group. This has one entry that is the number of interfaces and a table that contains the information about each specific interface.
3. Address Translation Group. This consists of a single table.
4. Internet Protocol Group. This consists of 20 individual items and 3 tables of items.

The row of a table or some other collection of individual items can be treated in a way similar to the attributes and other characteristics of a managed object in a CMIP model. Such an SNMP entity can also have pointers to other SNMP entities so a structure could be built up similar to that of a CMIP management model. However, this is not done for the standard groups used to manage Internet devices. There is also a more subtle difference between the two approaches due to the different nature of the communications networks.

The functional architecture that is the basis of much of the telecommunications management modeling is appropriate for transmission systems and cross-connections, but it is less appropriate for exchanges in which the management system is not directly aware of the on-demand connections, and it is even less appropriate for connectionless packet networks such as the Internet in which each packet may be routed differently. To have integrated management of the Internet and of telecommunications networks, a new approach may be needed.

The Open Distributed Processing Approach

The goal of the ODP approach is to provide independence from the technology and distribution of computing resources in a management system. This is achieved through a structured framework that captures the requirements of the management application, relates the information to the managed resources, and defines how this information may be accessed and manipulated. This framework consists of five viewpoints (Fig. 24), and, for a particular management application, there are five corresponding specifications. The ODP methodology consists of the clarification and development of the specifications associated with these viewpoints.

1. The enterprise viewpoint is concerned with the capture of requirements. These are expressed in terms of the contract between the client and the provider of the ODP system. The enterprise specification is a formal statement of the purpose and scope of the management systems and the requirements of the client.
2. The information object types, their relationships to each other and to the managed resources, and their states and permitted changes of state are the concern of the information viewpoint. These are formally expressed in the information specification, which defines the meaning of the information and how it is processed.
3. The computational viewpoint is concerned with the formation of the fundamental building blocks within the distributed management system. The computational specification refines the information objects into computa-

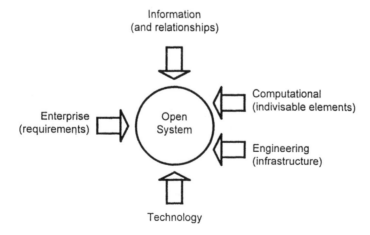

FIG. 24   The open distributed processing (ODP) viewpoints.

tional objects with defined interfaces. The computational objects supported by the distributed system are like atoms that cannot be subdivided unless new computational interfaces are defined.

4. The engineering viewpoint is concerned with the way in which the computational objects are allocated to the distributed components of the management system. In particular, the engineering specification determines the infrastructure of the management system and the mapping of computational objects onto its components.

5. The technology viewpoint is concerned with the implementation technology of the management system defined in the technology specification.

In terms of the ODP approach, the management models described so far are a mixture of computational and engineering specifications, with the distinction between these sometimes being ambiguous. In particular, it is not always clear when the management models only define the indivisible "atomic" objects of the management system and when they also define the physical interface between distributed physical components.

An enterprise specification is particularly useful when developing a management system because it should provide a clear statement of requirements; this is often only implicit in the formal definitions. Likewise, an information specification is a useful addition to the existing formal definitions of management models because relationships other than containment can easily become lost in the formal definition of the attributes of the managed objects. The enterprise and information specifications should be independent of both the particular approach used to define the computational objects and their mapping onto physical components, eliminating the need for these specifications to change as technologies and implementations evolve. In particular, as new forms of management modeling are developed, the ODP methodology allows a clear distinction to be made between those specifications that need to be changed and those that are not affected. If procurement specifications take the form of enterprise and information specifications, then it may be easier to take advantage of future developments.

Summary

Standard management models have been defined for most of the components of the telecommunications network. The models for SDH, PDH, exchanges, V5, and ATM are introduced here, together with network-level modeling that complements and integrates the separate element-level models. The challenges of developing a coherent approach to the management of both the telecommunications network and the Internet and of developing specifications that are futureproof still remain.

Bibliography

Abbott, E. A., *Flatland — A Romance of Many Dimensions*, Dover, New York, 1992.

Fatato, M., Modeling Telecommunications Networks' Transmission Systems, *IEEE Commun. Mag.*, 34(3):40–47 (March 1996).

Gillespie, A., and Rees, S., Access Network Management Modeling, *IEEE Commun. Mag.*, 34(3):62–72 (March 1996).

International Telecommunication Union, *Definition of Management Information*, ITU Recommendation X.721, ITU-T, Geneva, 1992.

International Telecommunication Union, *Architectures of Transport Networks Based on the Synchronous Digital Hierarchy (SDH)*, ITU Recommendation G.803, ITU-T, Geneva, 1992.

International Telecommunication Union, *Stage 2 and Stage 3 Description for the Q3 Interface: Alarm Surveillance*, ITU Recommendation Q.821, ITU-T, Geneva, 1992.

International Telecommunication Union Telecommunications Standardization Sector (ITU-T), *Basic Reference Model of Open Distributed Processing, Part 1: Overview and Guide to Use*, ITU-T Recommendation X.901, ITU-T, Geneva, 1993.

International Telecommunication Union Telecommunications Standardization Sector (ITU-T), *Common Elements of the Information Viewpoint for the Management of a Transport Network*, ITU-T Recommendation G.853-01, ITU-T, Geneva, 1996.

International Telecommunication Union Telecommunications Standardization Sector (ITU-T), *Stage 1, Stage 2 and Stage 3 Description for the Q3 Interface: Performance Management*, ITU-T Recommendation Q.822, ITU-T, Geneva, 1993.

Petermueller, W. J., Q3 Object Models for the Management of Exchanges, *IEEE Commun. Mag.*, 34(3):48–60 (March 1996).

Sexton, M., and Reid, A., *Transmission Networking: SONET and the Synchronous Digital Hierarchy*, Artech House, Boston, 1992.

Stallings, W., *SNMP, SNMPv2, and CMIP — The Practical Guide to Network-Management Standards*, Addison-Wesley, Reading, MA, 1993.

References

1. International Telegraph and Telephone Consultative Committee, *Principles for a Telecommunications Management Network*, CCITT Recommendation M.3010, ITU-T, Geneva, 1992.

2. International Telecommunication Union Telecommunications Standardization Sector (ITU-T), *Generic Functional Architecture of Transport Networks*, ITU-T Draft Recommendation G.805, ITU-T, Geneva, 1996.

3. Halsall, F., and Modiri, N., An Implementation of an OSI Network Management System, *IEEE Network Mag.* (July 1990).

4. International Telecommunication Union Telecommunications Standardization Sector (ITU-T), *Generic Managed Object Class Library*, ITU-T Recommendation M.3100, ITU-T, Geneva, 1995.

5. Gillespie, A., *Access Networks: Technology and V5 Interfacing*, Artech House, Boston, June 1997.

ALEX GILLESPIE

Routing in Telecommunications Networks

Introduction

Telecommunications networks are now ubiquitous. They come in all sizes. While some public telephone networks are made of tens of telephone exchanges, the popular Internet has million of nodes attached to it. An important problem to solve in public telephone networks, Internet, and actually in all telecommunications networks is how to establish routing paths. Routing is indeed a challenging issue and remains a vibrant area of research.

How is data or voice routed in a telecommunications network from a source node S to a destination node T? There exist myriad algorithms for routing data or voice in telecommunications networks. In the case of data, for instance, these algorithms range from best-path selection, by which the data are sent on a unique path (the "best" path), to flooding, by which data are sent on all possible paths between the source and the destination.

This article is devoted to routing in telecommunications networks. It presents the key techniques and backs the presentation by real-life examples. As routing goes hand in hand with switching and also as there exists a plurality of routing classification schemes, we start by relating routing to switching techniques and by briefly introducing the most popular routing classification schemes.

The table-driven approach to routing is presented after that since the *table-driven versus table-free* classification scheme was selected for this article. The table-free approach is studied afterward. We end by giving an overview of how the routing problem is concretely solved in selected real-life telecommunications networks.

Packet Switching, Circuit Switching, and Routing

Telecommunications networks are driven by two main switching techniques: packet switching and circuit switching. In this section, routing is first related to packet switching. It is subsequently related to circuit switching. The routing classification schemes are finally briefly introduced. Most textbooks on computer networks are good sources of information on switching techniques and routing classification schemes. Examples include Refs. 1 and 2.

Although packet switching is illustrated in this section by data transmission and circuit switching by voice transmission, the reader should be aware that

In this article, the term *telecommunications networks* is used in a broad sense and includes computer networks.

205

packet-switching techniques can also be applied to voice transmission. Telephony over the Internet is a good example of packet-switching-based-voice transmission. By the same token, circuit-switching techniques can also be applied to data transmission, although this seldom happens in real life.

Packet Switching and Routing

In packet-switched networks, the data sent by a source node S to a destination node T is broken in smaller units known as packets. Packet-switched networks can be connectionless or connection oriented. In connectionless packet-switched networks, the packets that make a message are transmitted independently. In connection-oriented packet-switched networks, they are transmitted the same way.

As far as routing is concerned, in connectionless packet-switched networks, Node S takes the routing decision for each and every packet. This means that packets, from the same message sent to Node T, might make the journey using different paths. On the other hand, in connection-oriented packet-switched networks, Node S makes the decision once, and all packets from the message are sent on the same path. The Internet is the archetype of connectionless packet-switched networks. Telenet (3), an international, commercial, value-added network, is an example of a connection-oriented packet-switched network.

Circuit Switching and Routing

Circuit switching is a common practice in telephony. It consists of establishing a physical path between the telephone exchanges to which users are attached. This physical path usually goes through many intermediate telephone exchanges. It is used as a dedicated physical path during calls.

As far as routing is concerned, circuit switching functions as connection-oriented packet switching in the sense that the establishment of a path is required prior to transmission. However, circuit switching differs significantly from connection-oriented packet switching. Circuit switching does not require the packetization of what is transmitted. Besides this, it does require the establishment of a dedicated physical path, while connection-oriented packet switching does not.

Routing Classification Schemes

Many different routing classification schemes are used in telecommunications literature. Selected examples are given below.

- *Table driven versus table free.* In these schemes, routing techniques are either table driven or table free. In table-driven routing, the node consults a table in order to make the routing decision. In table free routing, the decision is made automatically, without the help of a table.

- *Static versus dynamic*. Routing techniques are either static or dynamic in this scheme. In static routing, the routing decision is independent of the traffic level, while in dynamic routing the traffic level is taken into account. In practice, this means that in dynamic routing the routing algorithm adjusts itself to the traffic level, while in static routing it does not.
- *Centralized versus distributed*. These schemes are based on the level of centralization (distribution) of routing decisions. In centralized routing, routing decisions are made by a dedicated supervisory node based on information collected from all the nodes of the network. In distributed routing, decisions are made autonomously by each node based on information collected by the node.

Routing classification schemes are not mutually exclusive. As the reader might have guessed, a table-driven (or a table-free) routing technique is either static or dynamic. By the same token, a static (or a dynamic) routing technique is either table driven or table free. The *table-driven versus table-free* approach is followed in the rest of this article.

Routing in Telecommunications Networks: The Table-Driven Approach

In this section, we start by presenting the general principles of table-driven routing. Dijkstra's algorithm, the most popular algorithm for populating routing tables, is then discussed. Clustering, a technique for reducing the size of routing tables, is finally presented.

General Principles

Routing tables are located at every and each node of routing-table-driven telecommunications networks. They contain, for each destination that can be reached from the node, the path(s) leading to it. They might contain as well other information such as the estimated time for reaching the destination or the number of intermediate nodes on the way to the destination.

Routing tables are static when the information they contain is stored at system-generation time and does not change during the normal operation of the network. They are dynamic otherwise. In centralized routing, the content of the tables is generated centrally by a supervisory node, while in distributed routing each node generates its table using information collected from the other nodes.

For each destination that can be reached from a node, more than one path is usually stored. This allows the continuation of the normal operation of the network even when a path becomes unavailable. Figure 1 depicts a hypothetical

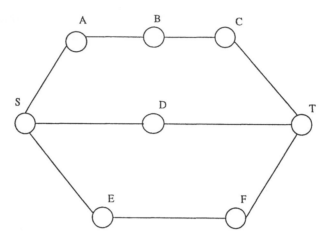

FIG. 1   A hypothetical telecommunications network.

telecommunications network. Table 1 shows a simplified version of the routing tables at Nodes S and D for Destination T. The simplified routing tables indicate how to reach destination T. From Table 1, it can be seen that SDT is the path for reaching T from S.

Populating Routing Tables

As can be seen from Fig. 1, besides the path indicated by the routing tables for reaching T from S (i.e., SDT), it is possible to reach T from S using other paths, namely, SABCT and SEFT. This raises an interesting question: How have the routing tables of Table 1 been populated? In other words, how has SDT been selected as the path for routing from S to T?

Best-path selection algorithms are popular techniques for populating routing tables. Their use implies a ranking among the possible paths leading from a given source to a given destination. A very simple criterion for ranking paths is their length. The chief assumption behind this criterion is that routing costs more on long paths than on short ones. Another assumption is that cost minimization is an objective.

A common measure of length is the number of intermediate nodes between the source and the destination. These intermediate nodes are also called hops.

TABLE 1   Simplified Routing Tables at Nodes S and D for Destination T

Source	Destination	Next Hop
S	T	D
D	T	T

Most best path selection algorithms are variants of a seminal algorithm invented by Dijkstra (4). Dijkstra's algorithm is rooted in the realization that if a given node N is known to be on the shortest path between a source node S and a destination node T, then the shortest path between N and S is known.

The realization is derived from the fact that the shortest path between S and T is equal to the sum of the shortest path between S and N and the shortest path between N and T. Knowing that N is on the shortest path between S and T implies, therefore, that either the shortest path between S and N or the one between N and T is known.

The algorithm uses the above to iteratively construct paths of increasing length from S to the other nodes of the network until the destination T is reached. During the construction, the nodes are divided in three sets. The first set, Set A, contains the nodes for which the path of minimal length from S is known. The second, Set B, contains the nodes from which the next node to be added to the first set will be selected, and the third, Set C, contains the remaining nodes.

Besides Sets A, B, and C, three other sets are considered during the construction. The first is Set I, the set of branches occurring in the minimal paths from Node N to the other nodes of Set A. The second, Set II, is the set of branches from which the next branch to be placed in Set I will be selected. The third, Set III, contains all the other branches. The reader should note that a branch is a path between two adjacent nodes.

Reducing the Size of Routing Tables

As stated above, routing tables are stored in each and every node of the network. They contain an entry for each and every destination that can be reached. In small- and medium-size telecommunications networks, these tables have tens or hundreds of entries. However, in large-size networks, they can have up to hundreds of thousands and even millions of entries, producing the need for reduction of the size of the routing tables.

Routing tables with hundreds of thousands or millions of entries are not practical for many reasons. An important portion of the memory of the nodes needs to be dedicated to the storage of these huge tables. Also, a lot of processing power is needed to scan and eventually update the tables. Schemes for reducing the size of routing tables are certainly a must for the smooth functioning of large telecommunications networks.

Kleinrock and Kamoun have systematically and thoroughly studied the problem (5,6). Their original idea consists of keeping at each node an entry in the routing table for the destinations close to the node and an entry per set of destinations for the destinations far from the node. The distance can be expressed in terms of hops or any other appropriate measurement.

This original idea can be realized through a hierarchical clustering of the nodes in the network. Figure 2 shows a 24-node network with a hierarchical clustering. The routing tables now have 10 entries, illustrated below by the example of the nodes in Cluster 1.1.

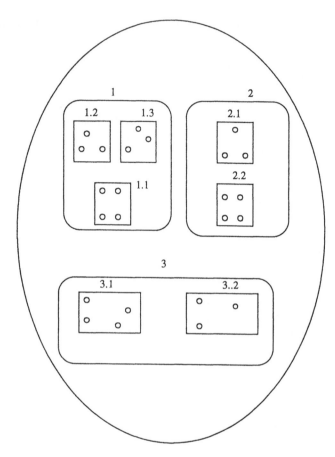

FIG. 2   A hypothetically clustered telecommunications network with 24 nodes.

Any node in Cluster 1.1 has in its routing table

- an entry for each of the four nodes in cluster 1.1, including itself
- an entry for each of the two other clusters at the same level (i.e., Clusters 1.2 and 1.3)
- an entry for each of the two clusters included in Cluster 2 (i.e., Clusters 2.1 and 2.2)
- an entry for each of the two clusters included in Cluster 3 (i.e., Clusters 3.1 and 3.2)

The routing tables would have had 24 entries if clustering was not used since the network has 24 nodes. As the reader can easily check, by merging Clusters 1.1 and 1.2, the tables' lengths will become 12.

For a given telecommunications network, there are many possible ways of hierarchically clustering the nodes. But, whatever way the nodes are hierarchically clustered, clustering has as a consequence an increase in the routes' lengths.

This is the penalty to pay for the gain in space and processing power linked to the decrease in the tables' lengths.

As there are many possible ways to cluster the nodes of a network hierarchically, an interesting issue is the determination of the optimal clustering, meaning the one for which the lengths of the routing tables are the shortest. Since the penalty to pay for the decrease in routing table length is the increase in route length, another interesting issue is how heavy the penalty is when the network is very large.

Kamoun and Kleinrock have solved both issues (5,6). The optimal number of clustering levels in a network with N nodes is $\ln N$. With that optimal clustering, each node has $e \ln N$ entries. Besides this, in very large networks these reductions may be achieved with essentially no increase in route lengths. This result has actually been qualified by the two researchers as intuitively pleasing, but possibly obvious.

Routing in Telecommunications Networks: The Table-Free Approach

Routing tables are not welcome in some networks. In the case of high-speed telecommunications networks, for instance, the delay caused by the consultation of a table prior to the switching of packets might not be acceptable. Table-free routing schemes have therefore been designed as alternatives to table-driven schemes.

We start this section by introducing flooding, a well-known scheme for table-free routing. After flooding, the table-free schemes for networks with regular topologies are discussed. Besides flooding and the schemes for networks with regular topologies, various other table-free routing techniques exist. The section ends by surveying them.

Flooding

The principles behind flooding are very simple. The approach does have merits. It, however, has major drawbacks. This explains why many improvements have been proposed over the years.

General Principles

In the purest form of flooding, when a source node S wants to send a packet to a destination Node T, it starts by sending the packet to all the neighboring nodes, meaning all the nodes to which it is directly connected. Each node, on reception of the packet, presents it to all neighboring nodes except the node from which the packet has been received. This process is reiterated until the destination T is reached.

As an illustration, when the source Node S of Fig. 1 wants to send a packet to the destination Node T, it starts by sending the packet to Nodes A, D, and E. Node A then sends it to Node B, which sends it in turn to Node C, which finally sends it to Node T. Node D, on reception of the packet, relays it to Node T. In the case of Node E, the packet is first transmitted to Node F, which forwards it to Node T.

Merits

Many features make flooding attractive. It is very simple to implement. All a node has to do when it receives a packet that is not for itself is to resend the packet to all its neighbors except the neighbor from which it has received the packet. Besides its simplicity, flooding has other distinct merits. Two are worth being mentioned.

1.   Although flooding does not calculate the shortest path between sources and destinations, it does guarantee that packets always arrive at their destination following the shortest path. This results from the fact that packets go to their destination following all possible paths.
2.   Flooding is very robust to node failures. As packets follow all possible paths, there is a very high probability that they arrive at the destination even when intermediate nodes fail. As an illustration, a packet sent from Source S to Destination T (Fig. 1) will reach T even when Nodes A and B fail at the precise moment S sends the packet. This makes flooding particularly suitable for applications such as military applications for which large numbers of intermediate nodes may be blown in pieces at any instant.

Drawbacks

Flooding does have many drawbacks. The most important is the overhead it generates. This overhead leads to very poor utilization of the network resources and also to a marked tendency to congestion. Since several copies of the packets are produced by every intermediate node, the total number of packets in transit in the network can be very important, even in small-size networks. The network can be literally flooded. As an example, a packet sent by the source Node S to the Destination T (Fig. 1) will be replicated nine times before reaching Node T. Node T will actually receive three copies.

Beyond Pure Flooding

As routing by flooding has distinct merits, many schemes have been designed to dam the flooding while keeping the advantages. The destination can, for instance, return by flooding an acknowledgment as soon as it receives the first copy of the packet. When this acknowledgment is received by a node, the node stops sending copies of the packet that has been acknowledged.

The scheme described above still has the disadvantage of producing substantial additional traffic. Another possibility is that nodes number sequentially packets they send. They then retransmit only packets they have not retransmitted previously. They recognize these packets by looking at the source address and the sequence number. In this scheme, routing by flooding becomes less simple as intermediate nodes must keep track of the packets they retransmit.

Routing by controlled flooding remains definitively the scheme that dams the flooding while keeping the advantages. It has been systematically studied by Lesser and Rom (7). Controlled flooding adds controls to the pure flooding in order to limit the extent of the flood.

Controlled flooding adds two basic controls to flooding. The first is the wealth of the message, which is assigned to the message by the source node. The second is the weight assigned to every link of the network. In order to traverse a link, a message must have a wealth equal to or greater than the weight of the link. When a message does traverse the link, its wealth is reduced by the link weight.

These two controls can clearly dam the flooding. By assigning, for instance, a wealth equal to the sum of the weight of all the links on the shortest path to a given destination, the destination will receive only one packet as all the other replicated ones will be discarded before they reach it. On the other hand, if the wealth is equal to the sum of the weight of all links on the longest path to a given destination, the destination will receive all copies.

The key issue is therefore the assignment of the two controls as different controls result in different routing patterns. If we assume that all links of the hypothetical network of Fig. 1 have the same weight of 1, by assigning 2 as the wealth only the packet following the shortest path between the Source S and the Destination T will reach the destination. On the other hand, if we assign 4 as the wealth, all replicates will reach the destination.

Table-Free Routing in Networks with Regular Topologies

If a telecommunications network has a highly regular topology, it is possible for any source node to compute the position of any destination node without having to resort to a routing table. There are many different techniques for routing in networks with highly regular topologies. In this article, Manhattan street networks (MSNs) are used to illustrate how routing is done in networks with highly regular topologies.

Figure 3 shows a simplified MSN. An MSN has unidirectional links arranged in a structure that resembles the streets and avenues in Manhattan. A natural addressing scheme for such a network is the use of the row and the column numbers as shown by Fig. 3. Besides this integer addressing scheme, a fractional addressing scheme has also been introduced by Maxemchuk (8).

The fractional addressing has two key advantages over the integer addressing:

1. It allows the additions of new rows and columns without changing the addresses of existing nodes.

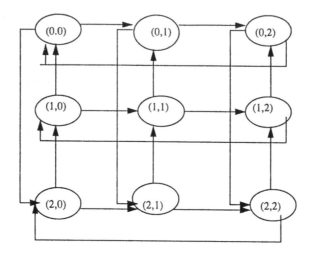

FIG. 3   A simplified nine-node Manhattan street network.

2.   It allows the nodes to make routing decisions without knowing the number of rows and columns in the network.

   A precise set of rules has been specified for ensuring that packets are routed on the shortest paths. The rules operate on the destination address of the incoming packet and the address of the node making the routing decision. The two addresses are first transformed with the assumption that the destination is located at the center of the network. For more details, the reader can consult Ref. 8.

Other Table-Free Routing Techniques

Among the other table-free routing techniques, two are worth mentioning: random routing and source routing.

Random Routing

In random routing, when a source node wants to send a packet to a destination node, the packet is first sent to a neighbor that has been randomly selected. On receipt of the packet, the receiving node randomly selects one of its neighbors (excluding the node from which the packet has been received) and forwards the packet to it. This process is reiterated until the destination is reached.
   Random routing is very easy to implement. The packet always reaches the destination, assuming of course that there is no link failure in the network. Random routing also avoids packet duplications since each time only one copy of the packet is sent. It, however, leads to long packet delays as the packets do not necessarily follow the shortest route.

Source Routing

In source routing, the route to follow from a source node to a destination node is determined by the source node. There is no need for routing tables as a packet carries in its header the complete list of nodes to cross in order to reach the destination. As an example, a packet sent by Node S to Node T (Fig. 1) will carry in its header SDT. Whenever any intermediate node receives it, it will forward it to the next node in the list.

The key issue in source routing is the computing of the list of nodes to be crossed on the way to the destination. This computing can be done in either a centralized or distributed manner. The centralized approach is used below to illustrate how the computing is done.

When the centralized approach is followed, the network is provided with a path server. This path server continuously collects routing information from all the nodes in the network. The information is used to compute continuously the shortest path between any two nodes of the network. When any node of the network wants to send a packet to a given destination, it gets the list of intermediate nodes from the path server and inserts it in the packet header.

Routing in Telecommunications Networks: Two Real-Life Examples

The two most popular telecommunications networks are certainly the public telephone system and the Internet. In this section, we use them to illustrate how routing is done in real-life telecommunications networks.

Routing in Public Telephone Systems

The telephone system was invented by Alexander Graham Bell in 1877. At the very beginning, every telephone was connected to every other telephone. There was no routing problem to solve at that time, but the system evolved very rapidly. It first evolved to a model in which all telephones were connected to a central switch. The routing decision was then taken by the central switch and consisted of merely connecting the calling telephone to the called telephone.

Routing became critical in public telephone systems when telephone switches started springing up everywhere. It was then possible to select the route to use for any given call since each call could be routed on different possible routes. Routing in public telephone networks is table driven.

The tables are stored at each switch and indicate the next switch on the way to the switch to which the called telephone is connected. We review below the different table-driven routing strategies used in public telephone systems. For a comprehensive overview of the public telephone system, the reader can consult Ref. 9.

Static Hierarchical Routing

Static hierarchical routing remains the prevalent routing strategy in public telephone systems, although the progressive introduction of dynamic schemes dates
to the late 1980s. In hierarchical routing, the switches are hierarchically clustered. We start by presenting the general principles of static hierarchical routing.
The principles are then illustrated by the specific case of the U.S. network. The
routing tables are finally studied.

In the simplest form of static hierarchical routing:

- There is no connection between switches at the same hierarchical level, except for the switches at the highest level of the hierarchy.
- A switch at any given level of the hierarchy, except the highest level, is
  connected to only one switch of the next hierarchical level.
- A switch at any given level of the hierarchy, except the lowest level, is
  connected to many switches of the previous level.

The simplest form of hierarchical routing is seldom found in real life. More
elaborate forms are rather common. In these elaborate forms, depending on the
traffic volume, there can be a direct connection between switches at the same
hierarchical level, and a switch at a given hierarchical level can also be connected
to more than one switch of the next hierarchical level.

Figure 4 shows a simplified version of how static hierarchical routing is done
in the U.S. network. There are five hierarchical levels. The switches at the lowest
level, meaning the switches to which the telephones are connected, are known
as Class 5 or end-office switches. The Class 4 or toll office switches are next,
followed successively by the Class 3 or primary switches, the Class 2 or sectional
switches, and the Class 1 or regional switches.

The final route is the only route that would have been used in the simplest
form of static hierarchical routing. However, as can be seen in Fig. 4, high-usage
routes have been added. High-usage routes connect any two switches that have
sufficient traffic between them to make a direct route economical.

The content of routing tables in static hierarchical routing is stored in the
switches at systems-generation time and does not change during the normal
operation of the public telephone systems. For each destination switch that can
be reached, the next possible hops are stored. The route type is also stored,
meaning final or high usage.

The switch uses well-defined rules to decide which type of route to route the
traffic. Final routes, for instance, are designed to handle their own direct traffic
and also overflow traffic from the high-usage routes. It should be noted that
faults such as call looping seldom happen in hierarchical static routing due to
the hierarchical structure of the network. Furthermore, the amount of switches
to be crossed on the way to any given destination is limited by the number of
hierarchies in the network.

Class

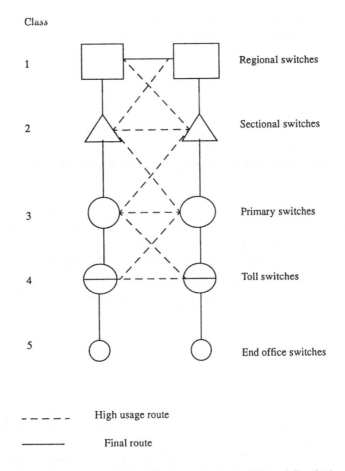

FIG. 4   A simplified version of hierarchical routing in the U.S. public telephone system.

Dynamic Routing

The key difference between dynamic routing and static hierarchical routing is that, in dynamic routing, the content of the routing tables changes. The changes can occur in a preplanned manner or even on line in real time. Dynamic routing has several advantages over static hierarchical routing. The most important is that it increases network utilization by varying routing patterns in accordance with traffic patterns.

There are two main types of dynamic traffic routing: time-sensitive dynamic traffic routing and real-time dynamic traffic routing. In time-sensitive dynamic traffic routing, the content of the routing tables is changed several times a day, typically every hour, to respond to known traffic shifts. The changes are preprogrammed and done without manual intervention. Reprogramming of the changes is typically done once every week.

In real-time dynamic traffic routing, the changes to the routing tables do not rely exclusively on precalculated routing patterns. Switches can sense immediate traffic load and, if necessary, initiate changes. In this case, the content of the routing tables may change many times every hour.

The Internet

The Internet grew out of ARPANET, a network funded by the U.S. Defense Advanced Research Projects Agency (DARPA). When it came into being in the 1970s, it initially connected a small number of computer networks and the associated hosts. It now connects a very large number of computer networks all over the world, and the combined network made by these interconnected computer networks is now known as the Internet.

Routing on the Internet is very interesting because many different techniques are used. Most of these techniques are table driven, but a table-free technique (i.e., flooding) is also used. In order to grasp how routing is done on the Internet, it is of prime importance to understand the Internet's key principles and architecture. These basic principles and architecture are presented first. As the architecture relies on the concept of an autonomous system, the intra-autonomous system routing and the interautonomous system routing are subsequently presented.

Key Principles

There are two key principles behind the Internet. The first one is the end-to-end control. This principle implies that very little control is done inside the network. It is rather up to the two ends exchanging packets to implement mechanisms such as packet-loss detection if deemed necessary. A consequence of this principle is that the rules for routing packets on the Internet are relatively simple.

The second principle is "Internet Protocol (IP) over everything." The IP is the routing protocol used on the Internet. For a survey, the reader can consult Ref. 10. In order to allow any and every computer network to connect to the Internet, the IP has been designed in such a way that it can be used for routing between any computer network and the rest of the Internet independently of the routing methods in use inside the computer network.

Architecture

From the architecture point of view, the Internet can be seen as a set of autonomous systems. An autonomous system can be loosely defined as a set of routers and networks under the same administration. It can be anything from a router directly attached to a local-area network to a corporate network with multiple local-area networks linked by a corporate backbone. The reader should note that routers are nodes to which no end user is attached. Their sole purpose is routing.

It can be easily guessed from the architecture described above that routing on the Internet is hierarchical. The highest level of the hierarchy is the inter-autonomous system routing. When the packet reaches the autonomous system to which the destination belongs, routing is done at the next level of the hierarchy in order to send the packet to the network to which the destination belongs. When the network is reached, routing is then done at the last level of the hierarchy in order to reach the destination.

Intra-Autonomous System Routing

Routers for routing inside autonomous systems are known as interior gateways. The routing tables in interior gateways are dynamic. They are populated by techniques known as interior gateway protocols (IGPs). Examples of IGPs are the routing information protocol (RIP) and the open shortest path first (OSPF). It should be stated that flooding plays a role in OSPF, although OSPF is mostly table driven.

The RIP belongs to the family of the distance vector protocol. For a comprehensive introduction to the distance vector protocol, the reader can consult Ref. 11. The RIP, as all distance vector protocols, allows a distributed computation of the shortest path between a given source node and any destination.

In the RIP, the routing table in any source node is empty when the node is powered up. It is then progressively built and constantly updated using information the neighbors send periodically. The information periodically sent by each neighbor includes the distance between the neighbor and other node(s) of the network.

The OSPF is much more complex than the RIP. In the OSPF, each node of the network maintains a database with information on all the other nodes of the network. The node constantly computes the shortest distance to any and every node that can be reached, using the information in the database in conjunction with the Dijkstra's algorithm. When a link goes up or down in the network, the router responsible for the link uses flooding to send the information throughout the network. All the nodes can then update their database to reflect the change.

Inter-Autonomous System Routing

Some of the routers of an autonomous system are selected by the administration of the system to act as routers for intersystem routing. They are known as external gateways and are entry points in adjacent autonomous systems. Inter-autonomous system routing is table driven. The tables are populated using techniques known as exterior gateway protocols (EGPs).

When first initialized, each exterior gateway receives the content of a routing table known as a reachability table. This table contains a list of potential neighbor exterior gateways. The potential neighbor exterior gateways are, of course, located in other autonomous systems. Three distinct procedures then follow: neighbor acquisition, neighbor reachability, and network reachability.

During the neighbor acquisition procedure, the exterior gateway exchanges information with its potential neighbors to find out whether they agree to act as neighbors. This agreement implies, among other things, that the neighbor has the soliciting external gateway listed in its reachability table as a potential neighbor. The potential neighbors that do not agree will be taken out of the exterior gateway reachability table.

The neighbor reachability procedure consists of constantly monitoring the links between the exterior gateway and the neighbor gateways with which neighboring agreements have been reached. The network reachability procedure, on the other hand, allows the exterior gateway to exchange reachability information with its neighbors.

The reachability information sent by an exterior gateway to a neighbor is the list of networks that can be reached through the exterior gateway. It should be stressed that these networks and the exterior gateway do not necessarily belong to the same autonomous system. An exterior gateway relays the reachability information received from its neighbors to all gateways belonging to its autonomous systems. The gateways use the information to update their routing tables.

Summary

Routing plays a vital role in telecommunications networks. Many routing techniques have been designed over the years. These techniques are classified using various schemes. In this article, we reviewed the routing techniques using a routing-table-based scheme. According to that scheme, routing is either table driven or table free. We have illustrated the techniques by two real-life examples: the public telephone system and the Internet.

In table-driven routing, nodes consult tables in order to determine, for packets to be sent, the next hop(s) on the way to the destination. These tables are most of the time populated using Dijkstra's algorithm and its variants. Thanks to this algorithm and its variants, packets between any given source and any given destination follow the shortest path between the destination and the source. As routing tables' sizes increase with network size, clustering techniques are used in large networks in order to reduce the size of the tables.

In table-free routing, nodes can determine the next hop(s) on the way to destinations without the support of tables. The most popular table-free method is certainly flooding. In pure flooding, nodes simply relay packets to all their neighbors. Pure flooding has evolved to methods by which a certain level of control is exerted. Besides flooding, table-free routing includes techniques such as random routing, source routing, and also techniques for routing in networks with highly regular topologies.

Routing in the public telephone systems is table driven. It evolved from a highly rigid structure in which the content of the tables is static, stored at system generation, to a highly flexible structure in which the content of the tables changes many times every hour. Routing in the Internet is also table driven. However, one of the Internet routing algorithms, the open shortest path first, combines table-driven routing with a form of flooding, showing that it is possible to get the best of two worlds.

Bibliography

Ash, G. R. (ed.), Dynamic Routing in Telecommunications Networks, *IEEE Commun. Mag.*, special issue, 33(7) (July 1995).

Barensel, C., Dobosiewicz, W., and Gburzynski, P., Routing in Multihop Packet Switching Networks: Gb/s Challenge, *IEEE Network*, 9(3):38–61 (May–June 1995).

Huitema, C., *Routing in Internet*, Prentice-Hall, Englewood Cliffs, NJ, 1995.

References

1. Black, U. D., *Data Networks, Concepts, Theory and Practice*, Prentice-Hall, Englewood Cliffs, NJ, 1989.
2. Tanembaum, A. S., *Computer Networks*, Prentice-Hall, Englewood Cliffs, NJ, 1996.
3. Roberts, L. G., Telenet Principles and Practice, *Communications Networks*, 315–329 (September 1975).
4. Dijkstra, E. W., A Note on Two Problems in Connexion with Graphs, *Numerishe Mathematik*, 1:269–271 (1959).
5. Kleinrock, L., and Kamoun, F., Hierarchical Routing for Large Networks — Performance Evaluation and Optimization, *Computer Networks*, 1:155–174 (1977).
6. Kleinrock, L., and Kamoun, F., Stochastic Performance Evaluation of Hierarchical Routing for Large Networks, *Computer Networks*, 1:155–174 (1977).
7. Lesser, O., and Rom, R., Routing by Controlled Flooding in Communications Networks, *INFOCOM*, 170–179 (August 1989).
8. Maxemchuk, N. F., Routing in the Manhattan Street Network, *IEEE Trans. Commun.*, COM-35(5) (May 1987).
9. Bellamy, Y. J., *Digital Telephony*, John Wiley, New York, 1991.
10. Stallings, W., IPv6: The New Internet Protocol, *IEEE Commun. Mag.*, 34(7) (July 1996).
11. Ford, L. R., and Fulkerson, D. R., *Flows in Networks*, Princeton University Press, Princeton, NJ, 1962.

ROCH H. GLITHO

RS-232 Interface (see Binary Serial Data Interchange: EIA-232-D Standard)

Sampling Theorem (see Electrical Filters: Fundamentals and System Applications)

Sarnoff, David

David Sarnoff was born February 27, 1891, in Uzlian, a Jewish stetl (village) in Minsk province, Russia. His father Abraham Sarnoff emigrated to America in 1896 to find some way to support his family; David and the rest of the family followed in 1900.

Arriving in New York City to find his father sick and the family with little, if any source of income, he later recalled asking himself, "If I don't help my family, who will?"

He went to work hawking newspapers, which added a small, but regular amount to the family till. He also sold snacks in theaters, ran errands for the butcher, and, on special occasions, sang in the synagogue—all of which contributed to his income. It also set the pattern for a long life of hard work and self-reliance, which contributed to his later success.

Young Sarnoff soon decided he could do better in the newspaper business with a base, and from somewhere—tradition says a friendly social worker—promoted the $200 necessary to purchase the newsstand at the corner of 46th Street and 10th Avenue. This work soon increased his income, but it put an end to his formal education after elementary school.

In 1906 Sarnoff decided to try for a position as messenger for a newspaper, but walked into the wrong office and ended up as messenger for the Commercial Cable Company, one of the first undersea cable companies. Although he did not keep the job long, it began his lifelong interest in telegraphy. Among his first purchases was a telegraph key, on which he practiced until he gained fluency. A contact from his amateur telegraphy introduced him to the Marconi Wireless Telegraph Company of America as a junior operator and office boy.

In 1901 Guglielmo Marconi first demonstrated the possibilities of wireless telegraphy (1); he did so again in 1903 to a delighted President Theodore Roosevelt, but the technology remained untried commercially. Marconi was eagerly trying to promote the idea of wireless telegraphy for ship-to-shore and ship-to-ship communications. Some maritime lines hired his operators, and Sarnoff took advantage of another operator's illness to work a stint on a ship, as well as filling in for the landside operators whenever they would let him.

The incident that made Sarnoff's name and Marconi's fortune occurred April 14, 1912—the sinking of the *Titanic*. Although it is unlikely that Sarnoff was at his post at the time of the first news, it did not take him long to get there. For 72 hours he occupied the post, copying the names of the survivors and other communications from the Cunard liner *Carpathia*, the first ship on the scene of the disaster. This incident proved the value of ship-to-shore wireless communications for news. It also proved beyond the shadow of a doubt the value of wireless communications for safety and led to laws requiring wireless equipment and trained operators on every ship.

Sarnoff soon realized that he would never make himself rich as a telegrapher. Although there is little doubt that money was not an end in itself to him, he realized its potential for improving his life and the lives of those around him. In an epochal meeting with some friends in 1913, he summed up the situation, "The place to make money is where the money is coming in," and forswore the life of the engineer for that of the salesperson.

As a salesperson for Marconi, he was able to visit his competition, compare their equipment with his, and learn their sales tricks and add them to his. He also maintained friendly relationships with operators from all over, whose advice and complaints he heeded and passed on as he could.

As he rose to junior levels of management, Sarnoff's advice was more and more frequently taken seriously. But in 1916 he wrote a memo to his supervisor, Edward Nally, that older and wiser heads pigeonholed as a crazy idea — a radio music box. In the memo, Sarnoff outlined a plan for the creation of wireless music services and pointed out that there was a huge audience for such services within a 50-mile radius of New York City. He also pointed out the usefulness of such a service for disseminating news, baseball scores, lectures, and so on. The service Sarnoff outlined in 1916 was, with only slight differences, exactly what we know today as radio. By that time Lee De Forest (2) (who held the contested patents on the audion tube, which made radio reception possible) had demonstrated that voice transmission services were both possible and practicable. Sarnoff's memo even pointed out that if only 1 in 15 American families purchased a proposed $75 receiver, that was nearly $75 million profit for American Marconi — an almost unheard of amount then. But the heads of the corporation had no confidence in the plan and shelved it.

At this point, the life of David Sarnoff became entwined with that of Edwin Howard Armstrong (3), a brilliant young inventor. In 1914, Sarnoff had visited Columbia University to see a demonstration of a new high-powered receiver developed by Armstrong. Impressed by it, Sarnoff championed its use by Marconi and eventually persuaded his bosses to adopt it. Armstrong became a frequent caller at Sarnoff's home and office, and eventually married Sarnoff's secretary. Armstrong and Sarnoff were friends for a time, but years later they became involved in a decade-spanning lawsuit.

Another influence that came into Sarnoff's life at this time was Doctor Alfred Goldsmith, a consultant to American Marconi and a genius inventor with more than 122 patents. Sarnoff's brilliance lay in his ability to manage the eccentricities of his inventive friends and in converting their ideas into profitable uses for the corporation.

World War I was the lever that created the Radio Corporation of America (eventually known as RCA), one of the corporate giants of our time, from American Marconi, a division of a British company. American Marconi eagerly submitted itself to Navy jurisdiction during the war as the Navy took control of all broadcasting and shut down all the amateurs. After the war, it seemed inconceivable that Britain should maintain control of the world's communications, so an effort was organized to break apart the Marconi companies. General Electric Corporation, at the behest of the U.S. government, purchased the entire stock of American Marconi from its British parent and created the Radio Corporation of America on December 1, 1919. RCA was to be a patriotic enterprise, with no thought (at least at first) of profit, but was to ensure that American inventions would be used by an American corporation to control America's airwaves. AT&T received a large block of RCA stock in return for its patents, as did Westinghouse. An odd fourth partner was the United Fruit Company, which bought in for $1 million in return for RCA taking over a fledgling telegraph company they had started in Central America.

As commercial manager, Sarnoff found himself in effective control of this patchwork corporation. He soon became general manager, and his presence was more and more necessary to defuse the feuds that constantly developed among the corporate partners over the direction in which RCA should proceed. As general manager of RCA, his salary was raised to $15,000 a year—a far cry from the $5.50 a week at which the messenger boy had started. Sarnoff consistently proved himself a good manager and a linchpin of the company. At first the company's efforts remained focused on transoceanic telegraphy, but Sarnoff had not forgotten his 1916 memo. By 1920 he had developed his radio music box idea still further.

After the war, Fred Cooper, an engineer for Westinghouse, had resumed his amateur broadcasting from station 8XK. A store in Pittsburgh, Pennsylvania, soon saw the possibilities and placed an advertisement for amateur radio sets so people could hear what was being broadcast. Another such station, 8MK, soon was set up in Detroit, Michigan, to broadcast election returns, news, and music. At this point, a Westinghouse executive named H. P. Davis saw the store's ad and put two and two together, foreseeing limitless opportunities. At Davis' urging, Cooper expanded his broadcasting. Station 8XK became KDKA, the country's first commercial radio station, which is still in operation to this day.

Sarnoff held his tongue and temper at having been "scooped." He went along with RCA's plans to manufacture radio receivers, but he wanted to get into broadcasting as well. He arranged for the broadcast of the Jack Dempsey/Georges Carpentier 1921 heavyweight prizefight, using new equipment designed by General Electric. Luck was with him. Only moments after Dempsey's fourth-round knockout of Carpentier ended the fight, the equipment literally melted into what Dr. Goldsmith described as a molten mass.

Westinghouse now established two new stations, including WJZ in Newark, New Jersey. Sarnoff tried briefly to compete with WJZ, then gave up and purchased the Newark station, shut it down, and moved it, after much struggle with his bosses, into Aeolian Hall in New York City. Between 1923 and 1926, WJZ and its twin WJY operated there, in full public view. A number of skilled engineers, along with announcer Milton Cross, for so many years familiar to American listeners as the voice of the Metropolitan Opera, came with WJZ from Newark.

At this point the competition began to come from a new direction. Long dissatisfied with its deal in the creation of RCA, AT&T went its own way, opening its own radio station, WEAF. This put RCA at a disadvantage because it could no longer use telephone lines to broadcast from outside the station and had to rely on less technically perfect Western Union lines. Another new idea first tried by WEAF was paid advertising, which also allowed WEAF to pay its entertainers, a practice not followed by Sarnoff's station until much later.

Sarnoff was able to pressure phone company executives into leasing lines for special events, but that backfired occasionally, as at a broadcast of a speech by British Prime Minister David Lloyd-George. The phone lines delivered flat, but recognizable speech. By the time it was broadcast, it was unintelligible, and many regular listeners called to complain. Sarnoff's legendary wrath was aroused, and at a meeting he assailed all his top managers about the failure. Carl Dreher, later one of Sarnoff's biographers, was one of the engineers of the event; he reports that, feeling he had nothing to lose, he offered no excuses, but merely pointed out that they had ventured into unknown territory. Now the territory was known, Dreher pointed out, and there could be no repetition of the event. Sarnoff said, "That was a good speech," and dismissed the meeting. He was not vindictive, says Dreher in his biography of Sarnoff, only efficient (4).

At the beginning of 1922, hostilities appeared between the radio and telephone sides of RCA. An agreement made in January 1921 had smoothed over the question of patent rights among the four companies founding RCA, but these founders had never envisioned the creation of the radio broadcasting business. By 1924 RCA was selling $50 million a year worth of receivers as part of a $358 million a year industry—larger now than its giant founder General Electric's other businesses together. The 1921 contract allowed RCA to sell only to "amateurs." At that time there was no "public." The question hinged on whether the public would be considered amateurs under the meaning of the contract.

Even then, Sarnoff's creativity was working. While radio was still a local business, Sarnoff had plans for a national broadcasting empire. He realized the professional aspects of broadcasting as entertainment called for a different set of skills than those possessed by the engineers who had created the field. He felt the time was right for a national solution, and he created it—NBC, the National Broadcasting Company.

Although a judge had ruled in favor of RCA in almost all of its disputes with its parent corporations, AT&T was still fighting the ruling. Sarnoff's plan for a national network, which included some specific concessions to AT&T regarding land links, seemed a perfect way for them to bow gracefully to the inevitable. Sarnoff, too, had to give up something. He was forced to abandon his notion that radio could be operated as a charitable operation. In 1926 NBC was created as the manager of two networks—the Blue network, which included WJZ, and the Red network, headed by WEAF, for which RCA had paid AT&T $1 million as part of the settlement.

Sarnoff never stopped planning and dreaming. He also never forgot anything he had seen. In the wireless telegraphy days, he had been offered a device, designed by the General Electric engineers, for putting the dots and dashes on film. He had rejected it for a more efficient ink-to-tape device designed by an in-house consultant. Now he saw a good use for it—adding voices to motion pictures. So RCA went to Hollywood and set up the Photophone Corporation.

As usual, top management staff with which Sarnoff dealt chose to favor his rival, this time Electronic Research Products, Incorporated, an arm of Western Electric. It was a wise decision, as it turned out, as the rivalry between General Electric and Westinghouse prevented any effective development of equipment by either, while Western Electric was all ready to begin.

Sarnoff was forced to start his own studio to use the Photophone process. In October 1928 he met with financier Joseph Kennedy, and the two created RKO Radio Pictures, Incorporated, by joining RCA with the Keith-Albee-Orpheum theaters.

In 1929, Owen Young, Sarnoff's boss at General Electric, was asked by U.S. President Coolidge to head a commission to meet with the German government and discuss revising the reparation plans made at the end of World War I. Young, impressed with Sarnoff's negotiating skills, brought him along. Sarnoff ended up doing most of the negotiating, and the treaty he worked out with Dr. Hjalmar Schacht won him a great deal of recognition.

At the end of 1929, Sarnoff was finally made president of RCA. In that capacity he oversaw one of his greatest triumphs, the merger between RCA and the Victor company on December 26, 1929. This merger finally gave RCA the manufacturing facility it needed to begin shaking off the yoke of General Electric and Westinghouse.

In May 1930, the federal government, which had called originally for the creation of RCA, now decided that the entire process had been illegal. The Justice Department sued General Electric, RCA, and all the other partners in a giant antitrust action. AT&T, in an attempt to make a separate peace with the government, exercised its contractual right to withdraw from the combination. Sarnoff's skills as a negotiator were called on, and he rose to the challenge. On November 21, 1932, two weeks before the trial was to begin, a consent decree was signed. Sarnoff had gotten almost everything he wanted for RCA in the exchange. The electrical companies withdrew from RCA's board and distributed their stock to their shareholders, leaving Sarnoff in full control of RCA for the first time. On the downside, the company accepted almost $18 million in additional debt, which eventually forced it to sell RKO.

The agreement also cleared the way for another great Sarnoff dream to become reality—television. The technology behind picture transmission was not new, having been first achieved by the German inventor Paul Nipkow in 1884, but the technology remained undeveloped for a long time. Sarnoff was aided in its development by Vladimir Zworykin, a brilliant inventor who immigrated to the United States in 1919 and went to work for Westinghouse. In 1923 Zworykin had applied for a patent for his "iconoscope," the pattern for a working television camera. Westinghouse proved uninterested in pursuing that line of research, so in 1929 Zworykin brought his ideas to RCA. He told an enthusiastic Sarnoff the system could be built and completed in a few months for $100,000. In fact, it took 10 years and more than $10 million. But it took a visionary of Sarnoff's caliber to authorize and back that expenditure during the depression, with profits minuscule for 6 of those 10 years and with a loss of more than $1 million in 1932.

RCA began limited television broadcasting at the time of the 1939 World's Fair. Standing before his iconoscopic camera at the opening of the World's Fair in New York City in April 1939, Sarnoff announced "a new art so important in its implications it is bound to affect all society" (5). He was a little bit premature. World War II intervened to delay further development of television.

Sarnoff, prescient as he often was, foresaw the coming of war and was a little bit ahead of the rest of us. When the Japanese attacked Pearl Harbor, he cabled immediately to President Roosevelt that RCA was "ready and at your instant service" (4). That service was called on, and RCA produced untold amounts of war materiel. Sarnoff also was called to service, although he spent only about a year in uniform.

Much of his tour of duty was spent in Washington, D.C., as a troubleshooter for the Signal Corps. When General Eisenhower asked for a top communications expert to be sent to him in London to oversee military communications and handle all the publicity and press arrangements for the coming invasion, Sarnoff was sent. As usual Sarnoff managed to make it all happen, even increasing the westbound transmission capacity sixfold to handle the expected D day news traffic.

The war may have taken up Sarnoff's time, but it never pulled his attention completely away from business. Lt. Col. Walter Brown, an RCA employee who also served under Sarnoff in London, recalls that at a dinner given by NBC's London manager for Sarnoff, he asked if there would be color TV in RCA's future. Sarnoff replied that color TV would be RCA's highest priority—before they had even established black and white TV as a going business (6).

By 1946 RCA was ready to go into the TV business. The first cables were laid for NBC's new network of TV stations, and RCA's manufacturing facilities were ready to handle the pent-up postwar demand. By 1950 more than seven million homes had TV sets: Sarnoff had recovered RCA's investment with a profit, and TV was a going and growing concern.

The rest of Sarnoff's career consisted of continuing to build RCA. As with any major company, there were triumphs and there were setbacks. RCA could not, for instance, ever make headway in either the home appliance or the computer industries; both proved expensive debacles. Sarnoff dealt with strikes and labor unrest the same way he met other problems—lay down the law when you can, negotiate when you have to.

Sarnoff's political life was not as independent as his business life. His naturally conservative tendencies made him an outspoken anti-Communist in the 1950s. He became friends with Richard Nixon and used his presidency at NBC to direct their policies in questionable directions during the red-baiting scares of that era. Seeing the failure of his hard-line cold war stance, he modified his positions somewhat, but received encouragement for his work from newly inaugurated President John F. Kennedy, who said he would welcome advice should Sarnoff care to give it.

Sarnoff's personal life was as quiet and unassuming as his business life was flamboyant. On July 4, 1917, Sarnoff married Lizette Hermant, a French Jewish girl from the Bronx, after a long courtship. They had three sons: Robert (1918), Edward (1921), and Thomas (1927). They lived comfortably but unostentatiously, first in a series of apartments, then, in 1937, in a house on East 71st Street where they spent the rest of their lives.

The one major disaster of Sarnoff's life was the legal entanglement with his old acquaintance E. Howard Armstrong. Among Armstrong's greatest developments was the perfection of frequency modulation (FM) for radio transmission. Armstrong received a basic patent for FM in 1933. While recognizing FM's superior quality, Sarnoff never cared to pursue it, feeling that TV was more important, indeed, that TV might supplant radio entirely. Armstrong was

forced to build his own radio station and faced constant litigation with companies, including RCA, which manufactured FM receivers without paying royal-
ties. Armstrong also sued RCA and NBC for allegedly influencing the Federal
Communications Commission (FCC) to obstruct issuing of FM broadcast licenses. Litigation continued, to Armstrong's financial ruin, until the end of
January 1954, when Armstrong committed suicide. Outraged at what she felt
was Armstrong's mistreatment at the hands of Sarnoff, Armstrong's widow
(Sarnoff's former secretary) continued with the suit for 13 years, rejecting settlement offers, until the suit was finally settled in Armstrong's favor.

In 1967 Sarnoff was honored with the opening of the David Sarnoff Library,
an adjunct of the RCA/David Sarnoff Research Center at Princeton, New
Jersey. In 1969 he retired as chairman of the board and president of RCA and
was succeeded by his son Robert.

David Sarnoff died December 12, 1971.

Sarnoff's career and ambitions might best be described as single minded.
Gifted with remarkably quick understanding and strong managerial skills, he
chose to use those gifts to pursue his company's goals. Had he chosen to become
wealthy, he might have built an astonishing fortune using his connections. Instead, although well paid, he chose to devote himself to improving the means of
communication he felt would benefit all humankind and to do that in a practical
way—by making them profitable.

While he did not invent radio, TV, or sound films, his contribution to
those industries was the irreplaceable impetus that kept people working on and
perfecting them. So doing, he left only a comparatively small fortune, but a
mighty monument.

Bibliography

Lyons, E., *David Sarnoff, a Biography*, Harper and Row, New York, 1966.
Sarnoff, D., *Looking Ahead, the Papers of David Sarnoff*, McGraw-Hill, New York,
  1968.

References

1. Bedi, J. E., Marconi, Guglielmo. In *The Froehlich/Kent Encyclopedia of Telecommunications*, Vol. 10 (F. E. Froehlich and A. Kent, eds.), Marcel Dekker, New
   York, 1995, pp. 361–368.
2. Dawson, D. K., de Forest, Lee. In *The Froehlich/Kent Encyclopedia of Telecommunications*, Vol. 5 (F. E. Froehlich and A. Kent, eds.), Marcel Dekker, New York,
   1993, pp. 285–289.
3. Armstrong, Edwin Howard. In *The Froehlich/Kent Encyclopedia of Telecommunications*, Vol. 1 (F. E. Froehlich and A. Kent, eds.), Marcel Dekker, New York,
   1991, pp. 335–340.

4. Dreher, C., *Sarnoff, An American Success*, Quadrangle/New York Times Book Company, New York, 1977.
5. *The Birth of an Industry* (as quoted in *David Sarnoff, a Biography* by E. Lyons).

    Text of address by David Sarnoff at the World's Fair, Flushing Meadows, New York, April 20, 1939.
6. Personal communication, Lt. Col. AUS (Ret.) W. R. Brown to C. Dreher (as quoted in Ref. 4).

MARC DAVIDSON

Satellite Communications
(see Overview of Satellite Communications)

Security of the Internet

Overview of Internet Security

As of 1996, the Internet connected an estimated 13 million computers in 195 countries on every continent, even Antarctica (1). The Internet is not a single network, but a worldwide collection of loosely connected networks that are accessible by individual computer hosts in a variety of ways, including gateways, routers, dial-up connections, and Internet service providers. The Internet is easily accessible to anyone with a computer and a network connection. Individuals and organizations worldwide can reach any point on the network without regard to national or geographic boundaries or time of day.

However, along with the convenience and easy access to information come new risks. Among them are the risks that valuable information will be lost, stolen, corrupted, or misused, and that the computer systems will be corrupted. If information is recorded electronically and is available on networked computers, it is more vulnerable than if the same information is printed on paper and locked in a file cabinet. Intruders do not need to enter an office or home and may not even be in the same country. They can steal or tamper with information without touching a piece of paper or a photocopier. They can create new electronic files, run their own programs, and hide evidence of their unauthorized activity.

Basic Security Concepts

Three basic security concepts important to information on the Internet are confidentiality, integrity, and availability. Concepts relating to the people who use that information are authentication, authorization, and nonrepudiation.

When information is read or copied by someone not authorized to do so, the result is known as *loss of confidentiality*. For some types of information, confidentiality is a very important attribute. Examples include research data, medical and insurance records, new product specifications, and corporate investment strategies. In some locations, there may be a legal obligation to protect the privacy of individuals. This is particularly true for banks and loan companies; debt collectors; businesses that extend credit to their customers or issue credit cards; hospitals, doctors' offices, and medical testing laboratories; individuals or agencies that offer services such as psychological counseling or drug treatment; and agencies that collect taxes.

Information can be corrupted when it is available on an insecure network. When information is modified in unexpected ways, the result is known as *loss of integrity*. This means that unauthorized changes are made to information, whether by human error or intentional tampering. Integrity is particularly important for critical safety and financial data used for activities such as electronic funds transfers, air traffic control, and financial accounting.

Information can be erased or become inaccessible, resulting in *loss of availability*. This means that people who are authorized to get information cannot get what they need. Availability is often the most important attribute in service-oriented businesses that depend on information (e.g., airline schedules and on-line inventory systems). Availability of the network itself is important to anyone whose business or education relies on a network connection. When a user cannot get access to the network or specific services provided on the network, they experience a *denial of service*.

To make information available to those who need it and who can be trusted with it, organizations use authentication and authorization. *Authentication* is proving that a user is whom he or she claims to be. That proof may involve something the user knows (such as a password), something the user has (such as a "smart card"), or something about the user that proves the person's identity (such as a fingerprint). *Authorization* is the act of determining whether a particular user (or computer system) has the right to carry out a certain activity, such as reading a file or running a program. Authentication and authorization go hand in hand. Users must be authenticated before carrying out the activity they are authorized to perform. Security is strong when the means of authentication cannot later be refuted—the user cannot later deny that he or she performed the activity. This is known as *nonrepudiation*.

Why Care About Security?

It is remarkably easy to gain unauthorized access to information in an insecure networked environment, and it is hard to catch the intruders. Even if users have nothing stored on their computer that they consider important, that computer can be a "weak link," allowing unauthorized access to the organization's systems and information.

Seemingly innocuous information can expose a computer system to compromise. Information that intruders find useful includes which hardware and software are being used, system configuration, type of network connections, phone numbers, and access and authentication procedures. Security-related information can enable unauthorized individuals to get access to important files and programs, thus compromising the security of the system. Examples of important information are passwords, access control files and keys, personnel information, and encryption algorithms.

Judging from CERT® Coordination Center (CERT/CC) data and the computer abuse reported in the media, no one on the Internet is immune. Those affected include banks and financial companies, insurance companies, brokerage houses, consultants, government contractors, government agencies, hospitals and medical laboratories, network service providers, utility companies, the textile business, universities, and wholesale and retail trades.

The consequences of a break in cover a broad range of possibilities: a minor loss of time in recovering from the problem, a decrease in productivity, a significant loss of money or staff hours, a devastating loss of credibility or market opportunity, a business no longer able to compete, legal liability, and the loss of life.

History

The Internet began in 1969 as the ARPANET, a project funded by the Advanced Research Projects Agency (ARPA) of the U.S. Department of Defense. One of the original goals of the project was to create a network that would continue to function even if major sections of the network failed or were attacked. The ARPANET was designed to reroute network traffic automatically around problems in connecting systems or in passing along the necessary information to keep the network functioning. Thus, from the beginning, the Internet was designed to be robust against denial-of-service attacks, which are described in a section below on denial of service.

The ARPANET protocols (the rules of syntax that enable computers to communicate on a network) were originally designed for openness and flexibility, not for security. The ARPA researchers needed to share information easily, so everyone needed to be an unrestricted "insider" on the network. Although the approach was appropriate at the time, it is not one that lends itself to today's commercial and government use.

As more locations with computers (known as *sites* in Internet parlance) joined the ARPANET, the usefulness of the network grew. The ARPANET consisted primarily of university and government computers, and the applications supported on this network were simple: electronic mail (E-mail), electronic news groups, and remote connection to other computers. By 1971, the Internet linked about two dozen research and government sites, and researchers had begun to use it to exchange information not directly related to the ARPANET itself. The network was becoming an important tool for collaborative research.

During these years, researchers also played "practical jokes" on each other using the ARPANET. These jokes usually involved joke messages, annoying messages, and other minor security violations. Some of these are described in Steven Levy's *Hackers: Heroes of the Computer Revolution* (2). It was rare that a connection from a remote system was considered an attack, however, because ARPANET users comprised a small group of people who generally knew and trusted each other.

In 1986, the first well-publicized international security incident was identified by Cliff Stoll, then of Lawrence Berkeley National Laboratory in northern California. A simple accounting error in the computer records of systems connected to the ARPANET led Stoll to uncover an international effort, using the network, to connect to computers in the United States and copy information from them. These U.S. computers were not only at universities, but at military and government sites all over the country. When Stoll published his experience in a 1989 book, *The Cuckoo's Egg* (3), he raised awareness that the ARPANET could be used for destructive purposes.

In 1988, the ARPANET had its first automated network security incident, usually referred to as "the Morris worm" (4). A student at Cornell University (Ithaca, NY), Robert T. Morris, wrote a program that would connect to another computer, find and use one of several vulnerabilities to copy itself to that second computer, and begin to run the copy of itself at the new location. Both the original code and the copy would then repeat these actions in an infinite loop to other computers on the ARPANET. This "self-replicating automated network attack tool" caused a geometric explosion of copies to be started at computers all around the ARPANET. The worm used so many system resources that the attacked computers could no longer function. As a result, 10% of the U.S. computers connected to the ARPANET effectively stopped at about the same time.

By that time, the ARPANET had grown to more than 88,000 computers and was the primary means of communication among network security experts. With the ARPANET effectively down, it was difficult to coordinate a response to the worm. Many sites removed themselves from the ARPANET altogether, further hampering communication and the transmission of the solution that would stop the worm.

The Morris worm prompted the Defense Advanced Research Projects Agency (DARPA, the new name for ARPA) to fund a computer emergency response team, now the CERT® Coordination Center, to give experts a central point for coordinating responses to network emergencies. Other teams quickly sprang up to address computer security incidents in specific organizations or geographic regions. Within a year of their formation, these incident response teams created an informal organization now known as the Forum of Incident Response and Security Teams (FIRST). These teams and the FIRST organization exist to coordinate responses to computer security incidents, assist sites in handling attacks, and educate network users about computer security threats and preventive practices.

In 1989, the ARPANET officially became the Internet and moved from a government research project to an operational network; by then, it had grown to more than 100,000 computers. Security problems continued, with both aggressive and defensive technologies becoming more sophisticated. Among the major security incidents (5) were the 1989 WANK/OILZ worm, an automated attack on VMS systems attached to the Internet, and exploitation of vulnerabilities in widely distributed programs such as the sendmail program, a complicated program commonly found on UNIX™-based systems for sending and receiving electronic mail. In 1994, intruder tools were created to "sniff" packets from the network easily, resulting in the widespread disclosure of user names and password information. In 1995, the method that Internet computers use to name and authenticate each other was exploited by a new set of attack tools that allowed widespread Internet attacks on computers that have trust relationships (see the section on exploitation of trust, below) with any other computer, even one in the same room. Today, the use of the World Wide Web and Web-related programming languages create new opportunities for network attacks.

Although the Internet was originally conceived and designed as a research and education network, usage patterns have radically changed. The Internet has become a home for private and commercial communication, and at this writing it is still expanding into important areas of commerce, medicine, and public service. Increased reliance on the Internet is expected over the next five years, along with increased attention to its security.

Network Security Incidents

A *network security incident* is any network-related activity with negative security implications. This usually means that the activity violates an explicit or implicit security policy (see the section on security policy). Incidents come in all shapes and sizes. They can come from anywhere on the Internet, although some attacks must be launched from specific systems or networks, and some require access to special accounts. An intrusion may be a comparatively minor event involving a single site, or a major event in which tens of thousands of sites are compromised. (When reading accounts of incidents, note that different groups may use different criteria for determining the bounds of an incident.)

A typical attack pattern consists of gaining access to a user's account, gaining privileged access, and using the victim's system as a launch platform for attacks on other sites. It is possible to accomplish all these steps manually in as little as 45 seconds; with automation, the time decreases further.

Sources of Incidents

It is difficult to characterize the people who cause incidents. An intruder may be an adolescent who is curious about what he or she can do on the Internet, a college student who has created a new software tool, an individual seeking personal gain, or a paid "spy" seeking information for the economic advantage of a corporation or foreign country. An incident may also be caused by a disgruntled former employee or a consultant who gained network information while working with a company. An intruder may seek entertainment, intellectual challenge, a sense of power, political attention, or financial gain.

One characteristic of the intruder community as a whole is its communication. There are electronic news groups and print publications on the latest intrusion techniques, as well as conferences on the topic. Intruders identify and publicize misconfigured systems; they use those systems to exchange pirated software, credit card numbers, exploitation programs, and the identity of sites that have been compromised, including account names and passwords. By sharing knowledge and easy-to-use software tools, successful intruders increase their number and their impact.

Types of Incidents

Incidents can be broadly classified into several kinds: the probe, scan, account compromise, root compromise, packet sniffer, denial of service, exploitation of trust, malicious code, and Internet infrastructure attacks.

Probe

A probe is characterized by unusual attempts to gain access to a system or to discover information about the system. One example is an attempt to log in to an unused account. Probing is the electronic equivalent of testing doorknobs to find an unlocked door for easy entry. Probes are sometimes followed by a more serious security event, but they are often the result of curiosity or confusion.

Scan

A scan is simply a large number of probes done using an automated tool. Scans can sometimes be the result of a misconfiguration or other error, but they are often a prelude to a more directed attack on systems that the intruder has found to be vulnerable.

Account Compromise

An account compromise is the unauthorized use of a computer account by someone other than the account owner, without involving system-level or root-level privileges (privileges a system administrator or network manager has). An account compromise might expose the victim to serious data loss, data theft, or theft of services. The lack of root-level access means that the damage can usually be contained, but a user-level account is often an entry point for greater access to the system.

Root Compromise

A root compromise is similar to an account compromise, except that the account that has been compromised has special privileges on the system. The term *root* is derived from an account on UNIX systems that typically has unlimited, or "superuser," privileges. Intruders who succeed in a root compromise can do just about anything on the victim's system, including run their own programs, change how the system works, and hide traces of their intrusion.

Packet Sniffer

A packet sniffer is a program that captures data from information packets as they travel over the network. That data may include user names, passwords, and proprietary information that travels over the network in clear text. With perhaps hundreds or thousands of passwords captured by the sniffer, intruders can launch widespread attacks on systems. Installing a packet sniffer does not necessarily require privileged access. For most multiuser systems, however, the presence of a packet sniffer implies there has been a root compromise.

Denial of Service

The goal of denial-of-service attacks is not to gain unauthorized access to machines or data, but to prevent legitimate users of a service from using it. A denial-of-service attack can come in many forms. Attackers may "flood" a network with large volumes of data or deliberately consume a scarce or limited resource, such as process control blocks or pending network connections. They may also disrupt physical components of the network or manipulate data in transit, including encrypted data.

Exploitation of Trust

Computers on networks often have trust relationships with one another. For example, before executing some commands, the computer checks a set of files that specify which other computers on the network are permitted to use those commands. If attackers can forge their identity, appearing to be using the trusted computer, they may be able to gain unauthorized access to other computers.

Malicious Code

Malicious code is a general term for programs that, when executed, would cause undesired results on a system. Users of the system usually are not aware of the program until they discover the damage. Malicious code includes Trojan horses, viruses, and worms. Trojan horses and viruses are usually hidden in legitimate programs or files that attackers have altered to do more than what is expected. Worms are self-replicating programs that spread with no human intervention after they are started. Viruses are also self-replicating programs, but usually require some action on the part of the user to spread inadvertently to other programs or systems. These sorts of programs can lead to serious data loss, downtime, denial of service, and other types of security incidents.

Internet Infrastructure Attacks

These rare but serious attacks involve key components of the Internet infrastructure rather than specific systems on the Internet. Examples are network name servers, network access providers, and large archive sites on which many users depend. Widespread automated attacks can also threaten the infrastructure. Infrastructure attacks affect a large portion of the Internet and can seriously hinder the day-to-day operation of many sites.

Incidents and Internet Growth

Since the CERT Coordination Center began operating in 1988, the number of security incidents reported to the center has grown dramatically, from less than

100 in 1988 to almost 2500 in 1995, the last year for which complete statistics are available as of this writing. Through 1994, the increase in incident reports roughly parallels the growth of the size of the Internet during that time. Figure 1 shows the growth of the Internet and the corresponding growth of reported security incidents.

The data for 1995 and partial data for 1996 show a slowing of the rate at which incidents are reported to the CERT/CC (perhaps because of sites' increased security efforts or the significant increase in other response teams formed to handle incidents). However, the rate continues to increase for serious incidents such as root compromises, service outages, and packet sniffers.

Incident Trends

In the late 1980s and early 1990s, the typical intrusion was fairly straightforward. Intruders most often exploited relatively simple weaknesses, such as poor passwords and misconfigured systems, that allowed greater access to the system than was intended. Once on a system, the intruders exploited one or another well-known, but usually unfixed, vulnerability to gain privileged access, enabling them to use the system as they wished.

There was little need to be more sophisticated because these simple techniques were effective. Vendors delivered systems with default settings that made it easy to break into the systems. Configuring systems in a secure manner was

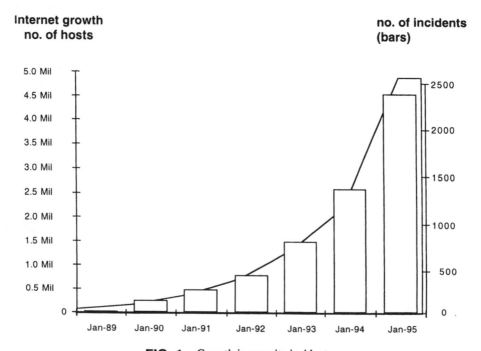

FIG. 1   Growth in security incidents.

not straightforward, and many system administrators did not have the time, expertise, or tools to monitor their systems adequately for intruder activity.

Unfortunately, all these activities continue in 1996; however, more sophisticated intrusions are now common. In eight years of operation, the CERT Coordination Center has seen intruders demonstrate increased technical knowledge, develop new ways to exploit system vulnerabilities, and create software tools to automate attacks. At the same time, intruders with little technical knowledge are becoming more effective as the sophisticated intruders share their knowledge and tools.

Intruders' Technical Knowledge

Intruders are demonstrating increased understanding of network topology, operations, and protocols, resulting in the infrastructure attacks described in the previous section on Internet infrastructure attacks. Instead of simply exploiting well-known vulnerabilities, intruders examine source code to discover weaknesses in certain programs, such as those used for electronic mail. Much source code is easy to obtain from programmers who make their work freely available on the Internet. Programs written for research purposes (with little thought for security) or written by naive programmers become widely used, with source code available to all. Moreover, the targets of many computer intrusions are organizations that maintain copies of proprietary source code (often the source code to computer operating systems or key software utilities). Once intruders gain access, they can examine this code to discover weaknesses.

Intruders keep up with new technology. For example, intruders now exploit vulnerabilities associated with the World Wide Web to gain unauthorized access to systems.

Other aspects of the new sophistication of intruders include the targeting of the network infrastructure (such as network routers and firewalls) and the ability to cloak their behavior. Intruders use Trojan horses to hide their activity from network administrators; for example, intruders alter authentication and logging programs so that they can log in without the activity showing up in the system logs. Intruders also encrypt output from their activity, such as the information captured by packet sniffers. Even if the victim finds the sniffer logs, it is difficult or impossible to determine what information was compromised.

Techniques to Exploit Vulnerabilities

As intruders become more sophisticated, they identify new and increasingly complex methods of attack. For example, intruders are developing sophisticated techniques to monitor the Internet for new connections. Newly connected systems are often not fully configured from a security perspective and are, therefore, vulnerable to attacks.

The most widely publicized of the newer types of intrusion is the use of the packet sniffers described in the section above on packet sniffers. Other tools are used to construct packets with forged addresses; one use of these tools is to mount a denial-of-service attack in a way that obscures the source of the attack. Intruders also "spoof" computer addresses, masking their real identity and successfully making connections that would not otherwise be permitted. In this way, they exploit trust relationships between computers.

With their sophisticated technical knowledge and understanding of the network, intruders are increasingly exploiting network interconnections. They move through the Internet infrastructure, attacking areas on which many people and systems depend. Infrastructure attacks are even more threatening because legitimate network managers and administrators typically think about protecting systems and parts of the infrastructure rather than the infrastructure as a whole.

In the first quarter of 1996, 7.5% of 346 incidents handled by the CERT Coordination Center involved these new and sophisticated methods, including packet sniffers, spoofing, and infrastructure attacks. A full 20% involved the total compromise of systems, in which intruders gain system-level, or root, privileges. This represents a significant increase in such attacks over previous years' attacks, and the numbers are still rising. Of 341 incidents in the third quarter of 1996, nearly 9% involved sophisticated attacks, and root compromises accounted for 33%.

Intruders' Use of Software Tools

The tools available to launch an attack have become more effective, easier to use, and more accessible to people without an in-depth knowledge of computer systems. Often, a sophisticated intruder embeds an attack procedure in a program and widely distributes it to the intruder community. Thus, people who have the desire but not the technical skill are able to break into systems. Indeed, there have been instances of intruders breaking into a UNIX system using a relatively sophisticated attack and then attempting to run DOS commands (commands that apply to an entirely different operating system).

Tools are available to examine programs for vulnerabilities even in the absence of source code. Though these tools can help system administrators identify problems, they also help intruders find new ways to break into systems.

As in many areas of computing, the tools used by intruders have become more automated, allowing intruders to gather information about thousands of Internet hosts quickly and with minimum effort. These tools can scan entire networks from a remote location and identify individual hosts with specific weaknesses. Intruders may catalog the information for later exploitation, share or trade with other intruders, or attack immediately. The increased availability and usability of scanning tools mean that even technically naive would-be intruders can find new sites and particular vulnerabilities.

Some tools automate multiphase attacks in which several small components are combined to achieve a particular end. For example, intruders can use a tool to mount a denial-of-service attack on a machine and spoof that machine's address to subvert the intended victim's machine. A second example is using a packet sniffer to get router or firewall passwords, logging in to the firewall to disable filters, then using a network file service to read data on an otherwise secure server.

The trend toward automation can be seen in the distribution of software packages containing a variety of tools to exploit vulnerabilities. These packages are often maintained by competent programmers and are distributed complete with version numbers and documentation.

A typical tool package might include the following:

- network scanner
- password cracking tool and large dictionaries
- packet sniffer
- variety of Trojan horse programs and libraries
- tools for selectively modifying system log files
- tools to conceal current activity
- tools for automatically modifying system configuration files
- tools for reporting bogus checksums

Internet Vulnerabilities

A vulnerability is a weakness that a person can exploit to accomplish something that is not authorized or intended as legitimate use of a network or system. When a vulnerability is exploited to compromise the security of systems or information on those systems, the result is a security incident. Vulnerabilities may be caused by engineering or design errors or faulty implementation.

Why the Internet Is Vulnerable

Many early network protocols that now form part of the Internet infrastructure were designed without security in mind. Without a fundamentally secure infrastructure, network defense becomes more difficult. Furthermore, the Internet is an extremely dynamic environment, in terms of both topology and emerging technology.

Because of the inherent openness of the Internet and the original design of the protocols, Internet attacks in general are quick, easy, inexpensive, and may be hard to detect or trace. An attacker does not have to be physically present to carry out the attack. In fact, many attacks can be launched readily from anywhere in the world — and the location of the attacker can easily be hidden. Nor is it always necessary to "break in" to a site (gain privileges on it) to compromise confidentiality, integrity, or availability of its information or service.

Even so, many sites place unwarranted trust in the Internet. It is common for sites to be unaware of the risks or unconcerned about the amount of trust they place in the Internet. They may not be aware of what can happen to their information and systems. They may believe that their site will not be a target or that precautions they have taken are sufficient. Because the technology is constantly changing and intruders are constantly developing new tools and techniques, solutions do not remain effective indefinitely.

Since much of the traffic on the Internet is not encrypted, confidentiality and integrity are difficult to achieve. This situation undermines not only applications (such as financial applications that are network based), but also more fundamental mechanisms such as authentication and nonrepudiation (see the section on basic security concepts for definitions). As a result, sites may be affected by a security compromise at another site over which they have no control. An example of this is a packet sniffer that is installed at one site but allows the intruder to gather information about other domains (possibly in other countries).

Another factor that contributes to the vulnerability of the Internet is the rapid growth and use of the network, accompanied by rapid deployment of network services involving complex applications. Often, these services are not designed, configured, or maintained securely. In the rush to get new products to market, developers do not adequately ensure that they do not repeat previous mistakes or introduce new vulnerabilities.

Compounding the problem, operating system security is rarely a purchase criterion. Commercial operating system vendors often report that sales are driven by customer demand for performance, price, ease of use, maintenance, and support. As a result, off-the-shelf operating systems are shipped in an easy-to-use but insecure configuration that allows sites to use the system soon after installation. These hosts/sites are often not fully configured from a security perspective before connecting. This lack of secure configuration makes them vulnerable to attacks, which sometimes occur within minutes of connection.

Finally, the explosive growth of the Internet has expanded the need for well-trained and experienced people to engineer and manage the network in a secure manner. Because the need for network security experts far exceeds the supply, inexperienced people are called on to secure systems, opening windows of opportunity for the intruder community.

Types of Technical Vulnerabilities

The following taxonomy is useful in understanding the technical causes behind successful intrusion techniques and helps experts identify general solutions for addressing each type of problem.

Flaws in Software or Protocol Designs

Protocols define the rules and conventions for computers to communicate on a network. If a protocol has a fundamental design flaw, it is vulnerable to exploitation no matter how well it is implemented. An example of this is the Network File System (NFS), which allows systems to share files. This protocol does not include a provision for authentication; that is, there is no way of verifying that a person logging in really is whom he or she claims to be. NFS servers are targets for the intruder community.

When software is designed or specified, often security is left out of the initial description and is later "added on" to the system. Because the additional components were not part of the original design, the software may not behave as planned, and unexpected vulnerabilities may be present.

Weaknesses in How Protocols and Software Are Implemented

Even when a protocol is well designed, it can be vulnerable because of the way it is implemented. For example, a protocol for electronic mail may be implemented in a way that permits intruders to connect to the mail port of the victim's machine and fool the machine into performing a task not intended by the service. If intruders supply certain data for the "To:" field instead of a correct E-mail address, they may be able to fool the machine into sending them user and password information or granting them access to the victim's machine with privileges to read protected files or run programs on the system. This type of vulnerability enables intruders to attack the victim's machine from remote sites without access to an account on the victim's system. This type of attack often is just a first step, leading to the exploitation of flaws in system or application software.

Software may be vulnerable because of flaws that were not identified before the software was released. This type of vulnerability has a wide range of sub-classes, which intruders often exploit using their own attack tools. For readers who are familiar with software design, the following examples of subclasses are included:

- race conditions in file access
- nonexistent checking of data content and size
- nonexistent checking for success or failure
- inability to adapt to resource exhaustion
- incomplete checking of operating environment
- inappropriate use of system calls
- reuse of software modules for purposes other than their intended ones

By exploiting program weaknesses, intruders at a remote site can gain access to a victim's system. Even if they have access to a nonprivileged user account on the victim's system, they can often gain additional, unauthorized privileges.

Weaknesses in System and Network Configurations

Vulnerabilities in the category of system and network configurations are not caused by problems inherent in protocols or software programs. Rather, the vulnerabilities are a result of the way these components are set up and used. Products may be delivered with default settings that intruders can exploit. System administrators and users may neglect to change the default settings, or they may simply set up their system to operate in a way that leaves the network vulnerable.

An example of a faulty configuration that has been exploited is anonymous File Transfer Protocol (FTP) service. Secure configuration guidelines for this service stress the need to ensure that the password file, archive tree, and ancillary software are separate from the rest of the operating system, and that the operating system cannot be reached from this staging area. When sites misconfigure their anonymous FTP archives, unauthorized users can get authentication information and use it to compromise the system.

Improving Security

In the face of the vulnerabilities and incident trends discussed above, a robust defense requires a flexible strategy that allows adaptation to the changing environment, well-defined policies and procedures, the use of robust tools, and constant vigilance.

It is helpful to begin a security improvement program by determining the current state of security at the site. Methods for making this determination in a reliable way are becoming available. Integral to a security program are documented policies and procedures and technology that supports their implementation.

Security Policy, Procedures, and Practices

Security Policy

A policy is a documented, high-level plan for organizationwide computer and information security. It provides a framework for making specific decisions, such as which defense mechanisms to use and how to configure services, and is the basis for developing secure programming guidelines and procedures for users and system administrators to follow. Because a security policy is a long-term document, the contents avoid technology-specific issues.

A security policy covers the following (among other topics appropriate to the organization):

- high-level description of the technical environment of the site, the legal environment (governing laws), the authority of the policy, and the basic philosophy to be used when interpreting the policy
- risk analysis that identifies the site's assets, the threats that exist against those assets, and the costs of asset loss
- guidelines for system administrators on how to manage systems
- definition of acceptable use for users
- guidelines for reacting to a site compromise (e.g., how to deal with the media

and law enforcement and whether to trace the intruder or shutdown and rebuild the system)

Factors that contribute to the success of a security policy include management commitment, technological support for enforcing the policy, effective dissemination of the policy, and the security awareness of all users. Management assigns responsibility for security, provides training for security personnel, and allocates funds to security. Technological support for the security policy moves some responsibility for enforcement from individuals to technology. The result is an automatic and consistent enforcement of policies, such as those for access and authentication. Technical options that support policy include (but are not limited to)

- challenge/response systems for authentication
- auditing systems for accountability and event reconstruction
- encryption systems for the confidential storage and transmission of data
- network tools such as firewalls and proxy servers

There are many books and papers devoted to site security policies, including requests for comments RFC 1244 (6) and RFC 1281 (7), guidelines written by the Internet Engineering Task Force.

Security-Related Procedures

Procedures are specific steps to follow that are based on the computer security policy. Procedures address such topics as retrieving programs from the network, connecting to the site's system from home or while traveling, using encryption, authentication for issuing accounts, configuration, and monitoring.

Security Practices

System administration practices play a key role in network security. Checklists and general advice on good security practices are readily available. Below are examples of commonly recommended practices.

- Ensure all accounts have a password and that the passwords are difficult to guess. A one-time password system is preferable.
- Use tools such as MD5 checksums (8), a strong cryptographic technique, to ensure the integrity of system software on a regular basis.
- Use secure programming techniques when writing software. These can be found at security-related sites on the World Wide Web.
- Be vigilant in network use and configuration, making changes as vulnerabilities become known.

- Regularly check with vendors for the latest available fixes and keep systems current with upgrades and patches.
- Regularly check on-line security archives, such as those maintained by incident response teams, for security alerts and technical advice.
- Audit systems and networks and regularly check logs. Many sites that suffer computer security incidents report that insufficient audit data is collected, so detecting and tracing an intrusion is difficult.

Security Technology

A variety of technologies have been developed to help organizations secure their systems and information against intruders. These technologies help protect systems and information against attacks, detect unusual or suspicious activities, and respond to events that affect security. In this section, the focus is on two core areas: operational technology and cryptography. The purpose of operational technology is to maintain and defend the availability of data resources in a secure manner. The purpose of cryptography is to secure the confidentiality, integrity, and authenticity of data resources.

Operational Technology

Intruders actively seek ways to access networks and hosts. Armed with knowledge about specific vulnerabilities, social engineering techniques,* and tools to automate information gathering and systems infiltration, intruders can often gain entry into systems with disconcerting ease. System administrators face the dilemma of maximizing the availability of system services to valid users while minimizing the susceptibility of complex network infrastructures to attack. Unfortunately, services often depend on the same characteristics of systems and network protocols that make them susceptible to compromise by intruders. In response, technologies have evolved to reduce the impact of such threats. No single technology addresses all the problems. Nevertheless, organizations can significantly improve their resistance to attack by carefully preparing and strategically deploying personnel and operational technologies. Data resources and assets can be protected, suspicious activity can be detected and assessed, and appropriate responses can be made to security events as they occur.

One-Time Passwords. Intruders often install packet sniffers to capture passwords as they traverse networks during remote log-in processes. Therefore, all passwords should at least be encrypted as they traverse networks. A better solution is to use one-time passwords because there are times when a password is required to initiate a connection before confidentiality can be protected.

One common example occurs in remote dial-up connections. Remote users,

---

*Social engineering refers to the use of human interaction (social skills) for getting or compromising information about the organization or its computer systems (e.g., by posing as a new employee or a repair person).

such as those traveling on business, dial in to their organization's modem pool
to access network and data resources. To identify and authenticate themselves
to the dial-up server, they must enter a user ID and password. Because this
initial exchange between the user and server may be monitored by intruders, it
is essential that the passwords are not reusable. In other words, intruders should
not be able to gain access by masquerading as a legitimate user using a password
they have captured.

One-time password technologies address this problem. Remote users carry a
device synchronized with software and hardware on the dial-up server. The
device displays random passwords, each of which remains in effect for a limited
time period (typically 60 seconds). These passwords are never repeated and are
valid only for a specific user during the period that each is displayed. In addition, users are often limited to one successful use of any given password. One-
time password technologies significantly reduce unauthorized entry at gateways
requiring an initial password.

Firewalls.   Intruders often attempt to gain access to networked systems by pre-
tending to initiate connections from trusted hosts. They squash the emissions of
the genuine host using a denial-of-service attack and then attempt to connect
to a target system using the address of the genuine host. To counter these
address-spoofing attacks and enforce limitations on authorized connections into
the organization's network, it is necessary to filter all incoming and outgoing
network traffic.

A firewall is a collection of hardware and software designed to examine a
stream of network traffic and service requests. Its purpose is to eliminate from
the stream those packets or requests that fail to meet the security criteria established by the organization. A simple firewall may consist of a filtering router,
configured to discard packets that arrive from unauthorized addresses or that
represent attempts to connect to unauthorized service ports. More sophisticated
implementations may include bastion hosts, on which proxy mechanisms operate on behalf of services. These mechanisms authenticate requests, verify their
form and content, and relay approved service requests to the appropriate service
hosts. Because firewalls are typically the first line of defense against intruders,
their configuration must be carefully implemented and tested before connections are established between internal networks and the Internet.

Monitoring Tools.   Continuous monitoring of network activity is required if a
site is to maintain confidence in the security of its network and data resources.
Network monitors may be installed at strategic locations to collect and examine
information continuously that may indicate suspicious activity. It is possible to
have automatic notifications alert system administrators when the monitor detects anomalous readings, such as a burst of activity that may indicate a denial-
of-service attempt. Such notifications may use a variety of channels, including
electronic mail and mobile paging. Sophisticated systems capable of reacting to
questionable network activity may be implemented to disconnect and block
suspect connections, limit or disable affected services, isolate affected systems,
and collect evidence for subsequent analysis.

Tools to scan, monitor, and eradicate viruses can identify and destroy malicious programs that may have inadvertently been transmitted to host systems. The damage potential of viruses ranges from mere annoyance (e.g., an unexpected "Happy Holidays" jingle without further effect) to the obliteration of critical data resources. To ensure continued protection, the virus identification data on which such tools depend must be kept up to date. Most virus tool vendors provide subscription services or other distribution facilities to help customers keep up to date with the latest viral strains.

Security Analysis Tools. Because of the increasing sophistication of intruder methods and the vulnerabilities present in commonly used applications, it is essential to assess periodically network susceptibility to compromise. A variety of vulnerability identification tools are available, which have garnered both praise and criticism. System administrators find these tools useful in identifying weaknesses in their systems. Critics argue that such tools, especially those freely available to the Internet community, pose a threat if acquired and misused by intruders.

Cryptography

One of the primary reasons that intruders can be successful is that most of the information they acquire from a system is in a form that they can read and comprehend. When you consider the millions of electronic messages that traverse the Internet each day, it is easy to see how a well-placed network sniffer might capture a wealth of information that users would not like to have disclosed to unintended readers. Intruders may reveal the information to others, modify it to misrepresent an individual or organization, or use it to launch an attack. One solution to this problem is, through the use of cryptography, to prevent intruders from being able to use the information that they capture.

Encryption is the process of translating information from its original form (called *plaintext*) into an encoded, incomprehensible form (called *ciphertext*). Decryption refers to the process of taking ciphertext and translating it back into plaintext. Any type of data may be encrypted, including digitized images and sounds.

Cryptography secures information by protecting its confidentiality. Cryptography can also be used to protect information about the integrity and authenticity of data. For example, checksums are often used to verify the integrity of a block of information. A checksum, which is a number calculated from the contents of a file, can be used to determine if the contents are correct. An intruder, however, may be able to forge the checksum after modifying the block of information. Unless the checksum is protected, such modification might not be detected. Cryptographic checksums (also called message digests) help prevent undetected modification of information by encrypting the checksum in a way that makes the checksum unique.

The authenticity of data can be protected in a similar way. For example, to transmit information to a colleague by E-mail, the sender first encrypts the information to protect its confidentiality and then attaches an encrypted digital signature to the message. When the colleague receives the message, he or she checks the origin of the message by using a key to verify the sender's digital signature and decrypts the information using the corresponding decryption key. To protect against the chance of intruders modifying or forging the information in transit, digital signatures are formed by encrypting a combination of a checksum of the information and the author's unique private key. A side effect of such authentication is the concept of nonrepudiation. A person who places their cryptographic digital signature on an electronic document cannot later claim that they did not sign it, since in theory they are the only one who could have created the correct signature.

Current laws in several countries, including the United States, restrict cryptographic technology from export or import across national borders. In the era of the Internet, it is particularly important to be aware of all applicable local and foreign regulations governing the use of cryptography.

Information Warfare

Extensive and widespread dependence on the Internet has called new attention to the importance of information to national security. The term *information warfare* refers to the act of war against the information resources of an adversary. Like warfare on land or in the air, information warfare is one component of a range of attack strategies for dominating an adversary in order to gain or maintain an objective.

Information warfare is divided into two categories: offensive and defensive. The purpose of offensive information warfare is to attack the information resources of an adversary to gain dominance. Defensive information warfare is the protection of your information assets against attack.

Information assets can take many forms, from messages sent by courier in diplomatic bags to the computers used to analyze enemy positions based on satellite data. In computer security, information assets include digital information, the computers that process them, and the networks that transmit the digital information from place to place. Computer security is a key element for protecting the availability, integrity, and confidentiality of all these information assets.

Internet security protects information assets consisting of computers, information, and networks that are part of the Internet. Internet security is related to information warfare when the Internet contains information assets that are important to the information warfare objective. For example, if an adversary can use the Internet to access battle plans, the Internet is being used for information warfare.

Internet security is important to both offensive and defensive information warfare because the Internet is a global and dependable resource on which many countries rely. Historically, military networks and computers were unreachable by nonmilitary participants. The Internet, however, provides a cost-effective way for military and government units to communicate and participate in achieving objectives. Use of the Internet means that individuals, multinational companies, and terrorist organizations all can gain access to important information resources of governments and military forces. Thus, it is important to address Internet security concerns as a key component of defensive information warfare.

Because the Internet is global, it can be an avenue of attack for offensive information warfare by many governments. One of the battlefields for a future military offensive could very well involve the Internet. Intruder technology (as described in a separate section above) could be used by a government as a weapon against information resources or used randomly by a terrorist organization against civilian targets.

In the study of information warfare, there are many new problems to solve that are not evident in other forms of warfare. These problems include identifying the enemy, responding without making your systems vulnerable to attack, and gathering intelligence on the Internet about preparations for a military exercise. These and other problems are likely to be the subject of discussion and investigation for some time to come.

The Future

Research and development efforts are underway to allow critical applications to operate in the future in a more secure environment than exists today.

Internetworking Protocols

Most of the network protocols currently in use have changed little since the early definitions of the ARPA research and education network when trust was the norm. To have a secure foundation for the critical Internet applications of the future, severe weaknesses must be addressed: lack of encryption to preserve privacy, lack of cryptographic authentication to identify the source of information, and lack of cryptographic checksums to preserve the integrity of data (and the integrity of the packet routing information itself). New internetworking protocols are under development that use cryptography to authenticate the originator of a packet and to protect the integrity and confidentiality of data.

The IETF (Internet Engineering Task Force) Proposed Standard for the Next Generation Internet Protocol (IPng) is being designed to cope with the vastly increased addressing and routing needs associated with the exponential growth of the Internet. IPng provides integral support for authenticating hosts and protecting the integrity and confidentiality of data.

The first release of IPng is officially termed IPv6 (Internet Protocol version 6). Since it is impractical to replace the existing protocol instantly and simultaneously throughout the Internet, IPv6 is designed to coexist with the current version of IP, allowing for a gradual transition over the course of years. Implementations of IPv6 for many routers and host operating systems are underway.

In the future, authentication protocols will increasingly be supported by technology that authenticates individuals (in the context of their organizational or personal roles) through the use of smart cards, fingerprint readers, voice recognition, retina scans, and so forth.

Protocol design, analysis, and implementation will be the subject of continued research. A primary goal is 100% verifiably secure protocols (that is, protocols as provably secure as the cryptographic algorithms supporting them), but researchers are nowhere near attaining this goal.

Intrusion Detection

Research is underway to improve the ability of networked systems and their managers to determine that they are, or have been, under attack. Intrusion detection is recognized as a problematic area of research that is still in its infancy. There are two major areas of research in intrusion detection: anomaly detection and pattern recognition.

Research in anomaly detection is based on determining patterns of "normal" behavior for networks, hosts, and users and then detecting behavior that is significantly different (anomalous). Patterns of normal behavior are frequently determined through data collection over a period of time sufficient to obtain a good sample of the typical behavior of authorized users and processes. The basic difficulty facing researchers is that normal behavior is highly variable based on a wide variety of innocuous factors. Many of the activities of intruders are indistinguishable from the benign actions of an authorized user.

The second major area of intrusion detection research is pattern recognition. The goal here is to detect patterns of network, host, and user activity that match known intruder attack scenarios. One problem with this approach is the variability that is possible within a single overall attack strategy. A second problem is that new attacks, with new attack patterns, cannot be detected by this approach.

Finally, to support the needs of the future Internet, intrusion detection tools and techniques that can identify coordinated distributed attacks are critically needed, as are better protocols to support traceability.

Software Engineering and System Survivability

Current software engineering methods and practice have had only limited success in managing the intellectual complexity of designing and implementing software. Moreover, in the design of software systems, security concerns are typically an afterthought (addressed through add-ons and software patches) rather than being an integral part of the overall design. This means that software systems of any significant size and complexity are likely to have exploitable security flaws. Because managing the intellectual complexity of software is difficult, up-front security design in products is rare, and detailed knowledge about systems is widespread, systems will be breached in spite of our best efforts to make them invulnerable. Therefore, the concept of information systems security must encompass the specification of systems that exhibit behaviors that contribute to survivability in spite of intrusions. Only then can systems be developed that are robust in the presence of attack and are able to survive attacks that cannot be completely repelled.

System survivability is the capacity of a system to continue performing critical functions in a timely manner even if significant portions of the system are incapacitated by attack or accident. We use the term *system* in the broadest possible sense, which includes networks and large-scale "systems of systems."

Although the concepts and practices associated with system survivability are embryonic, they include (but are not limited to) traditional areas of software engineering and computer science such as reliability, testing, dependability, fault tolerance, verification of correctness, performance, and information system security. Promising research in survivability encompasses a wide variety of research methods in software engineering. Inoculation tools may be developed that will automate the distribution of security fixes, throughout an entire network infrastructure, to provide comprehensive protection from a newly discovered security flaw. The concept of inoculation may be further generalized to encompass adaptive networks, which consist of distributed cooperative network elements that exchange information on security problems and actively change and adjust in response to security threats.

Web-Related Programming and Scripting Languages

Downloading interesting, informative, or entertaining "content" from a remote site to a user's local machine is central to the activity of Web browsing (or "net surfing"). The content getting the most attention from Web users and the greatest concern from security experts is executable content, code to be executed on the local machine on download. This executable content may provide live audio of a conference in progress, a jazz tune, three-dimensional (3-D) animation effects, or hostile code that destroys the local file system. Executable code is authored using one or more Web-related programming or scripting languages designed specifically for the production of platform-independent executable content. Languages in this category include JAVA and ActiveX. Executable content is called an "applet" in JAVA and a "control panel" in ActiveX.

Web-related programming languages pose new security challenges and concerns because code is downloaded, installed, and run on a user's machine without a review of source code (the recommended practice for secure use of publicly available software). These activities can be triggered by following any hypertext link or opening any page while browsing. A user may not even be aware that code has been downloaded and executed. Some Web-related programming languages, most notably JAVA, have built-in security features, but security experts are concerned about the adequacy of these features.

As executable content makes Web browsing even more alluring, further research in software engineering and greater user awareness will be necessary to counter security risks. Presently, the security of executable content depends on the correctness of multiple vendors' implementations, the inherent security of platform-independent "virtual machines," and the safety of the source code that is executed. In the foreseeable future, users need to be educated about the risks so they can make informed choices about where to place their trust.

Intelligent Autonomous Agents—A New Computing Paradigm

The future Internet environment is likely to be increasingly dependent on an agent-based model of computing, with significant implications for Internet security. Agents are executable software objects with executions that are not tied to any specific host or computing resource or to any geographical or logical network location. Agents perform computation and communication defined by a user, but the execution platforms are typically outside the user's administrative control (and outside the administrative control of the user's organization). The conceptual model of agent operation is one in which an intelligent agent, at the request of a user, goes to one or more remote hosts to perform a computation or gather information and then returns to the user with the result. An agent's mode of operation may range from partially to fully autonomous, and the degree to which an agent is autonomous may vary throughout the life of that agent.

A future agent-based computing environment may include features such as these:

- Agents share information and cooperate to complete the user's task.
- Agents protect themselves with intrinsic security mechanisms but also depend on some measure of extrinsic security provided by the infrastructure and cooperating agents.
- Since most of an agent's activity takes place outside the user's domain of administrative control (and hence outside any firewall designed to protect the user), the traditional firewall has little to contribute to security.
- Replication and agent diversity provide increased survivability while under attack and under conditions of degraded or uncertain infrastructure support.
- Agents communicate to enhance the detection of threats. Specialized sensor agents are specifically designed to detect particular types of threats, and groups of diverse sensor agents provide the entire agent "collective" with a comprehensive profile of current threats.
- The agent-supported infrastructure protects itself and takes defensive action without user intervention.

Acknowledgments: CERT is a registered trademark and service mark of Carnegie Mellon University. UNIX is a registered trademark of AT&T Bell Laboratories.

List of Acronyms

ARPA   Advanced Research Projects Agency
CERT/CC   CERT® Coordination Center

DARPA   Defense Advanced Research Projects Agency
FIRST   Forum of Incident Response and Security Teams
FTP   File Transfer Protocol
IETF   Internet Engineering Task Force
IP   Internet Protocol
IPng   Next Generation Internet Protocol (official name is IPv6)
IPv6   Internet Protocol version 6 (also informally called IPng)
NFS   Network File System
RFC   Request for Comments

Bibliography

Caelli, W., Longley, D., and Shain, M., *Information Security Handbook*, Stockton
    Press, New York, 1991.
Chapman, D. B., and Zwicky, E. D., *Building Internet Firewalls*, O'Reilly & Associates,
    Sebastopol, CA, 1995.
Cheswick, W. R., and Bellovin, S. M., *Firewalls and Internet Security: Repelling the
    Wily Hacker*, Addison-Wesley, New York, 1994.
Comer, D. E., *Internetworking with TCP/IP*, 3 vols., Prentice-Hall, Englewood Cliffs,
    NJ, 1991 and 1993.
Garfinkel, S., *PGP: Pretty Good Privacy*, O'Reilly & Associates, Sebastopol, CA, 1995.
Garfinkel, S., and Spafford, G., *Practical UNIX and Internet Security*, 2d ed., O'Reilly
    & Associates, Sebastopol, CA, 1996.
Kaufman, C., Perlman, R., and Speciner, M., *Network Security: Private Communica-
    tion in a Public World*, PTR Prentice-Hall, Englewood Cliffs, NJ, 1995.
McGraw, G., and Felten, E. W., *Java Security*, John Wiley & Sons, New York, 1996.
National Research Council, *Computers at Risk: Safe Computing in the Information
    Age*, National Academy Press, Washington, D.C., 1991.
Schneier, B., *Applied Cryptography: Protocols, Algorithms, and Source Code in C*, 2d
    ed., John Wiley & Sons, New York, 1996.

References

1. Network Wizards. Data is available on line: http://www.nw.com/.
2. Levy, S., *Hackers: Heroes of the Computer Revolution*, Anchor Press/Doubleday,
    Garden City, NY, 1984.
3. Stoll, C., *The Cuckoo's Egg: Tracking a Spy Through the Maze of Computer
    Espionage*, Doubleday, New York, 1989.
4. Denning, P. J. (ed.), *Computers Under Attack: Intruders, Worms, and Viruses*,
    ACM Press, Addison-Wesley, New York, 1990.
5. CERT Coordination Center, CERT® advisories and other security information,
    CERT/CC, Pittsburgh, PA. Available on line: http://www.cert.org/ and ftp://
    info.cert.org/pub/.
6. Internet Engineering Task Force, Site Security Policy Handbook Working Group,

Site Security Handbook, RFC 1244. Available on line: ftp://info.cert.org/pub/ietf/sshpwg/.

7. Internet Engineering Task Force, Network Working Group, *Guidelines for the Secure Operation of the Internet*, RFC 1281. Available on line: ftp://info.cert.org/pub/ietf/sshpwg/.

8. Internet Engineering Task Force, Network Working Group, *The MD5 Message-Digest Algorithm*, RFC 1321. Available on line: ftp://info.cert.org/pub/tools/md5/.

THOMAS A. LONGSTAFF
JAMES T. ELLIS
SHAWN V. HERNAN
HOWARD F. LIPSON
ROBERT D. McMILLAN
LINDA HUTZ PESANTE
DEREK SIMMEL

Self-Routing—Cell Based Switching

(see Fast, High-Performance Local-Area Networks, see Fast Packet Network: Data, Image, and Voice Signal Recovery, see High-Performance Packet Switching with Deflection Networks)

Services and Network Interworking in Wide-Area Networks

Introduction

There has never been such a proliferation of options and choices to the user while selecting a wide-area network (WAN). Traditionally, WANs have been based on leased or dial-up lines with routers located at strategic locations, running, among others, the Internet Protocol (IP), Internet packet exchange (IPX), Systems Network Architecture (SNA), AppleTalk, DECnet, and X.25. New transport technologies have led to new network services being offered by the service providers in the WAN, such as asynchronous transfer mode (ATM) cell relay service, frame relay service, and the Internet. Pricing, applications requirements, and capabilities of individual services have led them to coexist, necessitating one type of network to interwork with another type so that a user on one network can communicate with another user on a different network. Figure 1 illustrates some commonly used networks and the various interworking scenarios. Two types of interworking typically take place: network and service (Fig. 2). In network interworking, the same protocol is used on each end station, but a different protocol is used between them. The protocol conversion in both directions takes place inside the network, and the presence of another network protocol in the middle is transparent to end users. On the other hand, in service interworking, the two end stations can use different protocols but communicate with each other as peers. The protocol conversion is done inside the networks.

This article describes the various interworking functions (IWFs) and protocols for scenarios illustrated in Fig. 1. It is assumed that the reader has some familiarity with network protocols and services. The reader is advised to review Ref. 1 for a selection of articles on network and service interworking in the wide-area network.

Frame Relay and Asynchronous Transfer Mode Cell Relay Interworking

Frame relay (FR) is a fast, variable-size packet technology derived from X.25, with its origins in the work done for the Integrated Services Digital Network (ISDN) standards. Frame relay relies on the fact that most of the physical transmission medium is fiber based and generally error free, and much of the delivery assurance can be pushed to the sending and receiving devices since these are becoming increasingly more powerful than before. Therefore, frame relay removed the link-by-link flow control at the data-link layer of the X.25 Protocol and Link Access Procedure-Balanced (LAP-B). Most implementations of frame relay have used permanent virtual connection (PVC) service, but recently efforts

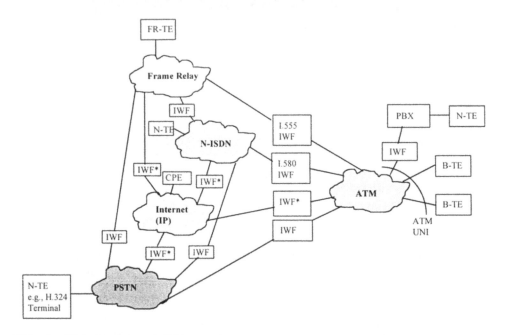

FIG. 1 Wide-area network (WAN) and access network interworking scenarios (ATM = asynchronous transfer mode; B-TE = broadband terminal equipment; CPE = customer premises equipment; FR-TE = frame relay terminal equipment; IWF = interworking function; NISDN = Narrowband Integrated Services Digital Network; N-TE = narrowband terminal equipment; PBX = private branch exchange; PSTN = public switched telephone network). *The IWF is typically a router.

have been under way to also implement switched virtual connection (SVC) FR service. For more details on FR, the reader should see Refs. 2–5.

On the other hand, ATM is a fixed, 53-byte packet, commonly referred to as cell, transport, and multiplexing technology, and the user service is typically referred to as cell relay service (CRS). For more details on ATM and CRS, the reader should see Refs. 6–9.

Both frame relay and ATM are fundamentally connection-oriented technologies, but the ATM also supports connectionless data transfers. The key difference between the frame relay service (FRS) and CRS is that FR is a data-link layer, nonguaranteed-delay, packet-switching technology, whereas ATM is a packet (cell) switching technology capable of delay guarantees in data transfers. Thus, while FR is best suited for non-real-time traffic, ATM can support both non-real-time and real-time traffic. Figures 3 and 4 illustrate the frame relay and ATM cell structures, respectively. For a brief overview of FR and ATM and a detailed discussion of FR/ATM interworking, the reader is referred to Ref. 2.

Frame Relay/Asynchronous Transfer Mode Network Interworking

In the FR/ATM network interworking scenario, two FR networks are interconnected via an ATM network with an FR/ATM IWF on either side. The IWF

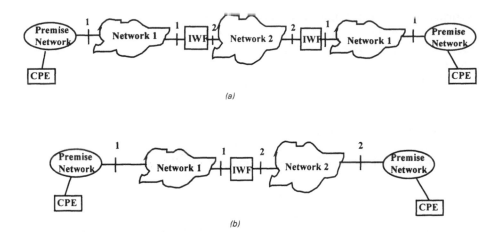

(a)

(b)

FIG. 2 Types of interworking: *a*, network interworking scenario; *b*, service interworking scenario. A Network 1 type interface is indicated by 1, and a Network 2 type interface is indicated by 2. The interworking function (IWF) can be implemented as a stand-alone device or integrated with either Network 1 or Network 2. If integrated with the network, the IWF need not support the interface of that network since it is internal to the terminating network element in that network.

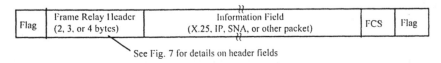

See Fig. 7 for details on header fields

FIG. 3 Frame relay structure (FCS = frame check sequence; SNA = System Network Architecture).

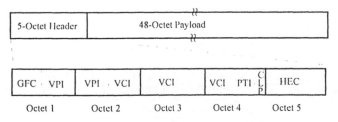

FIG. 4 Asynchronous transfer mode (ATM) cell structure at the user–network interface (UNI). At the network–node interface (NNI), the generic flow control (GFC; 4 bits) field is used with the virtual path identifier (VPI; 8 bits) field, making it 12 bits long (CLP = cell loss priority; HEC = header error control [8 bits]; PTI = payload type identifier [3 bits]; VCI = virtual channel identifier [16 bits]).

conforms to International Telecommunication Union Telecommunications Standardization Sector (ITU-T) Recommendations (Recs.) I.555 (10) and I.365.1 (11). Based on these recommendations, the ATM Forum and the Frame Relay Forum developed a set of specifications for network interworking (7,12). Figure 5 shows three FRS access configurations with six interconnection possibilities.

The ITU-T Rec. I.555 defines two network interworking scenarios that cover all six interconnection possibilities. Scenario 1 connects two FR networks or customer premises equipment (CPE) using Broadband Integrated Services Digital Network (BISDN) and Scenario 2 connects an FR network or CPE with an ATM CPE using BISDN. The reference configuration A3-B3 is not supported in ITU-T Rec. I.555. Therefore, no IWF exists in the network for this configuration. The FR-SSCS (frame relay service-specific convergence sublayer) constitutes the upper part of the ATM adaptation layer (AAL) that sits on top of the common part of the AAL5 (CPAAL5). Its purpose is to emulate FR bearer service in the ATM CPE connected to an ATM network and to provide interworking between an ATM network and an FR network. The IWF maps the following features between the FR service functions and the ATM cell relay functions:

Variable-length protocol data unit (PDU) formatting and delimiting
Error detection
Connection multiplexing
Loss priority
Congestion indication
PVC status management

Figure 6 illustrates the network interworking protocol stacks for the two scenarios described above, and Fig. 7 shows the structure of the FR-SSCS PDU with the three specified FR header formats. From the FR frame, the flag, FCS, and flag fields are stripped, and the remainder of the frame (header and information field) are encapsulated in the CPAAL5. Forward explicit congestion indication/backward explicit congestion indication (FECN/BECN) and

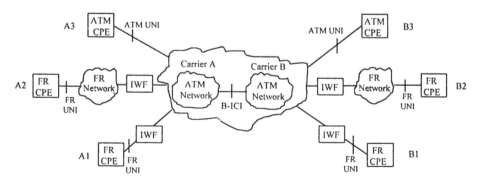

FIG. 5   Frame relay/asynchronous transfer mode network interworking scenarios.

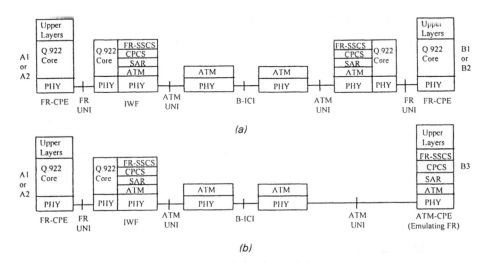

(a)

(b)

FIG. 6 Frame relay/asynchronous transfer mode interworking protocols: *a*, Scenario A1 or A2 interworking with B1 or B2; *b*, Scenario A1 or A2 interworking with B3 (B-ICI = Broadband Integrated Services Digital Network intercarrier interface; CPCS = common part convergence sublayer; FR-SCCS = frame relay service-specific convergence sublayer; PHY = physical layer; SAR = segmentation and reassembly).

discard eligibility (DE) fields of the FR header are mapped to (or from) explicit forward congestion indication (EFCI) and cell loss priority (CLP) fields of the ATM cell header.

The ATM and FR-SSCS layer management operate independently. The FR PVC status management procedures are transparently carried over the ATM layer via the FR-SSCS. Operations and management functions at the frame level (above the FR-SSCS) remain the same as in an FR network. However, at the AAL, the performance management is done at both the AAL5 CPCS and FR-SSCS sublayers.

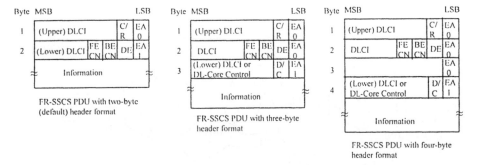

FIG. 7 Frame relay header, asynchronous transfer mode header, and frame relay service-specific convergence sublayer protocol data unit structures (BECN = backward explicit congestion indication; C/R = command/response; DC = DLCI or data-link core control indication; DE = discard eligibility; DLCI = data-link connection identifier; EA = extended address; FECN = forward explicit congestion indication).

Frame Relay/Asynchronous Transfer Mode Service Interworking

In service interworking, the FR service users and the ATM service users connected to FR and ATM networks, respectively, have no knowledge of the type of equipment at the other end, and their equipment does not perform any service-specific functions of the other users' service. Therefore, the FR and ATM networks and services can grow independently while being interconnected by IWFs. The IWFs also do not implement the FR-SSCF function; instead, they perform protocol translation in both directions. The Frame Relay Forum and ATM Forum have specified the Interworking Implementation Agreements (13) for Service Interworking. Figure 8 illustrates the service interworking reference configuration, and Fig. 9 shows the service interworking protocol stack (13).

Both the ATM CPE and the IWF implement null SSCS and AAL5 CPCS, and the BISDN Class C nonassured AAL5-based message protocol. Within the IWF, the Q.922 and AAL5 primitives are mapped from one to the other depending on the direction of the traffic. The reader is referred to Refs. 2 and 13 for a detailed mapping of the various fields in the FR header and the ATM header. For FR/ATM PVC management, two different management procedures are involved, one for the FR side and the other one for the ATM side of the IWF.

In order to transport upper-layer user protocols, two encapsulation methods are defined: transparent mode and translation mode. The type of mode used is selected for each PVC at the configuration time. In the transparent mode, the FR header is removed, and the entire user payload is encapsulated in an AAL5 PDU and segmented as ATM cells without regard to the type of protocol in use. There is no further mapping or fragmentation/reassembly required. This method is best suited for the transport of proprietary protocols.

In the translation mode, the upper-layer bridging and routing protocols are mapped between FR and ATM PVCs in the IWF. Which mode to use depends largely on the specific network topology. For example, when the two routers at either end communicating over an ATM WAN use the same encapsulation method, the transparent mode would work fine since the routers will handle the

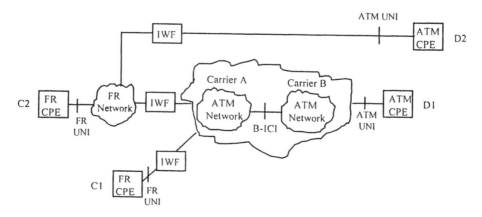

FIG. 8   Frame relay/asynchronous transfer mode service interworking scenario.

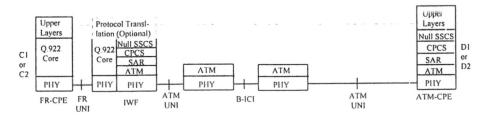

FIG. 9   Service interworking protocol stack.

upper-layer protocols appropriately. On the other hand, if a router at an FR site is communicating with an ATM edge device at the other side, then the translation mode might well be needed because of the need to transfer the upper-layer protocol information. One main advantage of the translation mode is that multiple protocols can coexist in a single PVC, whereas in the transparent mode each different protocol requires a separate PVC.

Narrowband Integrated Services Digital Network and Broadband Integrated Services Digital Network/ Asynchronous Transfer Mode Interworking

The asynchronous transfer mode was developed as a fixed packet-switching and transport technology for BISDN, with the goal of integrating all types of communications services and applications over a single network infrastructure. Prior to the standardization of BISDN, Narrowband ISDN (NISDN) was developed with the same goal of integrating all types of applications and services using circuit switching and $N \times 64$ kilobit per second (kb/s) circuits. The NISDN could support services that required bandwidths up to 2 megabits per second (Mb/s). Over the last five years, however, new and higher resolution multimedia applications have emerged requiring much larger bandwidth, thus providing impetus to the development and implementation of BISDN. Although NISDN had a slow start, its growth rate has picked up significantly.

Therefore, BISDN and NISDN will likely coexist for a significant period of time, necessitating the standardization and development of the interworking function between these two types of networks. Moreover, although ATM as a networking technology is capable of transporting voice from one ATM end station to another, its use has been limited due to the ubiquity requirement.

To address this, the industry, including the standards organizations, has been developing (1) service layer requirements and (2) interworking requirements and specifications with the existing telephony network. On the service side, work is being done on voice and video CODECs, ATM adaptation layer, end-user performance or quality of service (QoS), and functions that a native ATM terminal must perform to be able to interwork with a NISDN or PSTN customer device (e.g., phone, facsimile, multimedia terminals). On the interworking side, specifications are being developed for a BISDN/NISDN interworking function, including signaling. Figure 10 illustrates some of the main interworking configurations that have been the focus of industry standardization, including

1. ATM trunking between two narrowband networks (including PBXs)
2. ATM trunking for mobile narrowband services between a base station and a mobile switching center
3. A BISDN terminal attached to a BISDN communicating with a terminal equipment connected to a public switched telephone network (PSTN) or an NISDN network

The ATM Forum and ITU have been very active in developing service interworking specifications across narrowband and broadband (ATM) networks. Next, we briefly describe the protocol and procedures for each one of the above interworking scenarios.

Asynchronous Transfer Mode Trunking Between Two Narrowband Networks

Three separate specifications have been developed for the interconnection of narrowband networks using ATM to support 64-kb/s voice and telephony ser-

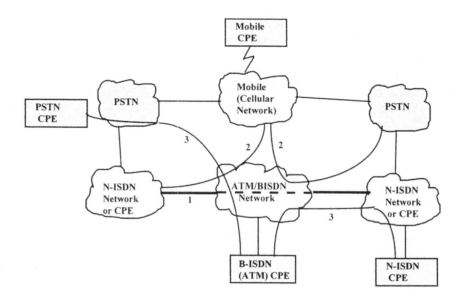

Scenarios 1, 2, and 3 are described in the text.

FIG. 10   Narrowband Integrated Services Digital Network (NISDN) and Broadband Integrated Services Digital Network (BISDN) interworking scenarios.

vices: (1) circuit emulation service (CES), (2) circuit emulation with dynamic bandwidth utilization based on activity detection on DS-1/E-1 channels, and (3) ATM trunking using ATM virtual connections.

Circuit Emulation Service

In CES, constant bit rate (CBR) traffic in a physical trunk is transported across an ATM network using AAL1 such that the performance is comparable to that experienced with the current plesiochronous digital hierarchy/synchronous digital hierarchy (PDH/SDH) technology, the bit integrity is retained (i.e., the analog signal loss is not inserted and voice echo control is not performed), and the structured or unstructured nature of the service in the narrowband network is preserved. The CES service interoperability specification is defined by the ATM Forum (14). The CES specification specifies support for the following types of CBR services:

1. Structured DS-1/E-1 $N \times$ 64 kb/s (fractional DS-1/E-1) service
2. Unstructured DS-1/E-1 (1.544 Mb/s, 2.048 Mb/s) service
3. Unstructured DS-3/E-3 (44.736 Mb/s, 34.368 Mb/s) service
4. Structured J2 $N \times$ 64 kb/s (fractional J2) service
5. Unstructured J2 (6.312 Mb/s) service

Figure 11 provides a reference model for circuit emulation services. As described in Ref. 14, the IWFs must be connected via physical interfaces defined in the ATM Forum user–network interface (UNI) specification. The CES IWFs are also connected to standard CBR circuits (e.g., DS-1/DS-3, J2, or E-1/E-3). The job of the two IWFs is to extend transparently the constant bit rate circuit to which they are connected across the ATM network such that the bit integrity is preserved. Thus, for facilities intended to carry voice or multimedia services, any required echo control must be performed either by the terminal equipment or before the ATM CES IWF is encountered. Note that the CES IWFs do not require explicit external physical interfaces to the CBR equipment and can be

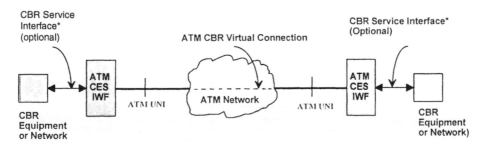

FIG. 11   Circuit emulation service (CES) interworking functions with the specified physical DS-1/E-1, J2, or DS-3/E-3 interfaces (CBR = constant bit rate). *The interworking function need not interface with the external interfaces; it can provide a "logical" structured or nonstructured DS-1/DS-3/E-1/E-3/J2 service and function as customer premises equipment.

used to provide a "logical" structured or unstructured DS-1/DS-3/E-1/E-3/J2 service, even if that service is not used to bridge between existing traditional circuits based on DS-1/DS-3, J2, or E-1/E-3. It is also possible that one IWF would interface to a circuit based on DS-1/DS-3, J2, or E-1/E-3 and the other would not. Both partial cell fill and cell fill formats are permitted. With respect to signaling, there is no specific mapping specified between signaling that is used in traditional DS-1/E-1/J2/DS-3/E-3 and $N \times 64$ kb/s services and ATM UNI 3.1/4.0 signaling. Both PVC and SVC (based on UNI 3.1) are supported between the CES IWFs.

In the structured mode, the CES is intended to support $N \times 64$ kb/s applications (e.g., voice, videoconferencing, local-area network [LAN] interconnection) such that a point-to-point fractional DS-1, E-1, J2, DS-3, and E-3 circuit is emulated. With respect to signaling, both channel-associated mode and unassociated (basic) mode are supported. Since the $N \times 64$ kb/s service can be configured to use only some of the time slots on a service interface (e.g., physical port), several independent emulated circuits, each on a separate ATM virtual channel connection (VCC) using the structured data transfer (SDT) mode of AAL1, can share the same service interface. This enables functional emulation of a DS-1/DS-0, E-1/DS-0, or J2/DS-0 digital cross-connect switch. The mapping function is responsible for assigning the bitstream to and from the SAR process to specific time slots in the service (Fig. 12).

In the unstructured CES, a point-to-point DS-1/E-1/J2/DS-3/E-3 circuit is emulated. The service is defined as a "clear channel pipe." Only one AAL1 or ATM VCC is provided in the ATM interface. The reader is advised to see Ref. 14 for more details on service description, AAL1 requirements, AAL user entity, clock recovery, ATM VCC requirements, signaling, call initiation procedures, and management functions (management information bases, MIBs).

Dynamic Bandwidth Circuit Emulation Service

In the CES V2 specification (14), the ATM VCCs corresponding to mapped time slots remain up for the duration of the call irrespective of their activity

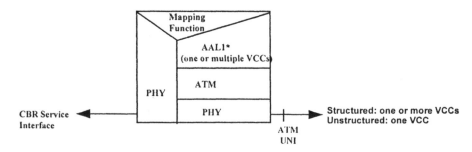

FIG. 12   The DS-1/E-1/J2 structured service interworking function and DS-1/E-1/J2/DS-3/ E-3 unstructured service interworking function (AAL = ATM adaptation layer; VCC = virtual channel connection). *In unstructured mode only one VCC is mapped, therefore there is only one AAL1 entity at the AAL.

level. A new specification has been developed that enables dynamic bandwidth utilization in an ATM network based on detecting which time slots of a given TDM trunk are active and which are inactive (15). When a nonactive state is detected in a specific time slot, the time slot is dropped from the next ATM structure, and the bandwidth it was using may be reused for other applications. Any method of time-slot activity detection can be applied (e.g., channel associated signaling [CAS], common channel signaling [CCS]) and is not specified in the specification. The active time slots are transported using the structured DS-1/E-1 $N \times 64$ kb/s service specified in the CES V2 specification (14). The dynamic bandwidth circuit emulation service (DBCES) IWF will typically perform the following functions: CES SDT DS-1/E-1 $N \times 64$ kb/s service as defined in Ref. 14, time-slot activity detection, Dynamic structure sizing (DSS) of the AAL1 structure that correlates to the active time slots in the TDM-to-ATM direction, recovering the active time slots from the AAL1 structure in the ATM-to-TDM direction and placing them in the proper slots in the TDM stream, and placing the proper signals in each of the time slots of the recovered TDM stream. Figure 13 shows a reference configuration depicting how dynamic bandwidth utilization will be used (15).

The ATM buffering functional block is responsible for multiplexing the ATM streams from various interfaces and serving those streams onto a common ATM interface. In a typical scenario, the DBCES IWF assigns a fixed bandwidth to the maximum configured structure size expected to be handled. (The configured structure is predetermined by the maximum number N of the 64-kb/s time slots provisioned on the trunk at the time of configuring the IWF. The N may represent a full or fractional DS-1/E-1 frame.) When some of the provisioned time slots in the IWF are not active, the IWF dynamically reduces the size of the structure, thus reducing the cell rate to the ATM queue. The dynamic bandwidth utilization (DBU) function in the ATM buffering functional block temporarily assigns this freed-up bandwidth to other unspecified-bit-rate

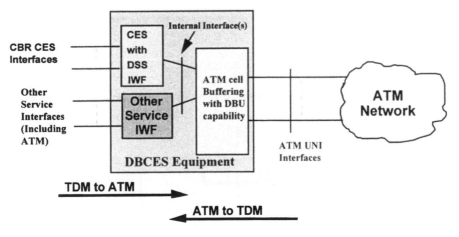

FIG. 13 Reference configuration for trunking with dynamic bandwidth circuit emulation service (DBCES) (DBU = dynamic bandwidth utilization; DSS = dynamic structure sizing).

(UBR) type service(s). When the inactive time slots become active again in the DBCES IWF, the bandwidth temporarily assigned to other services is given back to its proper owner.

Asynchronous Transfer Mode Trunking of N × 64 kb/s Channels Over Asynchronous Transfer Mode Virtual Channel Connection(s)

Specifications for ATM trunking of narrowband services are being developed by the ATM Forum (16). The specification provides a means for the interconnection of two narrowband networks through an ATM network using IWFs and extends the scope and capabilities of the CES V2 and DBCES specifications by providing more flexibility and efficiency in the use of ATM backbone resources. An ATM trunk is defined here as one or more ATM VCCs that carry a number of 64-kb/s narrowband channels and the associated signaling on demand. The number of narrowband channels in an ATM trunk can vary from one to some arbitrary number N.

The ITU-T Recommendation I.580 defines the following three interworking scenarios for ATM networks and narrowband networks (Fig. 14) (17). Note that the focus of I.580 recommendation is on public BISDN and public NISDN interworking.

1. Interconnection of an ATM network/terminal and a narrowband network/terminal
2. Interconnection of two narrowband networks via an ATM network
3. Interconnection of two ATM networks via a narrowband network

Scenario 1 requires mapping of both the user information and signaling protocols between BISDN and NISDN. The IWF that is required can implement the specification described in the section, "ATM Terminal Equipment Attached

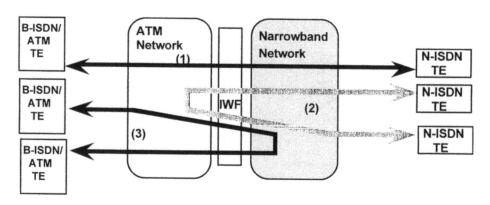

FIG. 14 Interworking scenarios between asynchronous transfer mode networks and narrowband networks (TE = terminal equipment).

to the BISDN Interworking with NISDN Terminal Equipment Attached to an NISDN," even though its focus is on the interworking of a private BISDN with a public or private NISDN.

Scenario 2 uses BISDN as a backbone network for the transport of NISDN information. The ATM trunking specification addresses Scenario 2 only and defines the requirements and capabilities of the IWF by which two narrowband networks are interconnected by a private, public, or private/public combination network (16). The traffic carried by the trunking method may be speech and voiceband telephony services (e.g., facsimile, modem) and the narrowband signaling associated with these services. Voice compression and silence removal are not supported.

Scenario 3 provides for a BISDN user to communicate with another BISDN user via an NISDN. This scenario is not very likely and has received limited attention in the industry. The IWF requirements for this scenario can be best met by combining the requirements of Scenarios 1 and 2.

Next, we briefly describe the interworking requirements and protocols for Scenario 2. Figure 15 shows the various ATM trunking mechanisms for Scenario 2.

Two mechanisms are supported: one-to-one mapping and many-to-one mapping. In one-to-one mapping, each 64-kb/s channel is mapped into one ATM VCC. In many-to-one mapping, multiple 64-kb/s channels are mapped into one ATM VCC. Both mechanisms use one IWF-to-IWF signaling connection to transport narrowband digital subscriber signaling number 1 (DSS1) or private integrated subscriber signaling system number 1 (PSS1) signaling messages between the IWFs within an SSCF-UNI/service specific connection-oriented pro-

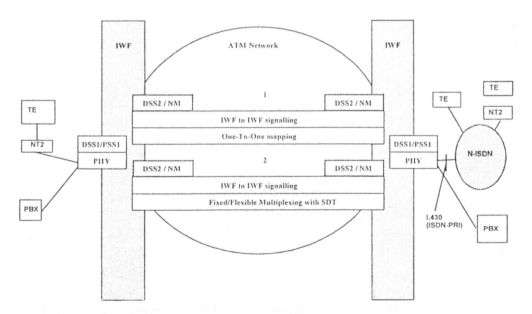

FIG. 15 Asynchronous transfer mode trunking configurations for Scenario 2 (PRI = Primary Rate Interface; SDT = structured data transfer).

tocl (SSCOP)/AAL5 cell stream. Narrowband CAS is carried in the same VCC as the 64-kbps channel. The narrowband networks connect to the IWF over DS-1, E-1, or 64-kbps trunks. The IWF provides a call-by-call switched service to the narrowband network. The NISDN (out-of-band) signaling is supported by terminating it in the IWF and transporting all narrowband messages in an already established signaling connection to the remote IWF. For the support of CAS, the IWF terminates the narrowband signaling and transports all the signaling information in the same ATM VCC that carries the voice information. Any services provided on the narrowband network are independent of their availability in the ATM network. Any service-specific information is transparently transported across the ATM network. The functionality of the IWF is illustrated in Fig. 16. The AAL1 is specified for the voice channel and AAL5 for the signaling channel.

Mobile Trunking Over the Asynchronous Transfer Mode

There has been an industry effort to develop specifications for using ATM to transport efficiently data of low bit rate (LBR), mainly compressed voice, typically for trunking in cellular networks (18). In cellular systems, the voice coding rates are typically 4 to 16 kb/s and could be variable. At these low rates, packetization delay becomes a problem. This has resulted in the need for the definition of a new minicell structure (new AAL) that will enable multiplexing

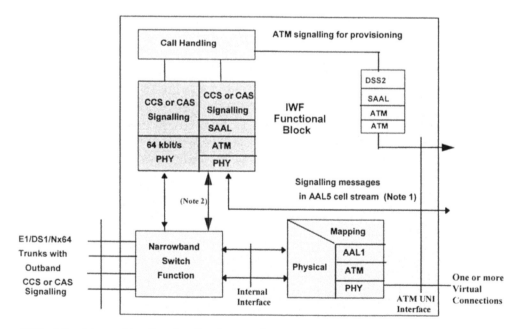

FIG. 16 Interworking functionality for asynchronous transfer mode trunking of narrowband networks. Note 1: This path is valid in the case of Narrowband Integrated Services Digital Network (CCS) signaling. Note 2: This path is valid in the case of CAS signaling only. (DSS2: digital subscriber signaling number 2; SAAL: signaling AAL).

of several channels at the ATM layer (multiple channels in one ATM cell) in the same ATM VCC. Thus, the new AAL will transport packets with all the packet sizes given by the low-bit-rate applications. Although the new voice AAL can be used in many different application environments, one area that is specifically targeted is mobile trunking (between the base station and the mobile station) (Fig. 17). The new AAL, formerly known as AAL-CU (AAL composite user), is now called AAL2.

The mobile trunking is targeted between the base station (BS) and the mobile switching center (MSC). A fundamental requirement is to allow multiple LBR voice channels to be multiplexed over a single ATM VC between the ATM end points. One or more voice channels would be mapped to a single ATM cell payload of 48 octets. Standardization of AAL2 is a very important activity in both the ITU and ATM Forum. At the recent meeting in February 1997 in Seoul, Korea, the ITU-T Study Group 13 (Q.6/SG13) published an AA2 specification with a three-octet minicell header (19). The same format has now been adopted by the ATM Forum. The AAL2 format is shown in Fig. 18.

Both the ITU and the ATM Forum have also been developing the AAL2 negotiation procedure (ANP) for using a dedicated AAL2 connection in an ATM VCC for the transfer of messages defined for ANP. The ANP controls the AAL2 connections multiplexed onto the same ATM VC as the AAL2 connection dedicated for ANP. The negotiation procedures are required to support on-demand AAL2 connections and include functions and messages for establishment and release of AAL2 (voice) connections. The negotiation procedures allow for symmetric operations, and the characteristics of the underlying ATM connections are managed by the normal ATM connection procedures invoked at the ATM connection setup. Additional functions that an AAL2 and/or an ANP may need to support are voice activation/deactivation; silence removal; operations, administration and maintenance (OAM) functions; timing synchronization; and other, yet-to-be-defined functions. The mobile trunking over the ATM specification has a considerable amount of work yet to be done, especially on ANP. A major portion of the specification is being done in the ITU and is being adopted by the ATM Forum. The work is not expected to be completed until early 1998.

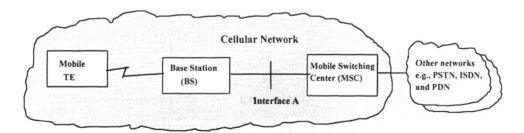

FIG. 17 Network reference configuration (interworking functions are in the base station [BS] and the mobile switching center [MSC]) (PDN = public data network). Interface A is between the BS and the MSC, where the asynchronous transfer mode adaptation layer 2 protocol is intended to be applied.

Octet
Number

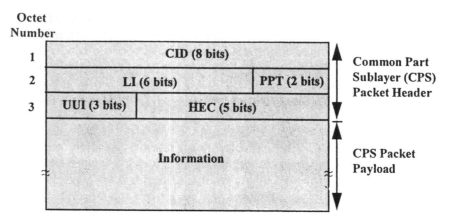

FIG. 18 Format of asynchronous transfer mode adaptation layer 2 common part sublayer (CPS) packet; actual information for the payload is 1 to 45/64 octets (CID = channel identifier [8 bits]; HEC = header error control [5 bits]; LI = length indicator [6 bits]; PPT = payload packet type [2 bits]; UUI = user-to-user indication [3 bits]).

ATM Terminal Equipment Attached to the B-ISDN Interworking with N-ISDN Terminal Equipment Attached to an N-ISDN

Specifications (20) have been developed at the ATM Forum to enable a native ATM terminal connected to a private or public BISDN to communicate with a private or public NISDN-attached terminal. The scenario addressed is one in which an ATM desktop accesses NISDN services offered by an NISDN through an IWF. Specifically, functions and requirements have been specified both for the native ATM terminal and the IWF to enable transport of a single 64-kb/s A-law or μ-law voiceband signal over a single VCC on demand. Specifications to support alternative low-bit-rate or high-quality coding schemes (including silence removal) are planned for development in later phases. Signaling interworking and support for UNI 4.0 supplementary services are also provided. Signaling interworking with an NISDN (DSS2) and at a Q.SIG (PSS1) interfaces is also included. The specification itself relies heavily on other standards (e.g., ATM Forum private network-to-network interface [PNNI] 1.0, UNI 4.0, ITU-T I.580, Q.2931, Q.2952, Q.2957). The reference configuration in which Scenario 1 (described in the section on ATM trunking of $N \times 64$ kb/s channels over ATM VCC[s]) will occur is shown in Fig. 19 (20). From a public/private standpoint, three interworking scenarios are considered: private BISDN to public NISDN, private BISDN to private NISDN, and ATM or broadband terminating equipment to private or public BISDN. The interworking between a public BISDN and a public NISDN is specified in I.580 and therefore is not covered in this specification. The signaling between a public BISDN and a private B-ISDN is specified in the ATM Forum UNI 4.0 specification (21).

The protocol reference model at the interface between the B-TE and the BISDN network is shown in Fig. 20. Any physical layer defined by the ATM Forum can be used. The B-TE has the option to support either the AAL1 specification for voice as described in ITU-T I.363.1 or the AAL5 as specified

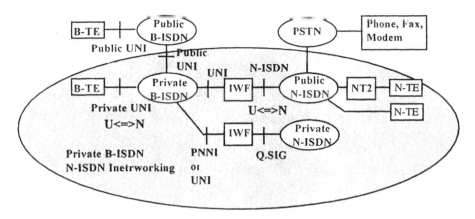

FIG. 19 B-TE and N-TE interworking reference configuration (B-TE = broadband terminal equipment; N-TE = narrowband terminal equipment; PNNI = private network-to-network interface).

in ITU-T I.363.5. It is the currently held view that a B-TE connected to a public BISDN must support AAL1 to conform to the public BISDN offerings. When using AAL5, the SSCS is null, and the AAL-SDU is mapped to one CPCS-SDU (no SSCS and no SSCOP). There is also a limit on the length of the CPCS-PDU, which may be up to 40 octets in increments of 8 octets. Thus, one AAL5 CPCS-PDU fits in one ATM cell payload of 48 octets. To enable error concealment and error correction methods to work at the application layer, the specification provides for the passing of the corrupted AAL5 CPCS-PDU, along with the error indication, to the higher layers according to the procedures of ITU-T I.363.5 as long as the length check passes. To support communication between an AAL1-based B-TE and an AAL5-based B-TE, an AAL1/AAL5 interworking function must be performed (not specified anywhere), which has the consequence of increasing delay due to reassembly and packetization functions in both directions. In addition, the signaling information must also be converted.

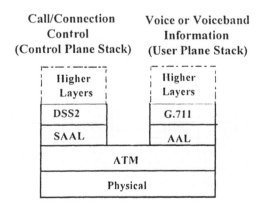

FIG. 20 A signaling and voiceband information protocol reference model.

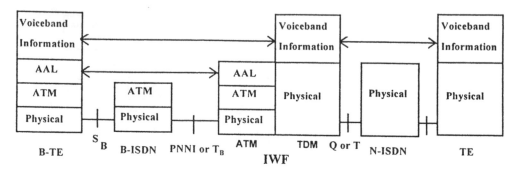

FIG. 21 User plane protocol reference model (TDM = time division multiplexing).

In the IWF, one ATM VCC is mapped to one NISDN channel on a per-call basis. On both sides, the voice traffic is 64 kb/s. For setting up the end-to-end connection, the IWF maps the BISDN signaling information (on VC = 5) to NISDN signaling information (on the D channel). When an IWF connects a private BISDN to a public NISDN, the signaling protocol to the left of the IWF is DSS2 (UNI 4.0) and to the right is DSS1 (Q.921). However, when an IWF connects a private BISDN to a private NISDN, the signaling protocol to the left of the IWF is PNNI 1.0 and to the right is PSS1.

The protocol reference models for the user plane and the control plane are shown in Figs. 21 and 22, respectively. Again, the interim local management interface (ILMI) as defined in the ILMI 4.0 (22) and PNNI 1.0 (23) specifications are specified for the UNI and PNNI, respectively.

Internet/Internet Protocol (IP) and Asynchronous Transfer Mode Interworking

As the Internet evolves into a QoS-enabled infrastructure, its backbone may very likely be ATM based. Furthermore, as ATM becomes deployed in the

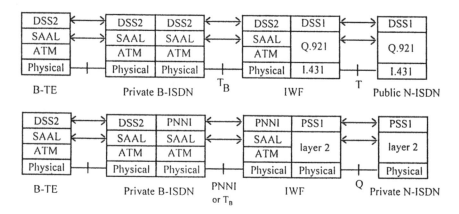

FIG. 22 Control plane protocol reference model.

enterprise networks and as a public cell relay service, there will be a greater need for the Internet and ATM to interoperate. The two most likely Internet/ATM Internetworking scenarios are (1) when an Internet user communicates with an ATM user and (2) when an Internet user communicates with another Internet user via an ATM network as a backbone to the Internet. The two scenarios are illustrated in Fig. 23. Note that the Internet is defined here as a network of router-based (nonguaranteed connectionless) networks running traditional Internetwork layer protocols (e.g., IP, IPX, DECnet, AppleTalk) by which routers enable communication across subnet boundaries. Subnets are typically built using LAN technologies (e.g., Ethernet, Token Ring, Fiber Distributed Data Interchange [FDDI]). The data flow in the traditional Internet is connectionless, whereas in the ATM network it is connection oriented. Therefore, some of the critical functions that the IWF performs are address mapping derived from the topology and reachability information, traffic mapping, QoS mapping, and other signaling and network management protocols.

The interworking function (IWF) can either be implemented in the router, in the ATM switching node, or as a stand-alone function. Several standards have been developed to support Internet/ATM interworking, namely, classical IP and Address Resolution Protocol (ARP) over ATM (24) and multiprotocol encapsulation over AAL5 (25). These have typically relied on preprovisioned ATM PVCs. Recently, efforts have been under way in the Internet Engineering Task Force (IETF) and the ATM Forum to develop a Multi Protocol Over ATM (MPOA) specification (26). The MPOA provides end-to-end Internetwork layer connectivity across an ATM network, including the case in which some hosts are connected directly to the ATM network and some are connected to LANs and WANs via an MPOA-capable edge device. Except for configuration information flows, Logical Link Control/Subnetwork Access Protocol (LLC/SNAP) encapsulation over ATM VCCs is used for all information flows (25).

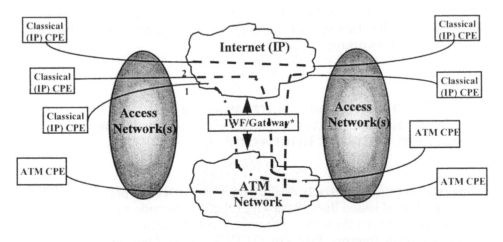

FIG. 23 Internet/asynchronous transfer mode interworking scenarios. Notes: 1 denotes scenario 1 where a classical CPE communicates with an ATM CPE; 2 denotes scenario 2 where a classical CPE communicates with another classical CPE via an ATM backbone. *The IWF/Gateway is typically an ATM-enhanced router.

Prior to the work on MPOA, the ATM Forum developed the LAN Emulation (LANE) specification that emulates Ethernet and Token Ring LANs over an ATM infrastructure (27). Therefore, LAN emulation provides a general solution for any protocol that runs over Ethernet or token ring to work transparently over ATM. Emulated LANs can be interconnected via routers to run Internetwork layer protocols in much the same way as they operate over Ethernet or Token Ring LANs. The IETF is developing the Next Hop Resolution Protocol (NHRP) and multicast address resolution server (MARS) protocols that will allow Internetwork layer protocols to operate over an ATM network. The MPOA development effort in the ATM Forum integrates LANE and NHRP, allowing intersubnet, Internetwork layer protocol communication over ATM-switched VCs without requiring routers. The LANE capability is preserved in the specification. One major goal of MPOA is to ensure interoperability with the existing infrastructure of routers by running standard Internetwork layer protocols such as open shortest path first (OSPF).

Public Switched Telephone Network/ Asynchronous Transfer Mode Interworking

Fundamentally, the PSTN has two logical styles. In the first, the application information is transported digitally in the last mile; in the other, the application information is transported in the analog domain over the final drop. The digital loop enables NISDN access and various other DSL (digital subscriber loop) services. It is the second type that is commonly associated with the PSTN. In both cases, though, the underlying infrastructure remains the same, except for some differences in the signaling procedures and protocols. There are no ATM UNI standards for analog termination in the ATM network. However, at the trunk level the ATM networks and the PSTN can be interconnected via any of the interworking function specifications described in the section on NISDN and BISDN/ATM interworking.

Frame Relay, Internet, Narrowband Integrated Services Digital Network, and Public Switched Telephone Network Interworking

Frame relay is being extensively used by the Internet service providers (ISPs) to replace traditional leased lines. Basically, FR PVCs provide a logical equivalent of leased lines between the nodes. The upper-layer protocols (e.g., Transmission Control Protocol/Internet Protocol [TCP/IP], User Datagram Protocol/Internet Protocol [UDP/IP]) remain the same and are transported as described in the section on frame relay and ATM cell relay interworking. Similarly, to take advantage of the higher throughput of ISDN, FR over ISDN has been gaining in popularity in the access network for ISP access. The NISDN/PSTN interworking standards have been in place for a long time, and public carrier networks that make use of them are in existence in large numbers and growing. Analog modems and NISDN are quite popular for accessing the Internet, and for access to corporate LANs in telecommuting and for connecting to content providers.

Evolution of Wide-Area Network Services

Driven by the growth in electronic commerce, electronic mail, multimedia, and a fundamental cultural shift toward an information society all over the world, the wide-area networks and services will continue to grow. In the near term, we are likely to witness explosive growth in the Internet and corporate intranets. It is expected that all the WAN services and interworking technologies described in this article will continue to grow because of varying user requirements, technology capabilities, pricing, and availability. With the maturing of the security protocols and standards, an increasing number of network providers are building intra- and extranets using the private and public Internet. With the rapid user acceptance of multimedia, efforts are under way to provide real-time transport capabilities in the Internet. This will probably lead the network providers to use ATM and FR in the backbone of the Internet, with attractiveness to the subscriber depending on the availability of the SVC capability. Simultaneously, standards work is progressing in the IETF to define protocols that will enable the Internet to support QoS. It is early to predict which technology(ies) will eventually dominate the Internet and the WAN a decade from now. Nevertheless, from a technological standpoint, some clear patterns are emerging. First, the distinction between routing and switching is disappearing, with routing used for address and route determination and the actual routing and transport done using packet/cell switching. Second, several different types of WANs will coexist, requiring interworking between them and with the existing networks. Third, it is unclear what role ATM will play in the real-time Internet of the future. Last, the business and regulatory issues will play a dominant role in the evolution of the WANs.

Conclusions

With ever-growing choices and options and varying user requirements, service and network interworking is becoming an important requirement for both the network operators and the end users. Because of a significant base of NISDN, PSTN, and router-based Internet equipment and services, new network services (e.g., FR, ATM cell relay) will coexist with them for a significant period of time. Standards organizations and industry forums have been actively cooperating to develop standards and interoperability specifications to meet these market demands. We hope that this will lead the equipment manufacturers to provide equipment sooner rather than later, thus enabling seamless interworking across different networks and technologies at the various layers of the user plane, control plane, and the management plane.

Acknowledgments: The author acknowledges the various standards and industry forum specifications that served as a rich source of information for the writing of this article. The reader is strongly advised to review these documents. A partial list is included in the list of references.

References

1. Dixit, S. (ed.), *IEEE Commun. Mag.*, 34:64–106 (June 1996); special issue, Service and Network Interworking in a WAN Environment.
2. Dixit, S., and Elby, S., Frame Relay and ATM Interworking, *IEEE Commun. Mag.*, 34:64–82 (June 1996).
3. International Telecommunication Union Telecommunications Standardization Sector, *Frame Mode Bearer Services*, ITU-T Rec. I.233, 1992.
4. International Telecommunication Union Telecommunications Standardization Sector, *ISDN Data Link Layer Specification for Frame Mode Bearer Services*, ITU-T Rec. Q.922, 1992.
5. International Telecommunication Union Telecommunications Standardization Sector, *Congestion Management for the ISDN Frame Relaying Bearer Service*, ITU-T Rec. I.370, 1991.
6. ATM Forum, *ATM User Network Interface (UNI) Specification Version 3.1*, Prentice-Hall, Englewood Cliffs, NJ, 1994.
7. ATM Forum, *BISDN Inter-Carrier Interface (B-ICI) Specification Version 1.1*, ATM Forum, Mountain View, CA, 1994.
8. ATM Forum, *Traffic Management Specification Version 4.0*, ATM Forum, Mountain View, CA, April 1996.
9. Bell Communications Research, *PVC Cell Relay Core Features*, Bellcore SR-3330, Bellcore, Red Bank, NJ, December 1995.
10. International Telecommunication Union Telecommunications Standardization Sector, *Frame Relaying Bearer Service Interworking*, ITU-T Rec. I.555, Com13, R2-E, July 1994.
11. International Telecommunication Union Telecommunications Standardization Sector, *Frame Relaying Service Specific Convergence Sublayer (FR-SSCS)*, ITU-T Rec. I.365.1, 1993.
12. Frame Relay Forum, *Frame Relay/ATM PVC Network Interworking Implementation Agreement*, FRF.5, December 1994.
13. Frame Relay Forum, *Frame Relay/ATM PVC Service Interworking Implementation Agreement*, FRF.8, April 1995.
14. ATM Forum af-vtoa-0078.000, *Circuit Emulation Services Version 2*, January 1997.
15. ATM Forum af-vtoa-0085.000 Ballot Final, *Specifications of (DBCES) Dynamic Bandwidth Utilization—In 64 kb/s Time Slot Trunking Over ATM—Using CES*, May 1997.
16. ATM Forum af-vtoa-0089.000 Letter Ballot, *Voice and Telephony Over ATM—ATM Trunking for Narrowband Services*, Version 1.0, June 1997.

17. International Telecommunication Union Telecommunications Standardization Sector, *General Agreement for Interworking Between B-ISDN and 64 kb/s ISDN*, ITU-T Rec. I.580, 1995.

18. ATM Forum BTD-VTOA-MOB-01.02, *Voice Over ATM—Mobile Trunking Specification*, April 1997.

19. International Telecommunication Union Telecommunications Standardization Sector, *BISDN ATM Adaptation Layer Type 2 Specification*, ITU-T Rec. I.363.2, Seoul, February 1997.

20. ATM Forum af-vtoa-0083.000 Ballot Final, *Voice and Telephony Over ATM to the Desktop Specification*, March 1997.

21. ATM Forum, *ATM User-Network Interface (UNI) Signaling Specification, Version 4.0*, July 1996.

22. ATM Forum, *Integrated Local Management Interface (ILMI) Specification, Version 4.0*, September 1996.

23. ATM Forum, *Private Network-Network Interface Specification, Version 1.0 (PNNI 1.0)*, March 1996.

24. Internet Engineering Task Force, *Classical IP and ARP Over ATM*, IETF RFC 1577, 1994.

25. Internet Engineering Task Force, *Multiprotocol Encapsulation Over AAL5*, IETF RFC 1483, 1993.

26. ATM Forum BTD-MPOA-MPOA-01, *MPOA Baseline Version 13*, February 1997.

27. ATM Forum, af-lane-0084.000 Letter Ballot, *LAN Emulation Over ATM*, Version 2, LUNI Specification, July 1997.

SUDHIR S. DIXIT

Shockley, William

William Shockley was born February 13, 1910, in London. His father was a mining engineer, and his mother was a mineral surveyor (1). He was brought up in Palo Alto, California, where his parents felt the local educational opportunities were inadequate, and he began his education with home study. The Shockleys had a neighbor, Perley A. Ross, who was a physicist at Stanford; Ross contributed significantly to the boy's interest in electricity, radio, and related topics.

After a brief period enrolled in the Palo Alto Military Academy, Shockley enrolled in Hollywood High School, Los Angeles, and also attended Los Angeles Coaching School, where he received instruction in physics.

In 1927 he entered the University of California at Los Angeles but stayed only a year before transferring to the California Institute of Technology in Pasadena. There he earned a bachelor's degree in physics in 1932 and was awarded a fellowship to teach at the Massachusetts Institute of Technology. His doctoral work was in applied solid-state physics, which would eventually lead him into his transistor-related activities.

In 1936, Shockley turned down a variety of job offers in order to join Bell Laboratories in Murray Hill, New Jersey, which offered him the opportunity to work with C. J. Davisson, who at that time led vacuum tube research at the laboratories.

During World War II, Shockley was on leave from the laboratories and served as director of research for the Anti-Submarine Warfare Operations Research Group at the War Department. He also served as a special consultant to the War Department and contributed to research into radar.

Returning to Bell Laboratories after the war, Shockley returned to solid-state physics and in 1945 was made codirector of that department within the laboratories. He worked with John Bardeen and Walter Brattain (2) on the development of the transistor, which was successfully demonstrated during 1947.

The original transistor, rigged from gold foil and plastic pressed against a block of germanium, showed instant improvements over other "valves," which allowed significant improvements over earlier devices for the amplification of electrical flow. This compact device, which revolutionized electronics around the world and enabled the miniaturization of components that continues to this day, drew its name from the phrase *transfer resistor*. The tiny device would eventually replace the large, clumsy, and very fragile vacuum tube, altering the field of electronics forever.

Shockley left Bell Laboratories in 1954. He founded the Shockley Semiconductor Laboratories, one of the most powerful of the modern corporations in what would come to be known as the Silicon Valley in California. Shockley Laboratories focused on a particular commitment of Shockley, to the utilization of germanium in the development of transistors. He would eventually hold more than 90 patents.

Shockley's laboratories nourished the knowledge, creativity, and eventual careers of many pioneers in electronics, among whom were an increasing number who determined that silicon was a better semiconducting material than germanium. With little opportunity to explore their preference, a group of seven of those engineers left the company. With backing from the Fairchild Camera and Instrument Company, they created Fairchild Semiconductor in Mountain View, California, in 1958. Pioneers from this group would later break away from Fairchild and form a company called Intel, which would become one of the most important manufacturers of components during the computer revolution.

In 1956, Shockley, Bardeen, and Brattain shared the Nobel Prize in physics, awarded to them for creating the transistor. Shockley became a lecturer at Stanford University in 1958. In 1960 his firm was sold to the Clevite Transistor Company, and he devoted his professional time almost entirely to teaching. In 1963 he became the first Alexander M. Poniatoff Professor of Engineering and Applied Science at Stanford, where he continued to teach through 1975.

In 1973 Shockley began to explore genetics, an area that would continue to interest him and to embroil him in controversy for the rest of his life. He ran for the U.S. Senate as a Republican in 1982, explaining that he would use his time in office to explain "dysgenics" — his theory that some other races have not evolved as fast as whites.

Shockley died August 12, 1989. He left a widow, Emmy, as well as children by his first wife, sons William and Richard and a daughter Alison Ianelli.

Bibliography

McGrath, M., William Shockley, Nobel Winner Who Stirred Racial Controversy, *Boston Globe*, August 15, 1989.
Polkinghorn, F., Interview of Warren P. Mason, *IEEE Oral History Transcripts*, IEEE Electronic Library. Available on line: http://www.ieee.org/history_center.
Rogers, E., and Larsen, J., *Silicon Valley Fever*, Basic Books, New York.

References

1. Slater, R., William Shockley, Co-Inventor of the Transistor. In *Portraits in Silicon*, MIT Press, Cambridge, MA, 1987, p. 141.
2. Sheddan, M. K., Brattain, Walter H. In *The Froehlich/Kent Encyclopedia of Telecommunications*, Vol. 2 (F. E. Froehlich and A. Kent, eds.), Marcel Dekker, New York, 1992, pp. 131–134.

MARYLIN K. SHEDDAN

Signal Processing (see Digital Signal Processing)

Signal Recovery for Data, Image, and Voice in Fast Packet Networks
(see Fast Packet Network: Data, Image, and Voice Signal Recovery)

Signal Theory

Introduction

Signal theory addresses problems in the representation of signals and the characterization of signal properties. The conceptually simplest methods for signal representation may, in fact, be too complex and unwieldy for useful signal processing. Two familiar concepts are representation by a graph of instantaneous values of the signal, the "oscillographic" representation, and by a graph of the Fourier transform of the signal, the "spectrographic" representation. Both of these are abstractions. Strictly speaking, neither comes even close to providing a universal framework for the variety of signals typically encountered in studies of physical and biological systems. It is generally well understood that some form of band limiting and duration limiting* are simultaneously necessary in order to yield a manageable (finite) representation that will accommodate a nontrivial class of signals.

An early, and still highly useful, technique applicable to signals of a periodic or a duration-limited nature is the Fourier series expansion. This leads to a discrete representation in terms of the sequence of Fourier coefficients, which can often be truncated to a finite sequence by invoking physical considerations that limit how rapidly the signal can fluctuate over the basic time interval. In many fields, but perhaps most noticeably in the field of speech processing, it became evident that neither a pure "time-domain" nor a pure "frequency-domain" viewpoint was adequate to capture the essential features of the signals under consideration. In 1946, a paper by Dennis Gabor addressed exactly this point (1). Many regard this paper as the genesis of modern signal theory.

In establishing a precise "time-frequency" framework for signal representation, Gabor made accessible, perhaps for the first time, several mathematical concepts to engineers. First, he made it clear that the analyst, be it machine or human, was free to choose the most suitable orthonormal basis for signal expansions. The familiar cosine/sine or cisoid (exponential) functions may not be the most suitable for some applications. In fact, Gabor's development is directed toward the goal of establishing that the cisoid multiplied by a Gaussian "probability" pulse with discrete values of time-offset and frequency-offset parameters formed a more useful basis for auditory signals. This approach alerted engineers to the utility of the concept of a signal space, that is, signals can be interpreted as vectors. Then, using standard linear space concepts such as inner products, norms, and projection operations, an immense variety of signal representation schemes is immediately available. The signal space idea also popularized the concept of a metric or "distance" between different signals, an idea that is central to the design of signal detectors, discriminators, and decoders.

---

*Many signals are referenced to one or more spatial coordinates rather than a single time coordinate. For this discussion, the techniques are the same in either case, and we use the time-coordinate terminology for convenience.

Second, Gabor made a convincing case for doing analysis with complex-value signals, an approach often regarded with skepticism by some engineers. He showed how Hilbert transforms could be used to form an analytic signal, which has a one-sided Fourier transform, thereby simplifying the methods by which certain signal characteristics, such as bandwidth and instantaneous frequency, could be expressed in general terms. The closely related concept of the complex envelope for band-pass signals is now standard terminology in all modulation systems, such as digital data modems.

Finally, Gabor presented the famous uncertainty principle applied to communication signals, showing that there exists a fundamental limit on how small the duration-bandwidth product of any signal can be. There followed several decades of activity in various pulse optimization problems solved by vector space methods such as gradient minimization or the Schwarz inequality. A typical problem of this nature might be to find the pulse with a given peak value and a given bandwidth that has the smallest energy.

Roughly three decades after Gabor's trend-setting paper, there was a major resurgence of interest in the theoretical aspects of time-frequency analysis, partly due to the fact that digital signal processing devices had advanced so much in speed and complexity that much more sophisticated signal processing became feasible. A related idea is that of multiresolution analysis and wavelet expansions, which has had a major impact on the way engineers think about signal representation problems. Wavelets are now being used in many applications for transmission and storage of audio and video signals. In the following sections of this article, we examine some of the foregoing ideas at an elementary level.

Signal Spaces

We can regard the continuous-time signal, or waveform $x(t)$, where t ranges over the entire real line R as a vector x in $L^2(R)$, the infinite-dimensional Hilbert space consisting of all square-integrable functions of time. The basic vector operation between two elements of this space is simply pointwise addition of the two waveforms. A Hilbert space also has an inner product (also called a scalar product or dot product) for any two vectors, denoted by (x,y). The inner product for $L^2(R)$, from which most signal properties of interest can be derived, is defined as*

$$(x,y) = \int x(t)y^*(t)dt \qquad (1)$$

Note that complex-value signals, and hence complex-value inner products (scalars), are accommodated in this definition.

A signal norm is induced by the inner product, namely, $\|x\| = (x,x)^{1/2}$. The physical quantity associated with the square of the norm is the energy of the

---

*All integration and summation limits in this article are $-\infty,\infty$ unless otherwise stated; $y^*(t)$ denotes the complex conjugate of $y(t)$.

signal, $E_x = \|x\|^2 = \int |x(t)|^2 dt$. The distance between any two signals in $L^2(R)$ (a non-negative, real quantity) is taken to be the norm of the difference, that is, $d(x,y) = \|x - y\|$. The distance measure is often used to characterize the error in a certain signal processing operation or to classify signals by a "nearest neighbor" criterion by which signals are grouped together when they are closest to one of a set of representative signals in the signal space. For example, consider a binary signaling system using transmitted waveforms s_0 and s_1. Because of additive noise, the received signal r could be anywhere in $L^2(R)$. A realistic, and in some cases optimum, detection scheme is to partition the signal space into two regions: the set of points closest to s_0 and another set that is its complement. Then, the decision as to which signal was transmitted is made according to which region contains the received signal r. Another important application for the distance metric is discussed below in terms of the time ambiguity of a signal.

The inner product (x,y) can be regarded as a linear functional of the signal x for any given "reference" signal y. The physical implementation of this operation can be accomplished in a straightforward manner. As shown in Fig. 1, there are two different approaches to this: (1) the multiplier-integrator and (2) the filter-sampler. In either case, it must be assumed that either x or y is essentially of limited duration so that the integral can be evaluated at some finite time. The multiplier-integrator version has the advantage that the reference function can be readily changed by simply modifying the y input to the multiplier. However, this requires relatively expensive (analog or digital) multiplier components. The more economical alternative is to make the inner product a

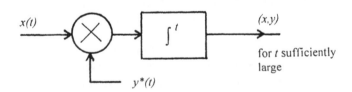

(a) Multiplier-integrator configuration.

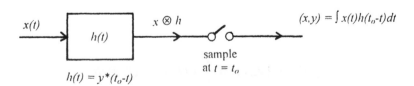

(b) Filter-sampler configuration.

FIG. 1 Implementation of the $L^2(R)$ inner product functional $(x,y) = \int x(t)y^*(t)dt$: *a*, multiplier-integrator configuration; *b*, filter-sampler configuration.

particular instantaneous value of a convolution integral so that, by making the impulse response of the filter in Fig. 1-*b* the time reverse of the *y* waveform, the sampled output corresponds to the desired inner product. Such a filter is sometimes said to be *matched* to the *y* waveform.

One of the most useful aspects of the signal-space viewpoint is in approximating an arbitrary signal as a linear combination of a prescribed set of basic waveforms. The set of expansion coefficients then becomes the discrete representation for a continuous-time waveform. The approach is succinctly stated by the orthogonal projection theorem: the linear combination $\hat{x}$ of basis waveforms $\{\phi_n\}$, which best represents *x* in the sense that $d(x,\hat{x}) = \|x - \hat{x}\|$ is minimized, is that which makes the error signal $x - \hat{x}$ orthogonal to every linear combination of the basis set $\{\phi_n\}$. Any two vectors are said to be *orthogonal* if their inner product vanishes. If we further specify that $\{\phi_n\}$ be an *orthonormal set*, that is, the basis vectors are pairwise orthogonal and each has unit norm, then the projection theorem can be stated somewhat more explicitly as: For $\hat{x} = \sum_{n=1}^{N} c_n \phi_n$, with $(\phi_n,\phi_m) = 0$ for $n \neq m$ and $(\phi_n,\phi_n) = 1$, the choice $c_n = (x,\phi_n)$ minimizes $\|x - \hat{x}\|$. Hence, $\hat{x} = \sum_{n=1}^{N} (x,\phi_n)\phi_n$ is the expression for the best approximation when an orthonormal basis is used.

The requirement that the basis be an orthonormal set is more a convenience than a restriction on the type of approximation that is employed. This is because, for an arbitrary basis set, an orthonormal basis can be formed using only linear combinations of vectors in the arbitrary set. The two bases are said to span the same subspace of $L^2(R)$, and the same approximating vector $\hat{x}$ results in either case. The Gram–Schmidt procedure is one technique for generating the orthonormal basis (2). Another technique is shown in a section below. There are numerous "classical" sets of orthonormal functions that have been used in signal analysis, such as Legendre, Laguerre, Chebyshev, Hermite, Walsh, prolate spheroidal, to name a few, in addition to the traditional sine/cosine functions for Fourier analysis (2).

Another important signal characteristic that can be defined in terms of an inner product is the time ambiguity of the signal, which also characterizes the energy distribution of the signal in the frequency domain. Consider the signal space separation of *x* and x_τ where x_τ is a time-shifted (by τ seconds) version of *x*, that is, $x_\tau(t) = x(t + \tau)$. Then, since $\|x - x_\tau\|^2 = (x - x_\tau, x - x_\tau) = (x,x) - (x_\tau,x) - (x,x_\tau) + (x_\tau,x_\tau)$; $\|x\|^2 = \|x_\tau\|^2 = (x,x)$ and $(x,x_\tau) = (x_\tau,x)^*$, the square of the distance between *x* and x_τ can be written as

$$d^2(x,x_\tau) = 2r_x(0) - 2Re\, r_x(\tau) \tag{2}$$

where

$$r_x(\tau) = (x_\tau,x) = \int x(t + \tau)x^*(t)dt \tag{3}$$

is defined as the time-ambiguity function for *x*.

This terminology results from the fact that time-shifted versions of the signal are easily distinguishable (nonambiguous) when the signal-space distance is large. Hence, a narrow $r_x(\tau)$ function is desirable when it is necessary to distinguish closely spaced versions of the same signal, as in radar systems, for example. The Fourier transform† of $r_x(\tau)$ is given by $R_x(f) = |X(f)|^2$, which is called the *energy spectral density* of the $x(t)$ signal. Clearly, a narrow time ambiguity function corresponds to a broad bandwidth signal. The frequency-domain phase-shift function of a signal does not affect its time ambiguity. Hence, there exist various "chirplike" signals that can have a very long duration, yet exhibit narrow time-ambiguity functions and a good time resolution capability (3).

Multiresolution Analysis and Wavelets

The type of basis set most commonly used in signal processing is generated from time-translated versions of a single basic waveform. Thus, the set $\{\phi(t - nT)\}$, where n takes on all integer values, spans a particular subspace of $L^2(R)$. Projection of an arbitrary continuous-time signal onto this subspace can be regarded as equivalent to mapping the signal into a discrete-time signal. It is a generalization of uniform sampling, and the familiar sampling theorem is a special case of this kind of projection. The N-dimensional subspaces spanned by the classical orthonormal sets as discussed above can be regarded as a kind of "block processing" of the signal. The finite-term expansion is applicable to a finite time interval, and the expansion process is repeatedly applied to successive time intervals. The time series representation, on the other hand, is more like "convolutional processing" in that elements of the discrete-time approximation are generated continually and sequentially for an indefinite time period. The implementation complexity of this approach is usually considerably less than that of the block processing approach.

As before, it is more convenient to perform the orthogonal projection if the $\{\phi(t - nT)\}$ were to form an orthonormal set. The condition of orthonormality is easily expressed in terms of the time-ambiguity function of the basic waveform, since

$$\int \phi(t - nT)\phi^*(t - mT)dt = r_\phi(mT - nT) \qquad (4)$$

Thus orthogonality requires that the $r_\phi(\tau)$ function have zeros at all integer multiples of T except $r_\phi(0) = 1$, that is, $r_\phi(nT) = \delta(n)$, the "unit-impulse" sequence. This is a condition familiar to data-transmission engineers — specifying a data pulse that produces no intersymbol interference. The frequency-domain version of this condition is known as the *Nyquist criterion*, and it is stated as

$$\sum_\ell R_\phi(f - \ell/T) = \sum_\ell |\Phi(f - \ell/T)|^2 = T \qquad \text{for all } f \qquad (5)$$

for the case under consideration here. It also follows directly from a relation known as the Poisson sum formula (2). Using Eq. (5), we can see that, for an

---

†We use $X(f) = \int x(t)e^{-j2\pi ft}dt$ and $x(t) = \int X(f)e^{j2\pi ft}df$ as the notations for a Fourier transform and its inverse, respectively.

arbitrary basic waveform $\omega(t)$ with Fourier transform $\Omega(f)$, the inverse transform of

$$\Phi(f) = \Omega(f)\left[\frac{1}{T}\Sigma_\ell|\Omega(f - \ell/T)|^2\right]^{-1/2} \tag{6}$$

makes $\{\phi(t - nT)\}$ an orthonormal set that spans the same subspace of $L^2(R)$ that $\{\omega(t - nT)\}$ does. This method of orthogonalization of the basis is an attractive alternative to the Gram–Schmidt procedure when dealing with time series representations. An example illustrating this procedure is shown in Fig. 2b.

Because of the above considerations, we are imposing no restrictions by assuming that $\{\phi(t - nT)\}$ is an orthonormal set and the best (minimum norm) time series approximation to an arbitrary waveform $x(t)$ is given, using the projection theorem, by

$$\hat{x}(t) = \Sigma_n c(n)\phi(t - nT); \qquad r_\phi(nT) = \delta(n) \tag{7}$$

where

$$c(n) = \int x(t)\phi^*(t - nT)dt \tag{8}$$

Note that with $h(t) = \phi^*(-t)$, the inner product that generates the discrete-time elements $c(n)$ is really a convolution integral. Therefore, the $c(n)$ are simply the uniformly sampled (every T seconds) output values of the matched filter $h(t)$ for the basic waveform $\phi(t)$ when the signal $x(t)$ is the input to the filter.

The familiar sampling theorem uses $\phi(t) = \text{sinc}(t/T) \triangleq [\sin(\pi t/T)]/(\pi t/T)$, which is an orthogonal set that could be normalized by scaling the basis functions by $\sqrt{1/T}$. The matched filter for $\text{sinc}(t/T)$ is an ideal low-pass filter with bandwidth $1/2T$, so the output of this filter is just a band-limited version of $x(t)$. Samples of the output yield the $c(n)$ for this particular representation. If $x(t)$ happens to be band limited to the frequency interval $|f| < 1/2T$, then the filter output is the same as $x(t)$, and under this condition we have $c(n) = x(nT)$ and $\|x - \hat{x}\| = 0$, and we can write

$$x(t) = \Sigma_n x(nT)\text{sinc}(t/T - n) \tag{9}$$

for all values of t.

When the signal $x(t)$ does not satisfy such restrictive, and physically unattainable, conditions as strict band limitation, then there will be some residual error, usually called an *aliasing error*, in the time series approximation. This error can be made small by making T small, at the obvious expense of increasing the rate of the resulting discrete-time signal. For a given sample interval T, one might try to reduce the error through appropriate selection of the basic waveform $\phi(t)$. It should be kept in mind, however, that no time series representation can give a close approximation to all signals in $L^2(R)$. This is because, for any $\phi(t)$, we can always find a signal that is orthogonal to the subspace spanned by $\{\phi(t - nT)\}$. In fact, $\psi(t) = \phi(-t)\sin(2\pi t/T)$ is such a signal.

The usual approach in signal processing is to assume the signals of interest in a particular application lie in some loosely defined class of essentially band-limited signals and then to choose a sample interval parameter T that gives an acceptably small error for all signals in the class. An important consideration is that different users of the same application (e.g., a medical imaging system) may impose different requirements on the acceptable error level or the same user may want to have control of the resolution level in order to "zoom in" on certain portions of the signal.

This has given rise to the concept of multiresolution analysis in the design of modern communication and signal processing systems (4,5). The basic idea is to allow a low-resolution rendition of a signal to be extended to the next higher level of resolution by adding a minimal amount of data to the already existing data at the current resolution level. At each resolution level, the signal is approximated by a time series representation but, by indefinitely increasing the resolution level, any signal in $L^2(R)$ can be closely approximated. Another way of saying this is that $L^2(R)$ can be spanned by a doubly indexed set of basis functions, $\phi_{mn}(t) = \phi_m(t - nT_m)$, where n indexes the time position as in the normal time series representation and m indexes the level of resolution by indicating the sample interval. In the conventional multiresolution analysis scheme, the $\phi_m(t)$ are time-scaled versions of a single basic waveform $\phi(t)$, in that $\phi_m(t) = (T/T_m)^{1/2}\phi(T/T_m)$. The usual arrangement is to make the scaling factors T/T_m equal to successive powers of 2, so that $T_m = 2^m T$, $\phi_m(t) = 2^{-m/2}\phi(2^{-m}t)$ and, for the mth level of resolution, the projection of $x(t)$ onto the subspace spanned by $\{\phi_m(t - nT_m)\}$ is†

$$\hat{x}_m(t) = \sum_n c_m(n)2^{-m/2}\phi(2^{-m}t - nT) \tag{10}$$

where

$$c_m(n) = \int x(t)2^{-m/2}\phi^*(2^{-m}t - nT)dt \tag{11}$$

since $\{\phi(t - nT)\}$ is assumed to be an orthonormal set as in Eqs. (7) and (8).

Note that in this traditional indexing scheme, increasing resolution corresponds to decreasing values of m. A key feature of multiresolution analysis is that the lower resolution subspaces are always contained within the higher resolution subspaces. Thus, it is possible to express the projection at level $m - 1$ as a sum of the projection at level m and a signal orthogonal to it:

$$\hat{x}_{m-1}(t) = \hat{x}_m(t) + w_m(t); \qquad (\hat{x}_m, w_m) = 0 \tag{12}$$

and

$$w_m(t) = \sum_n d_m(n)2^{-m/2}\psi(2^{-m}t - nT) \tag{13}$$

†The $2^{-m/2}$ amplitude scaling factor makes $\|\phi_m\| = 1$ for all m. The orthonormality of the mth level basis $\{\phi_m(t - nT_m)\}$ follows directly from the orthonormality of $\{\phi(t - nT)\}$.

The $\{2^{-m/2}\psi(2^{-m}t - nT)\}$, like the $\{2^{-m/2}\phi(2^{-m}t - nT)\}$, forms an orthonormal set for each level of resolution m. Also, the basic signal $\psi(t)$ is taken to be orthogonal to each signal in $\{\phi(t - nT)\}$, so that $w_m(t)$ is always orthogonal to $\hat{x}_m(t)$. The expression for $w_m(t)$ in Eq. (13) is called a *wavelet expansion*, and $\psi(t)$ is the basic wavelet-generating function. The complementary function $\phi(t)$ is called the *scaling function* for the particular version of multiresolution analysis.

Because of the "nested" nature of the subspaces used in multiresolution analysis, the scaling function is not an arbitrary function that generates an orthonormal set as in Eq. (7) and (8), but it must also be expressible as a linear combination of the functions spanning the next higher resolution. Since $\phi_0(t) = \phi(t)$ and $\phi_{-1}(t) = 2^{1/2}\phi(2t)$, we can write

$$\phi(t) = \sum_n 2h(n)\phi(2t - nT) \tag{14}$$

where

$$h(n) = \int \phi(t)\phi^*(2t - nT)dt \tag{15}$$

which is called the *dilation equation*; it represents a constraint that must be satisfied by a scaling function. Another constraint that is usually imposed is that at any resolution level, $\hat{x}_m(t)$, not $w_m(t)$, should contain the lower frequency components of the signal, specifically, the direct current (DC) component. It can be shown that this requires that $\int\phi(t)dt = \sqrt{T}$. The constraints specify conditions that the $\{h(k)\}$ sequence, or its discrete-time Fourier transform, $H(f) = \sum_k h(k)\exp(-j2\pi kTf)$, must satisfy

$$\sum_k h(k) = 1; \qquad H(0) = 1 \tag{16}$$

$$\sum_k h(k)h^*(k - 2n) = \delta(n)/2;$$

$$|H(f)|^2 + |H(f + 1/2T)|^2 = 1 \tag{17}$$

In a similar manner, the complementary wavelet function can also be expanded at the next higher level of resolution, giving an expression like the dilation equation

$$\psi(t) = \sum_n 2g(n)\phi(2t - nT) \tag{18}$$

where the $\{g(k)\}$ sequence, which specifies a particular type of wavelet, must satisfy a similar set of constraints:

$$G(0) = 1 \tag{19}$$

$$|G(f)|^2 + |G(f + 1/2T)|^2 = 1 \tag{20}$$

$$H^*(f)G(f) + H^*(f + 1/2T)G(f + 1/2T) = 0 \tag{21}$$

The first constraint Eq. (19) results from $\int \psi(t)dt = 0$, which comes about since $w_m(t)$ should not contain any part of the DC component of the signal. Equation (20), like Eq. (17), is due to $\{\psi(t - nT)\}$ forming an orthonormal set. The additional condition Eq. (21) results from making $\psi(t)$ orthogonal to every element in $\{\phi(t - nT)\}$. Given an $H(f)$ that satisfies the conditions for a scaling function, the corresponding wavelet conditions are always satisfied by

$$G(f) = \pm e^{-j2\pi Tf} H^*(f + 1/2T) \qquad (22)$$

or equivalently by

$$g(k) = \pm(-1)^{1-k} h(1 - k) \qquad (23)$$

Some examples of simple scaling functions and corresponding wavelets are shown in Fig. 2. The first example corresponds to projections representing piecewise constant approximations to a signal, the constant value in each sample interval being the average value of $x(t)$ over that interval, for a minimum norm approximation. Note that this differs from the familiar "sample-and-hold" operation, which also produces a piecewise constant approximation, but not one having the minimum error norm. It is easily seen that $\{h(0) = h(1) = 1/2;$ $h(k) = 0$ otherwise$\}$ satisfies the dilation equation for the rectangular scaling function. Then, using Eq. (23), the resulting wavelet is the rectangular doublet, or Haar function, shown in Fig. 2-*a*. The second example corresponds to minimum-norm piecewise linear approximations. The scaling function is obtained by applying Eq. (6) with $\Omega(f) = T\text{sinc}^2(Tf)$, the Fourier transform of a unit triangular waveform with a base width of $2T$. Numerical evaluation of the inverse transform of Eq. (6) gives the orthonormal equivalent of the triangular waveform (see Fig. 2-*b*). The wavelet function is obtained using Eq. (23). The third example illustrates the low-pass/high-pass nature of scaling functions and wavelets. In this case, the $H(f)$ is chosen to be an idealized half-band low-pass function with $H(f) = 1$ in $|f - \ell/T| \leq 1/4T$ for any integer ℓ and $H(f) = 0$ elsewhere. Then, $G(f)$ is a high-pass function complementary to $H(f)$, that is, $G(f) = 0$ in the pass-band region of $H(f)$ and $G(f) = 1$ elsewhere. The scaling function for this example is the sinc(t/T) function (see Fig. 2-*c*).

The discrete-time signals resulting from representing multiresolution waveforms exhibit a simple first-order recursion between adjacent levels of resolution. Starting from Eq. (12) and using the relations between scaling and wavelet functions at adjacent levels, we find

$$c_{m-1}(\ell) = 2^{1/2}\sum_n c_m(n)h(\ell - 2n) + 2^{1/2}\sum_n d_m(n)g(\ell - 2n) \qquad (24)$$

The two terms in Eq. (24) can be interpreted as discrete-time convolutions if the sequences are appropriately modified. Specifically, the first term is the convolution of the $\{h(n)\}$ sequence with a sequence derived from $\{c_m(n)\}$ by inserting zero elements between each element of the original sequence. This process is called *upsampling* (by a factor of 2) and is represented by the block with the ↑2 symbol in Fig. 3. Summing the two paths shown in Fig. 3 gives the discrete-time signal at the next higher level of resolution.

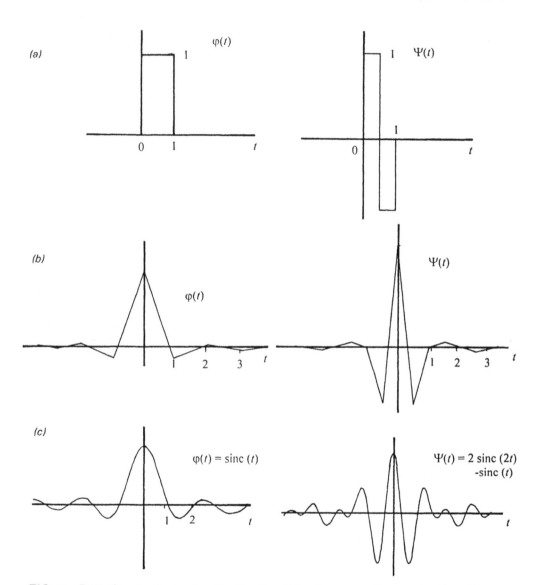

FIG. 2 Examples of orthogonal scaling function $\phi(t)$ and corresponding wavelet $\psi(t)$ (with $T = 1$): a, piecewise constant representation of arbitrary signal; b, piecewise linear representation; c, band-limited representation.

The inverse problem of resolving a given sequence $c_{m-1}(n)$ into its $c_m(n)$ and $d_m(n)$ components is handled by projections of $\hat{x}_{m-1}(t)$ onto subspaces spanned by $\{\phi_m(t - nT_m)\}$ and $\{\psi_m(t - nT_m)\}$, respectively. This results in the relations

$$c_m(n) = 2^{1/2}\sum_k h^*(k)c_{m-1}(2n + k) \qquad (25)$$

$$d_m(n) = 2^{1/2}\sum_k g^*(k)c_{m-1}(2n + k) \qquad (26)$$

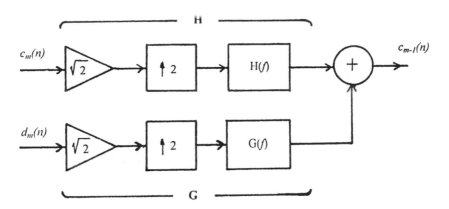

FIG. 3 Scalar-upsampler-filter configuration to reconstruct signal at next higher resolution level.

which can also be considered as discrete-time convolutions. Specifically, Eq. (25) is just the even-numbered samples of the convolution of the $\{c_{m-1}(n)\}$ sequence with $\{h^*(-n)\}$. Taking only the alternate samples of a signal is called *downsampling* (by a factor of 2), represented by the block with the $\downarrow 2$ symbol in Fig. 4.

The simple structures in Figs. 3 and 4 can be replicated to form an entire R-level multiresolution system as shown in Fig. 5. Letting operators $\tilde{H}$ and $\tilde{G}$ represent the filter-downsampler-scaler combinations of Eqs. (25) and (26), the left-hand structure in Fig. 5 performs the decomposition into a set of lower resolution signals. The right-hand part, consisting of operators H and G corresponding to Eq. (24), performs the reconstruction of the higher resolution signal by progressively summing in signals derived from the $d_m(n)$ sequences. In a typical application, flexibility is gained by the methods used to handle the intermediate signals $d_m(n)$. Some users will require only a subset of these signals to

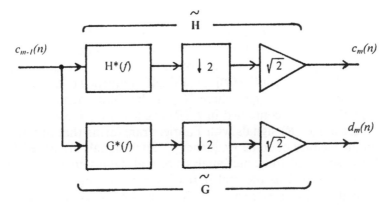

FIG. 4 Filter-downsampler-scaler configuration to decompose signal into its next lower resolution level components.

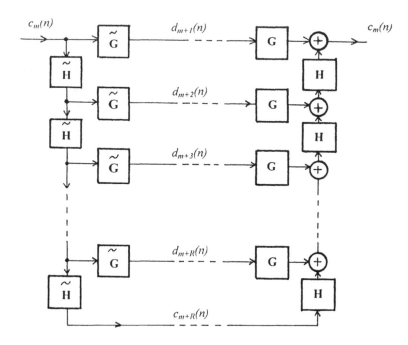

FIG. 5 R-level multiresolution system.

reconstruct lower resolution versions of the signals. In transmission or storage systems, the intermediate signals may be allocated different numbers of bits per sample, thereby minimizing the overall bit count. The scheme is somewhat like subband coding, except the intermediate signals represent information at different time scales instead of different frequency bands.

Analysis Using Complex-Value Signals

The concept of a "single-frequency" signal is often useful in signal analysis. The $\cos(2\pi f_0 t)$ and $\sin(2\pi f_0 t)$ functions are double-frequency signals, having components at $+f_0$ and $-f_0$. On the other hand, the cisoidal function $\exp(j2\pi f_0 t) = \cos(2\pi f_0 t) + j[\sin(2\pi f_0 t)]$ has a single-frequency component at f_0. Its Fourier transform is $\delta(f - f_0)$, where $\delta(\cdot)$ is the Dirac delta function. Note that such signals are not contained in $L^2(R)$ because they have infinite energy. However, there are $L^2(R)$ signals with Fourier transforms that are mainly concentrated in the vicinity of some frequency f_0. Such signals are called *band-pass signals*. Many aspects of frequency analysis of real signals are greatly facilitated by expressing the signal as the real part of a complex signal $x(t) = Re[z(t)]$. This is a straightforward extension of the familiar phasor representation for sinusoidal signals. Because of the many-to-one mapping nature of the $Re[\cdot]$ operator, there is no unique association of a complex $z(t)$ with a given real $x(t)$.

One of the most useful ways of making a one-to-one correspondence between $z(t)$ and $x(t)$ is to write $z(t) = x(t) + j[\hat{x}(t)]$, where $\hat{x}(t)$ is a linear transformation of $x(t)$. In particular, the usual assignment is to make $\hat{x}(t)$ the Hilbert transform of $x(t)$, that is,

$$\hat{x}(t) = \frac{1}{\pi} \int x(\tau)[t - \tau]^{-1}d\tau \qquad (27)$$

Equation (27) can be regarded as a convolution integral, so that $\hat{X}(f)$, the Fourier transform of $\hat{x}(t)$, can be expressed as $\hat{X}(f) = -j[\text{sgn}(f)]X(f)$ since $-j[\text{sgn}(f)]$ is the Fourier transform of the time function $1/\pi t$. Then, we have

$$
\begin{aligned}
Z(f) &= \{1 + j[-j \, \text{sgn}(f)]\}X(f) \\
&= 2X(f) \qquad \text{for } f > 0 \\
&= 0 \qquad \text{for } f < 0
\end{aligned}
\qquad (28)
$$

so that $Z(f)$ is always "one sided" in that it will always vanish for $f < 0$. The corresponding $z(t)$ is called the *analytic signal* or the *pre-envelope* for $x(t)$. Note that it is not necessary for $x(t)$ to be a band-pass signal or to invoke any "narrowband" approximations in order to derive its analytic signal equivalent. One of the first applications of the analytic signal was in defining an instantaneous frequency for arbitrary (not necessarily narrowband) signals. The definition for the time-dependent instantaneous frequency is

$$f_i(t) = \frac{1}{2\pi} Im \left\{ \frac{d}{dt} \ln [z(t)] \right\} = \frac{1}{2\pi} Im \left\{ \frac{\dot{z}(t)}{z(t)} \right\} \qquad (29)$$

The $1/2\pi$ factor is present to convert units from radians/second to hertz (Hz).

For signals of a definite band-pass character, such as carrier-modulated signals, with a $Z(f)$ concentrated in the vicinity of f_0, it is convenient to write $z(t) = w(t)\exp(j2\pi f_0 t)$, where $w(t)$ is called the *complex envelope* of $x(t)$ relative to a "center frequency" of f_0. Writing $z(t)$ in rectangular form, $z(t) = u(t) + jv(t)$, or, in polar form, $z(t) = A(t)\exp[j\Theta(t)]$, give the following expressions for the real band-pass signal:

$$
\begin{aligned}
x(t) &= Re[w(t)e^{j2\pi f_0 t}] \\
&= u(t) \cos(2\pi f_0 t) - v(t) \sin(2\pi f_0 t) \qquad (30) \\
&= A(t) \cos(2\pi f_0 t + \Theta(t)) \qquad (31)
\end{aligned}
$$

In conventional modulation terminology, regarding f_0 as the carrier frequency, the $u(t)$ and $v(t)$ signals are called the *in-phase* and *quadrature* components, respectively, of $x(t)$. The time functions $A(t) = |z(t)|$ and $\Theta(t)$ are the *envelope* and the *phase*, respectively, of the $x(t)$ signal. In terms of the complex envelope, the instantaneous frequency of $x(t)$ is

$$f_i(t) = f_0 + \frac{1}{2\pi} Im\left\{\frac{\dot{w}(t)}{w(t)}\right\} = f_0 + \frac{1}{2\pi}\dot{\Theta}(t) \qquad (32)$$

which gives the conventional definition of frequency deviation for a frequency modulation (FM) system.

The other main application of the analytic signal is to provide definitions for center frequency and bandwidth of an arbitrary signal. As explained in the Gabor paper (1), this can be done in terms of the first- and second-order moments of a normalized version of the $|Z(f)|^2$ function. Let $\omega(t) = z(t)/\|z\|$ be an energy-normalized version of the analytic signal. This normalization does not affect quantities such as bandwidth. Note that $\|z\|^2 = 2\|x\|^2$ and that $\int_0^\infty |\Omega(f)|^2 df = 1$. Now, define

$$f_0 = \int f|\Omega(f)|^2 df \qquad (33)$$

and

$$W_{rms}^2 = \int (f - f_0)^2 |\Omega(f)|^2 df \qquad (34)$$

as the center frequency and the square of the bandwidth of $x(t)$. These definitions are analogous to the mean and variance of a non-negative random variable with a probability density function that has the same form as $|\Omega(f)|^2$. In mechanics, f_0 and W_{rms} would correspond to the centroid and radius of gyration of a mass distributed according to $|\Omega(f)|^2$.

It is interesting to relate these frequency-domain parameters to the time domain through the instantaneous frequency concept. Let $a(t) = |\omega(t)|$ be the envelope of the energy-normalized signal, then we have the following relationships (2):

$$f_0 = \int a^2(t) f_i(t) dt \qquad (35)$$

$$W_{rms}^2 = \int a^2(t)[f_i(t) - f_0]^2 dt + (1/2\pi)^2 \|\dot{a}\|^2 \qquad (36)$$

This gives the intuitively satisfying result that the center frequency is the time average of the envelope-weighted instantaneous frequency. The mean-squared bandwidth expression has two parts; one is the time average of the envelope-weighted square of the instantaneous frequency deviation from f_0, while the other part is a term that depends only on the time fluctuations of the envelope. For a constant-envelope signal such as a standard FM signal, the second term in Eq. (36) vanishes. On the other hand, for a double-sideband amplitude modulation (AM) signal, there is no phase fluctuation, and the first term in Eq. (36) vanishes.

Using the same type of moment quantities to characterize the time position and the root-mean-squared (rms) duration of $x(t)$, we have

$$t_0 = \int t|x(t)|^2 dt / \|x\|^2 \tag{37}$$

$$T_{rms}^2 = \int (t - t_0)^2 |x(t)|^2 dt / \|x\|^2 \tag{38}$$

From these definitions, it is easy to derive an inequality relation for the duration-bandwidth product of any signal. The result is

$$T_{rms} W_{rms} \geq 1/4\pi \tag{39}$$

which is known as the *uncertainty principle* in signal analysis (1). It also follows easily that the signal that attains this lower bound on the duration-bandwidth product is the Gaussian envelope pulse waveform

$$x(t) = e^{-[((t-t_0)/2T_{rms})^2 + j2\pi f_0 t]} \tag{40}$$

which is centered at t_0 and f_0 in time and frequency with a bandwidth of $W_{rms} = [4\pi T_{rms}]^{-1}$.

Of course, alternative definitions for duration, bandwidth, or both will give different results for the optimum pulse exhibiting the minimum duration-bandwidth product. Gabor also considered the case of a strictly band-limited pulse having a minimum rms duration (1). The result is that

$$X(f) = \frac{\pi}{4W} \cos(\pi f/2W) \qquad \text{for } |f| \leq W$$

$$= 0 \qquad\qquad\qquad \text{otherwise} \tag{41}$$

resulting in $x(t) = [1 - (4Wt)^2]^{-1} \cos(2\pi Wt)$ as the band-limited pulse having the smallest value of T_{rms}.

In a later paper, Chalk considered the dual of the above problem (6), that is, what pulse, duration limited to the time interval $|t| \leq T$, loses the least fraction of its total energy in passing through a specified linear time-invariant filter? One of the cases Chalk considered for the filter was an ideal low-pass filter with cutoff frequency at W Hz. It is found, by application of standard techniques of the calculus of variations, that the optimum pulse in this case must satisfy the integral equation,

$$\int_{-T}^{T} \frac{\sin 2\pi W(t - \tau)}{\pi(t - \tau)} x(\tau) d\tau = \lambda x(t) \tag{42}$$

for $|t| \leq T$. Thus, the optimum pulse is seen to be an eigenfunction (the one with the largest eigenvalue) of a certain linear operator. This operator is composed of a gating operator that truncates the part of a signal lying outside the time interval $|t| \leq T$, followed by a band-limiting operator that suppresses all components outside the frequency interval $|f| \leq W$. Chalk demonstrated the solution to Eq. (42) using numerical techniques. Over a decade later, a series of papers showed that the eigenfunctions of Eq. (42) are the prolate spheroidal wave functions and extended Chalk's results by showing how these functions are used to form pulses that have a fraction of at least α^2 of their total energy in the time interval $|t| \leq T$ and at the same time a fraction of at least β^2 of the total energy in the frequency interval $|f| \leq W$ (7–9). Since these results were published, the prolate spheroidal wave functions have been extensively used as a basis for representing signals having an approximately band-limited character.

There is a broad variety of optimization problems that can be formulated and solved in essentially the same manner. One class of problem that received a lot of attention was in the design of pulses for data transmission for which the pulses satisfy a variety of constraints on peak value, energy, bandwidth, and so on while giving minimal intersymbol interference (2,10).

Time-Frequency Analysis

Having established that the elementary signal given in Eq. (40) is, in a particular sense, maximally concentrated at a point (t_0, f_0) in the time-frequency plane, Gabor (1) then proposed to use

$$\phi_{nk}(t) = e^{-\frac{\pi}{2}(t/T - n)^2 + j2\pi kWt} \tag{43}$$

as a doubly indexed set of functions as a basis to (approximately) span $L^2(R)$. Note that $\phi_{nk}(t)$ is centered at (nT, kW) in the time-frequency plane, $T = \sqrt{2\pi}\, T_{rms} \cong 2.506 T_{rms}$, and $W = 1/2T$. Then,

$$x(t) \cong \sum_n \sum_k c_{nk}\, \phi_{nk}(t) \tag{44}$$

and each element of the basis, called a *logon* by Gabor, covers an area $TW = 1/2$ in the time-frequency plane. Gabor acknowledged that finding the c_{nk} expansion coefficients is not trivial since the $\{\phi_{nk}(t)\}$ is not an orthonormal set.

Modern usage of time-frequency analysis tends to separate the problems of "discretization" of the signal (finding the c_{nk} values) and characterization of signal properties by the nature of the continuous distribution of energy density over the time frequency plane. The ideas of Gabor were carried over into the definition of the *Gabor transform* (also called the *short-time Fourier transform*):

$$X_g(t,f) = \int x(\tau)g^*(\tau - t)e^{-j2\pi f\tau}d\tau \tag{45}$$

The $g(\tau)$ function in Eq. (45) is usually duration limited and is regarded as a "sliding window" (positioned at time t) that selects a portion of the entire $x(\tau)$ waveform before the Fourier transform is taken. In this way, the changing spectral characteristics of the signal can be displayed as time progresses. It is helpful to think of speech signals as an example in which the time-varying spectral content has a fundamental importance. The Gabor transform retains all the information in a signal and, in fact, has an inverse given by

$$x(t) = \int \int X_g(\tau,\nu)g(t - \tau)e^{j2\pi\nu t}d\nu d\tau \qquad (46)$$

provided that $g(t)$ is normalized so that $\|g\| = 1$. In one-dimensional analysis, $|x(t)|^2$ and $|X(f)|^2$ are the energy temporal density and the energy spectral density, respectively. In time-frequency analysis, it is desired to use a two-dimensional density function $P(t,f)$, which is consistent with the one-dimensional concepts. In particular (11), we would like

$$P(t,f) \geq 0 \qquad (47)$$

$$\int \int P(t,f)dtdf = \|x\|^2 \qquad \text{(total energy)} \qquad (48)$$

$$\int P(t,f)df = |x(t)|^2 \qquad \text{(energy temporal density)} \qquad (49)$$

$$\int P(t,f)dt = |X(f)|^2 \qquad \text{(energy spectral density)} \qquad (50)$$

For a unit-energy signal, these conditions on $P(t,f)$ are the same as would be satisfied by the joint probability density function of two random variables. In that case, Eqs. (49) and (50) correspond to the marginal densities. The function $P(t,f) = |X_g(t,f)|^2$, commonly used in speech analysis, is the *spectrogram* of $x(t)$. It satisfies conditions in Eqs. (47) and (48), but

$$\int P(t,f)df = \int |g(\tau - t)|^2|x(\tau)|^2d\tau \qquad (51)$$

$$\int P(t,f)dt = \int |G(\nu - f)|^2|X(\nu)|^2d\nu \qquad (52)$$

for this case. We see that Eqs. (51) and (52) are the convolutions of energy temporal density and energy spectral density with $|g(-t)|^2$ and $|G(-f)|^2$, respectively. This reveals the limitation on the spectrogram for providing good resolution simultaneously in time and frequency. We would like both $|g(-t)|^2$ and $|G(-f)|^2$ to be "impulselike" in nature so that Eqs. (51) and (52) are approximately equal to $|x(t)|^2$ and $|X(f)|^2$, but the uncertainty principle will not permit this. If $g(t)$ is narrow compared to the interval of significant fluctuations in $x(t)$, then $G(f)$ is broad relative to the variations in $X(f)$, and Eq. (52) results in a highly smoothed version of $|X(f)|^2$. Conversely, if $G(f)$ is narrow, then Eq. (51) gives a highly smoothed version of $|x(t)|^2$.

We can look at the time-frequency resolution problem directly by extending the idea of the time ambiguity function discussed above. To this end, we examine the signal space separation between $x(t)$ and a time- and frequency-shifted version of $x(t)$. A slight modification on this is to define

$$x(t;\tau,\nu) = x(t - \tau)e^{j2\pi\nu t} \tag{53}$$

and then

$$d^2[x(t; -\tau/2, -\nu/2), x(t;\tau/2,\nu/2)] = 2w_x(0,0) - 2\,Re\,w_x(\tau,\nu) \tag{54}$$

where $w_x(\tau,\nu)$ is the inner product of the two versions of $x(t)$, first time shifted by $\pm\tau/2$ and then frequency shifted by $\pm\nu/2$. This symmetric shifting arrangement is used because then the resulting $w_x(\tau,\nu)$ function is independent of whether the time shifting precedes the frequency shifting or vice versa. The $w_x(\tau,\nu)$ function is called the *time-frequency ambiguity* function and is given by the inner product expression

$$w_x(\tau,\nu) = \int x(t + \tau/2)x^*(t - \tau/2)e^{-j2\pi\nu t}dt \tag{55}$$

The signal energy is given by $w_x(0,0)$ and $w_x(\tau,0) = r_x(\tau)$, the time-ambiguity function for $x(t)$. The function $R(\tau,\nu) = |w_x(\tau,\nu)|^2$ is called the *radar ambiguity function* and has been extensively used for the analysis and design of pulses for Doppler radar systems, which require both good range and velocity resolution in target returns (3). For unit-energy signals, $R(0,0) = 1$ and also

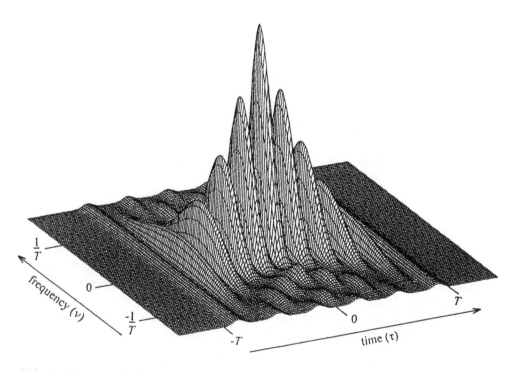

FIG. 6 Radar ambiguity function for rectangular envelope carrier burst. Duration of the burst is T seconds, and the burst consists of three complete cycles of the carrier. Secondary concentrations of ambiguity at plus and minus twice the carrier frequency are not seen in the figure because of the truncation of the frequency axis.

$$\int \int R(\tau,\nu)d\tau d\nu = 1 \tag{56}$$

which is another kind of expression for the uncertainty principle. For good time and frequency resolution, $R(\tau,\nu)$ should drop off rapidly from its peak value at $\tau = 0$, $\nu = 0$. However, the volume integral constraint of Eq. (56) means that the ambiguity cannot be made arbitrarily small over the entire $\tau\nu$ plane. If it is forced to be small in certain regions, then it must "pop up" in other regions of the $\tau\nu$ plane in order to satisfy Eq. (56). The radar ambiguity for a carrier burst with a rectangular envelope containing three complete cycles of the carrier is shown in Fig. 6. If the duration of the envelope is increased, then the constant-ν cross-sections of the ambiguity peak will be broader, while the constant-τ cross-sections will be correspondingly narrower.

The Wigner distribution is a popular alternative to the spectrogram for certain applications (11). It is given by the double Fourier transform of the time-frequency ambiguity function:

$$W_x(t,f) = \int \int w_x(\tau,\nu)e^{-j2\pi(f\tau - \nu t)}d\tau d\nu$$

$$= \int x(t + \tau/2)x^*(t - \tau/2)e^{-j2\pi f\tau}d\tau \tag{57}$$

Letting the energy density function be $P(t,f) = W_x(t,f)$, we find that all of the desired conditions on $P(t,f)$ except Eq. (47) are satisfied; there is no guarantee that $W_x(t,f)$ is non-negative or even real. Nevertheless, the Wigner distribution does not exhibit the resolution problems discussed above for the spectrogram. It has another kind of problem, however. Because of the quadratic nature of

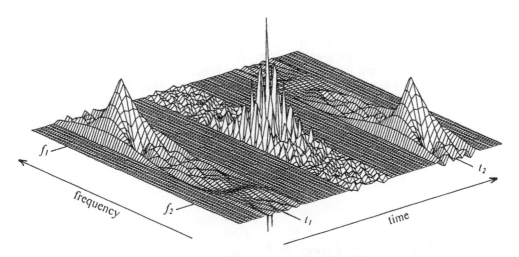

FIG. 7 Wigner distribution for the "beep-boop" signal: two rectangular envelope carrier bursts of frequency f_1 and f_2, centered at times t_1 and t_2, respectively. The burst durations are T seconds, and the frequency separation is $f_1 - f_2 = 5/T$. The time separation is $t_2 - t_1 = 4T$.

$W_x(t,f)$, when $x(t)$ is a multicomponent signal, "cross-product" terms appear indicating energy at times and frequencies not present in the $x(t)$ signal. These spurious terms in the Wigner distribution can be suppressed to some degree by suitably windowing the time-frequency ambiguity function before taking the Fourier transform as in Eq. (57).

This and many other aspects of time-frequency distributions are covered in the survey paper by Cohen (11). As an illustrative example, the Wigner distribution for the two-component "beep-boop" signal is shown in Fig. 7. This signal consists of two rectangular envelope carrier bursts at different frequencies, with the bursts being nonoverlapping in time. The spurious components of the distribution, centered at the mean value of the two frequencies and the mean value of the two times, are quite evident in the figure.

References

1. Gabor, D., Theory of Communication: Part 1. The Analysis of Information, *J. IEE*, 93(3):429–441 (November 1946).
2. Franks, L. E., *Signal Theory*, Prentice-Hall, Englewood Cliffs, NJ, 1969; reprint, Dowden and Culver, Stroudsburg, PA, 1981, 1988.
3. Cook, C. E., and Bernfeld, M., *Radar Signals*, Academic Press, New York, 1967.
4. Rioul, O., and Vetterli, M., Wavelets and Signal Processing, *IEEE Signal Proc. Mag.*, 14–38 (October 1991).
5. Mallat, S. G., A Theory for Multiresolution Signal Decomposition: The Wavelet Representation, *IEEE Trans. Pattern Analysis and Machine Intelligence*, 11(7): 674–693 (July 1989).
6. Chalk, J. H. H., The Optimum Pulse-Shape for Pulse Communication, *Proc. IEE*, 97(3):82–92 (March 1950).
7. Slepian, D., and Pollak, H. O., Prolate Spheroidal Wave Functions, Fourier Analysis, and Uncertainty—I, *Bell Sys. Tech. J.*, 40(1):43–63 (January 1961).
8. Landau, H. J., and Pollak, H. O., Prolate Spheroidal Wave Functions, Fourier Analysis, and Uncertainty—II, *Bell Sys. Tech. J.*, 40(1):65–84 (January 1961).
9. Landau, H. J., and Pollak, H. O., Prolate Spheroidal Wave Functions, Fourier Analysis, and Uncertainty—III: The Dimension of the Space of Essentially Time- and Band-Limited Signals, *Bell Sys. Tech. J.*, 41(4):1295–1336 (July 1962).
10. Franks, L. E., Further Results on Nyquist's Problem in Pulse Transmission, *IEEE Trans. Commun. Technol.*, COM-16(2):337–340 (April 1968).
11. Cohen, L., Time-Frequency Distributions—A Review, *Proc. IEEE*, 77(7):941–981 (July 1989).

LEWIS E. FRANKS

Signaling System #7   (see Common Channel Signaling)

SMDS—Switched Multimegabit Data Service

Introduction

This article is about *Switched Multimegabit Data Service* (SMDS). The name says it all:

- *Switched*: SMDS provides the capability for communications between any subscribers, just like the telephone network. In fact, SMDS even uses telephone numbers to identify subscribers (or at least their data communications equipment).
- *Multimegabit*: SMDS is intended for the interconnection of local-area networks (LANs) and therefore provides bandwidth similar to LANs. The multimegabit nature of SMDS makes it the first broadband (greater than 2.048 megabits per second [Mb/s]) public carrier service to be deployed.
- *Data*: SMDS is intended for carrying traffic found on today's LANs. This is generally called data, but in fact includes other types of traffic (e.g., images). The industry is rife with hyperbole about multimedia and video on demand. Yet, compared to the tens of millions of personal computers (PCs) currently connected to LANs, the market for these nondata services is a mere grease spot on the information highway. SMDS is meant to meet the substantial and growing need for high-speed LAN interconnection.
- *Service*: When it comes to public carrier data services, there is much confusion between technology and service.* SMDS is a service. It is not a technology or a protocol. It will follow the time-honored tradition of public carrier services; the service features will stay constant while the technology used in the carrier network is repeatedly improved. For example, SMDS is the first switched service based on asynchronous transfer mode (ATM) technology.

This article is excerpted from Ref. 1.† It contains an overview and a comprehensive treatment of SMDS.

This section describes some of the history behind SMDS, including the fundamental concepts and philosophy underlying the development of the service. It also enumerates sources of SMDS specifications.

The Problem: Market Perception of Telephone Companies

The history of SMDS can be viewed as beginning in July 1982 when the Institute of Electrical and Electronics Engineers (IEEE) Project 802 initiated a new work-

---

*This must be related to the data aspect since no such confusion exists for voice services. No one chooses their interexchange telephone service provider based on the protocols used to control the carrier's telephone switches.

† © 1994, *SMDS Wide-Area Data Networking with Switched Multi-megabit Data Service*, Prentice Hall, Englewood Cliffs, NJ.

ing group, IEEE P802.6, to develop standards for metropolitan-area networks (MANs). The key significance of the formation of IEEE P802.6 was that it reflected the realization among LAN vendors that there was a market for the connection of LANs across wide areas. Furthermore, it was perceived that existing standards activities were not adequately addressing this potential market. When considering wide-area connectivity, it was natural to think of telephone carriers. The 1982 statement of direction for the telephone industry was the integrated services digital network (ISDN).

Unfortunately, there was significant concern among LAN vendors about the adequacy of ISDN for meeting the LAN interconnection market. The greatest concern was the lack of bandwidth. At the time, the emphasis was on the 16-kilobits-per-second (kb/s) D channel for providing the data part of ISDN. Another possibility was to use the 64-kb/s B channel. Either approach would introduce a speed bottleneck. The time to transmit a 1000-byte packet on a Carrier Sense Multiple Access with Collision Detection (CSMA/CD) (10-Mb/s) LAN is 0.8 milliseconds (ms) (2). To do the same on a B channel takes 125 ms. It was hard to argue that using ISDN would provide transparent LAN interconnection.

Unfortunately for the telephone carriers, the lack of credibility for LAN interconnection was not limited to the dim view of ISDN. In 1982, telephone networks were considered to be slow and error prone. To a large extent, this view was accurate given the analog technology of the telephone networks and the modem technology of the day, which was the typical approach for both switched and private-line wide-area data connectivity. The only plausible carrier alternatives to the telephone companies in the early 1980s were the community antenna television (CATV) providers. As a result, more out of desperation than enthusiasm, the initial MAN protocol proposals were based on the use of CATV technology.

The formation of Working Group IEEE P802.6 stimulated work to begin within Bell Laboratories to develop high-speed data services that would be suitable for LAN interconnection. After the breakup of the Bell system, this work was continued by Bell Communications Research (Bellcore).* In 1985 and 1986, Bellcore, in cooperation with NYNEX and U.S. West, carried out successful concept trials of MAN token-ring technology to interconnect LANs across a metropolitan area.

As the MAN trials concluded in 1986, Bellcore work on defining a new, high-speed data service began. There were a number of goals:

- *The service should be aimed at LAN interconnection.* The LAN market had moved into a dynamic growth phase with ever-increasing numbers being installed, and market projections were highly optimistic. It did not take a

---

*Bellcore is a research and development company owned and principally funded by the Regional Bell Operating Companies (RBOCs). Among its activities is the definition of services that can be supported by the regulated networks of the RBOCs in order to foster nationwide service consistency. SMDS is the result of such service-definition work. In 1996 the RBOCs announced their plan to sell Bellcore to Science Applications International Corporation, a California-based consulting company.

great leap of faith to believe that a service that allowed the transparent interconnection of LANs across a wide area would find a significant market.

• *The service should offer more value than fast private lines.* By the time the MAN trials were completed, the position of the North American telephone carriers with respect to high-speed data communications had improved somewhat. The conversion of the telephone networks to digital transmission technology had made 1.544-Mb/s DS-1 (Digital Signal Level 1) private-line service viable. This service, in conjunction with the advent of remote bridges, made possible the interconnection of LANs with a bandwidth starting to approach that of the LANs. However, being simply a provider of private-line services for data was not considered a desirable business strategy (being in the private-line business is derisively referred to as being a WACCO [wire and cable company]). The inevitable arrival of local competition was expected to lead to private lines becoming a commodity. Thus, features and capabilities that provide customers with value beyond that available from private lines were considered vital.

• *The service should help Regional Bell Operating Companies (RBOCs) shed the image of being incapable of providing data communications services.* The demand for POTS (plain old telephone service) was expected to experience low, single-digit growth rates, whereas data communications (such as LAN interconnection) were expected to grow much faster. Therefore, it was vital that a useful data service be deployed quickly to build market presence and credibility for the RBOCs.

In addition to the above goals, there was a significant constraint. Because of the funding and ownership arrangements, Bellcore work on defining new services was constrained to focus on services that could be offered by the regulated network entities within the RBOCs. In this case, a key implication of this constraint was that the equipment that the regulated network provider owns, and that is on the customer's premises, is highly constrained in functionality. Another constraint was that the service cannot provide connectivity between local access and transport areas (LATAs) without the use of an interexchange carrier (IEC).

A Solution: Switched Multimegabit Data Service

The fundamental concept underlying SMDS is the exploitation of existing solutions for LAN interconnection. Figure 1 illustrates a typical approach to LAN interconnection in the local area. By using high-performance bridges or routers, LANs can be interconnected by means of backbone LANs. The approach for SMDS is to make the service look like a backbone LAN (see Fig. 2). This simple, but elegant, approach was conceived by Christine Hemrick while she was employed by Bellcore.

This basic approach to SMDS was established in late 1986. In early 1987, work began in earnest to flesh out the details of the service. In pursuing this work, several principles were followed.

The first principle was to be market driven. The intended market for SMDS

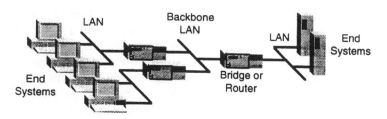

FIG. 1   Local-area interconnection of local-area networks (LANs).

high-speed interconnection of LANs was well defined from the beginning, and this was invaluable in keeping the work focused. It was believed that this market would soon blossom. This was of great concern because of the difficulty and long lead times involved in deploying new telephone-carrier-based services. Hence, the work was approached with a great sense of urgency. Furthermore, it was felt that the market opportunity, by the time the service could be deployed, would be for speeds above DS-1 (1.544 Mb/s). The assumption was that, once a DS-1 private-line data network was installed (along with the organization to maintain it), it would be difficult to induce a customer to migrate to the public network unless the customer needed a step up in performance.* By all indications, by the time SMDS could be deployed, DS-1 private-line data networks would be widely deployed. Consequently, the initial work was limited to defining DS-3 (Digital Signal Level 3; 44.736 Mb/s) interface rates (3). The addition of DS-1 interface rates was relatively easy and was done in 1989 (4). Low-speed (e.g., 56 kb/s), low-cost access to SMDS was defined in 1993 (5,6). The specification of DS-1 and slower interface rates was intended to make SMDS economically attractive to a larger share of the LAN interconnection market. The addition of European interface rates was straightforward and was done in 1993 (7).

The second principle was to respect customers' existing data network investment. No matter how wonderful the service offered by a carrier, if making use

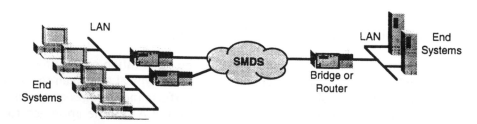

FIG. 2   Interconnection of local-area networks using switched multimegabit data service (SMDS).

*In retrospect, this assumption was not necessarily valid. The aggressive tariffs for DS-1 SMDS and the benefits of not having to manage a "rat's nest" of DS-1 private lines have proved to be strong inducements to customers to buy DS-1-based access to SMDS (see the overview sections of this article).

of it requires substantial changes to the customer's installed data network, the service will be unattractive. The basic concept of making SMDS look like a LAN addresses this principle. SMDS is synergistic with many existing data-networking architectures, and thus it has very little impact on the installed base of end systems, LANs, bridges, and routers. Group addressing (a form of multicasting, discussed in a separate section below) is an example of an element of SMDS that addresses this principle.

The third principle was to emphasize service attributes. A criticism that can be leveled at ISDN is that it was derived by looking at what was possible to implement in a telephone network without much consideration for the useful-ness of the resulting services (or even what the services might be). In defining SMDS, a conscious effort was made to ignore network implementation issues and instead focus on what is useful to the customer. For example, group ad-dressing is very useful to the customer, but its implementation details were not addressed by Bellcore until 1990 (8), three years after it was first specified as a service feature for SMDS.

The minimization of innovation was the fourth principle. Given the urgency of getting a useful LAN interconnection service deployed, it was imperative that unnecessary innovation be avoided. In particular, this meant avoiding the temptation to design the "glorious ultimate service" that would support all appli-cations using intelligent network concepts. By the time such a service could be deployed, one could expect completely different communications technologies to have taken over the market, perhaps of the type envisioned in *The President's Analyst*.*

The fifth principle was to "hit the dates." In early 1987, it was believed that 1991 was a realistic goal for the first deployment of SMDS. Earlier deployment was viewed as desirable from a market timing point of view. The realities of deploying a new service suggested that 1991 was optimistic, but realistic. In the end, Bell Atlantic derived revenue from a "pre-SMDS" service in December 1991. At the time of this writing, most of the RBOCs, GTE, and MCI are deploying SMDS; several IECs in addition to MCI have announced SMDS plans; telephone carriers in other parts of the world (e.g., Europe) have an-nounced SMDS plans; SMDS customer network management (CNM) has been deployed; and SMDS interest groups are operating in North America, Europe, and the Pacific Rim.

Sources of Specifications

A number of different organizations have been or are involved in activities related to SMDS specifications. Each is described below, including descriptions of the material that each organization developed.

---

*The President's Analyst is a movie from the late 1960s (predivestiture) in which TPC (The Phone Company) proposes injecting small wireless communications devices into the brains of all newborns as a way to save money on telephone-plant investment.

Bell Communications Research

Bellcore is a research and development company owned and principally funded by the RBOCs. One of the objectives of Bellcore is to provide specifications of services that can be offered by the RBOCs and thus promote consistency in telecommunications service throughout the United States. SMDS is one such service. The majority of the material in this article is drawn from Bellcore Technical Requirements and Technical Advisories (see Refs. 3–6 and 8–15).

SMDS Interest Group

The SMDS Interest Group (SIG) is an industry consortium that was formed in the United States in early 1991 to promote SMDS. By mid-1992, over 50 companies had joined. An example of one important activity of the SIG is the specification of interfaces to allow the use of a special SMDS channel service unit/data service unit (CSU/DSU) to allow access to the service (16,17). The SIG also drew up specifications for access to SMDS through frame relay and ATM (18,19). Another important focus of the SIG is the specification of ways to carry various network layer protocols over SMDS (20–22).

European SMDS Interest Group

The European SMDS Interest Group (ESIG) is the European counterpart to the SIG. It is promoting SMDS in Europe and to that end has developed specifications such as physical layer protocols for access to SMDS that are compatible with the European digital transmission hierarchy (7).

European Telecommunication Standardization Institute

The European Telecommunication Standardization Institute (ETSI) is developing standards for MANs and thus has developed physical layer convergence procedures for the European digital transmission hierarchy (23,24). These have been adopted for SMDS by the ESIG.

Pacific Rim Frame Relay/ATM/SMDS Interest Group

The Pacific Rim Frame Relay/ATM/SMDS Interest Group (PR FASIG) was formed in late 1992 to promote data services in the Pacific Rim. One PR FASIG activity may be the development of specifications for providing SMDS in Pacific Rim countries.

Internet Engineering Task Force

The Internet Engineering Task Force (IETF) looks at issues concerning the Internet and considers the use of SMDS in that context. For example, it has developed a specification for carrying the Internet Protocol (IP) over SMDS (25). The IETF is also responsible for the development of specifications based on the Simple Network Management Protocol (SNMP) (26,27), including some that are used for SMDS (28–30).

Institute of Electrical and Electronics Engineers Project 802

The SMDS Interface Protocol (SIP) has certain compatibility with the standard developed by IEEE P802.6 (31). The SIP (see the section, "Role of SMDS in a Data Communications Environment") is based on this standard.

Focus of This Article

This article looks at SMDS from customer and applications perspectives. This is in contrast to the Bellcore requirements documents that are related to SMDS. Because of the *Modified Final Judgment,** the RBOCs are not allowed to manufacture the equipment that they use in their networks, and they are not allowed to carry long-distance traffic. Consequently, Bellcore requirements are targeted at (potential) suppliers of equipment for RBOC networks and other network providers who must be involved if SMDS is to be widely offered in the United States.

In the United States, telecommunications regulation has evolved in such a way that may network providers coexist and compete in offering service. In particular, there are local exchange carriers (LECs) and interexchange carriers (IECs). LECs provide service within a single local access and transport area (LATA), while IECs are free to offer service between LATAs. Figure 3 shows some possible configurations involving LECs and IECs in providing SMDS. (The connection between IEC 1 and IEC 2 is shown in gray to emphasize the fact that the vigorous competition among interexchange carriers has inhibited the deployment of connections between IEC networks.) The complexity of these arrangements has resulted in a significant portion of Bellcore work being devoted to provision of SMDS by interconnected carrier networks.

The details of intercarrier provision of SMDS and the details of the equipment in the public network are of secondary importance to SMDS customers and their data communications equipment suppliers. Thus, this article focuses on the material contained in the SMDS service description requirements documents from Bellcore (9–11,14,15). In addition, material is presented that is not covered by the Bellcore requirements. For example, the SIG and the IETF have established specifications on how SMDS can be used in various important networking architectures (20–22,25), and this material is included here. Another example is customer premises equipment (CPE) implementation of the SIP.

**Modified Final Judgment* (MFJ) is the umbrella term used to refer to the consent decree that resulted in the breakup of AT&T in 1984. The MFJ was supplanted by The Telecommunications Act of 1996 passed by the U.S. Congress on February 1, 1996.

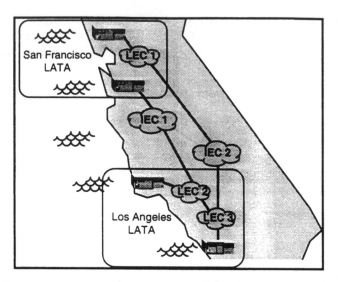

FIG. 3 Examples of interconnected networks providing switched multimegabit data service
(IEC = interexchange carrier; LATA = local access and transport area; LEC = local-
exchange carrier).

Overview

The sections below provide a high-level description of SMDS. Readers desiring
more detailed information are referred to Ref. 1.

The first section describes the general use of SMDS by explaining how SMDS
can be used in different networking architectures. The next section highlights
the key features of the service by making extensive use of examples. The rationale behind the features is also explained. The third section illustrates the economics of SMDS with a simple example. The fourth section describes the various access protocols for SMDS. A discussion of some technologies for providing
SMDS and comparisons of SMDS with some other services targeted at LAN
interconnection are also discussed.

Role in a Data Communications Environment

SMDS was designed with the expectation that it would be used as a physical
network in an internet. However, SMDS can play other data communications
roles as well. The various roles are described further in this section.

To understand better the potential roles of SMDS, it is helpful to take a
brief look at the SIP. The Level 3 Protocol Data Unit (L3_PDU) is the component of SIP that supports the SMDS features. All service features and performance objectives for SMDS are defined in terms of the transfer and treatment
of L3_PDUs. Figure 4 shows a simplified format of the L3_PDU. When
sending information, CPE that access SMDS construct the L3_PDU by filling

Destination Address	Source Address	Information
8 bytes	8 bytes	≤9188 bytes

FIG. 4   Simplified Level 3 Protocol Data Unit (L3__PDU) format.

in the destination address field, the source address field, and the information field. Each L3__PDU contains complete addressing information and thus can be delivered independently of other L3__PDUs. This type of service is called a datagram service. Unless there is an error, these fields are unchanged when the L3__PDU is delivered to the destination CPE.

Routing

The environment of a potential SMDS customer will probably contain a number of LANs and wide-area network (WAN) services. Examples of LANs include CSMA/CD (commonly called Ethernet) (2), token bus (32), token ring (33), and fiber distributed data interface (FDDI) (34). Examples of WAN services include DS-1 private lines and X.25 packet switching. The LANs and WAN services are integrated into a single network by an internetworking protocol implemented in routers and end systems. A router is a data packet store-and-forward device that operates at the network layer typically used to interconnect LANs and WANs. This configuration is generally referred to as an internet, and each LAN and WAN service is called a physical network. (Several LANs are often bridged, or interconnected, together to form a larger physical network.)

SMDS was explicitly designed to play the role of a physical network in an internet. An example of this role is portrayed in Fig. 5, which shows the use of SMDS by the Transmission Control Protocol/Internet Protocol (TCP/IP) (35,36). In this configuration, a router that is using SMDS to send an IP packet to another router (or an end system) must determine the proper SMDS Destination Address from the IP address of the recipient system. The details of this use of SMDS are described in Chapter 8 of Ref. 1 and in Ref. 25.

The TCP/IP architecture is very similar to many other Internet architectures, and the use of SMDS in these architectures is much like that shown in Fig. 5. Chapter 8 in Ref. 1 also describes the use of SMDS with the Open System Interconnection (OSI), AppleTalk, Xerox Network System (XNS), NetWare™ (Internet packet exchange [IPX]), 3Com's 3+ and 3+Open, VINES, DECnet Phase IV, and DECnet Phase V.

Bridging

As explained in the introductory sections, SMDS is designed to look like a backbone LAN. Thus, another possible way to use SMDS is with bridges. A bridge is a data packet store-and-forward device that operates at the data-link layer. It is typically used to interconnect multiple physical LANs into one logical

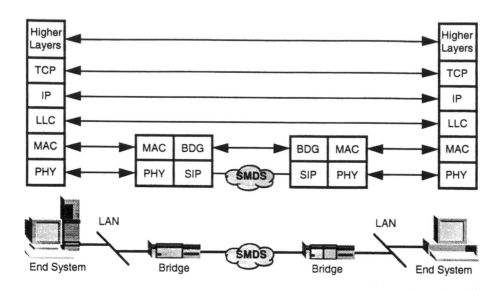

FIG. 5   Use of switched multimegabit data service as a physical network with the Transmission
Control Protocol/Internet Protocol (TCP/IP) (LLC = logical link control [IEEE 802.2
(46)]; MAC = media access control; PHY = physical layer; SIP = SMDS Interface
Protocol).

LAN. Figure 6 shows the use of SMDS with bridges and TCP/IP (35,36).
Virtually all IEEE 802 LANs and FDDI use 48-bit media access control (MAC)
addresses, while SMDS uses telephone numbers based on International Telegraph and Telephone Consultative Committee Recommendation (CCITT)

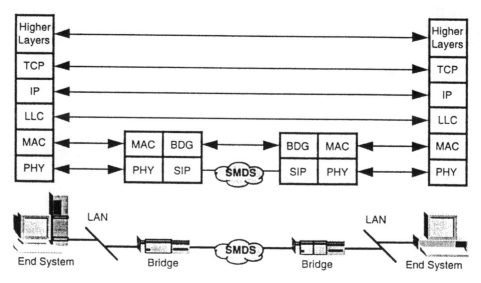

FIG. 6   The use of switched multimegabit data service with bridges (BDG = bridging functions).

F. 164 (37). The different address spaces will make translation bridging difficult, and thus it is expected that any use of SMDS with bridges will make use of encapsulation bridging. In other words, the MAC frame is copied into the SMDS information field by the bridge that is using SMDS to send to another bridge.

Multiplexing

Although switching is a major capability of SMDS, SMDS can be used to emulate private lines. As an example of this, Fig. 7 illustrates the use of SMDS for multiplexing. In Fig. 7, each service interface has two SMDS addresses associated with it. Each multiplexer has two ports, which is assumed to have a frame-based, link-level protocol such as High-level Data-Link Control (HDLC) Protocol or Synchronous Data-Link Control (SDLC) Protocol. When a frame is received on Port 1, it is encapsulated in the L3__PDU information field, and the destination address field is set to the transmit address (C). This L3__PDU is delivered to the multiplexer with Ports 3 and 4, and this multiplexer determines that Port 3 should receive the frame by matching C with the receive address of Port 3. Thus, in this example, Ports 1 and 3, 2 and 6, and 4 and 5 appear to be directly connected.

Many other variations of multiplexing are clearly possible, as well as various combinations of bridging, routing, and multiplexing. Like the paths worn in the

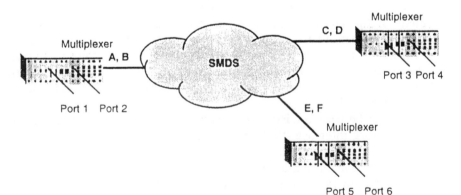

Port	Transmit address	Receive address	Connects to port
1	C	A	3
2	F	B	6
3	A	C	1
4	E	D	5
5	D	E	4
6	B	F	2

FIG. 7   The use of switched multimegabit data service for multiplexing.

grass of a new college campus, new ways of using SMDS will probably emerge as the service becomes widely deployed and used.

Overview of Key Features

The key features of SMDS are described in the following sections. The intent is to provide simple descriptions along with commentary on the rationale for their inclusion in the service.

Addressing

SMDS addresses use numbers structured according to CCITT Rec. E.164 (37). This is just a fancy way of saying "telephone numbers." The choice of the E.164 number structure is based on several considerations. First, the most prevalent network layer protocols, IP and IPX (38), are independent of the structure of physical network addresses and, as shown in Chapter 8 of Ref. 1, all of the internetworking protocols can work with E.164 numbers used as physical network addresses. Second, use of telephone numbers greatly facilitates deployment in the telephone networks. Modern telephone networks have extensive computerized operations support systems for such functions as service record inventory and billing. As might be expected, these operations support systems are typically keyed to telephone numbers. Therefore, the use of a different addressing scheme makes integrating SMDS into the current telephone network operations support systems difficult and in conflict with the goal of expeditious deployment of the service. Finally, a comprehensive administrative structure exists for telephone numbers — another plus for *quick* deployment of the service.

The format of an SMDS address is a 4-bit address type followed by up to 15 binary-coded decimal (BCD) digits structured according to E.164 specifications (37). This coding is chosen to facilitate routing within a service provider network as described in Ref. 8.

The interface between a CPE and the service provider's network is called the subscriber network interface (SNI). Up to 16 SMDS addresses can be assigned to a given SNI (at low-speed access to SMDS, this number is 2). One motivation for the ability to have more than one address per SNI is the support of multiplexing, as described in the section on multiplexing above. Another is the ability for a customer to have both public (well-known) and private addresses; this is described below in the section, "Address Screening."

Unicasting

SMDS unicasting occurs when an L3__PDU is transmitted across the SNI with an individual destination address (as indicated by the address type field). The L3__PDU is delivered to the SNI associated with the destination address. An exception to this occurs when the destination address is assigned to the source SNI. In this case, the L3__PDU is not delivered back to the source SNI. This exception accounts for the possibility of multiple CPEs attached to the SNI, as described in the section on distributed queue dual bus (DQDB) access. Since the full destination address is contained in every L3_PDU, there is no need for the CPE to be burdened with setting up and maintaining a connection before transmitting to any destination.

Group Addressing (Multicasting)

When an L3_PDU is transmitted across the SNI with a group destination address (as indicated by the type field), the L3_PDU is delivered to the SNIs associated with individual addresses that compose the group identified by the destination group address. Figure 8 shows an example of group addressing (multicasting).

Only one copy of the L3_PDU is delivered to an SNI, even if the SNI is identified by more than one individual address in the group address components. For example, the SNI with assigned addresses B and H in Fig. 8 receives only one copy of the L3_PDU even though both B and H are part of Group Address G. A copy of the L3_PDU is never delivered to the SNI from which it originated.

As shown in Fig. 8, the source address need not be a component of the group address. An example of the value of this is the ESHello (End System Hello) message of the OSI Connectionless Network Protocol (CLNP). Any OSI end system can send to a group address identifying all OSI intermediate systems even though the end system is not a member of the group. This distinguishes SMDS group addressing from the X.25 concept of a closed user group.

Source Address Validation

When an L3_PDU is sent to SMDS, the source address is checked to see if it is assigned to the originating SNI. If the source address is not assigned to the

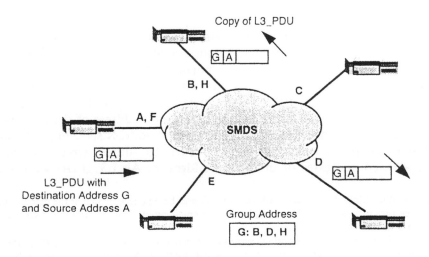

FIG. 8   Example of group addressing.

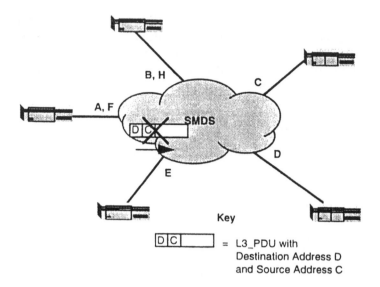

FIG. 9   Example of source address validation.

originating SNI, the L3__PDU is not delivered. An example of source address validation is shown in Fig. 9.

Source address validation is motivated by two considerations. First, in the case of charging on a per-L3__PDU basis, the source address *must* be validated to ensure that the proper customer is charged. Second, source address validation helps to prevent "spoofing" at the SIP level, by which the sender of an L3__PDU attempts to fool the receiver into believing that the L3__PDU was sent by a different source.

Address Screening

Address screening is a feature that lets a customer restrict communications based on SMDS addresses. The restrictions are defined by tables, called address screens, associated with each SNI. An individual address screen is a table consisting of individual addresses. A group address screen is a table consisting of group addresses. An L3__PDU is the subject of two types of address screening. With destination address screening, the destination address of the L3__PDU is compared with an address screen.* With source address screening, the source address is compared with an address screen.† An address screen can contain either allowed addresses or disallowed addresses. An L3__PDU with a com-

---

*It is natural to think of destination address screening as being implemented at the ingress point of the service provider network. However, it can be implemented anywhere in the network without any change to the service feature.

†It is natural to think of source address screening as being implemented at the egress point of the service provider network. However, it can be implemented anywhere in the network without any change to the service feature.

pared address (source or destination) that matches an entry in a screen with disallowed addresses is discarded, while an L3__PDU with a compared address that fails to match any entries in a screen with allowed addresses is discarded.

Figure 10 is an example of the use of address screens to limit communications across an SNI. The individual address screen and the group address screen are shown for the SNI with assigned addresses B and H. Both of these screens contain allowed addresses. The L3__PDUs sent from this SNI into the service provider network are subject to destination address screening. Examples in Fig. 10 are the L3__PDUs sent from B to G and from H to C. For the L3__PDU from B to G, the group address screen is used because the destination address is a group address. In this case, a match is found, and the L3__PDU is not discarded. For the L3__PDU from H to C, the individual address screen is used because the destination address is an individual address. In this case, a match is not found, and the L3__PDU is discarded.

The L3__PDUs addressed to be delivered to this SNI (i.e., with Destination Address B, H, or G) are subject to source address screening. Since a source address is always an individual address, the individual address screen for this SNI is used for source address screening. In Fig. 10, the L3__PDUs from C to G (discarded), from C to H (discarded), from C to B (discarded), and from E to B (not discarded) are all subjected to the source address screening for this SNI. The screens shown for this example SNI are those that would make this SNI part of a virtual private network consisting of A, B, D, E, and H. To complete the virtual private network, complementary screens are needed for the SNIs serving A, D, and E.

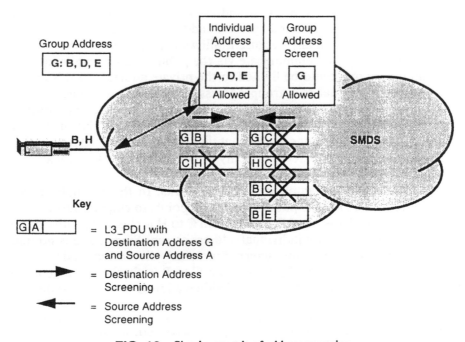

FIG. 10   Simple example of address screening.

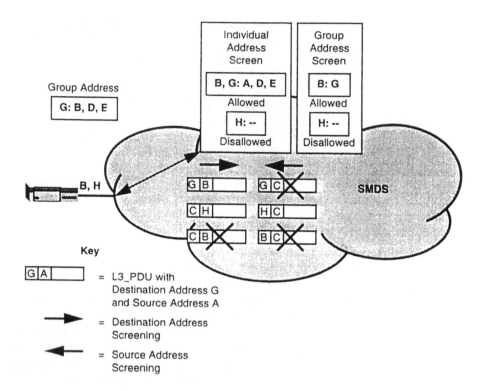

FIG. 11   Address screening with public and private addresses.

A more complex configuration of address screens can be used to effect the simultaneous use of "public" and "private" addresses on the same SNI.* This could be useful to a company because it allows the company's customers and prospective customers to communicate via the public address, while intracompany communications could be limited to the private address. In this arrangement, multiple individual and group address screens are used. Each screen is pointed to by addresses associated with the SNI. Individual address screens can be pointed to by either individual addresses assigned to the SNI or by group addresses that have at least one component that is an individual address assigned to the SNI. Group address screens can only be pointed to by individual addresses assigned to the SNI.

Figure 11 shows an example of how this works. In this example, B is the private address, and H is the public address. H points to empty address screens that are disallowed. Thus, the L3__PDU from C to H is checked (source address screened) against the empty individual address screen. Since there is no match (the screen is empty) and the screen "contains" disallowed addresses, the L3__PDU is delivered. The L3__PDU from C to B is checked (source address screened) against the {B, G: A, D, E} individual address screen and is discarded

---

*This more complex form of address screening is called out as an *objective* in Ref. 14, as opposed to a *requirement*. Consequently, it may not be supported in early offerings of SMDS. Check with your service provider before planning to use this feature.

because there is no match and the screen contains allowed addresses The L3_PDU from B to C is destination address screened using the {B, G: A, D, E} screen and discarded, while the L3_PDU from H to C is not discarded. The net result of the screening configuration in this example is that B can only exchange L3_PDUs with A, D, and E (and G), while H is unrestricted.

Notice that it is possible to have configuration conflicts between address screening and group addressing. For example, it is possible that group address screens could be such that all L3_PDUs sent to a given group address are always discarded. Another potential problem is a mistake in setting up the screens, thus preventing communications that are intended to be allowed. The use of the customer network management features, as described in a separate section below, aids in the timely detection and correction of such errors.

Access Classes

The bit rate of DS-3 (44.736 Mb/s) is approximately 28 times faster than the bit rate of DS-1 (1.544 Mb/s) and 672 times faster than SMDS based on DS-0 (Digital Signal Level 0) access (see the section, "Access Protocols"). This is a large difference when considering average throughput of potential SMDS customers. On the other hand, the higher rate of DS-3 can be very advantageous for bursts of data such as a large FDDI frame. The time to send a maximum-size FDDI frame (4500 bytes) on a DS-1-based SNI is approximately 31 ms. On a DS-3-based SNI, the same transmission takes only approximately 1.1 ms.*

It is clear that, if a service provider network is designed assuming full usage of all DS-3-based SNIs, then many more resources will be deployed in the service provider network than will be used. The access class concept is intended to allow a simple characterization of usage that will allow more efficient network implementation. The resulting economies can then be shared with the SMDS customers via low tariffs. Access classes are only defined for SNIs based on DS-3 and E-3 (European Transmission Level 3).

Access classes allow bursts of traffic but still limit the long-term sustained traffic rate. A burst is one or more L3_PDUs sent back to back at the full DS-3 bit rate. (Thus, the access class approach is not "fractionalizing" the DS-3.) An access class is described by a credit manager algorithm. An L3_PDU delivered to SMDS will be forwarded to its destination(s) if the available credit equals or exceeds the length of the L3_PDU information field. If the credit is not sufficient, the L3_PDU is discarded. The credit builds up continuously until it reaches a maximum. Whenever an L3_PDU is forwarded, the credit is reduced by the number of bytes in the L3_PDU information field. Figure 12 illustrates the credit calculation. Because the full DS-3 bit rate can be used if credit is available, the 4500-byte L3_PDU shown in Fig. 12 can traverse the SNI in approximately 1.1 ms.

The long-term maximum data transfer rate, called the sustained information rate (SIR), is just the rate of credit build-up. Five different access classes are

*These timing calculations are based on the fact that SIP on DS-3 carries 96,000 L2_PDUs per second, and each L2_PDU contains up to 44 bytes of an L3_PDU as payload.

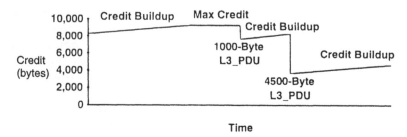

FIG. 12   Credit manager algorithm example.

defined for DS-3-based SNIs. Figure 13 shows the SIRs for these access classes. Four different access classes are defined for E-3-based SNIs. Figure 14 shows the SIRs for these access classes. In both cases, the first three classes correspond to the common LAN data rates of 4, 10, and 16 Mb/s. The other classes represent higher data rates.

From Fig. 12, it is clear that bursts of L3__PDUs can be transmitted when sufficient credit is available. The maximum burst is the number of information field bytes that can be transmitted in back-to-back L3__PDUs at the full DS-3 or E-3 bit rate without any of the L3__PDUs exceeding the available credit at the start of the burst when the credit is at the maximum value. The results of maximum burst calculations are presented in Figs. 15 and 16. The time to send such a burst is independent of the SIR because bursts are sent at the full DS-3 or E-3 bit rate. For example, the maximum burst of over 10 kb takes approximately 2.5 ms to send across the DS-3-based SNI.

There is some similarity between the concept of access class and the concept of committed information rate (CIR) of frame relay (39). (Access classes appear to predate CIR, having been first published in 1988; see Ref. 3.) However, a key difference is that when a frame relay frame exceeds the CIR, it can be marked for priority discard and then discarded if congestion is experienced in the network. With access classes, if an L3__PDU exceeds the available credit, it is

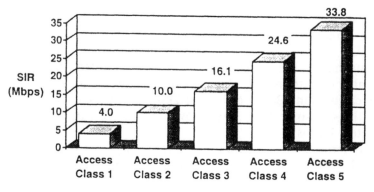

FIG. 13   Sustained information rates (SIRs) for subscriber network interfaces (SNIs) based on Digital Signal Level 3 (DS-3).

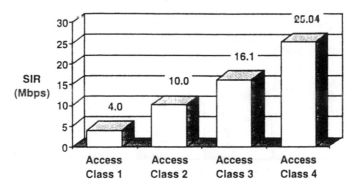

FIG. 14 Sustained information rates for E-3-based subscriber network interfaces.

always discarded. The motivation for this is to avoid unpredictable service performance for customers and simpler network implementation. By always discarding, SMDS keeps a customer from experiencing random, degraded performance as a result of having a poorly selected access class. Poor access class selection will cause high data loss, which will immediately manifest itself as high delay and low throughput or even broken transport layer connections. This should stimulate immediate corrective action, for example, through the use of customer network management (see the next section) and lead to more predictable future performance.

Customer Network Management

As discussed in the introductory sections, the underlying concept of SMDS is to provide a service that looks like a backbone LAN. One important property of today's LANs is that they can be monitored and configured from a remote network management station (NMS). As internets have grown larger with more physical networks, this network management capability has become critical to

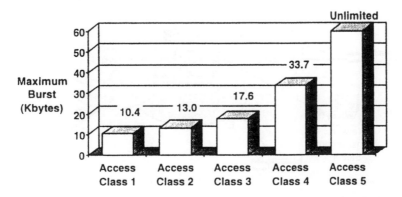

FIG. 15 Maximum burst sizes for subscriber network interfaces based on Digital Signal Level 3.

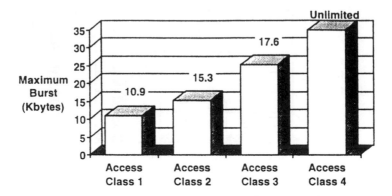

FIG. 16 Maximum burst sizes for E-3-based subscriber network interfaces.

the effective operation of the internets. Consequently, it is viewed as imperative that SMDS offer the customer the ability to manage and configure the service remotely. This service feature is called customer network management (CNM).

Because the majority of existing NMSs use the Simple Network Management Protocol (SNMP) (26,27), the SMDS network management feature uses this protocol to communicate with an agent within the service provider network. The concept is shown in Fig. 17. The customer's NMS sends requests to the agent, which in turn carries out the commands from the NMS. Two kinds of information are available: that dealing with SNIs and that dealing with subscription parameters. For example, the agent can retrieve performance measurements from various elements in the service provider network and forward the result to the NMS. Similarly, the agent can effect changes in the customer's service features, such as modifying an address screen.

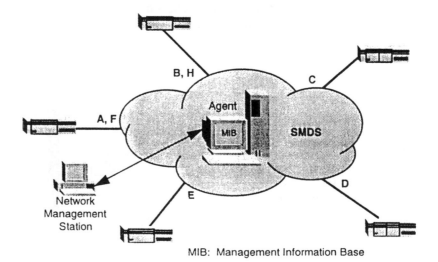

MIB: Management Information Base

FIG. 17 Customer network management concept.

TABLE 1   Assumptions for Economic Comparison (North America, 1993)

Item	First Cost ($)	Monthly Charge ($)	Equivalent Monthly Charge ($)
Per-mile charge for DS-1		20	20.00
Access charge per end for DS-1		340	340.00
CSU/DSU	2500		83.33
Router port	3000		100.00
SMDS tariff per SNI		700	700.00
CSU/DSU for SMDS	5000		166.67

Note: The equivalent monthly charge is computed from first cost by dividing by 30, which is based on a 3-year amortization and approximately a 13% interest rate.

In combination with the management of SMDS CPE, SMDS CNM service provides customers with a valuable tool to manage an internet that uses SMDS.

Economics

General statements about the environment in which a particular carrier service is economically attractive are very difficult to make. Tariffs, number of customer locations, and traffic patterns are but a few of the important factors. However, some of the dynamics can be illustrated by example.

In our example, the cost of SMDS is compared with that of dedicated DS-1 connections. The comparison is done in terms of equivalent monthly charges. The tariff and equipment cost assumptions are presented in Table 1. These numbers are representative for North America in 1993. They are not meant to be definitive statements of any tariffs or equipment costs. In fact, in general both equipment cost and tariffs are expected to fall over time. Tariffs for SMDS are still evolving. Most of the early RBOC intra-LATA offerings include a flat-rate tariff for each SNI. Tariffs based on traffic and/or distance may be offered in the future. For comparisons, a flat-rate tariff is assumed. Also assumed is the use of a special SMDS CSU/DSU,* which is a common configuration (see the section on DQDB access).

For each dedicated DS-1 connection, it is assumed that the length is 10 miles. Because the DS-1 tariff is assumed to be related to distance, the results of this comparison are highly dependent on this assumption. Also assumed is that the traffic is such that a single location will not overload a single DS-1 SNI. Finally, it is assumed that service installation charges are the same, therefore, they are not included in the comparison.

Because the DS-1 connections are assumed to be 10 miles, the monthly tariff is 10 times the per-mile charge plus twice the access charge (one access charge

---

*When a frame-based access protocol is used to access SMDS (see the section regarding data exchange interface access to SMDS), a special CSU/DSU is not needed, and the CSU/DSU cost is the same for both the dedicated DS-1 connection and SMDS.

for each end of the connection). The monthly equipment charge for each dedi-cated DS-1 connection (two ends) is two times the router-port charge plus two times the CSU/DSU charge. Thus, the total monthly charge for each dedicated DS-1 connection is 200.00 + 680.00 + 166.66 + 200.00 = $1246.66. For each SMDS SNI (one end), the monthly charge is the tariff plus the monthly charge for a router port plus the monthly charge for a CSU/DSU for SMDS, which equals 700.00 + 100.00 + 166.67 = $966.67.

The above calculations show that if just two locations, 10 miles apart, are to be communicating, the monthly charge for a single dedicated DS-1 connection ($1246.66) is less than the cost for the two-SMDS SNI arrangement ($1933.34). However, the comparison becomes more interesting as the number of locations increases.

If a full mesh of DS-1s is used, which maintains one hop between all loca-tions, the monthly charges are as shown in Fig. 18. Beginning with the addition of a third location, SMDS becomes significantly less expensive than dedicated DS-1 connections. As the number of locations increases, the SMDS advantage increases dramatically because the monthly charges for SMDS grow linearly, while the monthly charges for the full mesh of dedicated DS-1 connections grows quadratically.

Of course, it is unlikely that a full mesh of dedicated DS-1 connections will be used for a large number of locations. The least expensive topology for dedi-cated DS-1 connections is to connect all locations to a single hub location. Thus, $n - 1$ connections are used for n locations. The results of this comparison are shown in Fig. 19. In this case, SMDS and dedicated DS-1 connections are just about equal in monthly charges for four locations, with the SMDS having the advantage for larger numbers of locations and dedicated DS-1 connections hav-ing the advantage for two and three locations.

Actual dedicated DS-1 configurations are likely to fall somewhere between the two described above. In order to estimate the intermediate case, a partial-mesh topology is used. In this topology, each location has two dedicated DS-1 connections, one to each of two of the other locations. The results are shown in Fig. 20. Even for this moderate level of connectivity for the dedicated DS-1 connections, SMDS has lower monthly charges, beginning with three locations.

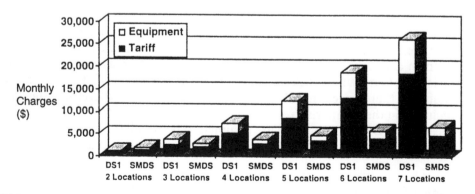

FIG. 18   Economics of private-line, full-mesh topology and switched multimegabit data service.

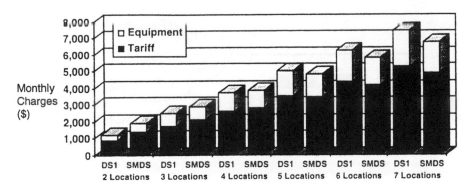

FIG. 19   Economics of private-line hub topology and switched multimegabit data service.

The above examples illustrate that SMDS looks more and more attractive as the number of communicating locations increases. This is exactly the behavior that is needed for a successful public switched service.

Access Protocols

In order to access SMDS, an access protocol that can carry the L3_PDU is required. Bellcore, in Ref. 14, specifies a protocol based on the DQDB standard (31). Subsequent to the publication of Ref. 14, access protocols based on the SMDS data exchange interface (DXI) (16) and on frame relay have been specified for access to SMDS (5,6,18,40). These three access protocols are briefly described in the following sections. Then, access to SMDS via the ATM is discussed.

Distributed Queue Dual Bus Access

DQDB is a protocol designed for a shared-medium network. It can be used for the equipment in a service provider network, and it is also the basis of SIP. The

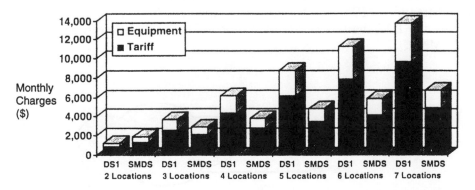

FIG. 20   Economics of private-line, partial-mesh topology and switched multimegabit data service.

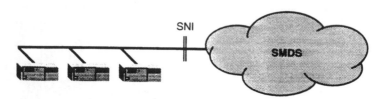

FIG. 21   Access of multiple customer premises equipment (CPE) to switched multimegabit data service.

section on technologies describes the former use, while in this section the focus is on the latter use.

One of the motivations for basing SIP on DQDB is that it allows multiple CPEs to access SMDS over a single physical access arrangement (see Fig. 21). However, typical configurations involve a single CPE (see Fig. 22). As shown in Fig. 22, SIP can be terminated in a single CPE device, such as a router, or it can be implemented in two devices using a special CSU/DSU. The interface between the CSU/DSU and the router is the SMDS DXI (16).

The SIP is said to be based on DQDB because a device that it conforms with, IEEE Standard 802.6-1990, can successfully access SMDS (31). However, for single-CPE access, a subset of that standard is all that is required. The SIP supports data rates of 1.544 Mb/s (DS-1), 2.048 Mb/s (European Transmission Level 1 [E-1]), 34.368 Mb/s (E-3), and 44.736 Mb/s (DS-3).

Data Exchange Interface Access

The SMDS Interest Group first developed the DXI (16) as an interface between a router or end system and a special CSU/DSU that would allow existing routers and end systems to access SMDS without the need for hardware upgrades. This was useful because hardware for SIP was not yet widely available. However, it requires that this new hardware be incorporated in the CSU/DSU (i.e., existing CSU/DSUs cannot be used). As a consequence, the SMDS Interest Group went on to specify the use of the DXI for direct-service access (40). MCI, in its HyperStream[SM] SMDS (41), has eliminated the need for a special CSU/DSU by making the DXI the service access protocol for speeds between 56 kb/s and 1.536 Mb/s.

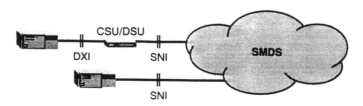

FIG. 22   Access of single customer premises equipment to switched multimegabit data service.

Flag	HDLC Header		HDLC Information			Flag
01111110_B	Address	Control	DXI Header	L3_PDU	CRC	01111110_B
1	1	1	2	0 – 9232	2 or 4	1

All lengths in bytes

FIG. 23 Encapsulation of the Level 3 Protocol Data Unit in the data exchange interface (DXI).

The DXI is based on HDLC procedures (42). The encapsulation of L3_PDUs in the HDLC frame is shown in Fig. 23.

Bellcore has published a technical service description for SMDS when the SNI is based on the DXI, which is referred to as low-speed SMDS (5). The SNI uses a DS-0 transmission path with bit rates of either 56 kb/s or 64 kb/s. Some service providers also offer a DS-1 transmission path with a bit rate of 1.536 Mb/s. Bellcore requirements (5) call for some differences from those in Ref. 14 with regard to the service features of low-speed SMDS. Most of the differences involve reduced addressing capabilities (e.g., the number of individual addresses that can be assigned to an SNI). The motivation for the reduced feature set is ease of implementation by the service provider. The SIG allows for use of the DXI at speeds of 56, 64, $n \times 56$, and $n \times 64$ kb/s, only observing that service feature implications for speeds above E-1 (2.048 Mb/s) are under study (40).

Frame Relay Access

Frame relay can be discussed as either a technology or a service. In this section, frame relay is discussed as a technology for supporting access to SMDS. (Frame relay is discussed as a service below.) The use of the Frame Relay Protocol to access SMDS is called SIP relay, and the service access interface is called the SIP relay interface (SRI) (6,18).

The Frame Relay Protocol is based on HDLC procedures (42), and thus it can be used to access SMDS much as described in the previous section. Figure 24 shows the encapsulation of the L3_PDU in the frame relay frame. The details of this encapsulation can be found in Chapter 5 of Ref. 1 and in Refs. 6 and 18. The data-link connection identifier (DLCI) in the frame header is set to a special value to identify the SIP relay interface permanent virtual circuit (SRI PVC). This value is not universally defined and thus will be defined by each service provider.

Flag	Frame Relay Header	Frame Relay Information		Flag
01111110_B	DLCI = SRI PVC	L3_PDU	CRC	01111110_B
1	2	0 – 9232	2	1

all lengths in bytes

FIG. 24 Encapsulation of the Level 3 Protocol Data Unit in the frame relay frame (CRC = cyclic redundancy check; DLCI = data-link connection identifier; SRI PVC = SIP relay interface permanent virtual circuit).

Given the similarity of frame relay encapsulation to DXI encapsulation, one might wonder why both protocols are needed. The answer is in the intended uses. The previous section discussed the motivation for the DXI. From the customer's point of view, an advantage of frame relay access to SMDS is that both frame relay service (see below) and SMDS can be accessed over a single interface. Since the tariff for a service includes a large component for the access line, using a single access line can yield significant savings compared to paying for two access lines, one for each service. The single access line can also yield savings by avoiding the use of a bridge or router port.

A service provider's motivation for the use of frame relay access to SMDS can be explained with the use of Fig. 25. Consider a carrier that has an existing network of frame relay switches. For simplicity, this network is represented by a single frame relay switch in Fig. 25. By using the encapsulation of the L3_PDU of Fig. 24, the existing frame relay network can transparently carry the L3_PDU to a location at which the SMDS service features and switching can be performed, shown as the combination of an interworking unit and an SMDS switching system in Fig. 25. Use of a frame relay permanent virtual circuit (FRPVC) allows the identification of the source service interface to allow support of source address validation and address screening. Thus, the use of frame relay access to SMDS allows the carrier to leverage the existing investment in frame relay technology in offering SMDS.

Bellcore has published a technical service description for SIP relay (6). The SMDS features supported by SIP relay differ from those in Ref. 14. Most of the differences involve reduced addressing capabilities. For example, only two individual addresses can be assigned to an SRI PVC.

Asynchronous Transfer Mode Access

ATM is a technology for switching and multiplexing based on 53-byte packets called *cells*. Multiplexing is accomplished by breaking each traffic stream into

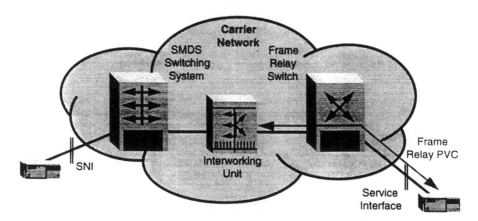

FIG. 25   Frame relay access to switched multimegabit data service.

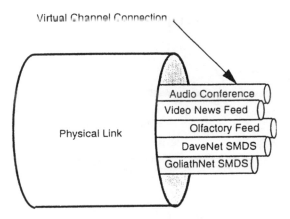

FIG. 26 Accessing switched multimegabit data service using asynchronous transfer mode.

cells and labeling the cells with a unique 24-bit code in the header of the cell. Each label corresponds to what is called a *virtual channel connection* (VCC). Thus, on a single physical interface to a network, multiple services can be accessed by using a different VCC for each service (see Fig. 26).

The protocols that are used for accessing SMDS over ATM are very similar to SIP. This follows from the cell orientation of SIP, in which L2__PDUs are also 53 bytes. One very important identical aspect of the protocol with SIP is that the service-feature-related fields of the L3__PDU are the same. The protocol layering for accessing SMDS over the ATM is shown in Fig. 27.

Technologies for Providing Switched Multimegabit Data Service

SMDS is a service, not a technology. In a properly designed service provider network, the customer should be unaware of and unconcerned with the technology being used by the service provider. The customer will only be concerned with the tariff, the features supported, and the performance. In principle, even

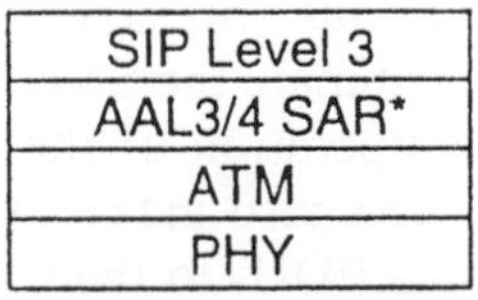

| SIP Level 3 |
| AAL3/4 SAR* |
| ATM |
| PHY |

FIG. 27 Protocol layering for accessing switched multimegabit data service using the asynchronous transfer mode (SAR = segmentation and reassembly). Note that AAL3/4 SAR refers to an asynchronous transfer mode protocol for carrying large, variable-length packets in multiple cells. *In 1996 the SMDS Interest Group approved the use of AAL5 SAR for carrying SMDS over ATM, in addition to the AAL3/4 SAR shown here.

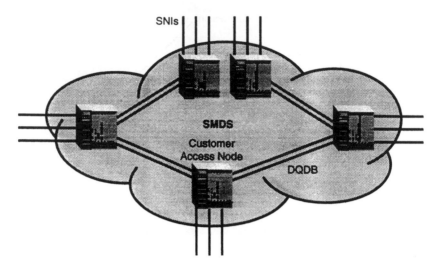

FIG. 28   The use of metropolitan-area network technology to provide switched multimegabit data service (DQDB = distributed queue dual bus).

supersonic carrier pigeons could be used to provide SMDS.* As the technology in the service provider network improves, the SMDS customer will not need to make any changes except to enjoy improved performance and lower tariffs.

Early implementations in service provider networks have used what can be described as MAN technology, which is typically dedicated only to SMDS. In the longer term, technology intended for supporting multiple services, such as ATM, may be used. The next sections describe these approaches.

Metropolitan-Area Network Technology

One approach to implementing SMDS in a service provider network is to use MAN technology. In the example illustrated in Fig. 28, a DQDB running on high-speed transmission systems is used to connect four central offices. In each central office are one or more customer access nodes (CANs). The CANs terminate SNIs and implement the various SMDS features such as address screening. (Note that because the CANs terminate the SNIs, aberrant behavior on the part of the CPE cannot disrupt the operation of the DQDB that connects the telephone central offices.) The CANs also make use of the connectionless MAC capability of DQDB to send L3__PDUs to the CAN(s) that terminate the destination SNI(s).

One of the reasons that MAN technology was considered attractive in the early work on SMDS was that this approach could be implemented quickly.

---

*In 1983, a short discussion actually occurred at an IEEE P802.6 meeting regarding the suitability of using supersonic carrier pigeons for MANs. After about two minutes with a calculator, it was concluded that the latency would be too great.

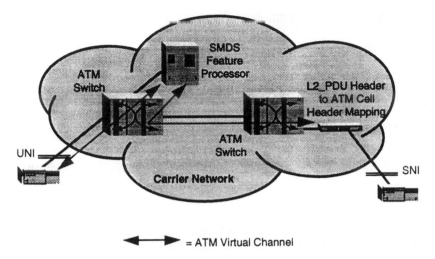

= ATM Virtual Channel

FIG. 29   Provision of switched multimegabit data service using asynchronous transfer mode (UNI = user network interface).

Indeed, the first trials and service implementations of SMDS by Bell Atlantic used this approach.

Broadband Integrated Services Digital Network

The broadband integrated services digital network (BISDN) is the concept that followed the ISDN. The BISDN differs from ISDN in that it is intended to provide much higher bandwidth, and that it will be based on ATM. Because ATM offers the promise of being able to support all forms of communications traffic, it is expected to provide the foundation on which telephone carriers will eventually implement all services.*

For SMDS, the multiplexing capability of ATM can be used in an arrangement like the one shown in Fig. 29. An ATM permanent virtual channel exists between the CPE and a device that provides the SMDS features, the SMDS feature processor. In effect, the ATM infrastructure is used to transfer L3_PDUs between points that can process them. The SMDS feature processor can use another permanent virtual channel to deliver the L3_PDU to the appropriate destination.

Figure 29 illustrates support of SMDS by both the SNI and the ATM user–network interface (UNI). For the UNI access, the ATM permanent virtual channel connects the CPE with the SMDS feature processor. With the SNI access, the permanent virtual channel can connect a simple mapping function to the SMDS feature processor. The mapping function need only map between the SIP L2_PDU network control information field and the ATM cell header.

*Clearly, this is a very long-term vision. The non-ATM-installed investment in telephone networks that is not written off is huge and will take a very long time to replace.

Comparison with Other Local-Area Network Interconnection Services

SMDS is not the only new carrier service that is targeted at LAN interconnection. In the past few years, there has been substantial interest in the industry concerning how these new services relate to each other.* In particular, the relationship of frame relay service and SMDS has been of high interest; this is discussed in the next section. Then, SMDS is compared with what here is called *native* LAN service. Finally, SMDS is compared with Cell Relay Service (CRS), which is expected to be one of the first public-carrier, ATM-based services.

Frame Relay Service

Above, frame relay was discussed as a technology. In this section, the concern is with frame relay in the context of a service. Frame relay service can be viewed as consisting of two different services: frame relay permanent virtual circuit service (FRPVC) and frame relay switched virtual circuit (FRSVC) service. Both are compared with SMDS in the next two sections.

Frame Relay Permanent Virtual Circuit Service.   With FRPVC, a customer can use a single service interface to exchange data with equipment connected to several other service interfaces. The bit rate of the service interface can range from 56 kb/s to 44.736 Mb/s. Connectivity between two service interfaces is established via a request to the service provider to configure a virtual connection between the service interfaces. Forms that the request can take include a telephone call, a paper mail message, or a facsimile transmission. The time from when the request is made to the time that the virtual connection is available can range from a few minutes to several weeks. Given this interval and the fact that there is usually a significant installation fee for each virtual connection, a customer will tend to leave the virtual connections in place once they are established. Hence, these connections are called permanent virtual circuits.

For each connection established at a service interface, there is (typically) a 10-bit identifier, a DLCI, assigned. To send data to a particular destination, the attached CPE forms a frame with a header that contains the DLCI value assigned to the PVC to the intended destination and transmits it across the service interface. The maximum frame size is usually 1600 bytes.

Because of the duration of the PVCs, the FRPVC is much like a private-line service and is ideal for private networks. Although SMDS can easily be used as a private-line replacement, with address screening providing similar levels of privacy, the ability to reach any SMDS subscriber's interface by simply using the proper E.164 address makes SMDS much better suited for public networking.

SMDS group addressing is another key difference. Such a multicasting capability is not available with FRPVC. The impact is that the methods of imple-

---

*In the recent past, some players in the industry have attempted to position these services as if they were in a competitive "fight to the death." These activities reminded us of professional wrestling promoters manufacturing nonexistent grudges in an attempt to sell more tickets, in this case to industry seminars and debates on the services.

menting routing and bridging are different with FRPVC. This is not to say that routing and bridging over FRPVC is not well understood. Bridging and routing implementations over private lines are very mature, and thus the private-line nature of FRPVC makes migration straightforward. However, as the number of permanent virtual circuits becomes large, the exchange of routing information on each connection becomes burdensome compared to the simplicity when group addressing is used to exchange routing information. For example, if a router is connected to a FRPVC with 100 PVCs across the interface, the router has to send 100 hello messages, one for each PVC, across the interface in order to maintain adjacency with the routers on the other end of each PVC. When a topology change occurs, swarms of update messages are exchanged on each PVC, which can look like an update message storm compared to the total bandwidth of the FRPVC interface.

Another key difference between SMDS and FRPVC is that the maximum user information in each data unit is 9188 bytes for SMDS versus 1600 bytes (typical) for FRPVC.

Frame Relay Switched Virtual Circuit Service. The FRSVC is like FRPVC with the additional capability for dynamic (e.g., subsecond) virtual connection setup and clearing. The setup and clearing are accomplished through a signaling protocol* that operates between the CPE and the network providing the FRSVC. This protocol is a variation on the connection-signaling protocol of ISDN. It is expected that the addressing used to identify end points for these connections will be based on E.164, and thus FRSVC will have public service capability similar to that of SMDS.

However, the signaling protocol makes the CPE implementation and operation more complex than for SMDS. First, there is the signaling protocol itself, which is nontrivial. Second, since FRSVC tariffs will probably include a connection holding time charge, the CPE will need to decide when to establish and when to clear a connection. Poor connection management can mean poor performance caused by excessive connection setups and clearings or excessive carrier charges for long-lived idle connections.

In other aspects, FRSVC and SMDS compare much the same as FRPVC and SMDS, as described in the previous section.

Native Local-Area Network Service

In a native LAN service, the subscriber network interface is a standard LAN such as Ethernet or FDDI. Several such interfaces are tied together as if the carrier network was a multiport store-and-forward repeater. Figure 30 shows how such a service might work for Ethernet. Any frame originated by A or B will be seen by C, D, E, and F. Thus, all end systems appear to be on a single

---

*It is hard to imagine a useful communications protocol that is not used to "signal" information between the communicating entities. However, the term *signaling* is deeply rooted in telecommunications culture predating "protocols" and even the telegraph. *Signaling protocol* usually refers to procedures to set up, maintain, and release connections.

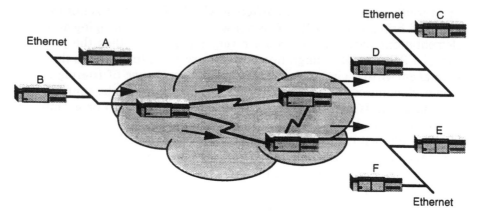

FIG. 30   Ethernet native local-area network service.

Ethernet except that end systems attached to different service interfaces will never collide with each other since the store-and-forward repeaters terminate the Ethernet collision domain on each port. This is the simplest implementation of a native LAN service. More sophisticated implementations could include filtering to reduce extraneous traffic and could also participate in the management of a bridging spanning tree. In all cases, the result is that all service interfaces are part of an MAC layer network.

A native LAN service has the advantage that it is accessed with the LAN MAC protocol. Thus, no special interface hardware is necessary to access the service. This is simpler than SMDS, but it also has some disadvantages. First, it is strictly intended for private network applications. Second, the number of devices attached to the service should be limited to prevent the management difficulties, such as broadcast storms, that can emerge with large, flat networks.

Cell Relay Service

In recent years, the communications industry has developed a huge interest in ATM. In all of the hoopla, ATM has come to stand for many things, including the following:

- a high-speed, scalable communications technology
- a galactic-scale communications architecture for the ultimate, ubiquitous, all-singing, all-dancing communications infrastructure
- a basis for the next generation of high-end public carrier services

ATM as the basis for new public carrier services is the focus here. Because of the ATM promise for very high speed and performance, it is natural to think of ATM as the basis for a LAN interconnection service. It is expected that SMDS will be one of these ATM-based services.

As of this writing, the details of these first ATM-based services are just emerging, although several public carriers have announced plans for such services (43). Therefore, educated guesses must suffice as to what the initial ATM-based services will entail. The conjecture about a generic ATM-based service is referred to as cell relay service (CRS).

CRS provides permanent virtual connections between service interfaces. It is very much like today's FRPVC, but with some important differences:

1. The bit rate of the service interface will be much higher, with 44.736 Mb/s (DS-3) being first offered and 155.52-Mb/s Synchronous Optical Network (SONET) Synchronous Transport Signal Level 3c (STS-3c) being offered later.
2. The unit of data transfer will be the ATM cell with a fixed-size payload of 48 bytes instead of a variable-length HDLC frame.
3. For virtual connection identification, 24 bits will be available instead of the 10 bits for FRPVC.
4. Instead of the CIR, CRS will use the generic cell rate algorithm (GCRA) as defined in Refs. 43 and 44.
5. Because of the need for hardware to convert between ATM cells and large, variable-length LAN frames, a special ATM CSU/DSU will be needed to access CRS during its early years of availability. The interface between the CSU/DSU and a router or end system is called the ATM data exchange interface (ATM DXI) (45).

The net effect is that CRS will be very much like FRPVC (an explicit goal of Mode 1A of the ATM DXI; see Ref. 45), except that it will have higher performance. Thus, the comparison of CRS with SMDS is identical to the comparison of FRPVC with SMDS, with the exception that CRS and SMDS will have similar bit rates for their service interfaces.

In 1997 carriers began to announce their intentions to provide CRS SVCs. When provided, CRS SVCs will be much like FRSVC except that they will have higher performance. Thus, the comparison of FRSVC and SMDS will also apply to this case.

Summary and Conclusion

In this article, the fundamental concepts underlying SMDS were examined. In the first part of the article, the problems faced by the public carriers in the mid-1980s, the motivation for creating SMDS, and its goals were reviewed. In the second part of the article, the role played by SMDS in a data communications environment as a physical network in an internet was reviewed. The features supported by SMDS such as unicast and multicast addressing, source address validation, address screening and SMDS access classes were described. How these features are designed to provide LAN-like service features and performance for end users over the wide area, as well as to provide some added security measures for the end users, were discussed. A few generic example calculations were given to compare SMDS's economics to that of dedicated DS-1 lines. The various access protocols that are defined for accesses to SMDS were described. Last, SMDS was compared to other popular or emerging public carrier data services, such as frame relay PVC or SVC services, native LAN service, and cell relay PVC or SVC services.

The mid-1990s witnessed the emergence of the public carriers as credible purveyors of wide-area data communications services, making higher speed, wide-area data connectivity more flexible and affordable to end users. SMDS is the first switched, broadband data service made available to end users. This and other public data services available or to be made available to the end users will undoubtedly have an impact on how wide-area networking is done for the remainder of the 1990s.

Acknowledgment: NetWare is a registered trademark and Internet Packet Exchange is a trademark of Novell, Incorporated. HyperStream is a service mark of MCI. 3Com is a registered trademark, and 3t and 3tOpen are trademarks of 3Com Corporation. AppleTalk is a registered trademark of Apple Computer, Inc. Xerox Network Systems and Ethernet are trademarks of Xerox Corporation. VINES is a registered trademark of Banyan Systems, Inc. DECnet Phase IV and DECnet Phase V are trademarks of Digital Equipment Corporation.

List of Acronyms

AAL 3/4	ATM adaptation layer 3/4
ANSI	American National Standards Institute
ATM	asynchronous transfer mode
BCD	binary-coded decimal
Bellcore	Bell Communications Research
BISDN	broadband integrated services digital network
CAN	customer access node
CATV	community antenna television
CIR	committed information rate
CLNP	connectionless network protocol
CNM	customer network management
CPE	customer premises equipment
CRS	cell relay service
CSMA/CD	carrier sense multiple access with collision detection
CSU	channel service unit
DLCI	data-link connection identifier
DQDB	distributed queue dual bus
DS-0	digital signal level 0
DS-1	digital signal level 1
DS-3	digital signal level 3
DSU	data service unit
DXI	data exchange interface

E 1	European transmission level 1
E-3	European transmission level 3
ESIG	European SMDS Interest Group
ETSI	European Telecommunication Standardization Institute
FDDI	fiber distributed data interface
FRPVC	frame relay permanent virtual circuit
FRSVC	frame relay switched virtual circuit
GCRA	generic cell rate algorithm
HDLC	high-level data link control
IEC	interexchange carrier
IEEE	Institute of Electrical and Electronics Engineers
IETF	Internet Engineering Task Force
IP	internet protocol
IPX	internet packet exchange
ISDN	integrated services digital network
L1__PDU	level 1 protocol data unit
L2__PDU	level 2 protocol data unit
L3__PDU	level 3 protocol data unit
LAN	local-area network
LATA	local access and transport area
LEC	local-exchange carrier
LLC	logical link control
MAC	media access control
MAN	metropolitan-area network
MFJ	Modified Final Judgment
MIB	management information base
NMS	network management station
OSI	Open System Interconnection
PDU	protocol data unit
PHY	physical layer
POTS	plain old telephone service
PR FASIG	Pacific Rim Frame Relay/ATM/SMDS Interest Group
PVC	permanent virtual circuit
RBOC	Regional Bell Operating Company
SDLC	synchronous data-link control
SIG	SMDS Interest Group
SIP	SMDS interface protocol
SIR	sustained information rate
SMDS	switched multimegabit data service
SNI	subscriber network interface
SNMP	simple network management protocol
SONET	synchronous optical network
SRI	SIP relay interface
STS-3c	synchronous transport signal level 3c
SVC	switched virtual circuit
TCP	transmission control protocol
UNI	user–network interface
VCC	virtual channel connection

WAN   wide-area network
XNS   Xerox Network System

References

1. Klessig, R. W., and Tesink, K., *SMDS Wide-Area Data Networking with Switched Multi-megabit Data Service*, Prentice-Hall, Englewood Cliffs, NJ, 1994.
2. Institute of Electrical and Electronics Engineers, *Carrier Sense Multiple Access with Collision Detection (CSMA/CD)*, IEEE Std 802.3-1985.
3. Bell Communications Research (Bellcore), *Metropolitan Area Network Generic Framework System Requirements in Support of Switched Multi-megabit Data Service*, Bellcore Technical Advisory TA-TSY-000772, Issue 1, February 1988.
4. Bell Communications Research (Bellcore), *Generic System Requirements in Support of Switched Multi-megabit Data Service*, Bellcore Technical Advisory TA-TSY-000772, Issue 2, March 1989.
5. Bell Communications Research (Bellcore), *Generic Requirements for Low Speed SMDS Access*, Bellcore Technical Advisory TA-TSV-001239, Issue 1, June 1993.
6. Bell Communications Research (Bellcore), *Generic Requirements for Frame Relay Access to SMDS*, Bellcore Technical Advisory TA-TSV-001240, Issue 1, June 1993.
7. European SMDS Interest Group, *SMDS Subscriber Network Interface Level 1 Specification*, Edition 1.0, ESIG-TS-002, June 1993.
8. Bell Communications Research (Bellcore), *Inter-Switching System Interface Generic Requirements in Support of SMDS Service*, Bellcore Technical Advisory TA-TSV-001059, December 1990.
9. Bell Communications Research (Bellcore), *Switched Multi-megabit Data Service Generic Requirements for Exchange Access and Intercompany Serving Arrangements*, Bellcore Technical Reference TR-TSV-001060, Issue 1, December 1991.
10. Bell Communications Research (Bellcore), *Generic Requirements for SMDS Customer Network Management Service*, Bellcore Technical Advisory TA-TSV-001062, Issue 2, February 1992.
11. Bell Communications Research (Bellcore), *Generic Requirements for Phase 1 SMDS Customer Network Management Service*, Bellcore Technical Reference TR-TSV-001062, Issue 1, March 1993.
12. Bell Communications Research (Bellcore), *Generic System Requirements in Support of Switched Multi-megabit Data Service*, Bellcore Technical Advisory TA-TSV-000772, Issue 3, October 1989.
13. Bell Communications Research (Bellcore), *Generic System Requirements in Support of Switched Multi-megabit Data Service*, Bellcore Technical Advisory TA-TSV-000772, Issue 3, Supplement 1, January 1991.
14. Bell Communications Research (Bellcore), *Generic Requirements in Support of Switched Multi-megabit Data Service*, Bellcore Technical Reference TR-TSV-000772, Issue 1, May 1991.
15. Bell Communications Research (Bellcore), *Local Access System Generic Requirements, Objectives, and Interfaces in Support of Switched Multi-megabit Data Service*, Bellcore Technical Reference TR-TSV-000773, Issue I, June 1991.
16. SMDS Interest Group, *SMDS Data Exchange Protocol, Revision 3.2*, SIG-TS-001/1991, October 22, 1991.
17. SMDS Interest Group, *SMDS DXI Local Management Interface Revision 2.0*, SIG-TS-002/1991, May 19, 1992.

18. SMDS Interest Group, *Frame Based Interface Protocol for Network Supporting SMDSCSIP Relay Interface*, Revision 1.0, SIG-TS-006/1993, February 1993.

19. SMDS Interest Group, *Protocol Interface Specification for Implementation of SMDS over an ATM-based Public UNI*, Revision 1.0, SIG-TWG-008/1994, May 3, 1994.

20. SMDS Interest Group, *Implementation of Phase IV DECnet over SMDS*, SIG-TS-003/1992, Revision 1.1, May 3, 1994.

21. Oppenheimer, A. B., *SMDSTalk: Apple Talk over SMDS*, SMDS Interest Group Informational Specification, SIG-TWG-019/1992, August 17, 1992.

22. Israel, J. E., *Transmission of Novell IPX Datagrams over the SMDS Service*, Version 1.0—SMDS Interest Group Informational Specification, SIG TWG-1993/042, July 1993.

23. European Telecommunication Standardization Institute, *Metropolitan Area Networks (MAN) Physical Layer Convergence Procedure for 2048 Mbit/s*, ETSI 300-213.

24. European Telecommunication Standardization Institute, *Metropolitan Area Networks (MAN) Physical Layer Convergence Procedure for 34,368 Mbit/s*, ETSI 300-214.

25. Piscitello, D. M., and Lawrence, J., *Transmission of IP Datagrams over the SMDS Service*, RFC 1209, InterNIC Information Services, March 1991.

26. Rose, M., *The Simple Book—An Introduction to Management of TCP/IP-Based Internets*, Prentice-Hall, Englewood Cliffs, NJ, 1991.

27. Case, J. D., Fedor, M., Schoffstall, M. L., and Davin, C., *Simple Network Management Protocol*, RFC 1157, InterNIC Information Services, May 1990.

28. Cox, T. A., and Tesink, K. (eds.), *Definitions of Managed Objects for the SIP Interface Types*, RFC 1304, InterNIC Information Services, May 1992.

29. Baker, F., and Watt, J., *Definitions of Managed Objects for the DS1 Interface Type*, RFC 1406, InterNIC Information Services, January 1993.

30. Cox, T. A., and Tesink, K. (eds.), *Definitions of Managed Objects for the DS3 Interface Type*, RFC 1407, InterNIC Information Services, January 1993.

31. Institute of Electrical and Electronics Engineers, *Distributed Queue Dual Bus (DQDB) Subnetwork of a Metropolitan Area Network*, IEEE Std 802.6-1990.

32. Institute of Electrical and Electronics Engineers, *Token-Passing Bus Access Method and Physical Layer Specifications*, IEEE Std 802.4-1985.

33. Institute of Electrical and Electronics Engineers, *Token Ring Access Method and Physical Layer Specification*, IEEE Std 802.5-1989.

34. American National Standards Institute, *Fiber Distributed Data Interface (FDDI)—Token Ring Media Access Control (MAC)*, ANSI X3.139-1987.

35. Postel, J., *Internet Protocol*, RFC 791, InterNIC Information Services, September 1981.

36. Postel, J., *Transmission Control Protocol*, RFC 793, InterNIC Information Services, September 1981.

37. International Telegraph and Telephone Consultative Committee (CCITT), Numbering Plan for the ISDN Era, CCITT Recommendation E.164. In *CCITT Blue Book*, International Telecommunication Union, Geneva, 1988.

38. Novell, *IPX Router Specification, Revision A*, Part Number 107-000029-001, June 16, 1992.

39. Frame Relay Forum, *Implementors Agreement*, FRF-1, January 1992.

40. SMDS Interest Group, *Frame Based Interface Protocol for SMDS Networks, Data Exchange Interface/Subscriber Network Interface Revision 1.0*, SIG-TS-005/1993, February 2, 1993.

41. MCI Telecommunications Corporation, *Hyperstream^{SM} Switched Multi-megabit Data Service*, May 1993 (brochure).

42. International Organization for Standardization (ISO) — Information Processing Systems, Data Communication, *High-Level Data Link Control Procedures — Frame Structure*, International Standard ISO 3309, 1984.
43. Bell Communications Research (Bellcore), *Generic Requirements for Exchange PVC Cell Relay Service*, Bellcore Technical Advisory TA-TSV-001408, Issue 1, August 1993.
44. Asynchronous Transfer Mode Forum, *ATM User-Network Interface Specification, Version 3.0*, September 10, 1993.
45. Asynchronous Transfer Mode Forum, *ATM Data Exchange Interface (DXI) Specification*, June 1993.
46. Institute of Electrical and Electronics Engineers, *Local Area Networks — Part 2: Logical Link Control*, IEEE Std. 802.2-1985.

ELYSIA C. TAN
ROBERT W. KLESSIG
KAJ TESINK

Space Communications

Introduction

Space communications are crucial to the manned spacecraft operations for transmissions of voice, command, telemetry, video, and data to and from the mission control centers. This article addresses the U.S. manned space communications and tracking systems. Other countries' space communications projects are not treated here. All satellite-related communications topics can be referred to other articles and also are not considered here.

The U.S. manned space communications programs include Space Shuttle *Orbiter* (SSO) and International Space Station (ISS). The SSO and ISS will utilize the National Aeronautical and Space Administration (NASA) Tracking and Data Relay Satellite System (TDRSS) to relay communications to and from the ground. In this article, the TDRSS is introduced first, followed by the descriptions of the communications and tracking system designs of SSO and ISS. The areas of requirements, configurations, operation, and performance for each system are discussed.

Tracking and Data Relay Satellite System

The present SSO and future ISS will be heavily dependent on NASA's TDRSS to provide the majority of the space-ground communications and tracking functions. So far, six TDRSs (F1, F3, F4, F5, F6, and F7) have been successfully deployed and are operational. Three new TDRSs (H, I, J) will be launched and become operational between the years 2000 and 2002.

Current Capabilities

The TDRSS is a space-based network that provides communications, tracking, and data acquisition and command/telemetry/audio/video services essential to the SSO, ISS, and low-earth orbital spacecraft such as the Compton Gamma Ray Observatory (GRO), Hubble Space Telescope, Earth Radiation Budget Satellite, Extreme Ultraviolet Explorer, and other NASA and non-NASA missions.

The TDRSS consists of two major elements: a constellation of at least two geosynchronous satellites (133° apart in longitude) at an altitude of 22,300 statute miles above the equator, with on-orbit spares, and a ground terminal complex located at White Sands, New Mexico. Figure 1 depicts the TDRS constellation and White Sands Complex (WSC) and Table 1 lists current TDRS utilization. Figure 2 shows the TDRS operational coverage.

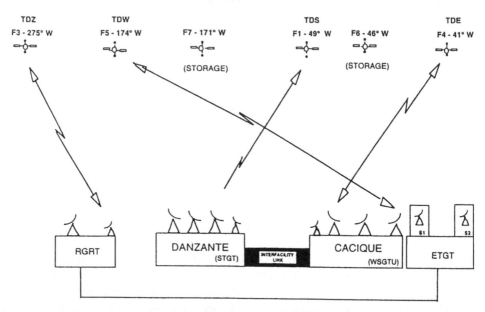

FIG. 1   Tracking and Data Relay Satellite System (TDRSS) constellation (ETGT = Extended TDRS Ground Terminal; RGRT = Remote Ground Relay Terminal; STGT = Second TDRSS Ground Terminal; WSGTU = White Sands Ground Terminal Upgrade).

TABLE 1   Tracking and Data Relay Satellite System Utilization

	Launched	Geosynchronous Orbit	In-Orbit Checkout	Utilization
TDRS-1	April 4, 1983 STS-6 (*Challenger*)	June 29, 1993	• December 28, 1983 • One-satellite system acceptance April 1985	• Currently at 49°W • Designated TDRS-Spare
TDRS-3	September 29, 1988 STS-26 (*Discovery*)	September 30, 1988	• January 15, 1989 • Two-satellite system acceptance July 1989	• Currently designated as TDRS-ZOE at 275°W
TDRS-4	March 13, 1989 STS-29 (*Discovery*)	March 14, 1989	June 9, 1989	• Currently designated as TDRS East at 41°W and providing full support
TDRS-5	August 2, 1991 STS-43 (*Atlantis*)	August 3, 1991	October 7, 1989	• Currently designated as TDRS West at 174°W and providing full support
TDRS-6	January 13, 1993 STS-54 (*Endeavor*)	January 14, 1993	March 4, 1993	• Currently at 46°W • Stored spare
TDRS-7	July 13, 1995 STS-70 (*Discovery*)	July 14, 1995	August 22, 1995	• Currently at 171°W • Stored spare

TDRS-2 lost January 28, 1986, aboard STS-51L (*Challenger*).

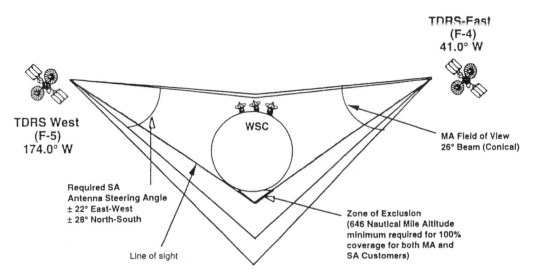

FIG. 2 Tracking and Data Relay Satellite System operational coverage (MA = multiple access; SA = single access).

The WSC is composed of the White Sands Ground Terminal (WSGT) and the Second TDRSS Ground Terminal (STGT), which contain the ground terminal communications relay equipment for the command, telemetry, tracking, and control equipment of the TDRSS. The NASA Ground Terminal (NGT) is co-located with the WSGT. The NGT is managed and operated by the NASA Goddard Space Flight Center (GSFC) and, in communication with the worldwide NASA Communications Network (NASCOM), is NASA's physical and electrical interface with the TDRSS. The NGT provides the interfaces with the common carrier, monitors the quality of the service from the TDRSS, and monitors remote data quality to the GSFC's Network Control Center (NCC). Figure 3 illustrates the TDRSS network configuration.

The TDRS can transmit and receive data and track a user satellite in a low-earth orbit at altitudes from 100 nautical miles (NM) to about 3000 NM for a minimum of 85% of its orbit. The TDRSS telecommunication services to and from the user's control and data-processing facilities operate in a real-time, bent-pipe mode at two frequency bands (Ku and S) and at S band in either a single-access (SA) or multiple-access (MA) mode.

The NASCOM network is composed of telephone, microwave, radio, submarine cables, and communications satellites. These various systems link the data flow through 11 countries with 15 foreign and domestic carriers and provide the required information between tracking sites and the Johnson Space Center (JSC) and GSFC control centers. Special wideband and video circuitry are used as needed.

Table 2 and Table 3 list the S-band forward- and return-link single-access services provided by the TDRSS, respectively (1,2). Table 4 and Table 5 list the Ku-band forward and return link single-access services provided by the TDRSS, respectively (1,2).

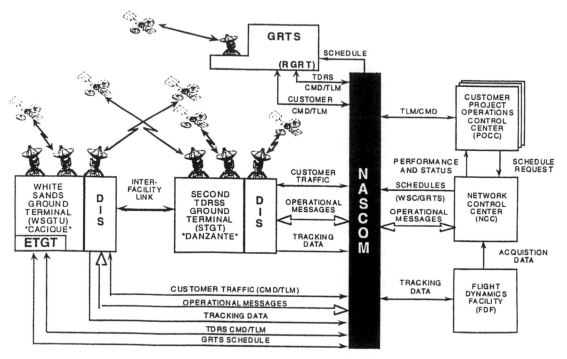

FIG. 3 Tracking and Data Relay Satellite System network configuration.

TABLE 2 S-Band Forward-Link Single-Access Services

Transmit frequency	2020.0 MHz to 2123.5 MHz, tunable in 0.05-MHz steps
Channel bandwidth	20 MHz
EIRP	48.5 dBW (SSO)
	46.4 dBW (end of life)
	46.2 dBW (high power)
	44.3 dBW (normal power)
	37.7 dBW (linear IF)
Data rates	100 b/s to 300 kb/s
Polarization	LHCP or RHCP Command selectable
Antenna gain	36 dB
Antenna pointing	Adjustable over ±22.5°E/W and ±31°N/S

EIRP = effective isotropic radiated power; RHCP = right-hand circular polarized; LHCP = left-hand circular polarized.

TABLE 3 S-Band Return-Link Single-Access Services

Return frequency	2200.0 MHz to 2300.0 MHz, tunable in 0.05-MHz steps
Channel bandwidth	10 MHz (3 dB)
Input user signal power range	−138 to −117 dBm
G/T	>8.9 dB/K
Data rates	1 kb/s to 6 Mb/s
Polarization	LHCP or RHCP Command selectable
Antenna gain	>36.8 dB
Antenna pointing	Adjustable over ±22.5°E/W and ±31°N/S

Two single-access users can be supported simultaneously.

Features of Tracking and Data Relay Satellite H, I, and J Series

The TDRS H, I, and J series (2) will be put into service starting between the years 2000 and 2002 and will provide customers the same S-band MA, S-band single-access (SSA), and Ku-band single-access (KuSA) services currently supported by the TDRS F3 through F7, but with system performance enhancement in terms of gain-to-system noise temperature (G/T) and effective isotropic radiated power (EIRP). Table 6 lists the MA, SSA, and KuSA performance comparison between the F3–F7 and the H, I, J. The deployment of H, I, and J architecture is shown in Fig. 4.

TABLE 4 Ku-Band Forward-Link Single-Access Services

Transmit frequency	13.775 GHz
Channel bandwidth	50 MHz (3 dB)
EIRP	49.4 dBW (end of life) 49.4 dBW (high power) 47.1 dBW (normal power) 41.3 dBW (linear IF)
Data rates	1 kb/s to 25 Mb/s
Polarization	LHCP or RHCP Command selectable
Antenna gain	>52.5 dB
Antenna pointing	Adjustable over ±22.5°E/W and ±31°N/S with autotrack capability

Two single-access users can be supported simultaneously.

TABLE 5   Ku-Band Return-Link Single-Access Services

Return frequency	15.0034 GHz
Channel bandwidth	225 MHz (3 dB)
Input user signal power range	−140 to −119 dBm
G/T	>23.8 dB/K
Data rates	1 kb/s to 300 Mb/s
Polarization	LHCP or RHCP Command selectable
Antenna gain	>53.54 dB
Antenna pointing	Adjustable over ±22.5°E/W and ±31°N/S with autotrack capability

Two single-access users can be supported simultaneously.

The TDRS H, I, and J, however, will also introduce new Ka-band single-access services to be time shared with existing Ku-band services in addition to upgrading the MA and SA performance. The Ka-band system will be a switched substitution for the current Ku-band customer services; it supports the same customer data rates as the present Ku-band services. The Ka-band will use the present space-to-ground (Ku-band) link design and will be essentially transparent to the ground station. The spacecraft Ka band will have the capability to support higher data rates than the current Ku-band system. Either the Ka band or Ku band can be selected independently for each single-access antenna by ground reconfiguration commands. Simultaneous S-band/Ku-band or S-band/Ka-band service is available for one user spacecraft or two co-orbiting user spacecraft.

Table 7 and Table 8 list TDRS H, I, J Ka-band forward-link and return-link single-access services, respectively. Table 9 lists the TDRS H, I, and J modes of operation.

TABLE 6   Performance Comparison of Tracking
and Data Relay Satellites F1–F7
and H, I, and J

	TDRS F3–F7	TDRS H, I, J
Forward-link EIRP		
MA	34.0 dBW	42.7 dBW
SSA	46.4 dBW	49.6 dBW
KuSA	49.4 dBW	49.7 dBW
Return-link G/T		
MA	0.5 dB/K	5.6 dB/K
SSA	8.9 dB/K	9.4 dB/K
KuSA	23.8 dB/K	24.4 dB/K

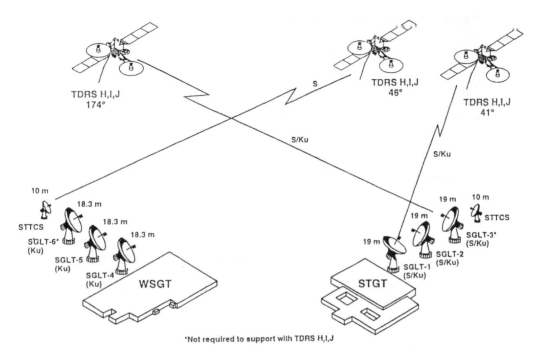

FIG. 4   Tracking and Data Relay Satellite H, I, and J architecture.

TABLE 7   Ka-Band Forward-Link Single-Access Services

Transmit frequency	22.555 to 23.545 GHz, tunable in 5-MHz steps
Channel bandwidth	50 MHz (3 dB)
EIRP	62.4 dBW (normal power)
Data rates	1 kb/s to 50 Mb/s
Polarization	LHCP or RHCP Command selectable
Antenna gain	> 52.3 dB
Antenna pointing	Adjustable over ±22.5°E/W and ±31°N/S with autotrack capability

Two single-access users can be supported simultaneously.

TABLE 8   Ka-Band Return-Link Single-Access Services

Return frequency	25.2534 to 27.4784 GHz, tunable in 25-MHz steps 25.550 to 27.200 GHz, tunable in 25-MHz steps
Channel bandwidth	225 MHz (25.2534 to 27.4784 GHz) (3 dB) 650 MHz (25.550 to 27.200 GHz) (3 dB)
G/T	>23.1 dB/K
Input user signal power range	−155 to −113 dBm
Data rates	1 kb/s to 300 Mb/s[a]
Polarization	LHCP or RHCP Command selectable
Antenna gain	>52.6 dB
Antenna pointing	Adjustable over ±22.5°E/W and ±31°N/S with autotrack capability

Two single-access users can be supported simultaneously.
[a]Spacecraft can accommodate 800 Mb/s.

Space Shuttle Communications and Tracking System Design

The design of the space shuttle communications and tracking system is more than 24 years old. It has served the program from the initial captive-active flights in 1977 to today with little change in the original design.

The advanced state-of-the-art digital communications techniques in the early 1970s such as voice digitization, channel coding, time division multiplexing, PSK (phase-shift keying) suppressed carrier modulation, and spread-spectrum modulation were necessarily employed due to use of low-gain, flush-mounted, wide-beam antennas and the TDRSS links (3). These advanced techniques, proven by the success of the space shuttle communications systems, led to and

TABLE 9   H, I, and J Modes of Operation

Forward-Link Services	Return-Link Services
2 S-band single access	2 S-band single access

2 Ku-band single access
  or
	2 Ku-band single access
	  or

1 Ku-band single access
  and
	1 Ku-band single access
	  and

1 Ka-band single access
  or
	1 Ka-band single access
	  or

2 Ka-band single access	2 Ka-band single access

became backbones of the modern communications systems, including satellite communications systems, wireless personal communications, and so on. Figure 5 depicts the overall space shuttle on-orbit communications links.

Even though the medium-scale, integrated, complementary metal oxide semiconductor (CMOS) circuitry that comprises most of the SSO network and signal processors like the network signal processor (NSP) and the payload signal processor (PSP), and custom large-scale integrated (LSI) devices are 1960s technology and became obsolete and are no longer used in today's modern communications system designs, the space shuttle communications and tracking system still functions well and will still operate beyond the year of 2000.

The SSO major space-ground operational communications and tracking systems are composed of the S-band network subsystem and the Ku-band subsystem.

S-Band Network Subsystem. The SSO S-band network subsystem provides integrated voice, command, ranging, and telemetry communications services with all three of the networks (NASA ground space flight tracking and data network [GSTDN], NASA TDRSS, and the U.S. Air Force [USAF] satellite control facility [SCF]) by way of two-way phase modulation (PM) links to the ground, directly or through the TDRSS. It also provides one-way transmission of wideband data directly to the ground by way of a frequency modulation (FM) link. The S-band network communication subsystem is the workhorse system

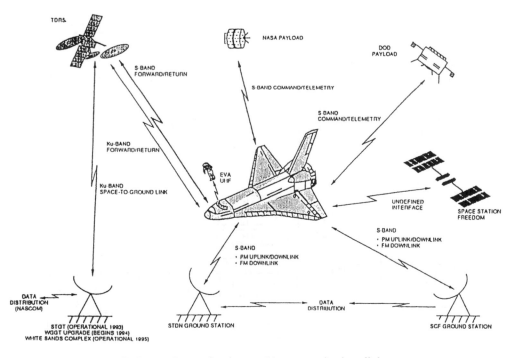

FIG. 5 Space shuttle on-orbit communications links.

that provides necessary flight operational data transmission and ranging capabilities. Figure 6 shows the S-band network subsystem block diagram.

Four flush-mounted antennas (quads) are used for the S-band PM operational links. The nominal coverage requirement for the antennas is 3 dBCi (decibels referenced to an isotropic radiator) at 87% of the sphere. Since the TDRSS link is marginal, the antenna performance is very critical. The four separate units are mounted on the side of the *Orbiter* forward fuselage, spaced approximately 90° apart. Selection of the proper antenna is controlled by the general-purpose computer. The switching algorithm incorporates a 2° dead band between the quadrants to avoid switch chatter. Within each quadrant, the antenna beam is switched forward or aft to enhance the system performance. Figure 7 illustrates SSO antenna locations, including S-band quads.

Two sections below illustrate the SSO S-band TDRSS relay forward-link and SSO S-band TDRSS relay return-link operations provided by the S-band network subsystem.

Ku-Band Subsystem. The Ku-band system can only operate during the on-orbit phase of the mission, when the SSO payload bay doors are open and the high-gain gimballed antenna reflector is deployed. In addition to the same normal operational services as provided by the S-band system, the main Ku-band subsystem is to provide wideband, high-rate data transmission via TDRS, which is not available on the normal S-band links and radar function and is for

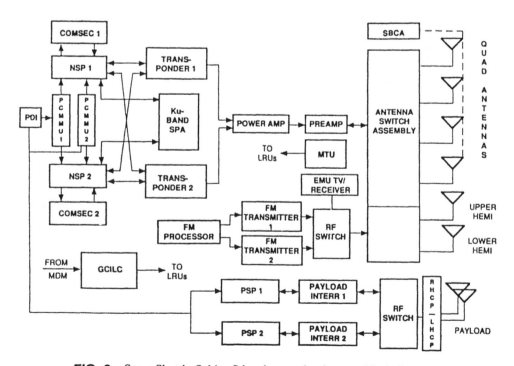

FIG. 6   Space Shuttle *Orbiter* S-band network subsystem block diagram.

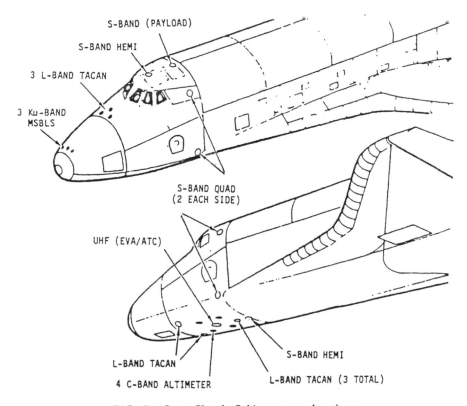

FIG. 7 Space Shuttle *Orbiter* antenna locations.

detecting and tracking detached payloads or other spacecraft during rendezvous maneuvers.

Separate sections below illustrate the SSO Ku-band TDRSS relay forward-link and SSO Ku-band TDRSS relay return-link operations provided by the Ku-band subsystem. Figure 8 is the Ku-band system functional block diagram.

Space Shuttle *Orbiter* S-Band TDRSS Relay Forward-Link Operation

In the SSO S-band TDRSS relay forward link, for a high-data-rate mode, two voice channels and one command channel are provided simultaneously. In the JSC Mission Control Center (MCC), each voice channel is digitized using delta modulation to 32 kb/s (kilobits per second). The MCC Bose–Chaudhuri–Hocquenghem (BCH) (127,50) [50 information bits, 77 parity check bits] encodes the 2.4-kb/s command messages into a 6.4-kb/s data sequence for command error detection, which is further processed into the permuted command form by exclusive-or bit by bit with a 128-bit Greenwich mean time (GMT) for command authentication. The permuted command channel is then time division multiplexed (TDM) with a 1.6-kb/s synchronization channel and the two 32-kb/s delta-modulated voice channels to form a 72-kb/s TDM data sequence, which is

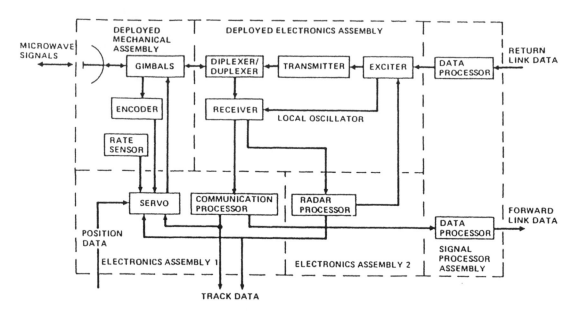

FIG. 8   Space Shuttle *Orbiter* Ku-band system functional block diagram.

encrypted to produce a 72-kb/s encrypted non-return-to-zero-level (NRZ-L) operational TDM data sequence. The 72 kb/s is transmitted to the TDRSS ground station (White Sands, NM), where it is encoded at a rate of 1/3 (1 bit in, 3 bits out) by a convolutional encoder. The 216-kb/s encoded sequence in bi-phase-L format is then multiplied by a pseudonoise (PN) NRZ-L code at the rate of 11.232 Mb/s (megabits per second). The resulting spread-spectrum-coded sequence is then used to PSK the carrier (2106.4 megahertz [MHz] or 2041.9 MHz). Figure 9 illustrates the TDM format.

In the SSO, one of the four quad antennas is selected and activated by the radio-frequency (RF) switch assembly to receive the PN/BPSK (binary phase-shift keying) modulated carrier and forward it to the preamplifier assembly for noise reduction. The S-band transponder then downconverts, despreads, and demodulates the received spread-spectrum signal. The NSP bit synchronizer produces a soft-decision version of the encoded 216-kb/s operational data, which is decoded using the Viterbi algorithm into the 72-kb/s encrypted operational TDM sequence.

The 72-kb/s operational data is then decrypted by a COMSEC (communication security) device. A frame sync demultiplexer in the NSP is used to identify the synchronization pattern and control encrypted bit inversion. The NSP will authenticate the permuted command data with the on-board 128-bit GMT code and the BCH command decoder and replace the 6.4-kb/s permuted command data with 6.4-kb/s encoded command data in the operational TDM sequence. The NSP will then separate the two voice channels and the encoded command channel. The NSP delta demodulators convert the two voice channels to analog voices, and the command decoder verifies the encoded command data sequence for any channel-induced error.

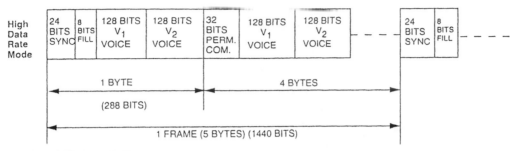

SYNC PATTERN: 76571440₈

SYNC PATTERN: 7657144$0_8$

1440 BITS/FRAME 50 FRAMES/SEC, TDM DATA RATE: 72 kbps
VOICE CHANNEL 1 **DATA** RATE : 72 x 5x128/1440 = 32 kbps, VOICE CHANNEL 2 **DATA** RATE : 72 x
5x128/1440 = 32 kbps, COMMAND CHANNEL DATA RATE : 72 x 4 x 32/1440 = 6.4 kbps, SYNC CHANNEL DATA
RATE : 72 x 32/1440 = 1.6 kbps

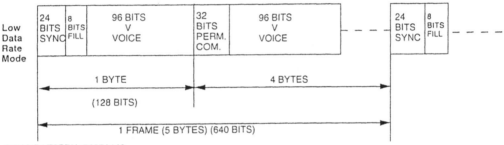

SYNC PATTERN: 7657144$0_8$

640 BITS/FRAME, 50 FRAMES/SEC, TDM DATA RATE: 32 kbps
VOICE CHANNEL **DATA** RATE : 32 x 5x96/640 = 24 kbps, COMMAND CHANNEL DATA RATE : 32 x 4 x
32/640 = 6.4 kbps, SYNC CHANNEL DATA RATE : 32 x 32/640 = 1.6 KBPS

FIG. 9 Space Shuttle *Orbiter* S-band forward-link operational time division multiplexed (TDM)
format.

For the low-data-rate mode, only one delta-modulated voice channel at the
rate of 24 kb/s is provided, and the resulting encrypted operational TDM channel is 32 kb/s instead of 72 kb/s. The encoded sequence is 96 kb/s. The operation of this low-data-rate mode is exactly the same as that of the high-data-rate
mode. This mode is used only when the link is deteriorated.

Figure 10 illustrates the S-band forward-link configuration, and Table 10
summarizes the S-band forward-link interface characteristics.

Space Shuttle *Orbiter* S-Band TDRSS Relay Return-Link Operation

In the SSO S-band TDRSS relay return link, for a high-data-rate mode, two
voice channels and one telemetry channel are provided simultaneously. In the
NSP, two voice channels, each digitized using delta modulation at 32 kb/s, is
TDM with the 128-kb/s telemetry channel to form a 192-kb/s operational TDM
signal. The operational TDM data sequence is then convolutionally encoded by
a rate 1/3 channel encoder to 576 kb/s (high data rate) or 288 kb/s (low data
rate) in the NSP. The NSP converts the encoded sequence to biphase-L format,
and the transponder is used to PSK modulate the link carrier, which is 240/221

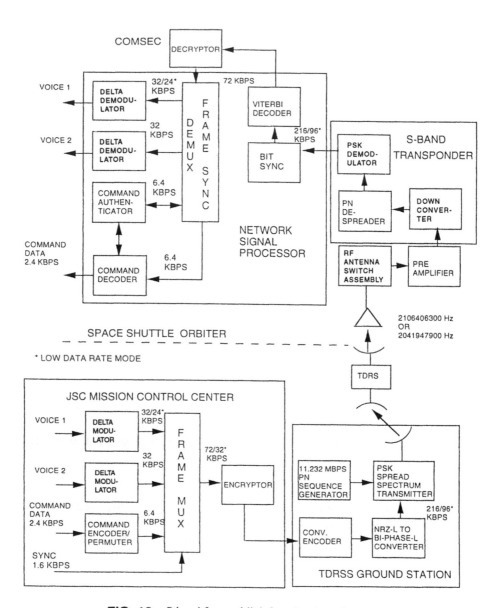

FIG. 10   S-band forward-link functional configuration.

times the received forward-link carrier frequency in the coherent turnaround mode. Figure 11 shows the S-band return-link TDM format.

When the auxiliary oscillator downlink carrier frequency (2287.5 MHz or 2217.5 MHz) is used, the 576-kb/s or 288-kb/s encoded data signal PSK modulates the carrier. The PSK-modulated carrier is amplified by the RF power amplifier and radiated to the TDRS from one of the quad antennas selected by the RF antenna switch assembly.

In the TDRSS ground station, a PSK receiver recovers the return-link carrier-modulating signal, and a soft-decision bit synchronizer and a Viterbi de-

TABLE 10 Space Shuttle *Orbiter* S-Band Forward-Link Interface Characteristics

Johnson Space Center Mission Control Center		
Information		
Channel	**Rate or Frequency**	**Premodulation Processing**
Voice 1	300–3000 Hz	Digitize voice channels to 32/24 kb/s[a] (delta modulation)
Voice 2	300–3000 Hz	BCH (127,50) encode and permute command data to 6.4 kb/s
Command	2.4 kb/s	Time division multiplex all channels to 72/32 kb/s[a] (sync pattern: 76571440_8)
Sync	1.6 kb/s	Encrypt the TDM channel Format encrypted data to NRZ-L

TDRSS Ground Station				
Channel Encoding	**Spectrum Spreading**	**Carrier Modulation Processing**	**Antenna Polariation**	**EIRP**
Convolutionally encode TDM data Code rate: 1/3 Constraint length: 7 Generator sequence: 7566127_8 Symbol rate: 216/96 kb/s ± 0.001% Symbol format: biphase-L	Exclusive or encoded TDM data with PN code PN chip rate: 11.232 mcps PN sequence length: 1023 PN sequence format: NRZ-L	PSK modulate carrier with PN-coded data Mod. index: $\pi/2$ ± 0.05 RAD Carrier frequency: 2106406300 Hz or 2041947900 Hz	RCP	46 dBW

(continued)

coder are used to provide the 192-kb/s or 96-kb/s operational TDM data sequence. A frame sync demultiplexer in the JSC MCC then identifies the synchronization pattern in the telemetry data and separates the two digital voice channels and the telemetry channel. The voice channels are routed to the delta demodulators for conversion to analog voices.

Figure 12 illustrates the S-band return-link configuration, and Table 11 summarizes the S-band return-link interface characteristics.

Space Shuttle *Orbiter* Ku-Band TDRSS Relay Forward-Link Operation

As in the case of the S-band TDRSS forward link, the Ku-band TDRSS forward link also employs a direct sequence, spread-spectrum signal design to reduce the power flux density as required by the International Telecommunication Union (ITU) International Radio Consultative Committee (CCIR) guideline. The TDM

TABLE 10 Continued

Space Shuttle *Orbiter* S-Band Pulse Modulation Subsystem

Min. G/T	Required P_{rec}/N_0	RF Antenna Switch Assembly	Preamplifier	S-Band Transponder	Network Signal Processor (NSP)
−28.0 dB/K	55.1 dBHz for BER 10^{-4} 51.6 dBHz[a]	Select one of four quad antennas	Amplify received RF signal and reduce front-end noise	Downconvert PN-coded carrier Despread carrier Demodulate carrier (COSTAS loop)	Bit sync demodulated baseband signal Viterbi decode baseband TDM data Decrypt TDM data[b] Frame sync and demultiplex TDM data Authenticate command data BCH decode command Delta demodulate voice channels

[a]Low-data-rate mode.
[b]By external COMSEC device.

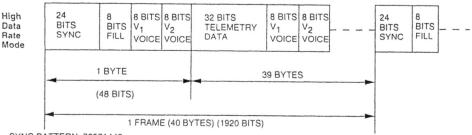

SYNC PATTERN: 76571440$_8$
1920 BITS/FRAME, 100 FRAMES/SEC, TDM DATA RATE: 192 KBPS
VOICE CHANNEL 1 DATA RATE : 192 x 40x8/1920 = 32 kbps, VOICE CHANNEL 2 DATA RATE : 192 x 40x8/1920 = 32 kbps,
TELEMETRY CHANNEL DATA RATE : 192 x 40 x 32/1920 = 128 kbps

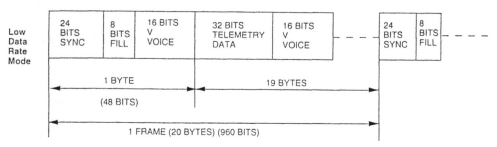

SYNC PATTERN: 76571440$_8$
960 BITS/FRAME, 100 FRAMES/SEC. TDM DATA RATE: 96 KBPS
VOICE CHANNEL DATA RATE : 96x 20X16/1960 = 32 kbps, TELEMETRY CHANNEL DATA RATE : 96 x 20 x 32/960= 64 kbps

FIG. 11 Space Shuttle *Orbiter* S-band return-link operational time division multiplexed format.

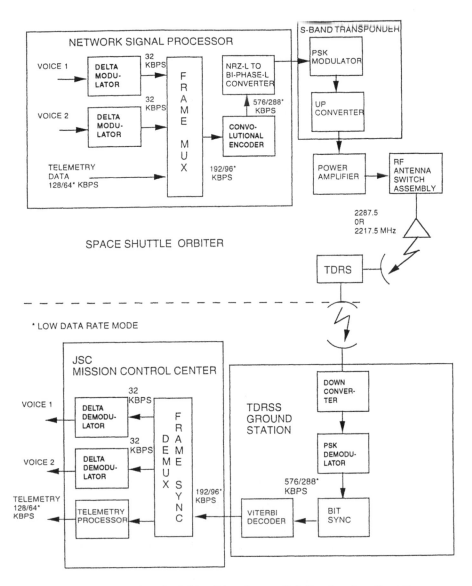

FIG. 12 Space Shuttle *Orbiter* S-band return-link functional configuration.

datastream is asynchronously modulo-2 added to the PN code prior to PSK modulating the carrier.

The PN chip rate of approximately 3.028 Mcps (mega chips per second) is coherently related to the 13.775-gigahertz (GHz) carrier frequency by the ratio of 31/(1469 × 96). The PN sequence is a 1023-chip-length gold code obtained by modulo-2 addition of two synchronous 1023-chip sequences obtained from two 10-stage linear shift registers with the feedback connection polynomial coefficients 3515 and 2011 in octal. The assigned gold code results from initializing the two shift register contents, respectively, at 0010000010 and 1001001000.

TABLE 11 Space Shuttle *Orbiter* S-Band Return-Link Interface Characteristics

Space Shuttle *Orbiter* S-Band Pulse Modulation Subsystem

Channel	Information Rate or Frequency	Network Signal Processor (NSP)	S-Band Transponder	RF Power Amplifier	RF Antenna Switch Assembly	EIRP
Voice 1 Voice 2 Telemetry	300–3000 Hz 300–3000 Hz 128/64[a] kb/s	Digitize voice channels to 32 kb/s (delta modulation) Time division multiplex all channels to 192/96 kb/s[a] (sync pattern: 76571440_8) Convolutionally encode the TDM channel to 576/288 kb/s[a] ±0.01% (biphase-L) Code rate: 1/3 Constraint length: 7 Generator sequence: 7566127_8	PSK modulate carrier with TDM-coded data Mod. index: $\pi/2$ ± 0.19 rad Carrier frequency: 2287.5 MHz or 2217.5 MHz ± 0.001% (auxilary or noncoherent mode) $240/221 \times$ received frequency (coherent mode)	Amplify PSK-modulated carrier 100 watts	Select one of quad antennas Antenna polarization RCP	17.0 dBW

TDRSS Ground Station

Required Received Power	Carrier Demodulation Processing	Baseband Processing
−177.7 dBW (for BER 10^{-4}) −180.7 dBW[a]	Downconvert channel-coded carrier Demodulate carrier (COSTAS loop)	Bit sync demodulated baseband signal Decode baseband TDM data (Viterbi soft decision)

Johnson Space Center Mission Control Center

Frame sync and demultiplex TDM Data

Delta demodulate voice channels

Process telemetry data

[a]Low-data-rate mode.

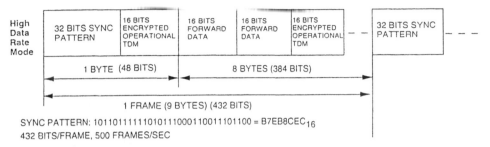

1 BYTE (48 BITS)   8 BYTES (384 BITS)

1 FRAME (9 BYTES) (432 BITS)

SYNC PATTERN: 10110111111010111000110011101100 = B7EB8CEC$_{16}$
432 BITS/FRAME, 500 FRAMES/SEC

TDM DATA RATE: 216 kbps, ENCRYPTED OPERATIONAL TDM CHANNEL **DATA** RATE : 216 x 9x48/432= 72 kbps
FORWARD DATA (TEXT AND GRAPHICS) DATA RATE : 216 x 8x32/1432 = 128kbps, SYNC CHANNEL DATA RATE :
216 x 32/432 = 16 kbps, IDLE FORWARD DATA CHANNEL IS REPLACED BY A REPEATED 127-BIT PN SEQUENCE.
THE PN BITS IN THE TDM FRAME REPEAT EVERY 127 FRAMES.

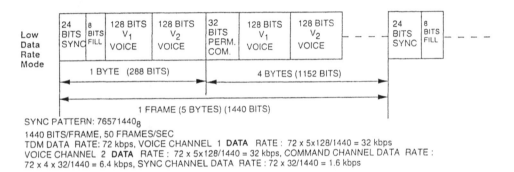

1 BYTE (288 BITS)   4 BYTES (1152 BITS)

1 FRAME (5 BYTES) (1440 BITS)

SYNC PATTERN: 7657144$_8$
1440 BITS/FRAME, 50 FRAMES/SEC
TDM DATA RATE: 72 kbps, VOICE CHANNEL 1 **DATA** RATE : 72 x 5x128/1440 = 32 kbps
VOICE CHANNEL 2 **DATA** RATE : 72 x 5x128/1440 = 32 kbps, COMMAND CHANNEL DATA RATE :
72 x 4 x 32/1440 = 6.4 kbps, SYNC CHANNEL DATA RATE : 72 x 32/1440 = 1.6 kbps

FIG. 13   Space Shuttle *Orbiter* Ku-band forward-link time division multiplexing format.

The Ku-band forward link has more capability than the S-band forward link and uplink. There are two modes of operation: the low-data-rate mode (Mode 2), which is the same as the high-data-rate mode of the S-band direct uplink and TDRSS forward link (72 kb/s, biphase-L), and the high-data-rate mode (Mode 1), which has an additional 128-kb/s data channel for text and graphics and a 16-kb/s sync channel (216 kb/s, biphase-L). Figure 13 illustrates the high-data-rate mode TDM format.

Forward error correction coding is not implemented due to high antenna gain provided by the 36-in diameter parabolic antenna. After the received signal is downconverted and despread, a COSTAS synchronization loop is used to demodulate the PSK carrier, and the recovered baseband signal plus noise are then detected by the bit synchronizer.

In the high-data-rate mode, two pairs of the bit synchronizers and demultiplexers are used. The first operates on the 216-kb/s TDM sequence, and the second set operates on the 72-kb/s operational data (voice and command). In the low-data-rate mode, only the second set bit sync/demux pair is employed. Forward-link encryption/decryption is only performed on the 72-kb/s operational data portion.

Figure 14 illustrates the Ku-band forward-link configuration, and Table 12 summarizes the Ku-band forward-link interface characteristics.

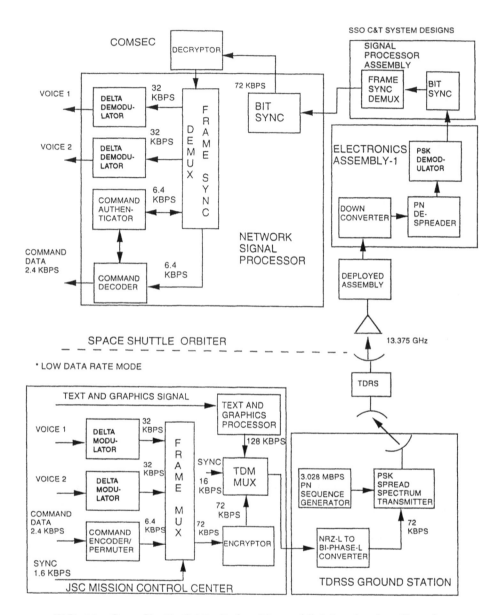

FIG. 14   Space Shuttle *Orbiter* Ku-band forward-link functional configuration.

Space Shuttle *Orbiter* Ku-Band TDRSS Relay Return-Link Operation

The Ku-band TDRSS relay return link provides the most throughput of any of the space shuttle communications links. Three channels are simultaneously transmitted on the return link. The modulation used depends on the mode of operation.

Mode 1 employs doubly unbalanced phase modulation, in that Channels 1 and 2 are used to quaternary phase-shift keying (QPSK) modulate an 8.5-MHz

TABLE 12 Space Shuttle *Orbiter* Ku-Band Forward Link High-Data-Rate Mode Interface Characteristics (Mode 1)

Johnson Space Center Mission Control Center			
	Information		
Channel	Rate or Frequency	Operational TDM Channel Processing	TDM Processing, 216 kb/s
Voice 1	300–3000 Hz	Digitize voice channels to 32 kb/s (delta modulation)	Time division multiplex operational TDM channel and forward data channel to 216 kb/s
Voice 2	300–3000 Hz		
Command	2.4 kb/s	BCH (127,50) encode and permute command data to 6.4 kb/s	Sync pattern: 26772706354_8
Sync	1.6 kb/s		
Forward data	128 kb/s	Time division multplex all channels to 72 kb/s (sync pattern: 76571440_8)	Format: biphase-L
Sync	16 kb/s	Encrypt the TDM channel Format encrypted data to NRZ-L	

TDRSS Ground Station			
Spectrum Spreading	Carrier Modulation Processing	Antenna Polarization	EIRP
Exclusive-or 216 kb/s TDM data with PN code	PSK modulate carrier with PN-coded data	RCP	Acquisition: 40.9–49.5 dBW
PN chip rate: $(31/141024) \times$ carrier freq. = 3.028 Mcps	Mod. index: $\pi/2$ rad $\pm$ 3°	Axial ratio $\leq$ 1 dB	Tracking: normal power: 46.5–49.5 dBW
PN sequence length: 1023	Carrier frequency: 13.775 GHz		High power: 48.5–51.5 dBW
PN sequence format: NRZ-L			

(continued)

square-wave subcarrier with 63° and 117° phase steps. The subcarrier and Channel 3 then QPSK modulate the 15.0034-GHz carrier with a 20/80 power ratio. Channel 3 performance is enhanced by employing rate 1/2 convolutional encoding and soft-decision Viterbi decoding.

On the ground, a QPSK receiver derives reference quadrature carriers for coherent QPSK demodulation to obtain Channel 3 encoded data and the QPSK 8.5-MHz subcarrier. A Viterbi decoder is used to decode the recovered soft-decision Channel 3 data from the bit synchronizer. An 8.5-MHz QPSK subcarrier demodulator is employed to recover Channel 1 and Channel 2, which are subsequently detected by the respective bit synchronizer. The three channel data signals are all output from the TDRSS ground station in NRZ-L format via digital transmission facilities. Figure 15 illustrates the Ku-band return-link

TABLE 12 Continued

			Electronics		
Min. G/T	Required P_{rec}/N_0	Deployed Assembly (DA)	Assembly-1 (EA-1)	Signal Processor Assembly (SPA)	Network Signal Processor (NSP)
6.0 dB/K	65.4 dBHz for BER 10^{-5} 64.0 dBHz for link acquisition threshold	Amplify received RF signal and reduce front-end noise Downconvert RF signal to first IF	Downconvert PN-coded carrier to 2nd IF Despread carrier Demodulate carrier (COSTAS loop)	Bit sync demodulated 216-kb/s baseband signal Frame sync and demultiplex 216-kb/s TDM data to 72-kb/s op. data and 128-kb/s forward data	Bit sync demodulated baseband signal Decrypt TDM data[a] frame sync and demultiplex TDM data Authenticate command data BCH decode command Delta demodulate voice channels

Space Shuttle Orbiter Ku-Band Subsystem

[a]By external COMSEC device.

Mode 1 configuration, and Table 13 summarizes the Ku-band return-link Mode 1 interface characteristics.

Mode 2 operation provides for Channels 1 and 2 identical to those used in Mode 1. Channel 3 can be from a variety of analog and digital sources. The modulation of Channels 1 and 2 on the subcarrier is again identical to that used in Mode 1, at a 27.5 : 72.5 power ratio. However, the subcarrier is converted to a sine wave, and Channel 3 is added linearly to produce a frequency division multiplexed (FDM) signal, which is then used to frequency modulate the 15.0034-GHz carrier.

On the ground, a linear FM demodulator recovers the FDM signal, which is routed to a 4.2-MHz low-pass filter to obtain the Channel 3 signal and to a coherent 8.5-MHz QPSK subcarrier demodulator to obtain Channels 1 and 2. The channels are routed from the TDRSS ground station by way of appropriate transmission facilities. Figure 16 illustrates the Ku-band return-link Mode 2 configuration, and Table 14 summarizes the Ku-band return-link Mode 2 interface characteristics.

International Space Station Communications and Tracking System Design

The ISS, which will be launched, assembled, and operational around the beginning of the next century, will be a permanent, earth-orbiting, manned space base. A large number of communications functions and services are required to

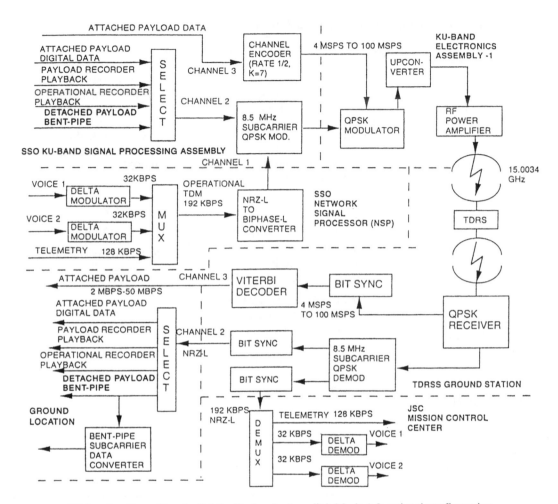

FIG. 15 Space Shuttle *Orbiter* Ku-band return-link Mode 1 functional configuration.

support not only basic space station operations, but also a wide variety of elements. The ISS Communications and Tracking (C&T) system designs, when implemented, will provide the required levels of system performance in terms of link margins and tracking system accuracy. High-quality video, audio, and digital data will be transmitted via the TDRSS, which will provide almost continuous coverage.

The basic ISS space-to-ground communications and tracking system is interfaced with the NASA Space Network (SN) via the TDRSS. The S-band single-access forward (SSAF) and return (SSAR) link services will provide tracking, telemetry, and command (TT&C) operational functions, while the Ku-band single-access return (KSAR) link services will provide the transmission of the scientific and video data from the station to the ground. The Electronic Systems Test Laboratory (ESTL) at the NASA JSC will be used to assure overall space station C&T system design verification, interface compatibility, and design certification from the breadboard phase to the final flight hardware phase.

TABLE 13 Space Shuttle *Orbiter* Ku-Band Return-Link Mode 1 Interface Characteristics

Space Shuttle *Orbiter* Ku-Band Subsystem		
Channel Data Selection Rate and Format	Signal Processing Assembly (SPA)	Deployed Assembly (DA): Electronics Assembly-1 (EA-1) and Antenna
Channel 1: Operational TDM 192 kb/s ± 0.01% bi-phase-L (From NSP)	Channels 1 and 2 QPSK modulate 8.5-MHz subcarrier power ratio: Channel 1: 27.5 ± 2.5% Channel 2: 72.5 ± 2.5%	Upconvert QPSK if carrier 1.87 GHz to 15.0034 GHz ± 0.0003%
Channel 2: Selected from attached payload (digital) 16 kb/s to 2 Mb/s NRZ-L, M, S or 16 kb/s to 1.024 Mb/s BI-ϕ-L, M, S	Convolutionally encode channel 3 to 4 to 100 Ms/s, NRZ-L Code rate: 1/2 Constraint length: 7 Generator sequence: 35707_8	RF amplify carrier (50 watts) Antenna: RCP Axial ratio ≤ 3 dB EIRP: 52.0 dBW minimum
Payload recorder playback 25.5 kb/s to 1.024 Mb/s BI-ϕ-L, M, S	QPSK modulate if carrier (1.87 GHz) with encoded channel 3 and 8.5-MHz QPSK subcarrier power ratio:	
Operational recorder playback 60 kb/s to 1.024 Mb/s BI-ϕ-L, M, S	Channel 3: 80 ± 5% 8.5-MHz QPSK subcarrier: 20 ± 5%	
Detached payload bent-pipe 16 kb/s to 2 Mb/s NRZ-L, M, S or 16 kb/s to 1.024 Mb/s BI-ϕ-L, M, S		
Channel 3: Attached payload digital 2 Mb/s to 50 Mb/s, NRZ-L, M, S		

TDRSS Ground Station		
Required Received Power	Carrier Demodulation Processing	Baseband Processing
− 160.9 dBW (For channel 3 BER 10^{-5})	Downconvert QPSK carrier Demodulate QPSK carrier	Demodulate 8.5-MHz QPSK subcarrier bit sync demodulated baseband signals Decode channel 3 (Viterbi soft decision)

Johnson Space Center Mission Control Center	Users
Frame sync and demultiplex operational TDM data	Process Channel 3 data
Delta demodulate voice channels	Process Channel 2 data
Process telemetry data	

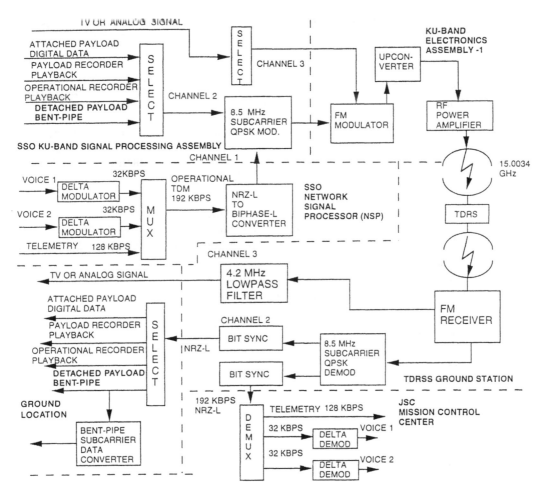

FIG. 16   Space Shuttle *Orbiter* Ku-band return-link Mode 2 functional configuration.

S-Band Single-Access Forward-Link Design

The S-band SA forward link from the ground, by way of the TDRSS to the space station, is composed of two channels: the command/audio channel and the range channel. The command/audio channel can accommodate one core data channel and up to two digital audio channels. Each voice channel is source encoded to 9.6-kb/s CCSDS (Consultative Committee for Space Data System) (5) compatible Bitstream Protocol Data Units (BPDUs) using the Motorola residually excited linear predictive (MRELP) algorithm.

The core data in the form of CCSDS source packets, which is connection-oriented command and session traffic, are formatted into Multiplexing Protocol Data Units (MPDUs) at a rate from 38.175 to 57.375 kb/s. Either BPDUs or MPDUs are encrypted with the Data Encryption Standard (DES) in the cipher feedback (CFB) mode and encapsulated into Virtual Channel Data Units (VCDUs), Reed-Solomon (R-S) (255, 223) encoded, and formatted into CCSDS

TABLE 14   Space Shuttle *Orbiter* Ku-Band Return-Link Mode 2 Interface Characteristics

Space Shuttle *Orbiter* Ku-Band Subsystem		
Channel Data Selection Rate and Format	Signal Processing Assembly (SPA)	Deployed Assembly (DA): Electronics Assembly-1 (EA-1) and Antenna
Channel 1: Operational TDM 192 kb/s ± 0.01% biphase-L (from NSP)	Channels 1 and 2 QPSK modulate 8.5-MHz subcarrier Power ratio: Channel 1: 27.5 ± 2.5% Channel 2: 72.5 ± 2.5%	Upconvert FM IF carrier 1.87 GHz to 15.0034 GHz ± 0.0003%
Channel 2: Selected from attached payload (digital) 16 kb/s to 2 Mb/s NRZ-L, M, S or 16 kb/s to 1.024 Mb/s BI-ϕ-L, M, S	Frequency modulate IF carrier (1.87 GHz) with Channel 3 and 8.5-MHz QPSK subcarrier	RF amplify carrier (50 watts) Antenna: RCP Axial ratio ≤ 3 dB EIRP: 52.0 dBW minimum
Payload recorder playback 25.5 kb/s to 1.024 Mb/s BI-ϕ-L, M, S	Peak frequency deviation: Channel 3: 11.0 MHz ± 10% 8.5 MHz QPSK subcarrier: 6 MHz ± 10%	
Operational recorder playback 60 kb/s to 1.024 Mb/s BI-ϕ-L, M, S		
Detached payload bent-pipe 16 kb/s to 2 Mb/s NRZ-L, M, S or 16 kb/s to 1.024 Mb/s BI-ϕ-L, M, S		
Channel 3: TV or attached payload analog (DC, 4.5 MHz); detached payload bent-pipe (1 KHz to 4.5 MHz)		

TDRSS Ground Station		
Required Received Power	Carrier Demodulation Processing	Baseband Processing
− 162.7 dBW (for Channel 3 SNR 26 dB [rms/rms] in 4.2-MHz lowpass filter)	Downconvert FM carrier Demodulate FM carrier	Demodulate 8.5-MHz QPSK subcarrier bit sync demodulated baseband signals (Channels 1 and 2) Restore Channel 3

Johnson Space Center Mission Control Center	Users
Frame sync and demultiplex operational TDM data	Process Channel 3 data
Delta demodulate voice channels	Process Channel 2 data
Process telemetry data	

Coded Virtual Channel Data Units (CVCDUs). Each CVCDU is then scrambled (exclusive-or'ed bit by bit) with a random sequence to provide the required bit transition density and forms a Channel Access Data Unit (CADU) by adding a 32-bit sync mark. The transmit data rate for the CADUs is 72 kb/s and is in the non-return-to-zero-mark (NRZ-M) format. This format is used to resolve possible channel bit phase ambiguity at the space station receiver side. Figure 17 illustrates the structure of the SSAF CADU.

The NRZ-M signal is modulo-2 added with the command/audio channel PN code at a nominal chip rate of 3 Mcps. The PN spread data on the command/audio channel and range channel PN code at a nominal chip rate of 3 Mcps are then used to binary phase-shift-key (BPSK) quadrature phases of the forward-link carrier (QPSK).

The command channel-to-range channel power ratio after carrier modulation is 10 decibels (dB). The resulting QPSK signal is converted to 2085.6875 MHz and transmitted to the space station via the TDRSS. In the space station, a directional high-gain horn antenna receives the SSAF signal and routes it to a low-noise amplifier (LNA), which outputs it to a transponder. In the transponder, the receiver acquires, tracks, despreads, demodulates, bit syncs, and detects the received command/audio channel signal into CADU forms.

A baseband signal processor is then used to R-S decode, DES decrypt, demultiplex, and deformat the data. The two 9.6-kb/s MRELP digital audio datastreams are processed and converted into two 192-kb/s linear pulse code modulation (PCM) digital audio datastreams prior to being transmitted to the audio distribution system.

Table 15 compares the design differences between the SSO and ISS. Figure 18 shows the SSAF link functional configuration, and Table 16 summarizes pertinent SSAF link design parameters.

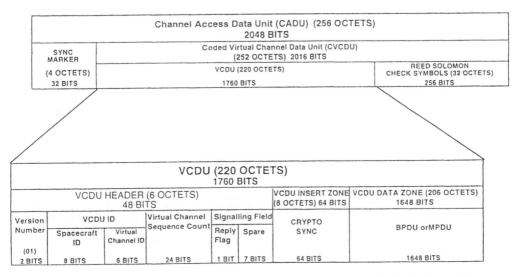

FIG. 17 International Space Station (ISS) S-band single-access forward (SSAF) Channel Access Data Unit (CADU) structure.

TABLE 15   International Space Station versus Space Shuttle *Orbiter*

Parameters	Space Station	Space Shuttle *Orbiter*
Data type	2 audio channels 1 command and data channel	2 audio channels 1 command and data channel
Frame format	CCSDS CADU (dynamic assignment)	TDM (fixed assignment)
Channel coding	Reed–Solomon (255, 223)	Convolutional (rate 1/3, constraint length 7)
Encryption	DES (cipher feedback)	Unique (stream cipher)
Data format	NRZ-M	Biphase-L
PN code rate	3.08 Mcps	11.232 Mcps
Ranging	PN code ranging channel	Tone ranging via GSTDN link or doppler tracking
Modulation	Unbalanced QPSK	BPSK
Bit error rate	10^{-5}	10^{-4}

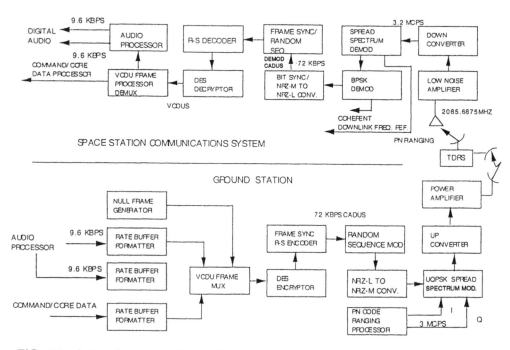

FIG. 18   International Space Station S-band single-access forward-link functional configuration.

TABLE 16 International Space Station S-Band Single-Access Forward-Link Design Parameters

Channel	Information Rate	Formatting/Encryption/Encoding	Spread Spectrum
Command Core data 1 channel No audio	57.375 kb/s	CCSDS CADU frame length: 256 octets Sync pattern: 4 octets (IACFFC1D$_{16}$)	I channel PN code chip rate: $\dfrac{31}{221 \times 96} \times F$
1 audio	47.775 kb/s	Header: 6 octets; crypto sync: 8 octets	(F = carrier fre-
2 audios	38.175 kb/s	Data fields: 206 octets (DES encrypted, CFB mode)	quency, 2085.6875 MHz)

Low data
rate (no
audio)
	4.781 kb/s	Reed-Solomon check symbols: 32 octets
		Transmitted data rate:
		  High data rate: 72 kb/s ± 8 ppm
		  Low data rate: 6 kb/s ± 8 ppm
		Data format: NRZ-M
			PN code format: NRZ-L
			PN code length: 1023 chips
			User PN code assignment
			number: 49

Ranging
  PN code epoch and chip rate synchronized to command channel
    PN code epoch and chip rate
  Truncated 18-stage shift register sequence: feedback connection
    polynomial, $x^{18} + x^{14} + x^9 + x^3 + 1$
  PN code length: 261,888 chips
  PN code epoch reference: All "1" state synchronized to command
    channel PN code generator initial condition

Q channel
PN code chip rate:
$$\frac{31}{221 \times 96} \times F$$
PN code format: NRZ-L

Carrier Modulation	Carrier Frequency	TDRS Minimum EIRP	Antenna Polarization	ISS Minimum G/T	ISS Required P_{rec}/N_0[a]
Unbalanced quadra- ture phase shift keying (UQPSK) I/Q (power ratio) = 10 dB ± 0.5 dB	2085.6875 MHz ± 1 Hz[b]	43.6 dBW	Right-hand circu- lar polarized (RHCP) Axial ratio TDRS: <1.5 dB ISS: <5 dB	72 kb/s −14.8 dB/K 6 kb/s −25.7 dB/K	72 kb/s 61.3 dBHz 6 kb/s 50.5 dBHz

[a]For channel BER of 10^{-5}.
[b]The ± 1 Hz frequency tolerance is the capability of the ground to initially set its transmit frequency.

S-Band Single-Access Return-Link Design

The SSA return-link carrier is either a coherent turnaround of the SSAF link for two-way Doppler extraction in the ground station or is generated by an on-board auxiliary oscillator. The transmitted frequency is 2265 MHz. As in the case of the forward link, two voice channels, each source encoded into 9.6-kb/s digital audio data using the MRELP algorithm, are processed into BPDUs. Telemetry data (core data) from 139.8 kb/s to 159 kb/s are processed into

MPDUs. All MPDUs or BPDUs are encapsulated into VCDUs and R-S (255, 223) encoded and formatted into CCSDS CVCDUs. Each CVCDU forms a CADU by adding a 32-bit sync mark. The formatted CADU is the same as in the SSAF link except that the crypto sync field is not used, and the transmission data rate is 192 kb/s instead of 72 kb/s. Figure 19 illustrates the SSAR CADU structure.

The CADUs at the fixed rate of 192 kb/s are transmitted on an alternate-bit basis, first onto the RF in-phase (I) channel at the rate of 96 kb/s and then onto the RF quadrature (Q) channel at the rate of 96 kb/s. The I and Q channel datastreams in the NRZ-L format are converted into the NRZ-M format prior to being rate 1/2 convolutionally encoded with constraint length $K = 7$.

The I and Q encoder output symbols are then modulated-2 added asynchronously, respectively, to the locally generated I and Q channel PN code at the chip rate of 3 Mcps, with the Q channel PN code offset by one-half chip relative to the I channel PN code. The resultant signals are used to BPSK quadrature phases of the IF carrier. This staggered quadriphase PN (SQPN) modulated IF carrier is upconverted and transmitted to the TDRSS through the directional high-gain horn antenna. On the ground, the PN receiver and QPSK demodulator acquire, track, despread, and demodulate the I and Q signals into baseband signals, which are detected and soft-decision decoded by the I and Q Viterbi decoders, respectively. The SN ground segment recombines the I and Q bitstreams to a 192-kb/s bit sequence after the NRZ-M-to-NRZ-M conversion and establishes CADU synchronization, Reed-Solomon decodes each CVCDU to VCDU, demultiplexes VCDUs, deformats, and routes data to ground-based users. The two 9.6-kb/s MRELP digital audio data are processed and converted into two analog audio channels prior to being distributed.

Figure 20 shows the SSAR link functional configuration, and Table 17 summarizes pertinent SSAR link design parameters.

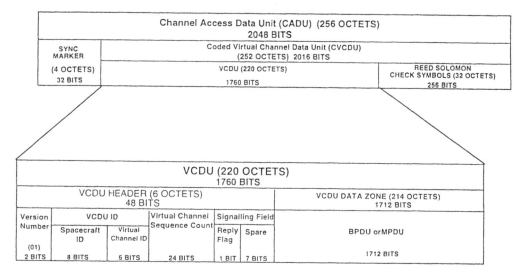

FIG. 19 International Space Station S-band single-access return Channel Access Data Unit structure.

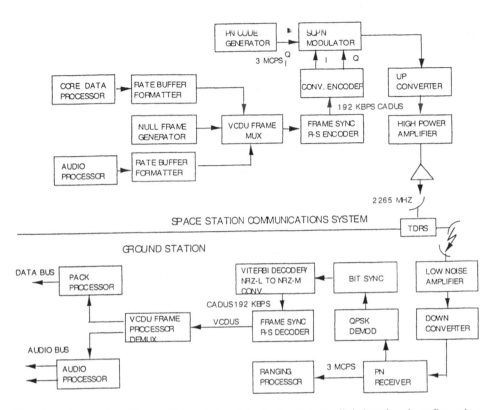

FIG. 20 International Space Station S-band single-access return-link functional configuration.

Ku-Band Single-Access Return-Link Design

Up to four channels of analog National Television System Committee (NTSC) composite video are processed and converted into digital video signals. The digital video signals, with associated audio, at the maximum rate of 43.196 Mb/s are formatted into CCSDS data packs, accumulated, rate buffered, and formatted into MPDUs. Up to eight channels of scientific or payload data in either bitstream format and/or CCSDS source packet data format are accepted, accumulated, rate buffered, and formatted into BPDUs and/or MPDUs. All MPDUs or BPDUs are encapsulated into VCDUs and R-S (255, 223) encoded with a symbol interleaving depth of 5 and formatted into CCSDS CVCDUs. Each CVCDU is then exclusive-or'ed bit by bit with a random sequence to provide the required bit transition density and a CADU is formed by adding a 32-bit sync mark. Figure 21 shows the CADU structure.

The CADUs at the fixed rate of 50 Mb/s are converted to NRZ-M from NRZ-L format prior to being BPSK modulated onto the carrier. The BPSK-modulated signal is transmitted via an RF high-power amplifier (HPA) to the TDRSS at a frequency of 15.0034 GHz. In the SN ground terminal, the BPSK demodulator acquires and tracks the carrier and provides demodulated base-band signals to the bit synchronizer, in which the data are detected and con-

TABLE 17 International Space Station S-Band Single-Access Return-Link Design Parameters

Information		Formatting/ Encoding	Channel Encoding/ Decoding	Spread Spectrum
Channel	Rate			
Digital audio (2 channel max)	9.6 kb/s	CCSDS CADU frame length: 256 octets Sync pattern: 4 octets (IACFFC1D$_{16}$) Header: 6 octets, Data fields: 214 octets Reed-Solomon check symbols: 32 octets Transmitted data rate: High-data-rate mode: 192 kb/s ± 8 ppm Low-data-rate mode: 12 kb/s ± 8 ppm Data format: NRZ-M	Convolutional code Constraint length: 7 Generator sequence: (3233013$_8$) Rate: 1/2 Symbol rate: 192 ks/s, 12 ks/s Symbol format: NRZ-L G2 inversion Decoding algorithm: Viterbi (maximum likelihood), 8-level soft decision	PN code chip rate: $\dfrac{31}{221 \times 96} \times F$ (F = carrier frequency, 2265 MHz) PN code format: NRZ-L Q-channel PN code delayed by one-half chip of I-channel PN code PN code length: DG1 Mode 1: 261888 chips[a] DG1 Mode 2: 2047 chips
Core data 1 channel No audio 1 audio 2 audio	159.0 kb/s 149.4 kb/s 139.8 kb/s			
Low data rate (no audio)	9.937 kb/s			

Carrier Modulation	Carrier Frequency	ISS Minimum EIRP	Antenna Polarization	TDRSS Required Total P_{rec}
Balanced QPSK I/Q = 0 dB ± 0.25 dB	DG1 Mode 1: 240/ 221 × SSAF carrier frequency DG1 Mode 2: 2265 MHz	High-gain antenna: 26.1 dBW Low-gain antenna: 14.3 dBW	RHCP Axial ratio TDRS < 1.5 dB ISS < 5 dB	−174.1 dBW[b] (192 kb/s) −186.1 dBW[b] (12 kb/s)

[a]The PN code sequences are gold codes obtained by modulo-2 addition of two synchronous 2047-bit maximal length sequences generated from three 11-stage feedback shift registers. The initial conditions for the shift registers are set by using the initial conditions required by PN Code No. 49.
[b]No user compliance loss for BER of 10^{-5}. No RFI degradation.

verted to the NRZ-L format from the NRZ-M format. The ground terminal then establishes CADU synchronization, removes the random sequence, R-S decodes and deinterleaves each CVCDU to VCDU, demultiplexes VCDUs, deformats, and routes data to ground-based users. The digital video is converted to the analog NTSC composite video format before distribution. Figure 22 illustrates the KSAR link functional configuration, and Table 18 summarizes pertinent link design parameters.

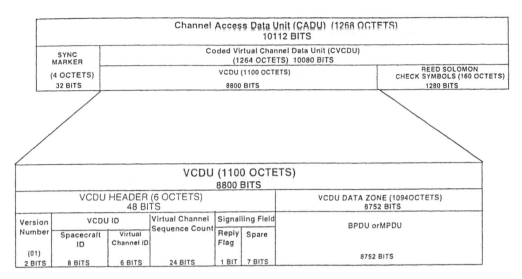

Channel Access Data Unit (CADU) (1268 OCTETS) 10112 BITS		
SYNC MARKER (4 OCTETS) 32 BITS	Coded Virtual Channel Data Unit (CVCDU) (1264 OCTETS) 10080 BITS	
	VCDU (1100 OCTETS) 8800 BITS	REED SOLOMON CHECK SYMBOLS (160 OCTETS) 1280 BITS

VCDU (1100 OCTETS) 8800 BITS					
VCDU HEADER (6 OCTETS) 48 BITS					VCDU DATA ZONE (1094OCTETS) 8752 BITS
Version Number (01) 2 BITS	VCDU ID		Virtual Channel Sequence Count	Signalling Field	BPDU orMPDU 8752 BITS
	Spacecraft ID 8 BITS	Virtual Channel ID 6 BITS	 24 BITS	Reply Flag / Spare 1 BIT / 7 BITS	

FIG. 21   International Space Station Ku-band single-access return (KSAR) Channel Access Data Unit structure.

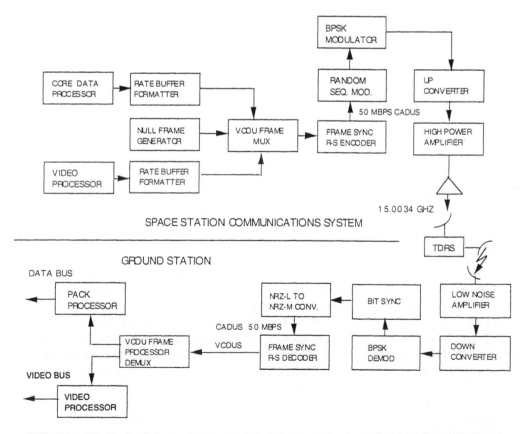

FIG. 22   International Space Station Ku-band single-access return-link functional configuration.

TABLE 18 International Space Station Ku-Band Single-Access Return-Link
              Design Parameters

	Information		
Channel	Number	Rate (Maximum per Channel)	Formatting/Encoding
High-data rate payload	8	43.196 Mb/s	CCSDS CADU frame length: 1264 octets
Digital video	4	43.196 Mb/s[a]	Sync pattern: 4 octets ($1ACFFC1D_{16}$) Header: 6 octets, Data fields: 1094 octets Reed-Solomon check symbols: 160 octets Transmitted data rate: 50 Mb/s $\pm$ 0.1% Data format: NRZ-M

Carrier Modulation	Carrier Frequency	ISS Minimum EIRP	Antenna Polarization	TDRSS Required Total P_{rec}
Binary phase-shift keying (BPSK)	15.0034 GHz	52.0 dBW	LHCP Axial ratio: TDRS < 1 dB ISS < 2 dB	-160.8 dBW[b]

[a]Data rate for 6-bit video 60 fields/s is 41.259 Mb/s.
[b]No user compliance loss for BER of 10^{-5}, including RFI degradation, user compliance loss, and polarization loss.

The Role of ESTL (Electronic Systems Test Laboratory)

The JSC ESTL is a unique C&T system integration and verification laboratory used to conduct detailed spacecraft-to-spacecraft and spacecraft-to-ground RF communications systems tests on an end-to-end system basis. The ESTL test concept and capabilities to perform end-to-end C&T systems test are illustrated in Fig. 23.

The ESTL provides the capability to interface operational space vehicle communications equipment with external interoperating elements in a laboratory environment under closely controlled conditions. Communications and tracking system compatibility and performance tests typically conducted in the ESTL can be divided into three categories. The first, referred to as system design evaluation tests, involves use of spacecraft subsystem breadboards that are developed early in the program. The second category is the system verification tests. These tests are performed using spacecraft prototype hardware and are intended to validate system design. The third category is referred to as system certification tests. These tests are conducted using spacecraft flight hardware to certify that the RF communications links between all elements of a space pro-

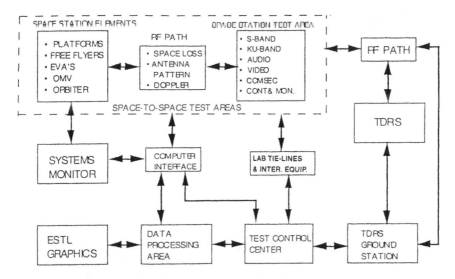

FIG. 23 Electronic Systems Test Laboratory (ESTL) test concept.

gram meet program requirements and that the system hardware and software are ready to support manned flight. The ESTL provides a continuing major contribution to the space shuttle program C&T system evaluation, verification, and certification process. The ESTL is also expected to play a similar important and active role to assure overall space station C&T system design verification, interface compatibility, and design certification.

Summary

In this article, the space communications for the U.S. manned spacecraft, Space Shuttle *Orbiter* and International Space Station (ISS), as well as the NASA Tracking and Data Relay Satellite System (TDRSS) were described.

Development of space shuttle communications and tracking systems required a very significant set of challenges. A large number of communications functions and services were required to support not only basic space shuttle vehicle operations, but also a wide variety of space shuttle cargo elements. The broad range of mission requirements and objectives dedicated a high degree of sophistication and complexity for the space shuttle C&T systems. In addition, C&T systems developed for the space shuttle were required to communicate with several classes of ground network terminals and spaceborne elements of the TDRSS. Development, for use on the space shuttle, of TDRSS-compatible systems that would satisfy data transmission requirements for launch, on-orbit operations, and entry proved to be one of the more significant challenges of the C&T system design and development.

Throughout the development phases, the various design challenges were met, and the resulting C&T system designs, when implemented, provided the required levels of system performance in terms of link margins and tracking system accuracy. High-quality video, audio, and digital data transmission capabilities are being provided over each of the space shuttle external RF links as applicable.

As in previous manned space programs, the space shuttle program clearly proved the importance of performing system-level analyses, simulations, and tests regarding the interface between space and ground networks. It was largely through the use of these tools early in the program that hardware that provided excellent performance for the space shuttle C&T tasks was developed.

However, the design of the space shuttle communications system is now over 24 years old. It has served the program from the initial captive-active flights in 1977 to today with little change in the original designs. Several system devices, including high-power traveling wave tubes (TWT), the antenna switch, the low-noise parametric amplifier, the medium-scale integrated CMOS circuitry that makes up most of the network and signal processors like the NSP and custom LSI devices, are 1960s technology and are no longer used in today's system designs. Some of these devices are no longer even procurable, and a repair would dictate a redesign to some level to accommodate today's technology.

The existing system has some design and operational deficiencies, such as PN code false lock, carrier false lock, modulation without data on, limited downlink telemetry channel utilization, large delay of command control for the payload interrogator, and so on; these require real-time operational work-arounds (S-band system) that thus add increasing cost to mission operations. Therefore, it is highly desirable to design a replacement S-band communications system for the SSO that eliminates obsolescence; reduces manufacturing, repair, maintenance, and operational cost; and minimizes the need for original equipment manufacture (OEM) support.

As for the ISS, a large number of communications functions and services are required to support not only basic ISS operations, but also a wide variety of elements. The ISS C&T system designs, when implemented, will provide the required levels of system performance in terms of link margins and tracking system accuracy. High-quality video, audio, and digital data are transmitted via the TDRSS, which will provide almost continuous coverage. In order to assure a space station C&T system design to meet program and mission requirements, the ETSL will be used to perform systems-level analyses, simulations, and tests regarding the interface between space and ground network systems. In this article, the space station C&T SSAF, SSAR, and KSAR system link designs were described. The role of the ESTL for the development of the space station C&T system was also emphasized.

References

1. *Space Network (SN) User's Guide*, STDN No. 101.2, Rev. 7, NASA Goddard Space Flight Center, November 1995.
2. *TDRSS Second Workshop Proceedings*, NASA Goddard Space Flight Center, Washington, D.C., June 1996.

3. Tu, K., et al., Space Shuttle Communications and Tracking System, *Proc. IEEE,* 75(3):356–370 (March 1987).
4. Tu, K., Space Station Space-to-Ground Communications and Tracking System, *Proceedings of 1994 International Conference on Communications Techniques (ICCT '94)*, Shanghai, China, June 1994.
5. *Recommendation for Space Data System Standards: Advanced Orbiting Systems, Networks and Data Links*, CCSDS 701.0-B-2, *Blue Book*, Issue 2, Washington, D.C., November 1992.

KWEI TU

Speech Processing
(see Digital Speech Processing)

Speech Recognition
(see Overview of Speech Recognition)

Spread-Spectrum Communications

Spread-spectrum modulation is one of many signaling methods for which there is a tradeoff between bandwidth and performance. For most other such signaling methods, the tradeoff is between the bandwidth and the probability of error for a channel in which the major disturbance is thermal noise. For example, M-ary orthogonal signaling provides such a tradeoff. By increasing M, the number of signals in the orthogonal signal set, the bandwidth requirement increases. If thermal noise is the only interference, and if the signal-to-noise ratio is sufficiently large, the error probability decreases to zero as M increases. Precise statements of this fact can be found in almost any book on digital communications (1).

For spread-spectrum modulation, the performance gains are for interference sources other than thermal noise. In fact, classical spread-spectrum modulation provides no performance improvement against thermal noise. The advantages of spread spectrum result from its ability to discriminate against narrowband interference, multipath interference, multiple-access interference, and other types of structured interference that arise in radio-frequency (RF) communication channels. Spread spectrum also provides some advantages for wire-line, optical, and other communication media, particularly if several transmitters have simultaneous access to the same frequency band.

In addition to protection against various types of interference, spread-spectrum signals are difficult for unauthorized receivers to demodulate or even to detect. This feature makes spread spectrum desirable for military systems and for civilian communication services that wish to offer a certain degree of privacy. Spread-spectrum modulation also distributes the transmitted power across a wide range of frequencies in a way that permits coexistence with other forms of communications that share the frequency band with the spread-spectrum system. Some of the advantages mentioned above can be illustrated conveniently if the spread-spectrum signal is modeled by a random process, and that is the approach taken in the next section.

Random Signal Model for Spread Spectrum

The modeling of a spread-spectrum signal as a random signal provides a simple framework in which to illustrate the basic concepts of spread-spectrum communications, especially the form of spread spectrum known as direct-sequence (DS) spread spectrum. In the random signal model, the spectral spreading signals are random processes. In addition to providing an illustration of the benefits of spread spectrum, the modeling of spread-spectrum signals by random processes has a historical connection to the early development of spread spectrum (2). Much of the early work on spread spectrum relates to the noiselike properties of DS spread-spectrum signals, and the sequences employed in DS spread-spectrum systems are sometimes referred to as pseudonoise sequences, even in current literature.

In the analysis of the random signal model, no attempt is made to be mathematically precise because random processes of the type used in the illustrations are not of interest in actual implementations of spread spectrum. The analysis presented in this section is only for the purpose of developing a physical insight into the basic concepts of spread-spectrum communications (3). In sections that follow, more precise analyses are presented, and these analyses are for models that are suitable for implementation. In a DS spread-spectrum signal, each data symbol is represented by a sequence of N elemental pulses, which are referred to as *chips*. If T is the symbol duration, we let $T_c = T/N$. In the most common example, the chip waveform is a rectangular pulse of duration T_c; consequently, T_c is often referred to as the *chip duration*. In general, however, the chip waveform can have any shape, and it need not be time limited.

The randomness in a spread-spectrum signal is primarily in its baseband spectral spreading signal. The spectral spreading signal is independent of the data to be transmitted, and its function is to spread the data signal over a wide range of frequencies. In this section, the spectral spreading signal is modeled as a random process, denoted $V(t)$. It is convenient for our illustration to let $V(t)$ be a zero-mean, stationary random process with autocorrelation function $R_V(\tau)$, given by

$$R_V(\tau) = \begin{cases} (T_c - |\tau|)/T_c, & |\tau| \leq T_c \\ 0, & |\tau| > T_c \end{cases} \tag{1}$$

The actual shape of the autocorrelation function is not particularly important, however. All that is required is that $R_V(\tau) \approx 0$ for $\tau > T_c$.

There are a number of different random signal models that can be used to illustrate the features of DS spread-spectrum communications. The essential requirement is that certain time averages can be accurately approximated by the corresponding probabilistic averages (i.e., the random process must be ergodic in the appropriate sense). If $V(t)$ is a Gaussian random process, for example, Eq. (1) is enough to guarantee that $V(t)$ is ergodic. But, certain other random processes are of even greater interest for modeling DS spread-spectrum signals. For example, a more accurate random signal model for DS spread spectrum is a sequence of pulses with random amplitudes and a suitable random time offset, such as

$$V(t) = \sum_{n=-\infty}^{\infty} A_n \psi(t - iT_c - U) \tag{2}$$

where $\psi(t)$ is the chip waveform, $\{A_n: -\infty < n < \infty\}$ is a sequence of independent random variables with $P(A_n = +1) = P(A_n = -1) = 1/2$ for each n, and U is a random variable that is uniformly distributed on the interval $[0, T_c]$ and independent of $\{A_n: -\infty < n < \infty\}$. If $\psi(t)$ is the unit-amplitude rectangular pulse of duration T_c, the resulting autocorrelation function is given by Eq. (1).

The power spectral density for a random process $V(t)$ that has the autocorrelation function of Eq. (1) is

$$S_V(f) = T_c \, [\text{sinc}(fT_c)]^2 \tag{3}$$

where $\text{sinc}(x) = \sin(\pi x)/(\pi x)$ for $x \neq 0$ and $\text{sinc}(0) = 1$. For purposes of this illustrative discussion, the *bandwidth* of a random process is defined as the first zero of its spectral density. More precisely, the bandwidth B of a baseband random process $V(t)$ is the smallest positive value of f for which $S_V(f) = 0$. For a random process that has an autocorrelation function given by Eq. (1) and a spectral density given by Eq. (3), the bandwidth is $B = 1/T_c$.

Let $d(t)$ be a binary data signal that consists of a sequence of rectangular pulses, each of which is of duration T and amplitude $\pm A$ (assume $A > 0$). Let $p_T(t)$ denote the unit-amplitude, rectangular pulse of duration T that begins at $t = 0$. Define

$$u_i(t) = (-1)^i A p_T(t)$$

for $i = 0$ and $i = 1$; that is, $u_0(t) = +A p_T(t)$ and $u_1(t) = -A p_T(t)$. Either

$$d(t) = u_0(t - nT)$$

or

$$d(t) = u_1(t - nT)$$

in the nth signaling interval, $nT \leq t < (n + 1)T$, depending on which binary digit, 0 or 1, is being sent over the channel in that interval. Analogous to the definition of the bandwidth of $V(t)$, the bandwidth of the data signal is defined as the first zero of the Fourier transform of $p_T(t)$, which is $1/T$.

The value of N is typically in the range of tens or hundreds for civilian systems, and it may be much larger for military systems. It follows from $N \gg 1$ that $B \gg 1/T$: the bandwidth of $V(t)$ is much greater than the bandwidth of the data signal $d(t)$. The transmitted signal is $V(t)d(t)$, which for most applications has a bandwidth approximately equal to the bandwidth of $V(t)$.

The channel model that is employed in this section is a baseband channel that has no thermal noise, but various other types of interference are present. This channel model is sufficient to illustrate several of the most important benefits of spread spectrum. The baseband receiver is a correlator that multiplies the received signal $r(t)$ by $V(t)$ and then integrates the product, as illustrated in Fig. 1.

For the data pulse transmitted during the 0th signaling interval, the decision statistic is

$$Z = \int_0^T V(t)r(t)dt$$

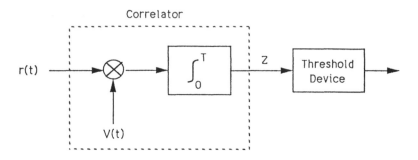

FIG. 1   Correlation receiver for $V(t)d(t)$.

The decision to be made is whether $d(t) = A$ or $d(t) = -A$ during a given signaling interval. If $Z > 0$, the receiver decides that a positive pulse was transmitted in the interval $[0,T]$; if $Z \le 0$, the receiver decides that a negative pulse was transmitted. In the absence of interference, the received signal is $r(t) = V(t)d(t)$, and so the decision statistic is

$$Z = \int_0^T V^2(t)d(t)dt$$

Denote by m the binary digit that is transmitted in the 0th signaling interval. The decision statistic for the 0th interval is given by

$$Z = (-1)^m A \int_0^T V^2(t)dt$$

for $m = 0$ or $m = 1$.

If $V(t)$ is ergodic in the appropriate sense, then, with high probability

$$T^{-1} \int_0^T V^2(t)dt \approx E\{V^2(t)\} = R_V(0) = 1$$

because the integration time is long compared with the inverse of the bandwidth of the random process $V(t)$ (i.e., $T \gg T_c$). As a result, if the binary digit m is sent over the channel, the decision statistic satisfies

$$Z \approx (-1)^m AT \tag{4}$$

It follows that if $m = 0$, then $Z > 0$, and the receiver correctly decides that a positive pulse was sent. If $m = 1$, then $Z \le 0$, and the receiver correctly decides that a negative pulse was sent. Thus, in the absence of any interference in the channel, the transmitted information is recovered from $V(t)d(t)$.

Notice that the receiver must know the random process $V(t)$ exactly, which is the reason that this conceptual model is not of interest for practical implementations. Many of the early attempts to develop spread-spectrum systems involved the use of a noise signal as a spectral spreading signal (2). In such an approach, it is necessary to have the same noise signal at the transmitter and receiver, and several mechanisms were invented to accomplish this. In practical implementations, the characteristics of $V(t)$ are approximated by a noiselike or "pseudonoise" waveform. For instance, many modern spread-spectrum systems use noiselike signals that are derived from shift-register sequences (e.g., see Refs. 4, 5, and 6). Such signals give the necessary randomness, yet they permit duplication of the spectral spreading signal at the receiver.

The analysis above and the final result given in Eq. (4) do not account for the possibility of interference in the channel. Such interference can result from authorized transmitters that operate in the same frequency band (e.g., in a multiple-access spread-spectrum system) or systems that emit energy in that frequency band (e.g., intentional jamming). The effects of these and other types of interfering signals can be illustrated by use of the random signal model.

Multiple-Access Capability

Suppose that there are K transmitted signals, and each is of the form described above. For simplicity, assume the K signals are synchronous in the sense that the data pulses for the different signals are aligned in time. The data signals are $d_k(t)$, $1 \leq k \leq K$, and each has the same form as the data signal $d(t)$ defined above. Each transmitter is assigned a spectral spreading signal $V_k(t)$, $1 \leq k \leq K$, and each spectral spreading signal is modeled as a zero-mean, stationary, random process with autocorrelation function $R_V(\tau)$. The processes are mutually independent. The received signal is now

$$r(t) = \sum_{k=1}^{K} V_k(t) d_k(t)$$

Under the assumption that the binary digit m is sent by the ith transmitter during the 0th signaling interval, the output of the ith correlation receiver is given by

$$Z_i \approx (-1)^m AT + \sum_{k \neq i} b_k A \int_0^T V_k(t) V_i(t)\, dt$$

where b_k is $+1$ or -1 depending on whether a 0 or 1 is sent by the kth transmitter in the 0th signaling interval.

Because $V_k(t)$ and $V_i(t)$ are independent, zero-mean, random processes for $k \neq i$, then

$$T^{-1} \int_0^T V_k(t) V_i(t) dt \approx E\{V_k(t) V_i(t)\} = 0 \qquad (5)$$

provided the random processes $V_k(t)$ and $V_i(t)$ are jointly ergodic in the appropriate sense. It follows that the decision statistic satisfies

$$Z \approx (-1)^m AT$$

which is the same as Eq. (4), even though other signals are now sharing the frequency band. To within the accuracy of the approximation given in Eq. (5), the other signals in the spread-spectrum system do not interfere with the transmission of the ith signal. This demonstrates the multiple-access capability of spread spectrum.

Interference Rejection Capability

Suppose that the received signal is

$$r(t) = V(t)d(t) + A'$$

that is, a dc signal has been added to the desired signal. The dc signal level A' can be either positive or negative. A dc signal in a baseband model corresponds to a sinusoidal signal with frequency equal to the carrier frequency in a RF system, so this is an appropriate baseband model for an interference tone at the carrier frequency. If the binary digit m is sent during the 0th interval, the output of the correlation receiver is

$$Z \approx (-1)^m AT + A' \int_0^T V(t)dt \tag{6}$$

The second term of the right-hand side of Eq. (6) represents the interference component at the output of the correlator. Assuming it is valid to equate the time average and the expectation, we see that

$$T^{-1} \int_0^T V(t)dt \approx 0$$

because $V(t)$ has zero mean and T is much larger than the inverse of the bandwidth of $V(t)$. The interference component in Eq. (6) is negligible if

$$(A'/A) \left\{ T^{-1} \int_0^T V(t)dt \right\} \approx 0$$

which is true provided that A' is not much larger than A. This demonstrates the dc interference rejection capability of a baseband spread-spectrum system. Sinusoidal interference rejection can be demonstrated via a similar argument, which uses

$$T^{-1} \int_0^T V(t)\cos(2\pi f_0 t)dt \approx 0$$

Sinusoidal interference at frequency f_0 in the baseband model corresponds to sinusoidal interference at frequency $f_c + f_0$ or $f_c - f_0$ in an RF system with carrier frequency f_c.

Antimultipath Capability

In a simplified baseband model of a multipath channel, the received signal might be

$$r(t) = V(t)d(t) - \beta V(t - \tau)d(t - \tau) \tag{7}$$

where β is in the range $0 \le \beta \le 1$ and $\tau > 0$. The second term on the right-hand side of Eq. (7) represents a multipath component that arrives with a delay of τ compared to the primary component. If, for example, $d(t) = +A$ for $-\tau \le t < T$, the decision statistic is

$$Z \approx AT - A\beta \int_0^T V(t - \tau)V(t)dt \approx AT\{1 - \beta R_V(\tau)\}$$

Recall that $R_V(\tau) = 0$ if $\tau \ge T_c$ so the multipath component does not interfere with the primary component if the delay in the multipath component is T_c or greater.

Time- and Frequency-Domain Interpretations of Despreading

Just as the modulation process in spread-spectrum communications is often referred to as *spreading* the data signal, the demodulation process is often referred to as *despreading*. For each of the types of interference considered above, the key feature is that the interfering signal has small correlation with $V(t)$. As a result, the interfering signal produces only a negligible change in the output of the correlator of Fig. 1. The integrator in the correlation receiver tends to smooth and "average out" the interfering signals. The effect of the correlator on the desired signal is quite different. It produces a large peak in response to the desired signal because it is matched to the shape of the rectangular data pulse. The output of the integrator is equivalent to an appropriate sample of the output of a time-invariant linear filter with a rectangular impulse response (1). Thus, from a time-domain point of view, the correlation receiver (or its equivalent matched-filter receiver) averages out the interference while enhancing the desired signal, thereby increasing the signal-to-interference ratio.

The despreading operation is simpler to illustrate in the frequency domain if $V^2(t)$ is approximately constant. Notice that if $V(t)$ is given by Eq. (2) and the chip waveform is the unit-amplitude rectangular pulse of duration T_c, then $V^2(t) = 1$ for all t. For such a spectral spreading signal, the multiplication by the local replica of the spectral spreading signal despreads (i.e., reduces the bandwidth of) the desired signal at the same time that it spreads the interference. The receiver "strips off" the spectral spreading signal from the data signal by multiplying the received signal by $V(t)$. The multiplication of the spread-spectrum signal $V(t)d(t)$ by $V(t)$ produces

$$V^2(t)d(t) = d(t)$$

which is the data signal, and it has a much narrower bandwidth than the original spread-spectrum signal.

After the multiplication process, the receiver passes the data signal through the integrator of Fig. 1, and this integration process has a bandwidth that is much smaller than B, the bandwidth of the spread-spectrum signal, but approximately the same as $d(t)$, the data signal. On the other hand, the result of the multiplication process on the interference signal is to increase the bandwidth of the interference. The product of the interfering signal and $V(t)$ has a bandwidth that is at least as large as B. The higher frequency components of this product signal are then filtered out by the integrator. The despreading process for the communication signal actually spreads the spectrum of the interference, and the subsequent narrowband filtering effect of the integrator therefore reduces the power in the interference.

Identification of Linear Systems

For many of the beneficial features of spread spectrum, the ideal choice for the transmitter's spectral spreading signal $V(t)$ would be white noise if it were not for the extreme difficulty (impossibility in most situations) of reproducing the noise waveform at the receiver. However, implementation issues are not of concern in this section since the objective is to develop an intuitive understanding of the benefits of spread spectrum. A white-noise spreading signal is a good test signal for linear system identification (i.e., estimation of the unknown impulse response of a linear system) (7). The spreading signal serves as the input signal, and standard methods for estimation of the crosscorrelation function for the input and output signals give an estimate of the impulse response (1).

One application of this approach to the estimation of an unknown impulse response is in mobile communications with fading channels (8), in which it may be necessary to estimate the channel's impulse response in order to perform equalization or multipath combining. In such an application, a wideband DS spread-spectrum signal is a good probe signal to transmit over the channel because such a signal can be designed to approximate a white-noise random process.

Direct-Sequence Spread-Spectrum Signals

As illustrated in the previous section, spread spectrum can provide a means of reliable communication in the presence of various types of interference. If a correlation receiver is employed, the goal in the design of DS spread-spectrum signals is to make the interference orthogonal to the communications signal. In many systems, true orthogonality is impossible to achieve, and the best that can be done is to make the interference approximately orthogonal to the communications signal. The greater the departure from orthogonality, the more that can be gained by using a receiver with more capability than a correlation receiver.

One segment $\xi(t)$ of a typical baseband spectral spreading signal is illustrated in Fig. 2. In this example, the segment is time limited to the interval $[0,T]$; that is, $\xi(t) = 0$ if $t < 0$ or $t \geq T$. Each of the seven shorter pulses in the segment is referred to as a *chip*. For the simple example shown in Fig. 2, the number of chips per segment is seven, but in practice this number is typically much larger.

If the waveform $\xi(t)$ is repeated every T seconds, a periodic signal $v(t)$ is generated. This signal, referred to as the *spectral spreading signal*, satisfies $v(t) = \xi(t - nT)$ for the nth signaling interval $nT \leq t < (n + 1)T$, and it can be represented as

$$v(t) = \sum_{n=-\infty}^{\infty} \xi(t - nT)$$

This last expression is a valid representation for a spectral spreading signal even if the original waveform $\xi(t)$ is not time limited to the interval $[0,T]$.

Notice that once the chip waveform is selected (in the present example, it is rectangular and has a duration T_c), the spectral spreading signal $v(t)$ is defined completely by specifying the sequence of pulse amplitudes for the chips in the time interval $[0,T]$. This sequence of pulse amplitudes is known as the *signature sequence* or *spreading sequence*. The former term is especially appropriate for multiple-access systems because the sequence is used to identify the signal (9).

The signature sequence for the waveform of Fig. 2 is $-1, -1, +1, -1, +1, +1, -1$. As an alternative to listing the sequence of polarities for the chips, it is common to use a binary notation in which the positive polarity is denoted by the binary 0 and the negative polarity denoted by the binary 1. In binary form, the signature sequence is written as 1101001.

If the sequence of chip amplitudes is denoted by $\mathbf{x} = (x_0, x_1, \ldots, x_{N-1})$, one segment of the spectral spreading signal is given by

$$\xi(t) = \sum_{i=0}^{N-1} x_i \psi(t - iT_c)$$

where $1/T_c$ is the chip rate and $\psi(t)$ is the chip waveform. Typical examples for the chip waveform are the rectangular pulse of duration T_c

$$\psi(t) = 1, \qquad 0 \leq t < T_c$$

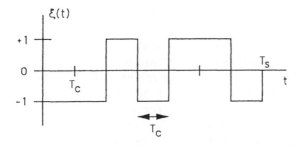

FIG. 2   Baseband spectral spreading signal with seven chips.

and the sine pulse of duration T_c

$$\psi(t) = \sin(\pi t/T_c), \qquad 0 \leq t < T_c$$

For the example illustrated in Fig. 2, the number of chips is $N = 7$, and $\psi(t)$ is the rectangular pulse of duration T_c.

A finite sequence $\mathbf{x} = (x_0, x_1, \ldots, x_{N-1})$ can be extended to an infinite periodic sequence by letting $x_{j+Nk} = x_j$ for each integer k and each integer j that is in the range $0 \leq j \leq N - 1$. We denote the periodic extension of $\mathbf{x}$ by (x_i), or, if explicit display of the index i is not necessary, we use the simpler notation x. The introduction of the infinite sequence permits us to write the spectral spreading signal as

$$v(t) = \sum_{i=-\infty}^{\infty} x_i \psi(t - iT_c) \tag{8}$$

This formulation is sufficiently general to include signature sequences that are not periodic; in fact, (x_i) can be any infinite sequence of real numbers. If (x_i) is periodic with period N, however, then $v(t)$ is periodic with period $T = NT_c$.

Let S be the *shift operator* on infinite sequences; specifically, for each sequence $x = (x_i)$, the infinite sequence Sx is the sequence $y = (y_i)$ defined by $y_i = x_{i+1}$ for $-\infty < i < +\infty$. More generally, for each integer j, define $S^j x$ to be the sequence y for which $y_i = x_{i+j}$ for $-\infty < i < +\infty$. The *period* of a sequence x is the smallest positive integer M for which $S^M x = x$. If the sequence x is the periodic extension of $\mathbf{x} = (x_0, x_1, \ldots, x_{N-1})$, then $S^N x = x$. Therefore, the period of the sequence x must be a divisor of N. For example, if N is even, the period of x could be $N/2$, as illustrated by the periodic extension of the vector

$$\mathbf{x} = (-1, +1, +1, -1, -1, +1, +1, -1)$$

which has length 8 but produces an infinite sequence of period 4. If

$$\mathbf{x} = (-1, +1, -1, +1, -1, +1, -1, +1)$$

the period of x is 2, and if

$$\mathbf{x} = (+1, +1, -1, +1, -1, -1, -1, +1)$$

the period is 8. All three vectors have length 8.

If the sequence x is the periodic extension of $\mathbf{x} = (x_0, x_1, \ldots, x_{N-1})$, we refer to the vector $(x_0, x_1, \ldots, x_{N-1})$ as the *segment* of the infinite sequence x. The period of x cannot exceed the length of the segment. If the signature sequence is periodic, the spectral spreading waveform $\xi(t)$ represents one segment of the baseband spectral spreading signal $v(t)$, and the baseband spectral spreading signal is a periodic continuous-time signal that has the period MT_c for some positive integer M that divides N.

Data can be communicated on the spectral spreading signal by superimposing amplitude modulation on $v(t)$. For purposes of illustration, consider a system with a data transmission rate of $1/T$ symbols per second, and suppose one data symbol is to be transmitted during each segment of the spectral spreading signal. If

$$\mathbf{b} = \ldots , b_{-1}, b_0, b_1, \ldots$$

is the sequence of data symbols to be sent to the receiver, the baseband spread-spectrum signal is

$$s(t) = A \sum_{n=-\infty}^{\infty} b_n \xi (t - nT)$$

where A is a fixed amplitude. Notice that if $\xi(t)$ is time limited to the interval $[0,T]$, the nth term of the sum is nonzero only if $nT \le t < (n + 1)T$. This formulation corresponds to a spectral spreading signal for which the duration of a segment is equal to the data symbol duration, and the period of the spectral spreading signal cannot exceed the duration of the segment. For some applications, it is necessary to use a spectral spreading signal with a longer period, in which case an alternative description of the spread-spectrum signal is preferred.

To obtain a more general formulation, and to develop an illustration of the bandwidth spreading accomplished by DS spread-spectrum modulation, define the data signal $d(t)$ as

$$d(t) = \sum_{n=-\infty}^{\infty} b_n p_T (t - nT) \tag{9}$$

where p_T is the rectangular pulse function defined by

$$p_T(u) = \begin{cases} 1, & 0 \le u < T \\ 0, & \text{otherwise} \end{cases}$$

The spread-spectrum signal can then be written as

$$s(t) = A v(t) d(t) \tag{10}$$

This formulation is valid even if the period of the spectral spreading signal is not the same as the duration T of the data pulse; in fact, the spectral spreading signal is not required to be periodic in this formulation.

The time- and frequency-domain relationships between the data signal and the resulting spread-spectrum signal are illustrated in Fig. 3. In this figure, $D(f)$ is the Fourier transform of the data pulse $d(t) = p_T(t)$, and $S(f)$ is the Fourier transform of the chip waveform, which for this illustration is a rectangular pulse of duration T_c. Only the main lobe and the two adjacent side lobes of $D(f)$ and $S(f)$ are shown. For some purposes, it is convenient to model the signature sequence and data sequence as independent sequences of independent

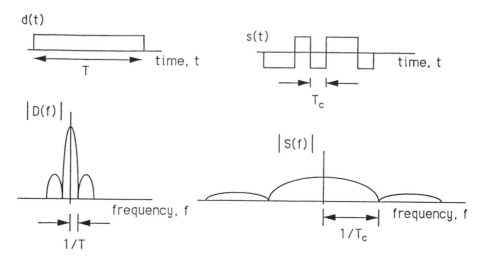

FIG. 3 Illustration of bandwidth expansion in direct-sequence (DS) spread spectrum.

random variables, each of which takes the value $+1$ with probability $1/2$ and the value -1 with probability $1/2$. If this random sequence model is adopted, and if an appropriate random time offset is included in the signals, then $D(f)$ and $S(f)$ are the power spectral densities for the data signal and spread-spectrum signal, respectively.

Notice that the first null in $D(f)$ is at $1/T$, and the spread-spectrum signal has its first null at $1/T_c$. The bandwidth of a DS spread-spectrum signal with N chips per data pulse is therefore N times the bandwidth of the data signal. Hence, the spectrum of the data signal is spread by a factor of N when it is multiplied by spectral spreading signal.

For transmission on RF or optical channels, it is necessary that the baseband spread-spectrum signal be modulated onto an appropriate sinusoidal carrier. Because both in-phase and quadrature modulation can be employed, two baseband DS spread-spectrum signals can be incorporated into a single DS spread-spectrum carrier-modulated signal. One form of DS spread-spectrum signal is given by

$$s(t) = Av_1(t)d_1(t)\cos(\omega_c t + \varphi) + Av_2(t)d_2(t)\sin(\omega_c t + \varphi) \qquad (11)$$

where $v_1(t)$ and $v_2(t)$ are baseband spectral spreading signals of the type defined in Eq. (8) and $d_1(t)$ and $d_2(t)$ are data signals of the type defined in Eq. (9). The baseband spectral spreading signals are obtained from two possibly different signature sequences, but we assume that they employ the same chip waveform $\psi(t)$ and therefore have the same chip duration. If the chip waveform is a rectangular pulse of duration T_c, the signal defined in Eq. (11) is a quadriphase-shift-key (QPSK) DS spread-spectrum signal. A time offset of $t_0 = T_c/2$ can be introduced in the in-phase component to give an offset QPSK DS spread-spectrum signal. If this time offset is employed and the chip waveform is the sine pulse, the resulting signal represents minimum-shift key (MSK) DS spread spectrum.

One example of DS spread spectrum is the low-cost packet radio (LPR) developed with the support of the Defense Advanced Research Projects Agency (DARPA) and the U.S. Army Communications and Electronics Command (CECOM). A description of this radio and a block diagram of the demodulator are given in Ref. 10. Some of the most important features are as follows. The chip rate is 12.8×10^6 chips per second, and the spread bandwidth is 20 MHz. The DS spread-spectrum modulation is MSK modulation, so the chip waveform $\psi(t)$ is the sine pulse. The RF signal is transmitted in 1 of 20 channels that span a frequency band from approximately 1.718 GHz to approximately 1.840 GHz. The LPR can transmit the encoded data at one of two transmission rates, 100 kilobits per second (kb/s) and 400 kb/s. The number of chips per bit is 128 at a transmission rate of 100 kb/s and 32 at a transmission rate of 400 kb/s.

A second example of DS spread spectrum is the wideband spread-spectrum cellular system known as the mobile cellular code-division multiple-access (CDMA) system. The technical requirements for the mobile cellular CDMA system are given in Ref. 11, which is the documentation for the interim standard known as IS-95. Additional information on this system can be found in Ref. 12.

The generation of the composite spread-spectrum signal for the forward link of the IS-95 system, which is the communications link from the base station to the mobile stations, is illustrated in Fig. 4. For simplicity, all forward-link channels in which any data modulation takes place, including the IS-95 paging and sync channels, are referred to as data channels in this figure. Also, the scrambling operation is not shown in Fig. 4. If scrambling is employed in a particular data channel, the corresponding data signal in Fig. 4 should be interpreted as a scrambled data signal.

In addition to the N data channels, there is a pilot channel that consists of a spectral spreading signal with no data modulation. The spreading sequence for the in-phase component of the pilot signal is the in-phase pilot sequence (I-pilot sequence), and the spreading sequence for the quadrature component of the pilot signal is the quadrature pilot sequence (Q-pilot sequence). The pilot signal provides the mobile stations with a convenient reference signal for use in demodulation of the forward-link data signals.

If the baseband spectral spreading signal derived from the I-pilot sequence is denoted by $x_1(t)$ and the baseband spectral spreading signal derived from the Q-pilot sequence is denoted by $x_2(t)$, the RF pilot signal can be expressed as

$$s_0(t) = A_0\{x_1(t)\cos(\omega_c t) + x_2(t)\sin(\omega_c t)\} \tag{12}$$

If $b_n(t)$ represents the nth baseband data signal ($1 \le n \le N$), the nth RF data signal is given by

$$s_n(t) = A_n w_n(t) b_n(t)\{x_1(t)\cos(\omega_c t) + x_2(t)\sin(\omega_c t)\} \tag{13}$$

where $w_n(t)$ is a baseband spectral spreading signal that is obtained from a special signature sequence that is referred to as a "Walsh function" in IS-95 (11). The Walsh functions of IS-95 are just the rows of the appropriate Hadamard

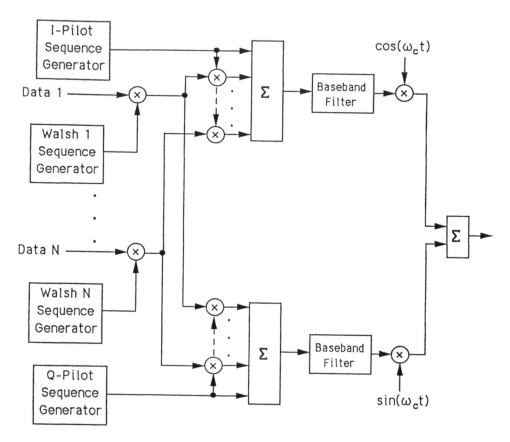

FIG. 4 Base station transmitted signal generation for direct-sequence spread-spectrum mobile cellular.

matrix (as described in the next section), so any two of them are orthogonal. The amplitudes $A_0, A_1, \ldots, A_N$ can be adjusted to give the desired amount of power in the data signals relative to the pilot signal.

The description of the nth data signal in Eq. (13) can be cast in the general form of Eq. (11) by letting the baseband spectral spreading signals be defined by

$$v_1(t) = w_n(t)x_1(t)$$

and

$$v_2(t) = w_n(t)x_2(t)$$

The baseband data signals of Eq. (11) are defined by $d_1(t) = d_2(t) = b_n(t)$. There is only one baseband data signal for each RF data signal in the IS-95 forward link.

The signals $s_0(t)$ and $s_n(t)$ of Eq. (12) and Eq. (13) can be written in a common form by defining the spectral spreading signals $g_{0,1}(t) = x_1(t)$ and $g_{0,2}(t)$

$= x_2(t)$ for the pilot signal and $g_{n,1}(t) = w_n(t)x_1(t)$ and $g_{n,2}(t) = w_n(t)x_2(t)$ for the data signals. The pilot signal can then be written as

$$s_0(t) = A_0\{g_{0,1}(t)\cos(\omega_c t) + g_{0,2}(t)\sin(\omega_c t)\}$$

and the nth RF data signal is given by

$$s_n(t) = A_n\, b_n(t)\{g_{n,1}(t)\cos(\omega_c t) + g_{n,2}(t)\sin(\omega_c t)\}$$

The transmitted RF signal on the forward link is

$$s(t) = \sum_{n=0}^{N} s_n(t)$$

If we define $b_0(t) = 1$ for all t, the forward-link transmitted signal can be expressed as

$$s(t) = \sum_{n=0}^{N} A_n b_n(t)\{g_{n,1}(t)\cos(\omega_0 t) + g_{n,2}(t)\sin(\omega_0 t)\}$$

We let $b_0(t) = 1$ for all t to reflect the absence of data in the pilot signal.

A third example of DS spread spectrum is the Global Positioning System (GPS), which consists of 24 satellites in 6 orbital planes and a ground control station (13,14). Each satellite transmits two DS spread-spectrum navigation signals. The first is an I-Q modulated DS spread-spectrum signal with carrier frequency $f_1 = 1575.42$ MHz. Two sequences are employed to spread the first navigation signal. One of the sequences is referred to as the *clear/acquisition* sequence by some authors and the *coarse acquisition* sequence by others. Fortunately, either can be abbreviated as C/A sequence, which is the most common designation in the literature. The second sequence also has two names: *precision* sequence and *protected* sequence. Each is abbreviated as P sequence.

Each satellite transmits a unique C/A sequence and a unique P sequence. The spreading sequences for the C/A components that are transmitted by different satellites are different sequences from a set of Gold sequences of period 1023 (6). The P sequences are very long segments of a single sequence that has a period in excess of 38 weeks (15). Each P sequence runs for 1 week without repeating, and it is reset at the end of each week.

If the spreading signal corresponding to the P sequence is denoted by $p(t)$ and the spreading signal corresponding to the C/A sequence is denoted by $x(t)$, then the first navigation signal transmitted by a particular satellite can be expressed as

$$s_1(t) = \sqrt{2}\, A_0\, d(t)p(t)\cos(2\pi f_1 t + \theta_1) + \sqrt{2}\, A_1\, d(t)x(t)\sin(2\pi f_1 t + \theta_1)$$

The baseband signal $d(t)$ represents a sequence of rectangular pulses that convey the navigation data from that satellite. The data rate is 50 bits per second (b/s).

The P component of the first navigation signal is

$$\sqrt{2}\, A_0\, d(t)p(t)\cos(2\pi f_1 t + \theta_1)$$

and the C/A component is

$$\sqrt{2}\, A_1\, d(t)x(t)\sin(2\pi f_1 t + \theta_1)$$

The power in the P component is half of that in the C/A component; that is, $A_0^2 = A_1^2/2$. As a result, the first navigation signal can be viewed as an imbalanced quadrature amplitude-shift-key (QASK) spread-spectrum signal or, equivalently, two unequal-power binary phase-shift keying (BPSK) spread-spectrum signals on quadrature carriers. The chip rate for the C/A spreading signal is 1.023 mega chips per second (Mc/s), and the chip rate for the P spreading signal is 10.23 Mc/s. Thus, the bandwidth of the P component is 10 times the bandwidth of the C/A component. Because the C/A spreading signal has the chip rate 1.023 Mc/s and the period of the C/A sequence is 1023, the waveform $x(t)$ repeats every 1 millisecond (ms). The data rate of 50 b/s corresponds to a data symbol duration of 20 ms, so there are 20 periods of the C/A spreading sequence per data pulse.

In order to reduce the range measurement errors that are caused by ionospheric refraction, a second navigation signal is transmitted at a different frequency than the first. The second navigation signal has the carrier frequency $f_2 = 1227.60$ MHz, and it is a standard BPSK DS spread-spectrum signal that normally carries the P sequence (14). When data is included in this signal, it can be expressed as

$$s_2(t) = \sqrt{2}\, A_2\, d(t)p(t)\sin(2\pi f_2 t + \theta_2)$$

The power in the second navigation signal is 7.9 decibels (dB) below the power in the C/A component of the first navigation signal (14).

All GPS receivers have access to the C/A component of the first signal, and even those receivers that have access to the P component typically use the C/A component to acquire an initial timing reference (i.e., coarse acquisition). The P sequence may not be available to all receivers at all times since it may be switched to a more secure sequence by the U.S. Department of Defense. For this and other reasons, some GPS receivers rely on the C/A component only. The P component can provide military receivers greater accuracy, more security, and better protection against jamming and incidental interference.

Sequences and Their Periodic Correlation Functions

A baseband spectral spreading signal $v(t)$ is a continuous-time waveform that is defined completely by its chip waveform $\psi(t)$ and its signature sequence (x_i). The mathematical relationship is given by

$$v(t) = \sum_{i=-\infty}^{\infty} \lambda_i \psi(t - iT_c)$$

If $v(t)$ has period T, its continuous-time periodic autocorrelation function is

$$r_v(\tau) = \int_0^T v(t)v(t + \tau)dt$$

Because $v(t)$ has period T, the only requirement on the limits for this integral is that the difference between the upper limit and the lower limit must be T. It is often convenient to normalize this autocorrelation function, so we define $R_v(\tau) = r_v(\tau)/T$ for all τ.

The inner product of two real vectors $\mathbf{x} = (x_0, x_1, \ldots, x_{N-1})$ and $\mathbf{y} = (y_0, y_1, \ldots, y_{N-1})$ of length N is defined by

$$(\mathbf{x}, \mathbf{y}) = \sum_{n=0}^{N-1} x_n y_n$$

The discrete periodic autocorrelation function for a sequence $x = (x_i)$ of period N is defined by

$$\theta_x(j) = \sum_{n=0}^{N-1} x_n x_{n+j}, \qquad 0 \le j \le N - 1$$

Notice that if the sequence y is defined by $y = S^j x$, and $\mathbf{y}$ is the segment $(y_0, y_1, \ldots, y_{N-1})$, then $\theta_x(j) = (\mathbf{x}, \mathbf{y})$.

The continuous-time periodic autocorrelation function can be determined from knowledge of the chip waveform and the discrete periodic autocorrelation function (3). For example, if $\psi(t) = 0$ if $t < 0$ or $t > T_c$ and

$$\lambda = \int_0^{T_c} \psi^2(t)dt$$

then $r_v(\tau)$ can be determined for $\tau = jT_c$ from the relationship $r_v(jT_c) = \lambda\theta_x(j)$, for $0 \le j \le N - 1$.

For a binary signature sequence, each element of the sequence (x_i) is either $+1$ or -1, so $x_i^2 = 1$ for each i. Thus, $\theta_x(0) = N$ for each binary signature sequence. For most applications, it is desirable for the discrete periodic autocorrelation function to be small for $j \ne 0$. The values of $\theta_x(j)$ for $j \ne 0$ are often referred to as the side lobes of the discrete periodic autocorrelation function. Similarly, the values of $r_v(\tau)$ and $R_v(\tau)$ for $\tau \ne 0$ are referred to as the side lobes of the continuous-time periodic autocorrelation function.

One family of sequences with a particularly good periodic autocorrelation function is the family of maximal-length linear feedback shift-register sequences, which are commonly referred to as *m*-sequences in the literature. The important properties of such sequences are given in Refs. 5 and 6. The period N of an *m*-sequence must be of the form $N = 2^n - 1$ for some integer n, and this period is the largest possible for any sequence generated by a linear feedback shift register with n storage elements.

An example of a linear feedback shift register that can generate an m-sequence is shown in Fig. 5. This shift register has three storage elements and a single modulo-2 adder (also referred to as an exclusive-OR gate). Modulo-2 addition is defined by

$$0 \oplus 0 = 1 \oplus 1 = 0 \quad \text{and} \quad 0 \oplus 1 = 1 \oplus 0 = 1$$

Suppose that initially each storage element holds a 1, as illustrated in Fig. 5. The next digit that is fed back and placed into the right-most register is $1 \oplus 1 = 0$, and the register contents become 110 as the left-most 1 shifts out of the register. Continuing in this fashion, we see that the sequence that shifts out of the register is

$$1110010111001011100.\ldots$$

This is a periodic sequence with period 7. The largest period that can be obtained from a linear feedback shift register with three storage elements is 7 because there are only seven different nonzero patterns of three binary digits, and each of these seven patterns can appear in the register at most once per period. The all-zeros pattern cannot appear in the register because if it were to appear it would produce a sequence consisting of all zeros (i.e., 000 in the register gives the output 00000000 . . .), and the period for the all-zeros sequence is 1 rather than 7. The sequence generated by the shift register is written in its ± 1 form by letting 0 be represented by $+1$ and 1 be represented by -1.

If x is an m-sequence of period $N = 2^n - 1$, its discrete periodic autocorrelation function is given by $\theta_x(0) = N$ and $\theta_x(j) = -1$ for $1 \le j \le N - 1$. The magnitudes of the periodic autocorrelation side lobes for an m-sequence are the smallest possible for any binary sequence of the same period (note that the period of an m-sequence is necessarily an odd integer). If the chip waveform is a unit-amplitude rectangular pulse of duration T_c, the corresponding continuous-time periodic autocorrelation function for a spectral spreading signal derived from an m-sequence of period N is given by

$$r_v(\tau) = \begin{cases} (NT_c - (N + 1)|\tau|), & |\tau| \le T_c \\ -T_c, & T_c < |\tau| \le (N - 1)T_c \end{cases}$$

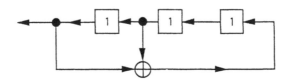

FIG. 5   A linear feedback shift register with initial loading 111.

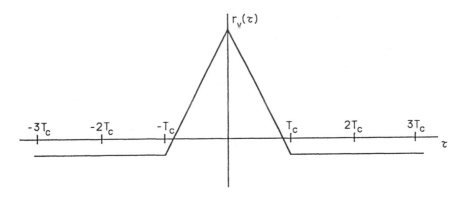

FIG. 6 Continuous-time periodic autocorrelation function for an m-sequence.

This autocorrelation function is illustrated in Fig. 6. Because $v(t)$ is periodic with period $T = NT_c$, the continuous-time periodic autocorrelation function satisfies $r_v(\tau) = r_v(\tau + iNT_c)$ for each integer i. In particular, $r_v(0) = r_v(iNT_c) = NT_c$ for each integer i. Notice also that $r_v(jT_c) = -T_c$ for $1 \le j \le N - 1$. The normalized continuous-time periodic autocorrelation function for a spectral spreading signal derived from an m-sequence of period N satisfies $R_v(0) = 1$ and $R_v(jT_c) = -1/N$ for $1 \le j \le N - 1$. Note the similarity between this normalized continuous-time periodic autocorrelation function and the autocorrelation function for the random process $V(t)$ considered in the first section (see Eq. (1) in particular).

The features of the autocorrelation function can be exploited in order to provide an appropriate error signal for a feedback system to track the spread-spectrum waveform. The receiver must know the timing of the incoming spectral spreading signal in order to demodulate the signal and recover the data. This timing changes because of timing drift in the clock at the transmitter, motion of the transmitter and receiver, and changes in the propagation conditions on the communication link. A system that can track these changes is shown in Fig. 7. This is a version of a delay-lock loop (16).

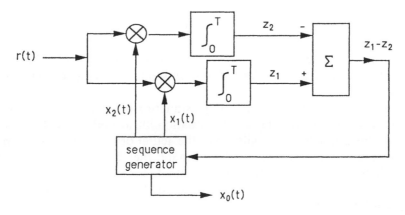

FIG. 7 Tracking loop for a direct-sequence spread-spectrum receiver.

Suppose that the spectral spreading signal has a time offset τ as it arrives at the receiver; that is, the received spreading signal is $v(t - \tau)$. The input $r(t)$ to the delay-lock loop normally consists of the sum of the desired signal $v(t - \tau)$, other communications signals, and noise. To illustrate the operation of the delay-lock loop, consider a situation in which the received signal is the desired signal only; that is, $r(t) = v(t - \tau)$. The signal $x_0(t)$ shown in Fig. 7 is known as the punctual signal. It is generated by the same sequence generator that generates $v(t)$, so $x_0(t) = v(t - \hat{\tau})$. The only possible difference between $x_0(t)$ and $v(t - \tau)$ is an offset in timing. If the loop is locked and tracking well, any timing offset is small: $\hat{\tau} \approx \tau$. In particular, the error $\tau_e = \tau - \hat{\tau}$ in the receiver's estimate of the timing offset should be small compared to the chip duration T_c.

The signal $x_1(t)$ is known as the *late signal*, and it is a delayed version of the punctual signal: $x_1(t) = x_0(t - \delta)$. The signal $x_2(t)$ is known as the *early signal*. It is an advanced version of the punctual signal: $x_2(t) = x_0(t + \delta)$. Thus, $x_1(t) = v(t - \hat{\tau} - \delta)$ and $x_2(t) = v(t - \hat{\tau} + \delta)$. The parameter δ is selected by the system designer to meet the requirements of the tracking system. A common choice is $\delta = T_c/2$. The outputs of the integrators are the correlations between the received signal and the early and late signals. Since all of these signals are just time offsets of the same spreading signal, these correlations are autocorrelations. In particular, the late sample is

$$z_1 = \int_0^T v(t - \tau)x_1(t)dt = \int_0^T v(t - \tau)v(t - \hat{\tau} - \delta) \, dt$$

and the early sample is

$$z_2 = \int_0^T v(t - \tau)x_2(t)dt = \int_0^T v(t - \tau)v(t - \hat{\tau} + \delta) \, dt$$

Because $v(t)$ has period T, it follows that the early sample is given by

$$z_1 = r_v(\tau - \hat{\tau} - \delta) = r_v(\tau_e - \delta)$$

and the late sample is given by

$$z_2 = r_v(\tau - \hat{\tau} + \delta) = r_v(\tau_e + \delta)$$

The resulting error signal is $z_1 - z_2 = r_v(\tau_e - \delta) - r_v(\tau_e + \delta)$. The early and late samples are illustrated in Fig. 8 for an m-sequence and a correlator spacing of $\delta = T_c/2$.

Notice that if the error in the timing estimate $\hat{\tau}$ is positive (i.e., $\tau_e > 0$), the error signal is also positive. This is the situation illustrated in Fig. 8. If the error is positive, the estimate $\hat{\tau}$ is smaller than the actual time offset τ, so the positive feedback results in an increase of the estimate, thereby reducing the error. On the other hand, if the error in the timing estimate $\hat{\tau}$ is negative ($\tau_e < 0$), the error signal is negative, and the estimate is reduced by the feedback mechanism. This in turn reduces the error.

For multiple-access communications and certain other applications, the

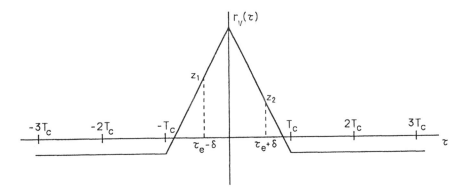

FIG. 8 Early and late samples for a delay-lock loop.

crosscorrelation functions for the sequences are at least as important as the autocorrelation functions. The discrete periodic crosscorrelation function for two sequences $x = (x_i)$ and $y = (y_i)$, each of period N, is defined by

$$\theta_{x,y}(j) = \sum_{n=0}^{N-1} x_n y_{n+j}, \qquad 0 \leq j \leq N - 1$$

For $j = 0$ the periodic crosscorrelation is the inner product

$$\theta_{x,y}(0) = (x,y) = \sum_{n=0}^{N-1} x_n y_n$$

If $\theta_{x,y}(0) = 0$, the sequences (x_i) and (y_i) are said to be *orthogonal*. Orthogonal sequences are of particular interest for code-division multiplexing (e.g., in the base station transmitter of the IS-95 mobile cellular communications system). If the sequence z is defined by $z = S^j y$, then

$$\theta_{x,y}(j) = \theta_{x,z}(0) = (\mathbf{x},\mathbf{z})$$

so the discrete periodic crosscorrelation for two sequences is the inner product for one of the sequences and a shifted version of the other.

Sets of orthogonal sequences can be obtained from the rows of Hadamard matrices (4). The simplest Hadamard matrix is the 2×2 matrix

$$H_1 = \begin{bmatrix} +1 & +1 \\ +1 & -1 \end{bmatrix}$$

for which the row vectors are the orthogonal vectors $(+1, +1)$ and $(+1, -1)$. The 4×4 Hadamard matrix is

$$H_2 = \begin{bmatrix} +1 & +1 & +1 & +1 \\ +1 & -1 & +1 & -1 \\ +1 & +1 & -1 & -1 \\ +1 & -1 & -1 & +1 \end{bmatrix}$$

which has orthogonal row vectors $(+1,+1,+1,+1)$, $(+1,-1,+1,-1)$, $(+1,+1,-1,-1)$, and $(+1,-1,-1,+1)$. Notice that H_2 can also be expressed in terms of the 2×2 Hadamard matrix H_1:

$$H_2 = \begin{bmatrix} H_1 & H_1 \\ H_1 & -H_1 \end{bmatrix}$$

For an arbitrary positive integer n, the relationship between H_{n+1} and H_n is

$$H_{n+1} = \begin{bmatrix} H_n & H_n \\ H_n & -H_n \end{bmatrix}$$

where H_{n+1} is a $2^{n+1} \times 2^{n+1}$ matrix and H_n is a $2^n \times 2^n$ matrix. Thus, if N is a power of 2, a set of N orthogonal vectors can be obtained from the rows of the Hadamard matrix H_m if $m = \log_2(N)$. A smaller number of orthogonal vectors can always be obtained by simply deleting rows of the matrix, so it is possible to obtain M orthogonal vectors of length N by this method as long as $M \le N$. Each vector is repeated periodically to form an infinite sequence of period N. For example, the vector $(+1,-1,-1,+1)$ produces the sequence

$$\ldots, +1,-1,-1,+1,+1,-1,-1,+1,+1,-1,-1,+1, \ldots,$$

and the vector $(+1,-1,+1,-1)$ produces the sequence

$$\ldots, +1,-1,+1,-1,+1,-1,+1,-1,+1,-1,+1,-1, \ldots$$

We refer to sequences generated in this manner as *Hadamard sequences*. Alternative names are Walsh functions (11) and Hadamard-Walsh sequences (12).

A similar set of sequences can be obtained from *m*-sequences. To obtain an example of the construction of orthogonal signals from *m*-sequences, return to the linear feedback shift register shown in Fig. 5. The sequence generated by this register is 1110010111001011100. . . . Next, form the array of all possible seven-bit vectors that come from the register. For the shift register in Fig. 5, these seven vectors are as follows:

1110010
1100101

$$1001011$$
$$0010111$$
$$0101110$$
$$1011100$$
$$0111001$$

These vectors can also be described as *left-cyclic shifts* of the initial pattern 1110010. At each step, the pattern is shifted one position to the left, and the left-most digit is moved to the right-most position to form the next row of the array.

The next step is to add a zero to the end of each vector, and then add the all-zeros sequence as the eighth row of the array. This gives the following array:

$$11100100$$
$$11001010$$
$$10010110$$
$$00101110$$
$$01011100$$
$$10111000$$
$$01110010$$
$$00000000$$

The final step is to replace each 0 by a $+1$ and each 1 by a -1. This gives a set of eight binary orthogonal vectors, each of length 8. For each positive integer n, a set of 2^n binary orthogonal vectors of length 2^n can be generated from a single m-sequence of period $2^n - 1$. These vectors can be repeated periodically to form infinite sequences.

The sets of Hadamard sequences and the sets of sequences generated from shift-register sequences in the manner described above are sets of orthogonal sequences. Thus, if (x_i) and (y_i) are two sequences from the same set of either type of sequences, then $\theta_{x,y}(0) = (\mathbf{x},\mathbf{y}) = 0$. However, it is not true, in general, that $\theta_{x,y}(j) = 0$ for $j \neq 0$. In fact, $\theta_{x,y}(j)$ can be very large if $j \neq 0$. For binary sequences of period 4, the largest possible value for $\theta_{x,y}(j)$ is 4, and this value is achieved if, and only if, $x = S^j y$ (i.e., x is a shifted version of y, shifted by j units). But, this is exactly the situation for the sequences we have considered thus far.

For example, suppose $x = (x_i)$ is the Hadamard sequence

$$\ldots, +1, -1, -1, +1, +1, -1, -1, +1, +1, -1, -1, +1, \ldots,$$

which is generated by the vector $(+1, +1, -1, -1)$, and $y = (y_i)$ is the Hadamard sequence

$$\ldots, +1, +1, -1, -1, +1, +1, -1, -1, +1, +1, -1, -1, \ldots,$$

which is generated by the vector $(+1, -1, -1, +1)$, then it is easy to see that $x = Sy$. The Hadamard sequence (x_i) is just a shifted version of the Hadamard sequence (y_i), shifted by one time unit. Consequently, for this example, $\theta_{x,y}(1)$

= 4. Alternatively, since the segments are of length 4, the definition of the periodic crosscorrelation function implies that

$$\theta_{x,y}(1) \ = \ \sum_{n=0}^{3} x_n y_{n+1}$$

which gives $\theta_{x,y}(1) \ = \ 4$ for these two sequences.

The orthogonal sequences generated from an m-sequence in the manner described also have poor periodic crosscorrelation functions. Consider the two vectors

00101110

and

01011100

Because these are cyclic shifts of each other, the two infinite sequences generated from these vectors (when converted to sequences of $+1$'s and -1's) are shifts of each other. Consequently, their periodic crosscorrelation function must take on the value 8 for some value of its argument (which is the largest possible value for sequences generated by vectors of length 8). In particular, if x denotes the infinite sequence obtained from the first vector and y denotes the sequence obtained from the second, then $\theta_{y,x}(1) = \theta_{x,y}(7) = 8$.

Fortunately, there are sequences that have periodic crosscorrelation functions $\theta_{x,y}(j)$ that are relatively small for all values of j. All of the sequences described in this section are generated by linear feedback shift registers, so the specifications of the sequences are given in terms of the feedback connections and initial conditions. A linear feedback shift register with n storage elements is illustrated in Fig. 9. The sequence generated by this register is denoted by $u = (u_i)$. Each of the tap coefficients $h_{n-1}, h_{n-2}, \ldots, h_1$ has value 0 or 1. If $h_k = 0$, there is no feedback through the kth tap, but if $h_k = 1$ there is feedback through the kth tap. The output sequence is described by the equation

$$u_{j+n} \ = \ u_j \oplus h_{n-1}u_{j+1} \oplus h_{n-2}u_{j+2} \oplus \ldots \oplus h_1 u_{j+n-1}$$

The two most common ranges for the integer j in this equation are $-\infty < j < \infty$, which gives a doubly infinite sequence, and $0 \leq j < \infty$, which gives a

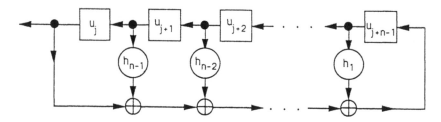

FIG. 9   General n-stage linear feedback shift register.

one-sided sequence that begins at time 0. Notice that u_{j+n} is the element that is being fed back into the right-most register as u_j is shifting out the left-most register.

For convenience, we define $h_n = 1$, which corresponds to having feedback from the left-most register, and $h_0 = 1$, which corresponds to having feedback into the right-most register, as shown in Fig. 9. Since these coefficients always take the value 1 in this section, they are not shown as variables in Fig. 9. A linear feedback shift register with n storage elements can be represented by the binary vector $h = (h_0, h_1, \ldots, h_n)$, where $h_0 = h_n = 1$ for all shift registers with n storage elements. This vector is referred to as the *feedback tap coefficient vector*.

In order to provide a compact designation for these shift registers, the feedback tap coefficient vectors are often represented in octal form (6). For example, if $n = 5$, the vector is $(h_0, h_1, h_2, h_3, h_4, h_5)$, which is grouped as $h_0 h_1 h_2$ and $h_3 h_4 h_5$. These two triples of binary numbers are then converted to octal. For example, the linear feedback shift register that has five storage elements and feedback taps given by the binary vector $(1,0,0,1,0,1)$ is represented in octal form as 45. The grouping process begins with the right-most three components of $(h_0, h_1, \ldots, h_n)$ and proceeds to the left. If the number of components is not a multiple of 3, the left-most group will not have the full three elements. For example, if $n = 4$, the vector is $(h_0, h_1, h_2, h_3, h_4)$ and the grouping is $h_0 h_1$ and $h_2 h_3 h_4$. Thus, the linear feedback shift register that has four storage elements and feedback taps given by the binary vector $(1,0,1,0,1)$ is represented in octal form as 25. As a final example, consider the linear feedback shift register of Fig. 5, which has three storage elements ($n = 3$). The feedback tap coefficient vector is $(1,0,1,1)$, which is expressed in octal as 13.

It is convenient to represent a linear feedback shift register with n storage elements by polynomial of degree n. The shift register illustrated in Fig. 9 corresponds to the polynomial

$$h(x) = h_0 x^n + h_1 x^{n-1} + h_2 x^{n-2} + \ldots h_{n-1} x + h_n$$

From the properties of the polynomial that represents a particular linear feedback shift register, certain properties can be determined for the sequence that is generated by that shift register. A simple example is the fact that, if the polynomial has the degree n, the period of the sequence generated by the corresponding shift register cannot exceed $2^n - 1$. Sequences of period $2^n - 1$ that are generated by a shift register with a polynomial that has degree n are precisely the m-sequences discussed in above, and the corresponding polynomial is a primitive polynomial (17). Tables of primitive polynomials are given in octal notation in (17).

Although each individual m-sequence has good periodic autocorrelation properties, not all pairs of m-sequences have good periodic crosscorrelation properties. Among the set of all m-sequences of a given period, however, certain pairs do have relatively small periodic crosscorrelation values, and these pairs are referred to as *preferred pairs* of m-sequences, and the associated primitive polynomials are referred to as *preferred pairs* of polynomials (6). For m-sequences of period 31, for example, the maximum periodic crosscorrelation for an arbitrary pair of m-sequences is 11, but the maximum periodic crosscorrelation for a preferred pair of m-sequences is only 9. The situation for m-sequences of period 127 is much more dramatic: for an arbitrary pair of m-sequences, the periodic crosscorrelation can be as large as 41, but for a preferred pair of m-sequences the maximum periodic crosscorrelation is 17. For m-sequences of period 4095, the maximum periodic crosscorrelation for an arbitrary pair of m-sequences is more than an order of magnitude greater than for a preferred pair of m-sequences.

The periodic crosscorrelation function for a preferred pair of m-sequences is also known to take on three values only. For example, if x and y are a preferred pair of m-sequences of period 31, $\theta_{x,y}(j)$ is equal to -1 for 15 values of j, equal to $+7$ for 10 values of j, and equal to -9 for 6 values of j. If the preferred pair of m-sequences has period 127, the three values taken on by the periodic crosscorrelation function are -1, $+15$, and -17. Preferred pairs of m-sequences are of interest for applications not so much because the number of different crosscorrelation values is small but primarily because the magnitudes of these values are relatively small compared with the sequence period and compared with the crosscorrelation values for pairs of m-sequences that are not preferred pairs. For each positive integer n, let $t(n)$ be defined by

$$t(n) = 1 + 2^{\lfloor (n+2)/2 \rfloor}$$

where $\lfloor (n+2)/2 \rfloor$ denotes the integer part of $(n+2)/2$. Notice that $t(5) = 9$ and $t(7) = 17$, and also notice that these are the maximum values of the magnitudes of the crosscorrelation functions for preferred pairs of m-sequences of periods 31 and 127, respectively. If the discrete periodic crosscorrelation function for two m-sequences of period $2^n - 1$ takes on the three values -1, $t(n) - 2$, and $-t(n)$ only, then it is said to be a *preferred three-valued crosscorrelation function* (6). A preferred pair of m-sequences of period $2^n - 1$ has a preferred three-valued crosscorrelation function.

A maximal connected set of m-sequences is a collection of m-sequences with the property that each pair in the set is a preferred pair. By requiring all of the sequences for a given application (e.g., spread-spectrum multiple-access communications) to be m-sequences from the same maximal connected set, one can be sure that no matter which two signals are considered, the discrete periodic crosscorrelation function for the two signals is a preferred three-valued crosscorrelation function. The reason that we cannot restrict attention to maximal connected sets for all applications is that the sizes of the sets are too small. For example, a maximal connected set of m-sequences of period 31 has only three sequences, and a maximal connected set of m-sequences of period 127 has only six sequences. For periods other than 127 that are of interest for most applications (e.g., from 7 to 65,535), none of the maximal connected sets has more than four sequences. Even though the total number of m-sequences of period N is quite large for large values of N, the size of a maximal connected set of such sequences remains small (e.g., for period 8191 there are a total of 630 m-sequences, but there are only four sequences in a maximal connected set).

Sets of Gold sequences (18) are much larger than maximal connected sets, and their crosscorrelation functions take on the same three values, but the periodic autocorrelation functions for most of the Gold sequences are not as good as the periodic autocorrelation functions for m-sequences. Also, Gold sequences of period $2^n - 1$ can be generated by linear feedback shift registers with $2n$ storage elements, so their periods are approximately the square root of the maximum periods for linear sequences generated with the same number of storage elements. However, if a large number of sequences is required for a given application, and if the crosscorrelation function is more important than the autocorrelation function, then the Gold sequences are much better than m-sequences for that application.

The simplest way to describe how to generate Gold sequences is to use the polynomial representation. If $f_1(x)$ and $f_2(x)$ are a preferred pair of primitive polynomials of degree n, the set of sequences generated by the product polynomial $f(x) = f_1(x)f_2(x)$ is a set of Gold sequences of period $2^n - 1$. Because the product polynomial has degree $2n$, the corresponding shift register has $2n$ storage elements. Alternative methods for generating Gold sequences are available, and there are many other classes of linear feedback shift register sequences that have different set sizes and different correlation properties (6).

Correlation Receivers for Direct-Sequence Spread Spectrum

Suppose the signal $s(t)$ is transmitted over a linear channel in which interference and noise are added, and let $r(t)$ denote the input to the communication receiver. For this channel model,

$$r(t) = s(t) + \eta(t) + X(t) \tag{14}$$

where $\eta(t)$ represents interference present on the channel and $X(t)$ represents thermal noise, which is modeled as a zero-mean, white Gaussian noise process. The signal $s(t)$ contains the information that is being conveyed to the receiving terminal, so it is referred to as the *desired signal*. It is a DS spread-spectrum signal of the type described in the previous sections. For example, the communication signal may be of the form

$$s(t) = A v(t)d(t)\cos(\omega_c t + \varphi)$$

where $v(t)$ is a baseband spectral spreading signal of the form given in Eq. (8), and $d(t)$ is a data signal defined by Eq. (9). The basic concepts can be illustrated with greater simplicity by considering baseband models for the communication signals and the receiver, in which case the desired signal is written as $s(t) = A v(t)d(t)$. The noise and interference are also described in terms of their baseband models.

In order to simplify the following presentation, we focus on chip waveforms that are time limited to the interval $[0, T_c]$. The baseband model for a correlation receiver is shown in Fig. 10. The correlation receiver has a replica of the spectral

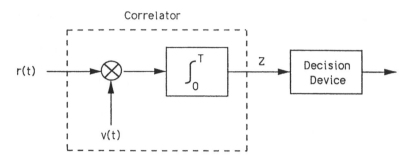

FIG. 10   Correlation receiver for the spread-spectrum signal $s(t)$.

spreading signal $v(t)$, and it processes the received signal by multiplying it by the spectral spreading signal and integrating the product of the two signals over the duration of the data pulse. The output of the integrator is Z, the statistic on which the receiver bases its decision as to which data symbol was transmitted (e.g., the decision may depend on whether Z exceeds some threshold).

The statistic Z has three components: a desired signal component μ, which is the part of the output that is due to $s(t)$; an interference component I, which is the output of the integrator if the input is $\eta(t)$ only; and the noise component, which is the result of thermal noise at the input to the receiver. The desired signal component for the interval $[0, T]$ is given by $\mu = b_0 A T$.

The influence of thermal noise is no different from in a baseband digital communications system that does not employ spread spectrum. This can be seen from an examination of the noise component at the output of the correlator. The noise component is the zero-mean Gaussian random variable defined by

$$W = \int_0^T X(t)v(t)dt$$

It is easy to show that W has the same probability distribution as

$$W' = \int_0^T X(t)dt$$

which is the output of the correlator or matched filter for a system with $d(t)$ as the transmitted signal (i.e., a system that does not use spread spectrum). As a result, the novel features of spread spectrum are demonstrated by focusing on the interference component of the decision statistic. The interference component is defined by

$$I = \int_0^T \eta(t)v(t)dt \tag{15}$$

If the interference is due exclusively to the presence of one or more additional DS spread-spectrum signals in the same frequency band as the desired signal, then the interference component I is referred to as the *multiple-access interference.*

Suppose that the only interference on the channel is another DS spread-spectrum signal $s'(t)$ with the same general characteristics as the desired signal (e.g., the same chip waveform, chip rate, and data rate). The multiple-access interference signal differs from the desired signal in that its amplitude is A' rather than A, its signature sequence is (y_i) rather than (x_i), and it arrives at the receiver with a time delay of τ units relative to the desired signal. The time delay is used to account for any offsets between the timing references for the two signals and any differences in the propagation times from the two transmitters to the common receiver. The amplitudes A' and A are used to account for the power levels in the transmitted signals and the propagation losses in the two communications paths. If A' is larger than A, the interference signal has more power than the desired signal. Such a situation can result if the source of the interfering signal is closer to the receiver than the source of the desired signal, so it is often referred to as the *near-far condition.* An unfavorable near-far condition, whether it is a result of greater power in the transmitted interfering signal or shorter range from the interfering transmitter to the receiver, can severely limit the multiple-access capability of a DS spread-spectrum system in which the demodulation is accomplished by correlation receivers.

Suppose that the signature sequences for the desired signal and the interference signal each have period N. Analogous to Eq. (8) and Eq. (9), the spectral spreading signal for the interference is expressed as

$$y(t) = \sum_{i=-\infty}^{\infty} y_i \psi(t - iT_c)$$

and its data signal is

$$d'(t) = \sum_{n=-\infty}^{\infty} d_n p_T(t - nT)$$

The interference signal is given by

$$s'(t) = A' y(t) d'(t)$$

The spectral spreading waveform for the interference is

$$\zeta(t) = \sum_{i=0}^{N-1} y_i \psi(t - iT_c)$$

and so the interference signal can also be written as

$$s'(t) = A' \sum_{n=-\infty}^{\infty} d_n \zeta(t - nT)$$

Notice that the nth term of this expression is nonzero only if $nT \le t < (n+1)T$.

Because $\eta(t) = s'(t - \tau)$, it follows from Eq. (14) that the received signal is given by

$$r(t) = s(t) + s'(t - \tau) + X(t) \tag{16}$$

and the interference component is

$$I = \int_0^T s'(t - \tau)v(t)dt \tag{17}$$

The signals are of infinite duration in the model, so it suffices for most purposes to restrict the time delay to the range $0 \le \tau < T$. This means that during the interval $[0,T]$, the only nonzero terms of the interference signal $s'(t - \tau)$ are $d_{-1}\zeta(t - \tau + T)$ and $d_0\zeta(t - \tau)$. If τ is in the interval $[0,T]$, the only values of t in $[0,T]$ for which $\zeta(t - \tau + T)$ can be nonzero are those that satisfy $0 \le t < \tau$. Similarly, $\zeta(t - \tau)$ can be nonzero only for $\tau \le t \le T$. By substituting for $s'(t - \tau)$ in Eq. (17), it is easy to see that

$$I = A'\{d_{-1}R_{y,v}(\tau) + d_0\hat{R}_{y,v}(\tau)\} \tag{18}$$

where the continuous-time partial crosscorrelation functions $R_{y,v}$ and $\hat{R}_{y,v}$ are defined by

$$R_{y,v}(\tau) = \int_0^\tau y(t - \tau)v(t)dt \tag{19}$$

and

$$\hat{R}_{y,v}(\tau) = \int_\tau^T y(t - \tau)v(t)dt \tag{20}$$

Notice that the amplitudes of two consecutive data symbols of the interference signal influence the output of the correlator. Even the magnitude of the interference depends on the relative polarities of these two pulse amplitudes. For example, if the sequence (d_n) is a binary sequence from the set $\{-1,+1\}$, then in general different interference magnitudes result from the two cases $d_{-1} = d_0$ and $d_{-1} = -d_0$.

Because of the timing offset between the two spread-spectrum signals, if $d_{-1} = -d_0$, the multiple-access interference cannot be expressed in terms of the periodic crosscorrelation function only. The appropriate function for characterizing the interference in an asynchronous DS spread-spectrum multiple-access system is the aperiodic crosscorrelation function (9) since it can handle both $d_{-1} = d_0$ and $d_{-1} = -d_0$.

The aperiodic crosscorrelation function for two vectors $\mathbf{u} = (u_0, u_1, \ldots, u_{N-1})$ and $\mathbf{v} = (v_0, v_1, \ldots, v_{N-1})$ of length N is defined by

$$C_{u,v}(i) = \sum_{j=0}^{N-1-i} u_j v_{j+i}, \qquad 0 \le i \le N - 1 \tag{21}$$

$$C_{u,v}(i) - \sum_{j=0}^{N-1+i} u_{j-i}v_j, \qquad -(N-1) \le i < 0 \tag{22}$$

and $C_{u,v}(i) = 0$ if $i > N$ or $i < -N$. The function $C_{u,v}(i)$ is also the aperiodic crosscorrelation function for the infinite sequences u and v that are the periodic extensions of **u** and **v**, respectively.

The sequence elements that are involved in the expression for the crosscorrelation function $C_{u,v}(i)$ for $0 \le i \le N - 1$ are illustrated in Fig. 11. In particular, the details in this illustration correspond to $i = 2$. Notice from the definition or from the illustration that $C_{u,v}(i) = C_{v,u}(-i)$. The aperiodic autocorrelation function is a special case of the aperiodic crosscorrelation function. If the vector **v** is replaced by the vector **u** in the above definition of the aperiodic crosscorrelation function, the result is the *aperiodic autocorrelation function $C_u(i)$* for the vector **u**.

It is shown in Ref. 3 that the continuous-time crosscorrelation functions can be determined from the aperiodic crosscorrelation function for the signature sequences x and y. In particular, the continuous-time crosscorrelation functions can be expressed in terms of functions that depend on the signature sequences only and functions that depend on the chip waveform only. The sequence-dependent parts can be expressed in terms of the partial autocorrelation functions for the chip waveform, which are

$$R_\psi(s) = \int_0^s \psi(t)\psi(t + T_c - s)dt, \qquad 0 \le s \le T_c \tag{23}$$

and

$$\hat{R}_\psi(s) = \int_s^{T_c} \psi(t)\psi(t - s)dt, \qquad 0 \le s \le T_c \tag{24}$$

These two functions are related by $R_\psi(s) = \hat{R}_\psi(T_c - s)$.

If τ is in the interval $iT_c \le \tau < (i + 1)T_c$ for $0 \le i \le N - 1$, the continuous-time partial crosscorrelation functions of Eq. (19) and Eq. (20) can be written as

$$R_{y,v}(\tau) = C_{y,x}(i - N)\hat{R}_\psi(\tau - iT_c) + C_{y,x}(i + 1 - N)R_\psi(\tau - iT_c) \tag{25}$$

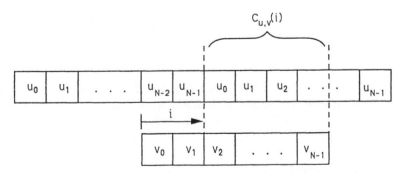

FIG. 11　Illustration of the crosscorrelation of $C_{u,v}(i)$ for $i = 2$.

and

$$\hat{R}_{y,v}(\tau) = C_{y,x}(i)\hat{R}_{\psi}(\tau - iT_c) + C_{y,x}(i + 1)R_{\psi}(\tau - iT_c) \qquad (26)$$

As illustrated by these expressions, it is the aperiodic crosscorrelation, not the periodic crosscorrelation, that determines the multiple-access interference between two DS spread-spectrum signals with symbol times that are not synchronized. Furthermore, they are impossible to synchronize in a distributed mobile, wireless communications network with multiple transmitters and multiple receivers (e.g., see the discussion of the distributed topology below).

In DS spread-spectrum multiple access, which is sometimes referred to as code-division multiple access (CDMA), there are a number of different spread-spectrum signals that occupy the same frequency band simultaneously (19). Rather than having just one interfering DS spread-spectrum signal, as we have discussed so far, there are several interfering signals. However, the interference caused by each signal is precisely of the type described above. If the system has K transmitted signals, the receiver that is matched to one of them experiences interference from the other $K - 1$ signals, and $K - 1$ aperiodic crosscorrelation functions are required to characterize this interference.

Not only is the multiple-access capability enhanced by careful selection of the spreading sequences, but the ability of the DS spread-spectrum system to provide protection against tone jamming and multipath is also improved by employing spreading sequences that have good correlation properties. The correlation properties that influence the performance of DS spread-spectrum communication systems are derived from the aperiodic autocorrelation and crosscorrelation functions for the sequences used by the transmitters in the system.

In order to examine the influence of sequence selection, we consider the processing that takes place in a correlation receiver for a spread-spectrum communication system. Suppose that a DS spread-spectrum system has N chips per data pulse. If binary data modulation is employed (e.g., binary PSK), each data pulse represents a binary channel symbol. Alternatively, each data pulse can represent a binary symbol sent on one component of a spread-spectrum signal that has both in-phase and quadrature modulation (e.g., QPSK or MSK).

Consider a communication receiver that is matched to the desired signal, and suppose the desired signal uses the signature sequence $\mathbf{x} = x_0, x_1, \ldots, x_{N-1}$ to spread the pulse that is being demodulated. Because of the asynchronism between the desired signal and the interference, two consecutive pulses of the interference signal affect the outcome of the demodulation of a single symbol of the desired signal. Suppose that the interference signal uses sequences $\mathbf{y} = y_0, y_1, \ldots, y_{N-1}$ and $\mathbf{z} = z_0, z_1, \ldots, z_{N-1}$ for these two consecutive pulses in the interference signal.

The signals seen by the correlator are illustrated in Fig. 12 for an offset of k times the chip duration between the desired signal and the interference ($0 \leq k < N$). It suffices to restrict attention to such offsets because the complete correlation function can be determined from the values of the correlations that correspond to time offsets that are integer multiples of a chip duration (3).

The receiver correlation process produces two correlations between segments of the sequences involved: a suffix of $\mathbf{y}$ is correlated against a prefix of $\mathbf{x}$, and a

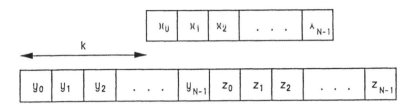

FIG. 12   Sequence alignment for an offset of k chips.

prefix of $\mathbf{z}$ is correlated against a suffix of $\mathbf{x}$. These two correlations are the two aperiodic crosscorrelations $C_{x,y}(k)$ and $C_{x,z}(k - N)$, respectively. Because the polarity of each pulse involved can be either positive or negative, the two aperiodic correlations may either add or subtract to produce the correlator output for the pulse being demodulated. The correlator output that results from the presence of interference is a constant times one of the quantities of the form

$$\pm C_{x,y}(k) \pm C_{x,z}(k - N)$$

and the choices for the two signs depend on the polarities of the pulses involved. Because we have no control over these polarities (they are determined by the data symbols being transmitted), it is desirable to choose the sequences in a way that guarantees both $|C_{x,y}(k) + C_{x,z}(k - N)|$ and $|C_{x,y}(k) - C_{x,z}(k - N)|$ are small.

Nearly all of the investigations of sequences with small correlation have been carried out for systems in which $\mathbf{z} = \mathbf{y}$; that is, the interference signal uses a single periodic sequence of period N, so each pulse of the interfering signal has the same spreading sequence. For investigations of DS spread-spectrum multiple-access capability in systems for which each signal has a signature sequence of period N, then $\mathbf{z} = \mathbf{y}$ and the correlations involved can be written in terms of the periodic crosscorrelation function $\theta_{x,y}$ and the odd crosscorrelation function $\hat{\theta}_{x,y}$ (6), which are related to the aperiodic crosscorrelation function by

$$\theta_{x,y}(k) = C_{x,y}(k) + C_{x,y}(k - N)$$

and

$$\hat{\theta}_{x,y}(k) = C_{x,y}(k) - C_{x,y}(k - N)$$

for $0 \leq k \leq N - 1$. For investigations of DS spread-spectrum acquisition and multipath communication in systems with sequences of period N, the sequences $\mathbf{x}$, $\mathbf{y}$, and $\mathbf{z}$ are all the same, and the correlations involved are the periodic and odd autocorrelations

$$\theta_x(k) = C_x(k) + C_x(k - N)$$

and

$$\hat{\theta}_x(k) = C_x(k) - C_x(k - N)$$

One of the choices to be made in sequence selection is the choice of the class of sequences to be considered. Maximal-length linear feedback sequences (*m*-sequences), Gold sequences, Kasami sequences, and several other classes of sequences have been investigated (6,20–23), primarily because of their good periodic autocorrelation and periodic crosscorrelation functions. Within each class, some sequences have better aperiodic correlation properties than others, and considerable reductions in the correlation values can be accomplished by careful selection of not only the class of sequences but also the particular sequences from that class that are employed as spreading sequences for a given system.

Even for a fixed set of sequences, there is more to be gained by optimizing the way in which the sequences are used in the spread-spectrum modulation. Consider the vectors that generate periodic sequences, and let T be the cyclic shift operator defined as follows (6,20): if $\mathbf{x} = x_0, x_1, x_2, \ldots, x_{N-1}$, then

$$T\mathbf{x} = x_1, x_2, x_3, \ldots, x_{N-1}, x_0$$

Similarly,

$$T^2\mathbf{x} = x_2, x_3, \ldots, x_{N-1}, x_0, x_1$$

$$T^3\mathbf{x} = x_3, \ldots, x_{N-1}, x_0, x_1, x_2$$

and so on.

For n in the range $0 \leq n \leq N - 1$, the vectors $T^n\mathbf{x}$ are the *phases* of $\mathbf{x}$ (6,20). For each choice of the pair of vectors $\mathbf{x}$ and $\mathbf{y}$, there are N different phases in which $\mathbf{x}$ can be placed and N different phases in which $\mathbf{y}$ can be placed, for a total of N^2 different orientations between the two vectors.

The operator T plays the same role for vectors as the operator S plays for infinite sequences. In particular, if the sequence x is the periodic extension of the vector $\mathbf{x}$, then the sequence $S^n x$ is the periodic extension of the vector $T^n\mathbf{x}$. As a result, if the sequence x is a periodic sequence with period N, the sequences $S^n x$, $0 \leq n \leq N - 1$, are referred to as the *phases* of x. In a DS spread-spectrum system for which there is one period of the signature sequence in each data symbol, the N different phases of $\mathbf{x}$ correspond to the N different possible orientations for the sequence within the data pulse. In effect, the N different phases represent N different starting points for the vector or N different orders in which the elements of $x_0, x_1, x_2, \ldots, x_{N-1}$ can be used within the data pulse.

Although the periodic correlation properties are the same for different phases of the sequences in a given set, the aperiodic (and odd) correlation properties differ greatly among the different phases (6). As a result, it is possible to select the phase of a sequence $\mathbf{x}$ to minimize some measure of the aperiodic autocorrelation or to select the phases of a pair of $\mathbf{x}$ and $\mathbf{y}$ to minimize some measure of the aperiodic crosscorrelation. The measures of aperiodic correlation that have been considered (6,20,21) include the maximum value of $|\hat{\theta}_{x,y}(k)|$ for all shifts k and all pairs of sequences $\mathbf{x}$ and $\mathbf{y}$, and the mean-squared value of $|\hat{\theta}_{x,y}(k)|$, averaged over all shifts k.

For a DS spread-spectrum multiple-access system with K transmitters, what is really desired is a set of K sequences for which any pair has a small aperiodic crosscorrelation. Just to illustrate the kind of performance improvements that can be obtained by sequence selection, the results of a performance evaluation are shown in Fig. 13 for a DS spread-spectrum multiple-access communications system with three equal-power received signals. The sequences employed are three nonreciprocal m-sequences of period 31 (6). The curve with the smallest error probabilities corresponds to having the three m-sequences in the phases

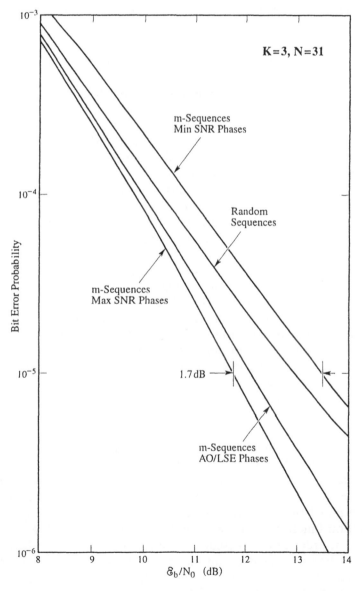

FIG. 13 Results of a performance evaluation for a direct-sequence spread-spectrum multiple-access communications system with three equal-power received signals.

that minimize a certain average interference parameter (20). These phases give the maximum value of a parameter known as the signal-to-noise ratio (SNR) parameter (3,9). The next lowest error probability is obtained for sequences in their AO/LSE phases, which are the phases that minimize a certain autocorrelation parameter (20).

The worst performance shown is for the three sequences in the phases that maximize the mean-square interference parameter (and therefore minimize the SNR parameter). The performance curve is included for these sequences as an illustration of how poor the performance can be if arbitrary phases are used for the m-sequences. For a bit error probability of 10^{-5} in a DS spread-spectrum multiple-access system that employs binary PSK modulation, the system with the sequences that minimize the SNR parameter requires that $\mathcal{E}/N_0$ be 1.7 dB larger than for the system with the sequences that maximize the SNR parameter. All three of these curves are for the same set of m-sequences, but the phases for the sequences are different. Presumably the performance gap would be even larger if the maximization and minimization of the SNR parameter were over a larger set of sequences.

The average performance for sequences selected at random is also shown in Fig. 13. The sequences are selected according to a uniform distribution on the set of all binary sequences of length N, and sequences for different transmitters are statistically independent. Investigations of the performance of DS spread-spectrum multiple-access communications with random spreading sequences have been published in Refs. 19, 24, and 25 and several other papers. As illustrated in Fig. 13, the average performance for a DS spread-spectrum multiple-access system that uses random sequences is neither as bad as for a system that employs m-sequences in their minimum SNR phases nor as good as for a system that employs m-sequences in their maximum SNR phases.

As mentioned above, the results of Fig. 13 are for a system in which the three received signals have equal power. The effects of multiple-access interference can be much more severe if an interfering signal has greater power than the desired signal, which is the unfavorable near-far condition mentioned previously. In DS spread-spectrum multiple access, it is important to control the power of the transmitted signals in order to limit the degradation caused by unfavorable near-far conditions.

Although there are some limited results on the performance of DS spread-spectrum multiple-access systems in which the sequence period is greater than the data pulse duration (26), very little optimization of sequences has been carried out for this situation. As a result, it is common to model the spreading sequences as random for systems with multiple data pulses per period of the spreading sequence. The performance obtained is then the average performance, averaged over the set of all sequences. It is also true that, as the number of sequences required by the system increases, the complexity of optimizing the crosscorrelation functions increases greatly. For a system with K transmitters, there are on the order of K^2 crosscorrelation functions that must be kept small. As a result, selections based on an autocorrelation criterion (e.g., AO/LSE) may be preferred or even necessary, even though the resulting choices will provide poorer performance than if the sequence selections are based on a crosscorrelation criterion. Of course, if the major source of interference is multipath rather than multiple-access interference, the optimization with respect to autocorrelation is preferred anyway (27–33). For protection against tone jamming, balance and other properties are also important, but many of these properties are more dependent on the class of sequences rather than the phases in which they are used. Additional information on the design of DS spread-spectrum waveforms and the selection of spreading sequences is given in Refs. 6, 19, and 23.

The selection of sequences also greatly influences the performance of DS spread-spectrum communications over a frequency-selective fading channel (8,34) for both a correlation receiver and a rake receiver (e.g., see Refs. 32 and 33). In a correlation receiver, the discrimination against multipath components is similar to the discrimination against multiple-access interference (28). If the received multipath components are of comparable power levels, use of a rake receiver should be considered. In a rake receiver, an attempt is made to utilize the energy in the multipath components rather than simply discriminate against them, but sequence selection is just as important. The rake receiver has multiple correlation receivers that track and demodulate different multipath components.

As illustrated in Ref. 32, proper sequence selection can permit a reduction in the number of demodulators in the rake receiver with no increase in error probability. Thus, sequence selection can result in reduced complexity requirements in the receiver, as well as decreased power requirements in the transmitter. For channels in which the maximum delay spread is less than the duration of the data pulse, a periodic sequence can be employed with one period per data pulse. If the multipath spread is larger than the data pulse duration, it is necessary for the sequence period to be longer than the data pulse, and this complicates the optimization.

As described in a previous section, the mobile cellular CDMA system (i.e., the IS-95 system) has the feature that the signature sequences in the composite signal transmitted by the base station are orthogonal. This orthogonality implies there is no multiple-access interference at the mobile station if channel conditions are perfect. The cellular system is an example of a *star topology*, which is the term used for a network in which all communication takes place between a single terminal and each of two or more other terminals. That is, one designated terminal is transmitting to multiple receivers or it is receiving from multiple transmitters. A radio network with star topology is illustrated in Fig. 14. A single cell of a cellular telephone system is an example of a star topology: the base station transmits to the mobile phones within the cell on the forward link and receives from the mobile phones on the reverse link.

If a network of terminals has a star topology and the channel is ideal (e.g., no multipath propagation), it is possible to control the arrival times of the received signals in such a way that the signature sequences for different signals are perfectly synchronized at each receiver. Since the data signals to be transmitted to the mobile phones of the cellular system are modulated and combined in a single system within the base station, it is easy to align the chips of the signature sequences for the signals transmitted by the base station. In the mobile cellular CDMA system, all of the data signals for the forward link are modulated onto the same RF carrier within the base station transmitter.

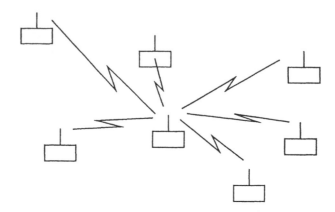

FIG. 14   A radio network with a star topology.

Mobile cellular CDMA systems typically must contend with multipath propagation on the channel. As a result of multipath propagation and other sources of channel distortion, the received versions of the different forward-link signals are not orthogonal, in spite of the star topology, so there is some multiple-access interference in the forward link of the mobile cellular CDMA system.

The ability of a DS spread-spectrum system to benefit from the use of orthogonal signature sequences depends critically on being able to maintain time synchronism among the different signature sequences during the modulation, transmission, propagation, and reception of the signals. As discussed above, in the absence of multipath and other channel disturbances, this time synchronism can be accomplished in the forward link of the mobile cellular CDMA system. For mobile radio systems and networks in which different data signals are modulated onto different RF carriers at different locations throughout the network, it is rarely feasible to provide the time synchronism needed to guarantee orthogonality among the transmitted spread-spectrum signals. Even if there is precise timing among the transmitted signals, it is still impossible to provide orthogonality among the received spread-spectrum signals at each of the terminals in the network unless the network has a very special topology, such as the star topology. Just the fact that the distances between the different pairs of transmitters and receivers can vary greatly throughout the network is enough to make it impossible to have all of the signature sequences aligned at each of the receivers.

A mobile, distributed, spread-spectrum multiple-access radio network is an example of an application in which precise synchronism among the signals arriving at each receiver is not possible. There is not a single base station (there may be no base stations at all), the network does not have a star topology, and different communications links have significantly different propagation delays. Furthermore, the delays for different links are changing independently of each other as a result of the movement of the terminals in the network. A distributed network topology is illustrated in Fig. 15.

In systems and networks for which precise synchronism among the received signals is impossible or impractical, the *aperiodic* correlation properties of the

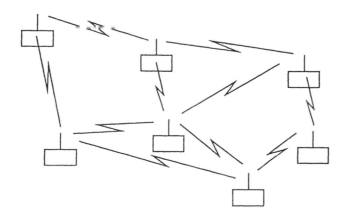

FIG. 15   A radio network with a distributed topology.

sequences determine the multiple-access performance of the system. Unfortunately, the standard sets of orthogonal sequences, such as the sequences employed in the forward link for the IS-95 mobile cellular CDMA system, have very poor aperiodic correlation properties. As a result, they are not good choices for signature sequences for other types of spread-spectrum systems, such as distributed radio networks.

For applications to GPS receivers, both the autocorrelation and crosscorrelation properties of the sequences are important. The autocorrelation functions for the sequences play a role in the acquisition performance of the receiver and in the ability of the receiver to discriminate against multipath interference. The crosscorrelation functions among the sequences employed by the different satellites determine the multiple-access capability of the GPS system. As mentioned in a previous section, all GPS receivers have access to the C/A component of the first navigation signal, whereas access to the P component may be restricted. For this reason, and in order to simplify the presentation, we limit the discussion of the operation of the GPS receiver to the C/A component only. The same principles apply to receivers that process and utilize the P component.

The C/A components of the navigation signals transmitted by different satellites employ different sequences from a set of Gold sequences of period 1023. As such, their periodic crosscorrelation functions take on the values -1, $+63$, and -65 only. The incoming signal at a GPS receiver consists of the sum of the navigation signals from all GPS satellites that are in view of the receiver. The first navigation signal from each satellite has the same carrier frequency f_1, so there is multiple-access interference at the output of each correlator whenever two or more satellites are in view. For the C/A component of the first navigation signal, the fact that there are 20 periods of each Gold sequence per data pulse means that the multiple-access interference due to C/A components from other satellites is determined largely by the periodic crosscorrelation functions. The correlation interval consists of 20 periods of the 1023-chip sequence, so for each interfering signal there are at least 19 periodic crosscorrelations and at most one odd crosscorrelation. If the appropriate pair of data bits agree in polarity, there are 20 periodic crosscorrelations and no odd crosscorrelations.

Similarly, multipath discrimination is determined largely by the periodic autocorrelation function for the Gold sequence used by the first navigation signal. The peak in the periodic autocorrelation function is 1023, of course, which occurs at the time offset of zero, and the maximum periodic autocorrelation side-lobe magnitude is the same as the maximum magnitude of the periodic crosscorrelation function, which is 65. The ratio of the maximum periodic autocorrelation side lobe to the peak for the periodic autocorrelation function is $65/1023 \approx 0.0635$, which is also the ratio of the maximum crosscorrelation to the peak autocorrelation.

In the GPS application, the *normalized multiple-access interference* is the ratio of the multiple-access interference to the desired signal if each is measured at the output of a correlator matched to the C/A spreading waveform of the desired signal. This ratio is equivalent to the maximum of the normalized multiple-access interference for two equal-power, first navigation signals arriving at a GPS receiver. Because this is the ratio of two voltages, and because

$$20 \log_{10}(65/1023) \approx -23.9 \, \text{dB}$$

the maximum multiple-access interference is nearly 24 dB below the desired signal.

The *pseudorange* is the product of the velocity of light and the difference between the time that a given chip is transmitted, as measured and reported by the satellite, and the time that it is received, as measured by the GPS receiver. For several reasons, including the fact that the receiver's clock is not in synchronism with the satellite's clock, this is not the actual range. Because the satellites' positions are known, and this information is obtained by the GPS receiver from the data signal, the pseudorange measurements to four satellites (which provide four equations) are sufficient to determine the three coordinates of the receiver's position and resolve the uncertainty in the receiver's clock (four unknowns).

If the error in the receiver's clock is denoted by Δ, the corresponding range error is $r = c\Delta$, where c is the velocity of light. If μ_i is the pseudorange to the ith satellite, which has reported it is at position (X_i, Y_i, Z_i), the four equations are

$$\mu_1 + r = [(X_1 - x)^2 + (Y_1 - y)^2 + (Z_1 - z)^2]^{1/2}$$

$$\mu_2 + r = [(X_2 - x)^2 + (Y_2 - y)^2 + (Z_2 - z)^2]^{1/2}$$

$$\mu_3 + r = [(X_3 - x)^2 + (Y_3 - y)^2 + (Z_3 - z)^2]^{1/2}$$

and

$$\mu_4 + r = [(X_4 - x)^2 + (Y_4 - y)^2 + (Z_4 - z)^2]^{1/2}$$

The four unknowns are r, the error in the receiver's clock, and x, y, and z, the coordinates of the receiver's position. In practice, there are several other variables that introduce errors in the determination of the receiver's position, including small errors in satellite clocks and satellite positions and the delays encountered as the signals propagate through the ionosphere (14,15).

The GPS receiver must track the spread-spectrum signals as the satellites move in their orbits (the receiver may be moving as well). This can be accomplished by a tracking subsystem that employs an appropriate modification of the baseband delay-lock loop illustrated in Fig. 7. The modification is necessary because the received signal is a data-modulated RF spread-spectrum signal, and the data symbols and the phase and frequency of the RF carrier are all unknown. Generally, this modification results in a noncoherent delay-lock tracking loop (16,35).

Frequency-Hop Spread Spectrum

Signals with very large bandwidths can be generated by a form of spread-spectrum modulation known as *frequency hopping*. The basic feature of frequency-hop spread spectrum is that the carrier frequency of a communications signal is changed or *hopped* over some predetermined set of frequencies. A communications signal with bandwidth W that is hopped over a set of q frequencies can occupy a total bandwidth of as much as qW. The resulting spread-spectrum signal is referred to as a *frequency-hop signal*. In the frequency-hop form of spread spectrum, the instantaneous bandwidth of the signal may be no more than the original communications signal, even though the long-term spectral occupancy increases by as much as a factor of q.

The RF band that is available for the frequency-hop spread-spectrum signal is divided into q subbands that are referred to as *frequency slots*. It is not necessary for the frequency slots to be contiguous, so it is possible to reserve some portions of the spectrum for use by other signals. The times at which the carrier frequency is changed are referred to as *hop epochs*, and the time intervals between consecutive hop epochs are called *hop intervals*. For the standard frequency-hop signal, the hop intervals all have the same length, but this is not necessary, and there are good reasons in some military systems for the use of variable-length hop intervals.

Data can be transmitted during all or part of each hop interval. The time interval during which a data signal is actually transmitted is called the *dwell interval* for the frequency-hop signal. There are various reasons, including protection against certain types of jamming, why it may be desirable to have the dwell interval smaller than the hop interval for some applications. In fact, it may be necessary to do so in order to provide time for the settling of the frequency synthesizer at the beginning of each hop interval.

In order to give a mathematical description of frequency-hop signaling, we consider a data communication signal of the form

$$v(t) = \sqrt{2}\, m(t)\cos[2\pi f_d t + \theta(t) + \varphi]$$

where $m(t)$ represents amplitude modulation and $\theta(t)$ represents phase modulation. One of the two modulation signals can be held constant in order to achieve amplitude or phase modulation alone. A frequency-hop signal that conveys the same information as $v(t)$ is obtained by replacing the fixed frequency f_d by a time-varying frequency. A system to accomplish this is illustrated in Fig. 16. In this model, the input to the frequency synthesizer is the sequence (ξ_j), and the output of the frequency synthesizer is the signal

$$c(t) = 2\cos[2\pi\psi(t)t + \alpha(t)]$$

where $\psi(t) = \xi_j$ during the jth hop interval. The signal $\alpha(t)$ represents the phase shifts that may be introduced by the frequency synthesizer as it switches frequencies at the beginning of each hop interval. The sequence (ξ_j) is referred to as the *hopping pattern* for the frequency-hop signal. In most applications, the hopping pattern is some type of pseudorandom sequence, but it is not necessarily generated by a linear feedback shift register.

The purpose of the band-pass filter in Fig. 16 is to remove the lower sideband of the signal generated by the mixer. The output of the band-pass filter is the frequency-hop signal

$$s(t) = \sqrt{2}\, m(t)\cos[2\pi\xi(t)t + \theta(t) + \varphi]$$

The time-varying carrier frequency for the frequency-hop signal is given by

$$\xi(t) = f_d + \psi(t)$$

In a frequency-hop system in which each hop interval is of duration T_h, the *hopping rate* is $1/T_h$. Frequency-hop spread-spectrum signals are generally divided into two categories according to the relationship between their hopping rates and data symbol transmission rates. If one or more data symbols are transmitted in each dwell interval, the frequency-hop signal is referred to as a *slow-frequency-hop* (SFH) signal; otherwise, it is a *fast-frequency-hop* (FFH) signal. In fast-frequency hopping, the transmission of each data signal requires at least two dwell intervals. Stated another way, the hopping rate is greater than the data symbol rate for FFH signals.

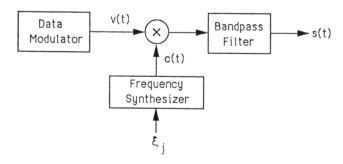

FIG. 16   Frequency-hop modulator.

On the other hand, it is not uncommon for tens or hundreds of data symbols to be transmitted in each dwell interval in a SFH system, and this corresponds to a hopping rate that is one or two orders of magnitude smaller than the data symbol rate. By definition, the hopping rate does not exceed the data symbol rate in an SFH system. The bandwidth of a FFH signal is determined primarily by the hopping rate and the number of frequency slots, but the bandwidth of a SFH signal is determined by the data symbol rate and the number of frequency slots. An SFH signal with six data symbols per hop is illustrated in Fig. 17.

The scenario illustrated in Fig. 18 exemplifies one of the advantages of frequency-hop modulation: several signals can coexist in the same frequency band, yet satisfactory performance can be obtained for each (36). Shown in Fig. 18 are two frequency-hop signals and a signal that occupies a fixed set of frequency slots. From the point of view of the receiver that is attempting to receive and demodulate the signal shown in Fig. 17, one of the frequency-hop signals of Fig. 18 represents the *desired signal*, the second represents *frequency-hop multiple-access interference*, and the third represents *partial-band interference*.

As illustrated in the second dwell interval of the desired signal, it may be that two frequency-hop signals occupy the same frequency slot for all or part of

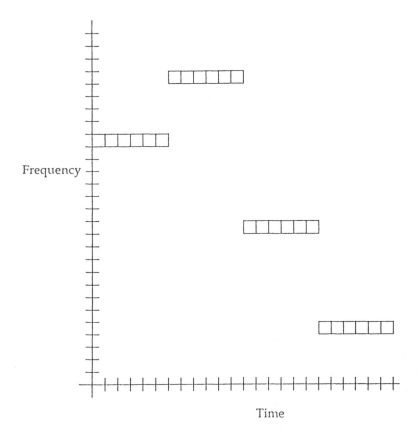

FIG. 17   A slow-frequency-hop (SFH) signal.

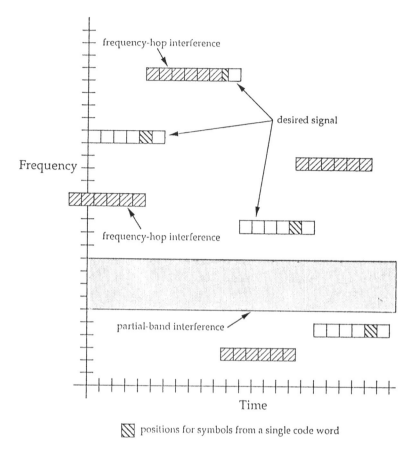

FIG. 18 Two slow-frequency-hop signals and partial-band interference.

one or more dwell intervals during the transmission of a message. In most
applications, the error probabilities are quite large for both signals during such
an event. The partial-band interference may represent a communication signal
operating at a fixed frequency, or, in a military system, it may represent partial-
band jamming. The frequency occupancy of the partial-band interference may
change from time to time, but such changes are normally much slower than the
hopping rate of the frequency-hop signals. If a symbol of the desired signal is
transmitted in a frequency slot that contains either multiple-access interference
or partial-band interference, we say that the symbol has been *hit*.

Frequency-hop signaling differs from direct-sequence spread spectrum in
that the former does not provide any protection for individual symbols that are
transmitted in a frequency slot that contains interference. As a result, frequency-hop signaling is vulnerable to various forms of partial-band interference
unless suitable error-control coding is employed (37). The primary benefit of
frequency-hop signaling is the frequency diversity that it provides, and this
frequency diversity can be exploited to compensate for the presence of interference in a fraction of the frequency slots.

The necessity for error-control coding in SFH transmission can be illustrated with a simple example. Suppose that for a particular binary data transmission system a bit error probability of 10^{-3} is required, and suppose the frequency band used by the SFH signal consists of 100 frequency slots. Assume each frequency slot is used with the same probability (i.e., the probability that the signal is in a particular slot at a given time is 0.01), and assume that very strong interference is present in one or more frequency slots (i.e., interference that has considerably more power than the desired signal). In this situation, the probability of error is no smaller than approximately 5×10^{-3} even if the interference is confined to a single slot and there is no interference at all in the other 99 frequency slots. The only way to improve this situation is to use error-control coding to correct some of the errors, because it is usually not possible to avoid use of the frequency slots that have interference. In most applications, the transmitter does not know the spectral occupancy of the interference, and this occupancy may be changing with time (e.g., the interference might be another SFH signal).

An error-correcting code may be used to permit correction of the errors that occur when data symbols are transmitted in frequency slots with severe interference. Several block codes and convolutional codes are suitable for this purpose. As illustrated in Fig. 18, the encoded symbols for a block code can be transmitted in such a way that no two code symbols from a given codeword are in the same dwell interval. In this way, one hit can cause at most one error in any single codeword.

One of the natural choices for error control in SFH systems is the Reed-Solomon code (38). A Reed-Solomon code of block length n that has k information symbols can correct any set of $(n - k)/2$ or fewer errors. If it is possible for the receiver to determine which symbols have been hit, the performance can be increased greatly by erasing some or all such symbols (39). The same Reed-Solomon code can correct $n - k$ or fewer erasures if there are no errors. More generally, it can correct any combination of t errors and e erasures provided $2t + e \leq n - k$. The decoder need not have perfect knowledge of which symbols are hit in order to obtain significant performance gains by erasing some of the received symbols (39). Many different methods have been developed for deciding which of the received symbols should be erased, and the performance of each of these methods has been evaluated (e.g., Refs. 39–41).

For fast-frequency hopping, it is possible to achieve acceptable performance in some communications environments without the use of error-control coding. However, two important points should be kept in mind: FFH can be viewed as a simple repetition code, and the use of more powerful error-control coding methods can improve performance greatly even for FFH systems.

The simplest method for achieving diversity in FFH systems is to send each symbol in each of several different frequency slots and use a simple majority vote among the corresponding decisions made by the demodulator. This is an example of a *hard-decision* combining scheme. *Soft-decision* combining schemes, such as summing the squares of the soft-decision demodulator outputs, have also been proposed (e.g., Ref. 42). The combination of error-control coding and simple frequency diversity provides tremendous protection against partial-band interference (42–45).

The SINCGARS mobile frequency-hop radio (46) operates in the portion of the very-high-frequency (VHF) band from approximately 30 MHz to approximately 88 MHz. The number of frequency slots available is 2320, and the width of each slot is 25 kilohertz (kHz). Digital data or digitized voice signals can be sent using a form of binary frequency-shift keying. The data rate is variable over a range from 75 b/s to 16 kb/s, and the hopping rate is approximately 100 hops per second. Enhanced versions of the SINCGARS radio use side information developed in the receiver to erase unreliable symbols, and Reed-Solomon coding is employed to correct errors and erasures. The original applications of SINCGARS include fully connected networks for voice communications. With the addition of suitable routing protocols, the SINCGARS radios can be employed in a fully distributed store-and-forward packet communication network.

References

1. Pursley, M. B., *Introduction to Digital Communications*, Addison-Wesley, Reading, MA, 1998.
2. Scholtz, R. A., The Origins of Spread-Spectrum Communications, *IEEE Trans. Commun.*, COM-30:822–852 (May 1982).
3. Pursley, M. B., Spread-Spectrum Multiple-Access Communications. In *Multi-User Communication Systems* (G. Longo, ed.), Springer-Verlag, Vienna, 1981, pp. 139–199.
4. Golomb, S. W., et al., *Digital Communications with Space Applications*, Prentice-Hall, Englewood Cliffs, NJ, 1964.
5. Golomb, S. W., *Shift Register Sequences*, Holden-Day, San Francisco, 1967.
6. Sarwate, D. V., and Pursley, M. B., Crosscorrelation Properties of Pseudorandom and Related Sequences, *Proc. IEEE*, 68:593–619 (May 1980).
7. Cooper, G. R., and McGillem, C. D., *Probabilistic Methods of Signal and System Analysis*, 2d ed., Holt, Rinehart, and Winston, New York, 1986, pp. 297–300.
8. Stein, S., Communication over Fading Radio Channels. In *The Froehlich/Kent Encyclopedia of Telecommunications*, Vol. 3 (F. E. Froehlich and A. Kent, eds.), Marcel Dekker, New York, 1992, pp. 261–303.
9. Pursley, M. B., Performance Evaluation for Phase-Coded Spread-Spectrum Multiple-Access Communication—Part I: System Analysis, *IEEE Trans. Commun.*, COM-25:795–799 (August 1977).
10. Fifer, W. C., and Bruno, F. J., The Low-Cost Packet Radio, *Proc. IEEE*, 75:33–42 (January 1987).
11. Telecommunications Industry Association (TIA), *TIA/EIA Interim Standard Mobile Station-Base Station Compatibility Standard for Dual-Mode Wideband Spread Spectrum Cellular System*, TIA, Washington, DC, July 1993.
12. Viterbi, A. J., *CDMA: Principles of Spread Spectrum Communication*, Addison-Wesley, Reading, MA, 1995.
13. Getting, I. A., The Global Position System, *IEEE Spectrum*, 36–47 (December 1993).
14. Enge, P. K., The Global Position System: Signals, Measurements, and Performance, *Intl. J. Wireless Information Networks*, 1(2):83–105 (1994).
15. Spilker, J. J., Jr., GPS Signal Structure and Performance Characteristics, *Navigation: J. Institute of Navigation*, 25(2):121–146 (Summer 1978).

16. Spilker, J. J., Jr., *Digital Communications by Satellite*, Prentice-Hall, Englewood Cliffs, NJ, 1977.

17. Peterson, W. W., and Weldon, E. J., Jr., *Error-Correcting Codes*, 2d ed., MIT Press, Cambridge, MA, 1972.

18. Gold, R., Optimal Binary Sequences for Spread Spectrum Multiplexing, *IEEE Trans. Information Theory*, IT-13:619–621 (October 1967).

19. Pursley, M. B., The Role of Spread Spectrum in Packet Radio Networks, *Proc. IEEE*, 75:116–134 (January 1987).

20. Pursley, M. B., and Roefs, H. F. A., Numerical Evaluation of Correlation Parameters for Optimal Phases of Binary Shift-Register Sequences, *IEEE Trans. Commun.*, COM-27:1597–1604 (October 1979).

21. Skaug, R., Numerical Evaluation of the Nonperiodic Autocorrelation Parameter for Optimal Phases of Maximal Length Sequences, *Proc. IEE*, Part F, 127:230–237 (1980).

22. Garber, F. D., and Pursley, M. B., Optimal Phases of Maximal-Length Sequences for Asynchronous Spread-Spectrum Multiplexing, *Elec. Lett.*, 16:756–757 (September 1980).

23. Skaug, R., and Hjelmstad, J. F., *Spread Spectrum in Communication*, Peregrinus, London, 1985.

24. Roefs, H. F. A., and Pursley, M. B., Correlation Parameters of Random Binary Sequences, *Elec. Lett.*, 13(16):488–489 (August 1977).

25. Lehnert, J. S., and Pursley, M. B., Error Probability for Binary Direct-Sequence Spread-Spectrum Communications with Random Signature Sequences, *IEEE Trans. Commun.*, COM-35:87–98 (January 1987).

26. Pursley, M. B., Sarwate, D. V., and Basar, T. U., Partial Correlation Effects in Direct-Sequence Spread-Spectrum Multiple-Access Communication Systems, *IEEE Trans. Commun.*, COM-32:567–573 (May 1984).

27. Turin, G. L., Introduction to Spread-Spectrum Antimultipath Techniques and Their Application to Urban Digital Radio, *Proc. IEEE*, 68:328–353 (March 1980).

28. Pursley, M. B., Effects of Specular Multipath Fading on Spread-Spectrum Communications. In *New Concepts in Multi-User Communications* (J. K. Skwirzynski, ed.), NATO Advanced Study Institute—Series E, Sijthoff and Noordhoff International Publishers, Alphen aan den Rijn, Netherlands, 1981, pp. 481–505.

29. Geraniotis, E. A., and Pursley, M. B., Performance of Coherent Direct-Sequence Spread-Spectrum Communications over Specular Multipath Fading Channels, *IEEE Trans. Commun.*, COM-33:502–508 (June 1985).

30. Lehnert, J. S., and Pursley, M. B., Multipath Diversity Reception of Spread-Spectrum Multiple-Access Communications, *IEEE Trans. Commun.*, COM-35:1189–1198 (November 1987).

31. Noneaker, D. L., and Pursley, M. B., On the Chip Rate of CDMA Systems with Doubly Selective Fading and Rake Reception, *IEEE J. Sel. Areas Commun.*, 12:853–861 (June 1994).

32. Noneaker, D. L., and Pursley, M. B., Selection of Spreading Sequences for Direct-Sequence Spread-Spectrum Communications over a Doubly Selective Fading Channel, *IEEE Trans. Commun.*, 42(12):3171–3177 (December 1994).

33. Noneaker, D. L., and Pursley, M. B., The Effects of Sequence Selection on DS Spread-Spectrum with Selective Fading and Rake reception, *IEEE Trans. Commun.*, 44(2):229–237 (February 1996).

34. Stein, S., *Communication Systems and Techniques*, Part 3 (M. Schwartz, W. R. Bennett, and S. Stein), McGraw-Hill, New York, 1966, pp. 561–584.

35. Holmes, J. K., *Coherent Spread Spectrum Systems*, Wiley, New York, 1982.

36. Pursley, M. B., Frequency-Hop Transmission for Satellite Packet Switching and

Terrestrial Packet Radio Networks, *IEEE Trans. Information Theory*, IT-32:652–667 (September 1986).

37. Viterbi, A. J., and Jacobs, I. M., Advances in Coding and Modulation for Noncoherent Channels Affected by Fading, Partial Band, and Multiple-Access Interference. In *Advances in Communication Systems*, Vol. 4, Academic Press, New York, 1975, pp. 279–308.

38. Berlekamp, E. R., The Technology of Error-Correcting Codes, *Proc. IEEE*, 68: 564–593 (May 1980).

39. Pursley, M. B., Reed-Solomon Codes in Frequency-Hop Communications. In *Reed-Solomon Codes and Their Applications* (S. B. Wicker and V. K. Bhargava, eds.), IEEE Press, New York, 1994, Chapter 8.

40. Baum, C. W., and Pursley, M. B., Bayesian Methods for Erasure Insertion in Frequency-Hop Communications with Partial-Band Interference, *IEEE Trans. Commun.*, 40:1231–1238 (July 1992).

41. Viterbi, A. J., A Robust Ratio-Threshold Technique to Mitigate Tone and Partial Band Jamming in Coded MFSK Systems, *1982 IEEE Military Commun. Conf. Record*, 22.4.1–22.4.5 (October 1982).

42. Simon, M. K., Omura, J. K., Scholtz, R. A., and Levitt, B. K., *Spread Spectrum Communications*, Vols. 1–3, Computer Science Press, Rockville, MD, 1985.

43. Pursley, M. B., Coding and Diversity for Channels with Fading and Pulsed Interference, *Proc. 1982 Conf. Info. Sci. Sys.* (Princeton University), 413–418 (March 1982).

44. Pursley, M. B., and Stark, W. E., Performance of Reed-Solomon Coded Frequency-Hop Spread-Spectrum Communications in Partial-Band Interference, *IEEE Trans. Commun.*, COM-33:767–774 (August 1985).

45. Torrieri, D. J., *Principles of Military Communication Systems*, Artech, Dedham, MA, 1981.

46. Kammer, D. E., SINCGARS—The New Generation Combat Net Radio System, *Proc. Tactical Commun. Conf.*, 64–72 (April 1986).

MICHAEL B. PURSLEY

Standard Interfaces for Wireless Communications

Introduction

Work on cellular and personal communications systems (or services) (PCS) in the United States has progressed in six different standards groups:

- *T1P1*. This committee is under the Alliance for Telecommunications Industry Solutions, Inc. (ATIS), and is responsible for PCS services requirements.
- *TR-45*. This committee is under the Telecommunications Industry Association (TIA) and is responsible for cellular services and protocols.
- *TR-46*. This committee is under the TIA and is responsible for PCS services and protocols.
- *T1S1*. This is an ATIS body responsible for the signaling protocols and is undertaking the role of the upper layer signaling protocols between various elements of the PCS system.
- *T1M1*. This is an ATIS body responsible for operations, administration, maintenance, and provisioning (OAM&P) and is undertaking the role of the OAM&P services and protocols for PCS.
- *The Joint Technical Committee of T1P1 and TR-46*. This committee is the result of a cooperative effort between T1P1 and TR-46 and has the responsibilities of developing the air interface requirements for PCS.

Clearly, with six U.S. standards groups examining the problem, there is overlap among the work.

In Europe, the work on the Global System for Mobile Communication (GSM) has progressed in a single group, European Telecommunication Standardization Institute (ETSI). The GSM is part of the Conference European Postal and Telecommunications Administrations (CEPT) and is based in Brussels. GSM commonly refers to both the group and the Pan-European cellular network project. Currently, all GSM activities occur under direction of the ETSI, located in Paris.

In this article, we examine three reference models for wireless communications (cellular and PCS) and describe the standard interfaces that they use.

Cellular and Personal Communications Services Reference Models in the United States

Key to the North American cellular and PCS systems is the use of a common reference model. Both the TIA (in TR-45 and TR-46) and Committee T1 (in

429

T1P1) have a reference model, but each model can be converted into the other one. All of the North American systems follow the reference model. The names of each of the network elements are similar, and some of the functionality is partitioned differently between the models. The main difference between the two reference models is how mobility is managed. Mobility is the capability for users to place and receive calls in systems other than their home system. In the T1 Reference Model, the user data and the terminal data are separate, thus users can communicate with the network via different radio personal terminals. In the TIA Reference Model, only terminal mobility is supported. A user can place or receive calls at only one terminal (the one the network has identified as owned by the user). The TIA functionality is migrating toward independent terminal and user mobility, but all aspects of it are not currently supported. Although these models are for the PCS, they apply equally well to cellular systems.

Telecommunications Industry Association Reference Model

The main elements of the TIA Reference Model (Fig. 1) are the personal station (PS), radio system (RS), mobile switching center (MSC), home location register (HLR), data message handler (DMH), visited location register (VLR), authentication center (AC), equipment identity register (EIR), operations system (OS), interworking function (IWF), and external networks (1).

The personal station (PS) terminates the radio path on the user side and

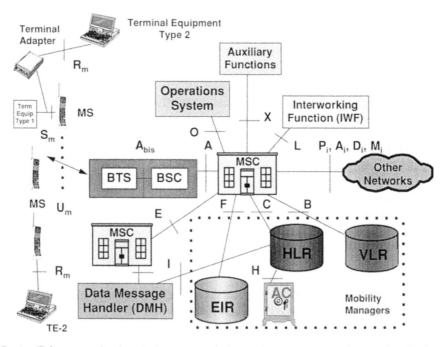

FIG. 1 Telecommunications Industry Association Reference Model (AC = authentication center; BSC = base-station controller; BTS = base transceiver system; EIR = equipment identity register; HLR = home location register; MSC = mobile switching center; VLR = visitor location register).

enables the user to gain access to services from the network. The PS can be a stand-alone device or can have other devices (e.g., personal computers, facsimile machines, etc.) connected to it.

The radio system (RS), often called the base station, terminates the radio path and connects to the personal communications switching center. The RS is often segmented into the base transceiver system (BTS) and the base-station controller (BSC). The BTS consists of one or more transceivers placed at a single location and terminates the radio path on the network side. The BTS may be co-located with a BSC or may be independently located. The BSC is the control and management system for one or more BTSs. The BSC exchanges messages with both the BTS and the MSC. Some signaling messages may pass through the BSC transparently.

The mobile switching center is an automatic system that interfaces the user traffic from the wireless network to the wireline network or other wireless networks. The MSC is also often called the mobile telephone switching office (MTSO) or the personal communications switching center (PCSC).

The home location register (HLR) is the functional unit used for management of mobile subscribers by maintaining all subscriber information (e.g., ESN [electronic serial number], DN [directory number], IMSI [international mobile subscriber identity], user profiles, current location, etc.). The HLR may be co-located with a MSC, be an integral part of the MSC, or may be independent of the MSC. One HLR can serve multiple MSCs, or an HLR may be distributed over multiple locations.

The data message handler (DMH) is used for generating data used for billing of the customer.

The visited location register (VLR) is linked to one or more MSCs. The VLR is the functional unit that dynamically stores subscriber information (e.g., ESN, directory number, user profile information, etc.) obtained from the user's HLR when the subscriber is located in the area covered by the VLR. When a roaming mobile station (MS) enters a new service area covered by a MSC, the MSC informs the associated VLR about the PS by querying the HLR after the PS goes through a registration procedure.

The authentication center (AC) manages the authentication or encryption information associated with an individual subscriber. As of this writing, the details of the operation of the AC have not been defined. The AC may be located within an HLR or MSC or may be located independently of both.

The equipment identity register (EIR) provides information about the personal station for record purposes. As of this writing, the details of the operation of the EIR have not been defined. The EIR may be located within an MSC or may be located independently of it.

The operations system (OS) is responsible for overall management of the wireless network.

The interworking function (IWF) enables the MSC to communicate with other networks when protocol conversions are necessary.

External networks are other communications networks, the public switched telephone network (PSTN), the Integrated Services Digital Network (ISDN), the Public Land Mobile Network (PLMN), and the packet-switched public data network (PSPDN).

The following interfaces are defined between the various elements of the system:

- *RS–MSC interface* (A interface). The interface between the radio system and the MSC supports signaling and traffic (both voice and data). The A interface protocols have been defined using SS7 (Signaling System No. 7), ISDN Basic Rate Interface/Primary Rate Interface (BRI/PRI) and frame relay.
- *BTS–BSC interface* (A-bis). If the radio system is segmented into a BTS and BSC, this internal interface is defined.
- *MSC–PSTN interface* (A_i). This interface is defined as an analog interface using either dual-tone multiple frequency (DTMF) signaling or multifrequency signaling.
- *MSC–VLR interface* (B interface). This interface is defined in the TIA Interim Standard 41 (IS-41) Protocol specification.
- *MSC–HLR interface* (C interface). This interface is defined in the TIA IS-41 Protocol specification.
- *HLR–VLR interface* (D interface). This interface is the signaling interface between an HLR and a VLR and is based on SS7. It is currently defined in the TIA IS-41 Protocol specification.
- *MSC–ISDN interface* (D_i interface). This is the digital interface to the public telephone network and is a T1 interface (24 channels of 64 kilobits per second [kb/s]) and uses Q.931 signaling.
- *MSC–MSC interface* (E interface). This interface is the traffic and signaling interface between wireless networks. It is currently defined in the TIA IS-41 Protocol specification.
- *MSC–EIR interface* (F interface). Since the EIR is not yet defined, the protocol for this interface is not defined.
- *VLR–VLR interface* (G interface). When communication is needed between VLRs, this interface is used. It is defined by TIA IS-41.
- *HLR–authentication-center interface* (H interface). The protocol for this interface is not defined.
- *DMH–MSC interface* (I interface).
- *MSC–IWF interface* (L interface). This interface is defined by the interworking function.
- *MSC–PLMN interface* (M_i interface). This interface is to another wireless network.
- *MSC–OS interface* (O interface). This is the interface to the operations systems. It is currently being defined in ATIS's standard body T1M1.
- *MSC–PSPDN interface* (P_i interface). This interface is defined by the packet network that is connected to the MSC.
- *Terminal adapter–terminal equipment (TA–TE) interface* (R interface). These interfaces will be specific for each type of terminal that will be connected to a PS.
- *ISDN–TE interface* (S interface). This interface is outside the scope of PCS and is defined within the ISDN system.
- *RS–PS interface* (U_m interface). This is the air interface.
- *PSTN–DCE interface* (W interface). This interface is outside the scope of PCS and is defined within the PSTN system.

- *MSC–AUX* (X interface). This interface depends on the auxiliary equipment connected to the MSC.

T1 Personal Communications Systems Reference Architecture

The T1 architecture (Fig. 2) is similar to the TIA model, but has some differences (2).

The following elements are defined in the model:

- *Radio personal terminal* (RPT). The RPT is identical to the PS of the TIA model.
- *Radio port* (RP). The RP is identical to the BTS of the TIA model.
- *Radio port intermediary* (RPI). The RPI provides an interface between one or more RPs and the RPC. The RPI allocates radio channels and may control handoffs. It is dependent on the air interface.

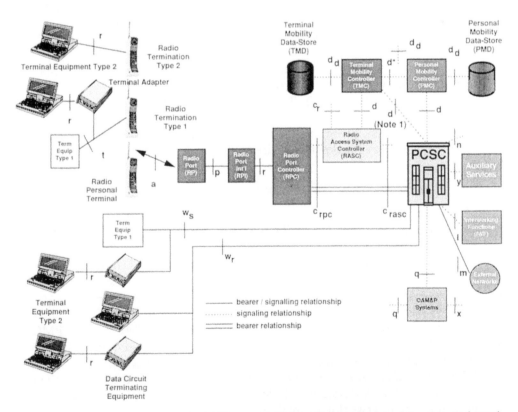

FIG. 2 T1 Reference Model (OAM&P = operations, administration, maintenance, and provisioning; PCSC = personal communications switching center). Note 1: The requirements associated with the reference point between the PSC and TMC are for further study. *This reference point may be between PMC and TMD and is for further study.

- *Radio port controller* (RPC). The RPC is identical to the BSC of the TIA model.
- *Radio access system controller* (RASC). The RASC performs the radio-specific switching functions of call delivery and origination, handoff control, registration and authentication, and radio access management (control of signaling channels).
- *PCS switching center* (PCSC). The PSC is similar to the MSC of the TIA Reference Model. Some of the functions of the MSC in the TIA model are distributed into other elements.
- *Terminal mobility controller* (TMC). The TMC provides the control logic for terminal authentication, location management, alerting, and routing to RPTs. This function is supported by the VLR and the MSC in the TIA Reference Model.
- *Terminal mobility data store* (TMD). The TMD maintains the associated terminal data and is similar to the VLR in the TIA Reference Model.
- *Personal mobility controller* (PMC). The PMC provides the control logic for user authentication, location management, alerting, and routing to users. This function is supported by the HLR and the MSC in the TIA Reference Model.
- *Personal mobility data store* (PMD). The PMD maintains the associated user data and is similar to the HLR in the TIA Reference Model.
- *OAM&P systems*. The OAM&P systems are identical to the OS in the TIA Reference Model.
- *Auxiliary services*. These represent a variety of services such as voice mail, paging, and so on that may be provided by the PCSC.
- *Interworking function* (IWF). The IWF is identical to the IWF of the TIA Reference Model.
- *External networks*. These include those networks (i.e., wired and wireless) that are not part of the described wireless network.

The following interfaces are described:

- *RP–RPT interface* (a interface). This interface is identical to the U_m interface of the TIA model.
- *PCSC–RPC interface* (c interface). This is similar to the A interface of the TIA model. If an RASC is used, the c_{rpc}, c_{rasc}, and c_r interfaces are defined.
- *TMC-to-other-elements interface* (d interface). The d interface is between the TMC and the RASC and between the PMC and the PSC. The d_d interface is between the TMC and the TMD and between the TMC and the PMC.
- *RPI–RPC interface* (f interface). This is between the radio port and the radio port intermediary; it may or may not be internal to a radio system.
- *PCSC-to-auxiliary-functions interface* (l interface). This is the same as the L interface in the TIA Reference Model. As in the TIA model, this interface is defined by the interworking function.
- *PCSC-to-external-networks interface* (m interface). This is the same as in the TIA Reference Model to external networks except that TIA segments this interface into type of network (A_i, D_i, M_i, and P_i).

- *PCSC–PCSC interface* (n interface). This interface is to other PCSCs.
- *RPI–RP interface* (p interface). This interface carries baseband bearer and control information, contained in the air interface, between the RPI and the RP.
- *OAM&P systems–PCSC interface* (q interface). This is the same as the O interface in the TIA Reference Model.
- *Terminal-adapter-to-terminal-equipment interface* (t interface). This interface depends on the type of equipment and is the same as the R interface of the TIA Reference Model.
- *PCSC-to-terminal-equipment interface* (w interface). This interface depends on the type of equipment and allows terminal equipment to be directly connected to the PCSC. There is no equivalent in the TIA Reference Model.
- *OAM&P-to-craft-terminal interface* (x interface). This interface provides capabilities to access the operations system to display, edit, and add/delete information.
- *PCSC-to-auxiliary-services interface* (y interface). This interface, which is the same as the X interface of the TIA Reference Model, depends on the auxiliary equipment connected to the PCSC.

Open Interfaces

Many of the interfaces in the reference architectures are closed and proprietary. Only the air interface (U_m or a interface) is fully open. However, the standards groups are in the process of publishing open standards for other interfaces. Some of the open interfaces are the interface between the switch and base-station controller (A interface or c interface), the air interface (U_m interface or a interface), the interface to the public telephone network, the interface for inter-MSC communications, and operations interfaces.

Work has progressed on several different protocols for the A interface based on different signaling protocols. For SS7-based signaling, the first open interface for the switch-to-base-station controller is based on the SS7 Signaling Protocol (3). A link between the two network elements uses one 56- or 64-kb/s channel in a T1 (1.544-megabits-per-second [Mb/s]) connection. The underlying protocol is then a point-to-point protocol that is consistent with the SS7 Protocol used for signaling between switches. For ISDN-based signaling, a signaling protocol that enables the MSC to be either a wireless switch separate from the wireline network or a wireline switch has also been developed (4). The protocol is based on ISDN BRI or PRI communications and is supported using one 56- or 64-kb/s channel in a T1 (1.544-Mbps) connection or 16 kb/s in a 144-kb/s BRI channel. For frame-relay-based signaling, as an interim step toward using asynchronous transfer mode (ATM) communications, and to support CDMA services, a frame relay signaling system is in the process of being standardized.

The air interface (U_m interface or a interface) is between the mobile station and the base-station transceiver. This is the interface that has been traditionally an open interface. In the United States, six air interfaces have been standardized for use with cellular/PCS systems. These interfaces are

- Advanced Mobile Phone System (AMPS) is an analog standard using frequency modulation for voice and Manchester frequency modulation (MFM) for data communications (5). Each AMPS channel occupies a bandwidth of 30 kilohertz (kHz).
- Combined code division multiple access/time division multiple access (CDMA/TDMA) (CCT) system is a digital standard that uses both time division multiplex and code division multiplex to assign 24 (or 32 with small cells) users to a 5.0-megahertz (MHz) system operating at a TDMA frame rate of 781.25 kb/s with a 5.0-Mbps spreading code or 16 users to a 2.5-MHz system operating at a TDMA frame rate of 468.25 kb/s with a 2.5-Mbps spreading code (6).
- Code division multiple access (CDMA) is a digital standard that uses spread spectrum and code division to distinguish multiple users on the same frequency (7,8). Each CDMA channel occupies 1.25-MHz bandwidth.
- Personal Access Communications System (PACS) is a digital standard that uses time division system to multiplex seven users into a 384-kb/s channel with a bandwidth of 300 kHz (9).
- Time division multiple access (TDMA) is a digital standard that uses time division to multiplex three users into one 30-kHz channel (10–15). It thus replaces one AMPS channel with three TDMA users.
- Wideband code division multiple access (W-CDMA) is a digital standard similar to CDMA, but using a wider bandwidth (16). Each W-CDMA channel occupies 5-, 10-, or 15-MHz bandwidth, depending on the parameters chosen for the system.

The interface to the public telephone network is a standard and open interface. Signaling connections are based on multifrequency signaling or on SS7 signaling (17). Voice transmission is based on μ-law pulse code modulation (PCM) or standard analog voice transmission.

For inter-MSC communications, when mobile telephones move from their home system to a visited system, communications must occur between the two mobile switches. Originally, this type of communication was based on proprietary signaling using private lines, modem connections, or X.25. An open interface based on SS7 signaling has been developed by the TIA (IS-41) (18). The SS7 signaling either can be on a private SS7 network linking multiple wireless service providers or can be on one of the many public SS7 networks owned and operated by the local- and interexchange carriers in the United States.

Major revisions were made in the existing standards, resulting in IS-41C and a new standard, IS-124, referred to as data message handling (DMH), was developed (19).

The DMH (IS-124) moves "near-real-time" messages between different places in the cellular network. The basic messages are related to call activities, when and where the call was made, length of the call, and so on. It fulfills the cellular/PCS operators' need for a way to transfer call detail records rapidly among various business entities involved in a call. Data message handling has critical features important to business, such as accountability, auditability, and traceability. The uses of DMH are for settlement, fraud monitoring, credit verification, credit limit monitoring, and real-time customer care in settling billing disputes. Data message handling provides a reliable data communications protocol to support applications that should not depend on the time-critical IS-41 signaling network. Future application of DMH could include

- Customer equipment configurations
- Short message (data component)
- E-mail and e-mail attachments
- E911 location data

Data message handling has very flexible message services, including

- *Certified delivery*. This service transfers the call detail information between network elements with a positive acknowledgment back to the sender. Thus, the sender knows when the data gets through.
- *Uncertified delivery*. This mechanism is called "send and pray." In contrast to certified delivery, the sender does not know if the receiver got the message. This message service has less overhead than the certified delivery.
- *Retransmission request*. This command is typically initiated by the receiver when he is sure data has been lost. The receiver knows something is missing by examining the message sequence numbers. This might be used to ask for the missing information, which was originally sent using uncertified delivery.
- *Record request*. This command is initiated by the receiving end when it needs a specific record or set of records to complete a job. For example, to complete a specific bill, an operator needs not only the network roaming bill, but the associated long-distance costs. The operator would use this message service to request that record.
- *Aggregate delivery*. This is similar to a batch header and trailer on an accounting tape and provides summary information on records previously sent.
- *Aggregate request*. This is how the receiver of records asks for summary information data that was previously sent. This message service allows the receiver to verify that all data sent has been received.
- *Rate request*. This is a request to an external service to put a dollar amount on the call based on the call detail record.

Data message handling has been designed to operate under many configurations, many modes, and many applications. It has been designed to be scalable and can migrate throughout the network. Further, it has been designed to allow operators to control the nature of the data they interexchange with other carriers and clearing houses.

Data message handling is compatible with the Open System Interconnection (OSI) seven-layer model. The DMH itself is an application service level in layer 7. The advantage of using the OSI Model is to allow decoupling of the software between layers. Using this model allows the operator to change networks while preserving investment in software.

Standard operations interfaces are based on the Telecommunications Management Network (TMN) (20–22). The TMN includes a logical structure that originates from data communications networks to provide an organized architecture for achieving interconnection among various types of operations systems and/or telecommunications equipment types. Management information is exchanged among operations systems and equipment using an agreed-on architecture and standardized interfaces, including protocols and messages. The Open System Interconnection (OSI) management technology has been chosen as the basis for TMN interfaces.

The principles of TMN provide for management through the definition of a management information model (Managed Objects) that is operated over standardized interfaces. This model was developed through the definition of a required set of management services divided into various service components and management functions. An information model represents the system and supports the management functions. Three important aspects of TMN that can be applied to manage a PCS network are

1.   A layered architecture with five layers: business management layer (BML), service management layer (SML), network management layer (NML), element management layer (EML), and network element layer (NEL). A layered architecture as applied to a PCS network is shown in Fig. 3.

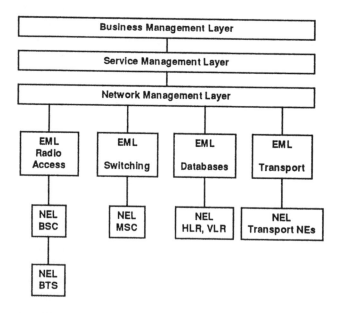

FIG. 3   Layered architecture applied to a personal communications systems network (BML = business management layer; EML = element management layer; NEL = network element layer; NML = network management layer; SML = service management layer).

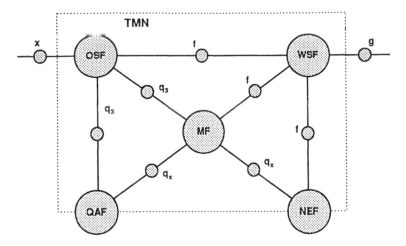

FIG. 4   Telecommunications Management Network (TMN) functional architecture and interfaces (MF = mediation function; NEF = network element function; OSF = operations systems function; QAF = Q adapter function).

2. A functional architecture that defines functional blocks: operations system function (OSF), mediation function (MF), workstation function (WSF), network element function (NEF), and Q adapter function (QAF) (see Fig. 4).

3. A physical architecture defining management roles for operations systems, communications networks, and network elements.

European Wireless Systems

The GSM is the global system for mobile communications (23–33) (Fig. 5); it is a Pan-European cellular mobile radio system. A GSM mobile user can receive services similar to those provided in the fixed public switched telephone network (PSTN). With a GSM system, it is possible to make and receive telephone calls in the same general manner as in fixed networks. Users are able to dial a phone number and be connected to the number of interest, irrespective of whether the dialed number corresponds to a mobile or fixed station. By using the same national and international dialing procedures that exist today, a mobile user can dial anyone's phone number in the world. Similarly, incoming calls from anywhere in the world can be received by a GSM user. Data and messaging services are also available in a GSM system.

The basic subsystems of the GSM architecture are the base-station subsystem (BSS), network and switching subsystem (NSS), and operation subsystem (OSS). The BSS provides and manages transmission paths between the mobile station and the NSS. This includes management of the radio interface between the mobile station and the rest of the GSM network. The responsibility of the

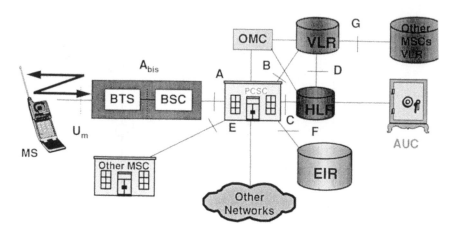

FIG. 5   Global System for Mobile Communication (GSM) Reference Model.

NSS is to manage communications and connect mobile stations to the relevant networks or other mobile stations. The NSS is not in direct contact with the mobile station; likewise, the BSS is not in direct contact with external networks. The operational part of the GSM system includes the mobile station, BSS, and NSS. The OSS provides a means for a service provider to control the operational part.

Major GSM system entities are the mobile station (MS), base-station subsystem (BSS), mobile switching center (MSC), home location register (HLR), visitor location register (VLR), equipment identity register (EIR), authentication center (AC), and operation maintenance center (OMC).

Standard open interfaces between network entities are defined in the GSM specifications. These interfaces are A, A-bis, B, C, D, E, F, G, and U_m (or air interface). Since interfaces between network elements are open and standardized, a given service provider has total flexibility in selecting and mixing different vendor's equipment in the network.

Base-Station Subsystem. The BSS consists of one base-station controller (BSC) and one or more base transceiver stations (BTSs). Each BTS serves a cell area. A BTS includes a controller, radio transmitter/receiver, radio-frequency (RF) amplifiers, RF combiners, an antenna, and much more. A BSC manages radio resources and executes procedures that are common to all BTSs. The BTSs are connected to a BSC over standard 32-channel digital facilities operating at 2.048 Mb/s. A BSC can be physically located anywhere between an MSC and its BTSs and can only be connected to one MSC. A BSC is connected to an MSC over standard 32-channel digital facilities.

Mobile Switching Center.   The functions of the MSC are

- Call handling that copes with the mobile nature of subscribers (e.g., paging)
- Management of logical radio-link channel during calls

- Management of MSC-BSS signaling protocol
- Control of inter-BSS handoffs
- Act as a gateway to interrogate the HLR for routing incoming calls to the called mobile station
- Exchange of signaling information with other system entities
- Other normal functions of a local-exchange switch in the fixed network

An MSC interfaces with other network elements, including

- An operation maintenance center (OMC).
- A visitor location register (VLR). All mobile stations that move around under base stations connected to the MSC are always managed by the same VLR.
- An equipment identity register (EIR). While it is possible for an MSC to communicate to multiple EIRs, it is highly unlikely since an EIR provides centralized and geographically independent functions.
- The MSC contacts a home location register (HLR) to determine how a call should be routed to a given mobile station. For incoming calls to a mobile station, the MSC would typically consult one HLR. For mobile-to-mobile calls in larger networks, an MSC could consult HLRs of other systems to help to minimize the trunk paths to other mobile stations.
- A given MSC can be interconnected to other MSCs to support inter-MSC handoffs. The E interface is only a signaling interface.

Home Location Register. The HLR contains the identities of mobile subscribers, their service profiles, and their location information. The location information is stored as a mobile station roaming number (MSRN), which is a directory number that the network uses to route calls to the MSC at which the mobile subscriber is located at the time of the call.

Visitor Location Register. The VLR contains the subscriber parameters and location information for all the mobile subscribers currently located in the geographical area controlled by that VLR. The VLR allocates the MSRN and (when required) a temporary mobile subscriber identity (TMSI) for secret identification of the mobile subscriber on the radio link.

Equipment Identity Register. The EIR contains a list of valid MS equipment identities, a list of mobiles under observation, and a list of mobiles for which service is barred. This database is accessed during the equipment validation procedure when a mobile accesses the system.

Authentication Center. The AC contains subscriber authentication data called authentication keys (K_i), generates security-related parameters required to authorize service using K_i, and a unique data pattern called a cipher key (K_c) required for encrypting user speech and data.

Operation Maintenance Center. The OMC monitors and controls the GSM Public Land Mobile Network (PLMN).

Mobile-Station-to-Base-Station Interface. The physical layer of the mobile-station-to-base-station (MS–BS) interface (U_m) is referred to as the radio subsystem layer. This layer interfaces to the data-link layer and radio resource management sublayer in the MS and base station and to other functional units in the MS and network subsystem (which includes the BSS and MSC) for supporting traffic channels. At the physical level, most signaling messages carried on the radio path are in 23-octet blocks. The data-link layer functions are link multiplexing, error detection and correction, flow control, and segmentation to allow for long messages on the upper layers. The protocol is similar to the ISDN Link Access Protocol, D Channel (LAPD) and is called LAPDm.

The radio resource layer manages the dialog between the MS and BSS concerning the management of radio connections, including connection establishment, control, release, and changes (e.g., during handoff). The mobility management layer deals with supporting functions of location update, authentication, and encryption management in a mobile environment. In the connection management layer, the call control entity controls end-to-end call establishment and management, and the supplementary services entity supports the management of supplementary services. Both protocols are very similar to those used in the fixed network. The short message service (SMS) protocol of this layer supports the high-level functions related to the transfer and management of short message services.

Base-Station-Controller-to-Base-Transceiver-System Interface. The A-bis interface between the BSC and BTS supports two types of communications links: traffic channels at 64 kb/s carrying speech or user data for a full- or half-rate radio traffic channel and signaling channels at 16 kb/s carrying information for BSC-BTS and BSC-MS signaling. There are two types of messages handled by the traffic management procedure part of the signaling interface, transparent and nontransparent. Transparent messages are between the MS and BSC-MSC and do not require analysis by the BTS. Nontransparent messages do require BTS analysis.

Base-Station-Subsystem-to-Mobile-Switching-Center Interface. The physical layer of the BSS-MSC (A) interface is a 2-Mb/s standard International Telegraph and Telephone Consultative Committee (CCITT) digital connection. The signaling transport uses the Message Transfer Part (MTP) and Signaling Connection Control Part (SCCP) of SS7. Error-free transport is handled by a subset of the MTP, and logical connection is handled by a subset of the SCCP. The application parts are divided between the BSS management application part (BSSMAP) and the BSS operation and maintenance application part (BSSOMAP). The BSSAP is further divided into the direct transfer application part (DTAP) and the BSS management application part (BSSMAP). The DTAP is used to transfer layer 3 messages between the mobile station and the MSC without BSC involvement. The BSSMAP is responsible for all aspects of the radio resource handling at the BSS. The BSSOMAP supports all the operation and maintenance communications of BSS.

Information transfer between GSM PLMN entities uses the MAP of SS7. This MAP contains a mobile application and several Application Service Elements (ASEs). It uses the service of the Transaction Capabilities Application Part (TCAP) of SS7. The mobile ASEs plus TCAP make up the mobile Application Entity (AE) of SS7. It uses the SCCP for routing, and only Class 0 (connectionless datagram) service is required. The MAP layers provide the necessary signaling functions required to provide services such as setting up mobile facilities for voice and nonvoice applications in a mobile network. The major procedures supported by MAP are

- location registration and cancellation
- handoff procedures
- handling supplementary services
- retrieval of subscriber parameters during call setup
- authentication procedures

List of Acronyms

AC	authentication center
AM	access manager
AMPS	Advanced Mobile Phone System
ASE	Application Service Element
ATIS	Alliance for Telecommunications Industry Solutions, Inc.
ATM	asynchronous transfer mode
AUC	authentication center
BML	business management layer
BS	base station; also known as a radio port or base transceiver system
BSC	base-station controller
BSS	base-station subsystem
BTS	base transceiver system; also known as a radio port
CCT	combined CDMA/TDMA
CDMA	code division multiple access
CO	central office
CPE	customer premises equipment
DECT	Digital European Cordless Telecommunications
DMH	data message handler
EIR	equipment identity register
EML	element management layer
ETSI	European Telecommunication Standardization Institute

FM	frequency modulation
GSM	Global System for Mobile Communication
HLR	home location register
ISDN BRI	Integrated Services Digital Network Basic Rate Interface (144 kb/s)
ISDN PRI	Integrated Services Digital Network Primary Rate Interface (1.544 Mb/s)
ITU	International Telecommunication Union
IWF	interworking function
MF	mediation function
MTP	Message Transfer Part
NEF	network element function
NEL	network element layer
NML	network management layer
NSS	network and switching subsystem
OAM&P	operations, administration, maintenance, and provisioning
OMC	operation maintenance center
OSF	operations systems function
OSI	Open System Interconnection
OSS	operation subsystem
OSS	operations support system
PACS	Personal Access Communications System
PCM	pulse code modulation
PCS	personal communications systems (or services)
PCSC	personal communications switching center
PLMN	Public Land Mobile Network
PMC	personal mobility controller
PMD	personal mobility data store
POTS	plain old telephone service
PS	personal station
PSPDN	packet-switched public data network
PSTN	public switched telephone network
QAF	Q adapter function
RASC	radio access system controller
RP	radio port; also known as base station or base transceiver system
RPC	radio port controller
RPI	radio port intermediary
RPT	radio personal terminal
RS	radio system
SCCP	Signaling Connection Control Part
SML	service management layer
SMS	service management system
SMS	short message service
SS7	Signaling System No. 7
TCAP	Transaction Capabilities Application Part
TDM	time division multiplexing
TDMA	time division multiple access
TIA	Telecommunications Industry Association

TMC	terminal mobility controller
TMD	terminal mobility data store
TMN	Telecommunications Management Network
VLR	visitor location register
WSF	workstation function

References

1. Telecommunications Industry Association Project PN-3169, *Personal Communications Services Network Reference Model for 1800 MHz*, Proposal, June 6, 1994.
2. Committee T1 – Telecommunications, *A Technical Report on Network Capabilities, Architectures, and Interfaces for Personal Communications*, T1 Technical Report #34, May 1994.
3. *MSC-BS Interface for Public 800 MHz*, TIA Interim Standard IS-634.
4. Telecommunications Industry Association TR-46, *ISDN Based A Interface (Radio System to PCSC) for 1800 MHz Personal Communications Systems*, Project PN-3344, Re-Ballot Submission, August 29, 1995.
5. Telecommuniations Industry Association, *Cellular System Mobile Station – Land Station Compatibility Specification*, TIA Interim Standard IS-91.
6. T1P1/94-089, PCS2000, *A Composite CDMA/TDMA Air Interface Compatibility Standard for Personal Communications in 1.8–2.2 GHz for Licensed and Unlicensed Applications*, Committee T1 Approved Trial User Standard, T1-LB-459, November 1994.
7. Telecommunications Industry Association, *Mobile Station-Base Station Compatibility Standard for Dual-Mode Wideband Spread Spectrum Cellular System*, TIA IS-95, July 1993.
8. T1P1/94-088, *Draft American National Standard for Telecommunications – Personal Station–Base Station Compatibility Requirements for 1.8 to 2.0 GHz Code Division Multiple Access (CDMA) Personal Communications Systems*, J-STD-008, November 1994.
9. *Personal Access Communications System, Air Interface Standard*, J-STD-014.
10. Telecommunications Industry Association, *Cellular System Dual-Mode Mobile Station Base Station Compatibility Standard*, TIA IS-54B, April 1992.
11. Telecommunications Industry Association, *800 MHz TDMA Cellular–Radio Interface – Mobile Station–Base Station Compatibility – Digital Control Channel*, TIA IS-136.1, December 1994.
12. Telecommunications Industry Association, *800 MHz TDMA Cellular–Radio Interface – Mobile Station–Base Station Compatibility – Traffic Channels and FSK Control Channel*, TIA IS-136.2, December 1994.
13. *PCS-1900 MHz, IS-136 Based, Mobile Station Minimum Performance Standards*, J-STD-009.
14. *PCS-1900 MHz, IS-136 Based, Base Station Minimum Performance Standards*, J-STD-010.
15. *PCS-1900 MHz, IS-136 Based, Air Interface Compatibility Standards*, J-STD-011.
16. *PCS-1900 Air Interface, Proposed Wideband CDMA PCS Standard*, J-STD-007.
17. *Compatibility Information for Interconnection of a Wireless Service Provider and an Local Exchange Carrier Network*, Bellcore Generic Requirements, GR-145-CORE, Issue 1, March 1996.

18. Telecommunications Industry Association, *Cellular Radio Telecommunications Intersystem Operations*, TIA Interim Standard IS-41C.
19. EIA/TIA *Cellular Radio Telecommunications Intersystem Non-signaling Data Communications (DMH)*, EIA/TIA IS-124.
20. T1M1.5/94-001R2, *Proposed Draft Standard—Operations, Administration, Maintenance, and Provisioning (OAM&P) Interfaces Standards for PCS*.
21. Klerer, System Management Information Modeling, *IEEE Commun. Mag.* (May 1993).
22. Hayes, S., A Standard for OAM&P of PCS Systems, *IEEE Personnel Commun. Mag.*, 1(4) (December 1994).
23. Aidarous, and Levyak, T. P., (Eds.), *Telecommunications Network Management into the 21st Century*, IEEE Press, Piscataway, NJ, 1994.
24. *GSM Overview, Glossary, Abbreviations, Service Phases*, GSM Specification Series 1.02–1.06.
25. *GSM Services and Features*, GSM Specification Series 2.01–2.88.
26. *GSM PLMN Functions, Architecture, Numbering and Addressing, Procedures*, GSM Specification Series 3.01–3.88.
27. *MS-BSS Interface*, GSM Specification Series 4.01–4.88.
28. *Radio link*, GSM Specification Series 5.01–5.10.
29. *Speech Processing*, GSM Specification Series 6.01–6.32.
30. *Terminal Adaptation*, GSM Specification Series 7.01–7.03.
31. *BSS-MSC Interface, BSC-BTS Interface*, GSM Specification Series 8.01–8.60.
32. *Network Interworking, MAP*, GSM Specification Series 9.01–9.11.
33. *ETSI Technical Specifications GSM 12 Series*.

VIJAY K. GARG
JOSEPH E. WILKES

Status of the Internet

Introduction

What is the Internet? How did it happen? How will it evolve? How will it influence society? How will it affect me? These questions are now common enough to be heard in various everyday contexts, even on the subway systems among urban commuters! This article attempts to provide some insights into where we are, how we got here, and where we may be headed. But, among the many lessons the Internet has taught us, one of the strongest is that we will probably not be able to predict very accurately future directions.

There are three basic techniques for establishing transfer of information in electronic systems: message switching, circuit switching, and packet switching. Message switching is exemplified from the earliest days of electrical communications by the telegraph, with messages, originally exclusively text, handled one at a time on a store-and-forward basis. In circuit switching, as in the plain old telephone system (POTS), a separate circuit is established for each call or conversation, originally exclusively voice oriented. In packet switching, a much more recent idea, messages are broken into segments called packets and routed over a network to be reassembled into the original format following reception at the destination; this operation is all digital and was originally intended exclusively for computer data.

In the telegraph system, the network intelligence was in the presence of human operators at interchange points; although the telephone system began with human operators as the primary means of control, a circuit-switching fabric was soon adopted. This has evolved to today's advanced intelligent network (AIN), which has many very useful features. In modern telephone systems, the intelligence for the system is embodied almost exclusively in the switching system, with the user terminals and telephones remaining dumb terminals for the most part. Packet-switching systems, on the contrary, were originally conceived as a means for transferring data between computers; as a consequence, the intelligence of the system is mostly concentrated at the periphery of the system in the user terminals, with the users taking much more responsibility for the operation of the system, including the validity of the data, and with no dedicated circuits established.

The predecessors of the Internet, generally identified as ARPANET (Advanced Research Projects Agency Network), BITNET (Because It's Time Network), CSNET (Computer Science Network), USENET (UNIX Users Network), and NSFNET (National Science Foundation Network), all utilized packet switching to handle data transfer, as does the Internet today.

Lawrence G. Roberts, in his alumni profile for the 50th anniversary of the Research Laboratory of Electronics at the Massachusetts Institute of Technology (MIT) (1), credits J. C. R. Licklider of MIT with the articulation in 1962 of

Although the author is an employee of the National Science Foundation, the views here are those of the author and are not intended to represent positions of the National Science Foundation.

the idea of a galactic network similar to what we now know as the Web. A careful analysis of the development of networking has been presented in Ref. 2; there are many key players and contributors. The astounding rate of growth of the area is one of the great stories of modern times. The passage of the Telecommunications Act in 1996 promised to enable a true revolution in telecommunications. Computers and communications are converging in many significant ways, as detailed by David Messerschmidt in Ref. 3. Electronic commerce is now a topic of study in business schools and a subject of everyday life and activity. A new body of rules, which may be referred to as Internet law, is just now beginning to emerge and is being developed day by day.

Predecessors

From early data transfer setups for time-sharing computers came ARPANET (4), the initial packet-switching network, arranged to permit collaboration among geographically dispersed computers. This network was developed under the auspices of the federal Defense Advanced Research Projects Agency (DARPA). It was followed by NSFNET (5), with the support of the National Science Foundation (NSF). The NSFNET spread to interconnect the United States and world wide research and education communities. By 1995, the interest in this network of networks worldwide was so strong that the term *Internet* had taken on an entirely new meaning, and the Internet was recognized and popularized.

Advanced Research Projects Agency Network

The ARPANET (one of the first computer networks, which led to the NSFNET, and thence to the Internet of today) was undertaken under the direction of the Defense Advanced Research Project Agency (DARPA) around 1968. The network was based initially on 50-kilobits-per-second (kb/s) transmission lines. The protocol suite was selected and implemented via a set of interface message processors (IMPs); each organized in a small minicomputer, the Honeywell DDP-516. (The first IMP now resides at UCLA as a historical instrument.) The software had to be designed to fit into 24 pages of 512 words per page in core storage. A quite detailed presentation of the IMP is found in Ref. 4. The net came up with 4 sites in 1969 and quickly grew to 10 sites with coast-to-coast coverage. Maximum packet size was approximately 1000 bits. Each packet was routed from IMP to IMP following a routing protocol designed to be so flexible as to not require the IMPs to know the network topology. Error control was performed at the IMPs. Communications protocols were developed to allow the IMPs to function as a collection of subnets.

This network was truly a marvel and well ahead of its time. It was so successful that it set the stage of development and launched the world on the path toward the Internet. In 1974, Kahn and Cerf published a description of the

Transmission Control Protocol/Internet Protocol (TCP/IP) for interconnecting networks following open architecture standards (6). This protocol has evolved over the years with a succession of implementations on an evolution of platforms, holding fast to the basic original principle of openness, into the foundation of the Internet of today.

National Science Foundation Network

The NSFNET developed from around 1985 as a network of networks designed to serve the interests of the research and education community. Early decisions were to use the TCP/IP protocol and to organize the government support as seed funding rather than continuing as a directly funded government infrastructure. Regional networks were encouraged, and this strategy initially gave rise to private network companies such as PSI, ANS CO+RE, UUNET, and, more recently, many Internet service provider companies (estimated to number 3400–3500 today).

The original NSFNET was composed of six nodes with routers referred to as "fuzzball" routers; these were connected with 56-kb/s links. This arrangement was quickly overwhelmed with traffic. A new-generation architecture was developed and put into operation by July 1988, with 13 nodes interconnected by T1, 1.5-Mb/s (megabits per second), links. This backbone network was developed and constructed by a team composed of Merit Network, Incorporated, IBM, MCI, and the University of Michigan. Regional networks and centers connected to the backbone included BARRNet, JVNCnet, Merit, MIDnet, NCAR/USAN, NCSA, NorthWestNet, NYSERNet (New York State Education and Research Network), Pittsburgh Supercomputing Center, San Diego Supercomputing Center, SESQUINET, SURAnet, and Westnet, later expanded to include NEARNET, Argonne National Laboratory, DARPA, FIX-East, FIX-West, and a second SURAnet node.

Traffic and network utilization grew at an almost overwhelming rate: in the first month of T1 operation, traffic doubled from previous levels. Demand continued to the point that a T3, 45-Mb/s backbone was commissioned. In fall 1991, the initial 16 T3 sites were operating, and the T3 backbone service connected 16 sites and 3500 networks. International interest and connectivity rose as well; by 1995, the NSFNET backbone was connected to 93 countries and over 50,000 networks.

As interest grew in the research and education community, it spread to the commercial sector. Commercial interests noted the value of a common network infrastructure, as opposed to completely private company network infrastructures. There was (is) enormous discussion and controversy as to how to mix requirements for research and education traffic with commercial traffic in constructive and equitable ways. The government, industry, and university partnerships that participated led to rapid transfer of the technology developed; this and, perhaps as important, the management/service model developed led to the creation of a new networking industry with companies such as PSI, ANS, CISCO, and many others. The area surrounding Washington, D.C., came to be identified as NETPLEX by *Fortune* magazine in March 1994. The growth of the networking industry and activity in the metropolitan Washington, D.C., area was compared to the very-large-scale-integrated (VLSI) silicon chip activity in California's "Silicon Valley."

As network activity grew at an astounding rate, the concept of "user service" took on an entirely new meaning: users, hosts, databases, and services all were now distributed across the entire globe and "finding it on the desktop" meant a truly global search. The Merit group established a Network Operation Center (NOC) that operated on a 24-hour, 7-day-a-week basis.

The NSFNET activity went beyond the development of the network itself to facilitate the development of tools and services needed to make the capabilities of the network accessible and useful to more and more segments of the community. This was necessary as the data available on the net grew more diverse in both disciplinary and geographic dimensions. Example research projects included the EDGAR (Electronic Data Gathering and Retrieval) services of the Securities Exchange Commission. This short-term project demonstrated that a complicated and time-sensitive regulatory database could be made available over the Internet; the activity continues as an ongoing activity begun as a research project.

The Clearinghouse for Networked Information and Data Retrieval (CNIDR) at MCNC developed tools for finding and retrieving data; it was instrumental in getting the U.S. Patent and Trademark Office information on line. The Internet Network Information Center (InterNIC) was organized to provide information services, directory services, and domain name registration. The Network Startup Resource Center (NSRC) has provided assistance both domestically and abroad to help organizations and countries undertake initial Internet activity. Not only data, but also video and audio tools have been developed. CU-SeeMe is a audio/videoconference tool originally developed at Cornell University and currently being commercialized by White Pine. This tool permits videoteleconferencing over the Internet at a performance level that has led to the creation of the Global Schoolhouse Project, which provides educational activity over the Internet primarily in K–12 schools.

Many organizations have become involved in networking in formal ways. EDUCOM and the Federation of American Research Networks (FARNET) are U.S. national organizations highly supportive and influential in the development of research and education networking. The Internet Engineering Task Force (IETF), which grew out of early DARPA network support programs, serves as a global networking development and coordination resource. The Internet Society (ISOC) is an international organization with the mission of serving the needs of the global on-line community. There are other support organizations in many nations around the globe.

Research-oriented organizations that support the development of networking capabilities include most of the telecommunications research organizations around the world. Some particularly notable projects have been undertaken at the Center for National Research Initiatives (CNRI) in Reston, Virginia. The CNRI leads the Cross Industry Working Team (XIWT), a multi-industry collaboration devoted to advanced networking research. The CNRI also led the Gigabit Testbed Project, with support from the NSF and DARPA. This project consisted of the development of five testbeds (Vistanet, Casa, Blanca, Aurora,

Nectar) operating at gigabit- or near-gigabit-per-second speeds. Although each of the testbeds had only a small number of nodes, they demonstrated that gigabit speeds lead to applications not possible at lower speeds. Bottlenecks to high-speed networking were identified, and substantial progress was made in resolving some of these. Another very advanced research testbed is the MAGIC testbed operated under DARPA support. The National Laboratory for Advanced Networking Research (NLANR) provides for cutting-edge research support of very high performance networking, including network measurement tools and applications.

Transition and Privatization

By mid-1992, it had become clear that the success of the NSFNET had led to commercial interest in the Internet at a level that was even then beginning to see unprecedented growth rates. The research and education community had developed a capability that promises to revolutionize the business and personal communications scene. The government support of the research and education community networks was neither sufficient nor appropriate for the support of commodity networking for the general public or for the continued support of what had become commodity networking for the research and education community. The sorting out of the transition of the Internet from an NSFNET backbone service to a privatized Internet took from 1992 to 1995. After extensive public input and debate, followed by solicitations, reviews, and competitive awards, a new support activity was commissioned, and the "old" NSFNET was phased out; it was shut down on April 30, 1995. Advanced or leading-edge activity in networking continued in the form of the new architecture.

The New Architecture

The new network architecture consisted of three main pieces or activities: the very-high-speed backbone network service (vBNS), the network access points (NAPs), and the routing arbiter (RA).

Very-High-Speed Backbone Network Service

The vBNS is a very-high-performance network built and operated by MCI under a cooperative agreement with the NSF (7). The network connects the five supercomputer centers with a network operating at 155 Mb/s initially, with planned upgrades to 622 Mb/s and then to 2.4 gigabits per second (Gb/s) over a five-year program that began on April 1, 1995. The vBNS also connects to the Internet at four NAPs at 45 Mb/s. This interconnection to the Internet will also be upgraded as technology permits.

A smaller network testbed enables testing of new concepts and equipment outside the main network. The vBNS is operated as an IP over the asynchronous transfer mode/Synchronous Optical Network (ATM/SONET). This activity is a partnership between the federal government, MCI, and the academic community. MCI had constructed the original configuration prior to the actual award, enabling basic operation to begin immediately with the award; it continues to devote substantial internal resources to the further development of the partnership. The vBNS ensures that advanced very-high-speed networking capabilities are available to research and education organizations that have developed and are developing applications requiring very-high-performance services. It also serves to develop very-high-speed, high-performance network capabilities.

Network Access Points and Routing Arbiter

The NAPs are points of interconnection and traffic exchange of Internet service providers. These exchange points were created to make sure the Internet did not break into subsets that could not communicate with each other in the transition to a commercial Internet. The RA provides routing information to the NAPs. Four priority NAPs were implemented under cooperative agreements with Metropolitan Fiber Systems (MFS), U.S. Sprint, and Bell Communications Research (Bellcore), and operated by MFS, U.S. Sprint, Ameritech, and PacTel (Pacific Telesis). By 1997, the commercial activity had progressed to the point that government participation in these facilities was no longer required. Numerous other exchange points are now active as well.

The domain name registration activity of the InterNIC, conducted by Network Solutions, Incorporated, is illustrative of the rapid growth of Internet interest in the commercial and personal arenas. In 1993–1994, the domain name database was composed primarily of .gov (government), .edu (education), and .org (organization) entries with some .com (commercial) entries. By September 1995, the number of names registered was increasing dramatically, with the growth taking place primarily in the .com area. The number of entries climbed from a base of about 2400 to 3000, with a growth rate of 3000/month in 1995. The increase in activity led to a decision to institute a fee for registration beginning in October 1995 as a means for supporting the extreme growth in registration activity, most of which was concentrated in the private sector. In early 1997, the number of registrants stood at about 900,000 and continues to increase at a rate of 30,000/month. The number of .gov, .edu, and .org entries is now only a fraction of 1% of the total. Today, the future of the activity is clouded by the desire for competition in the provision of the service and very significant concerns about the relationship between copyright and trademark rights and registration issues. As in the case of intellectual property rights in the Internet in other contexts, much remains cloudy in the legal framework of the Internet and electronic commerce.

The Web, Mosaic, and Netscape

One of the events that has defined the public interest in the Internet and led to the frenetic pace of expansion and excitement over the last three years has been the development of easy-to-use software for the location and downloading of major sources of information over the Internet. From the early applications such as E-mail and File Transfer Protocol (FTP), which required considerable computer skills for ready manipulation, came user-friendly mail-handler applications such as Eudora and Pine, and Archie and Veronica, then gopher tools for resource location over the Internet.

These things were immensely helpful, but the development of the World Wide Web approach with hypertext links from files in one computer to files in other computers located elsewhere on the Internet was definitive. This development came from Tim Berners-Lee at CERN, the European particle physics laboratory in Geneva (8). It was developed to enable researchers to work more efficiently and has spread broadly into the business world. The application Mosaic was conceived and created at the University of Illinois as an unanticipated side effect or by-product of computer and networking research there. It was highly successful when made widely available over the Internet to other researchers. Subsequently, it has been substantially improved and commercialized as Netscape, now reported to be used by 45 million users worldwide (9). Mark Andresson and his colleagues have become another major modern computer success story. Their company, Netscape, is now a market force.

Search engines such as AltaVista and Yahoo! have also contributed to the proliferation of home pages of information on the Internet. Tasks that were formerly quite time consuming, perhaps requiring several trips to a library and several telephone calls, can be accomplished quickly and easily by just "pulling it up on the web." Commerce and education are in a state of excitement and promise to undergo major changes in the way things are done on a day-to-day basis. The information revolution is well under way.

Problems and Issues

Performance

The success and popularity of the Internet have led to an explosion of user population and traffic. Access via dial-up modems has stressed the telephone switching centers through which the Internet traffic passes due to the substantial differences in holding times for Internet access compared to voice traffic. Telephone companies have begun to experiment with techniques for recognition of the Internet traffic and arrangements to take this traffic directly to the backbone, bypassing circuit switches as much as possible. Within the Internet service provider community, serious upgrades have been undertaken, such as the widely touted $300 million upgrade by America Online (AOL) to handle the volumes of traffic stimulated by their introduction of a flat rate for unlimited access. Backbone speeds of nationwide providers are all being upgraded.

However, the protocols in use do not permit an individual user to pay more for better service; the user may choose to upgrade to a T1 or T3 connection, not generally viable for a residential setting, but the protocols still prevent preferential treatment of one traffic stream over another. In addition, some implementations of the TCP/IP stacks running on personal computers (PCs) will not allow window sizes to adapt to give greater throughput than about 200 kb/s in some situations.

At busy times of the day, in particular, long waits for information, particularly if fancy graphics are featured, are commonplace. There have been "end of the world" predictions forecasting the collapse of the fundamental Internet infrastructure.

So, in a real sense the Internet is somewhat a victim of its own success. It is here with great flair and promise, but with some obvious performance deficits. The tendency to utilize more and more graphics and audio and video on the Internet leads to higher and higher bandwidth demand and indicates the congestion problems cannot be circumvented simply by "burying the problems with bandwidth."

Congestion and availability are performance issues that currently limit the utility of the Internet in some applications.

Security and Privacy

The Committee to Study National Cryptographic Policy, through the Computer Science and Telecommunications Board (CSTB) of the National Research Council, has published a report dealing with the role of cryptography in an electronic society (10). Issues involving privacy and security in electronic commerce, including arrangements for electronic payment, are reported in Ref. 11. There is a strong need for standards that will allow the general public to utilize electronic means of authentication, certification, nonrepudiation, protection, and secure sharing of information with controlled access by planned and unplanned means. Anonymity of transactions in some applications is also an issue, for instance, in circumstances for which electronic cash is the preferred means of exchange. The CSTB report contains a rather detailed set of recommendations for security and privacy policy requirements of the information society.

Major obstacles arise with conflicting requirements: individuals wish to conduct private exchanges, while law enforcement agencies wish, under appropriately controlled scenarios, to be allowed to monitor and understand these exchanges. Governments are reluctant to grant arbitrary access to information crossing their "boundaries." The United States, for example, controls the export of software and hardware designed to exceed certain levels of protection.

Standards must evolve to reconcile the differences in the conflicting requirements, to survive multinational scrutiny, and enable cross-border secure applications. Progress has been made, and applications such as Netscape now include the possibility of keyed security provisions. Much more is required to enable the spread of ubiquitous electronic commerce.

Intellectual Property Issues

Intellectual property issues and the right of an author to protect the value of contributions, both "original" contributions and "value-added" contributions, to the body of information in the age of electronic information access and exchange are complicated. The age of electronic access brings with it both enhanced opportunity and enhanced risk. The new medium provides for unparalleled opportunity in the distribution of information. For example, Netscape and other software companies have taken great advantage of the new Internet capability to allow clients to download new versions as required and at their own initiative. Collaborating groups utilize living documents on the Web to radically enhance the speed of interaction and scope of creativity.

The copyright protection of information is made more difficult by the ease of distribution. There is much to be done to implement procedures, practices, and controls to create a situation in which publishers will be eager to use the new medium. While the details of the medium are only beginning to take shape, some believe the medium will give rise to entirely new information exchange paradigms; a few believe the medium will prove to be as revolutionary as the printing press!

Some exciting possibilities for the protection of information include embedding identity information and tracking in the information format, with electronic marking and/or watermarking techniques (12). While not foolproof or undefeatable, these tools may provide protection levels more nearly comparable to or perhaps superior to those available in traditional scenarios.

Management and Administration

Who is in charge of the Internet? This is a frequently asked question. The answer is that there is no single, central control point. There is no governing body. There is no single person or agency in charge of the Internet — it is unofficially coordinated by the ISOC, a voluntary membership organization with the purpose of promoting global information exchange through Internet technology. A group of volunteers makes up the Internet Architecture Board, which meets regularly to approve standards and allocate resources. The ETF consists of volunteers and determines solutions to near-term technical problems of the Internet. But, there is really no central controlling body.

The current loose structure has served well to this point in time. As the Internet becomes more and more a vital tool for essential commerce, there is speculation that more definitive rules and procedures to ensure a minimum level of availability and accountability will evolve. Policy, standards, and law for the Internet will surely continue to develop and evolve for some time. Means for effecting resolution of conflicting interests will be required just as in other aspects of modern society.

One of the loudest current debates is centered on registration issues. An International Ad Hoc Committee (IAHC) under Internet Society leadership has released a proposed mechanism to create more top-level domains and more registries, thus providing for user choice in registry services. An open Internet process is proposing the creation of an American Registry for Internet Numbers (ARIN).

Interoperability

The telephone industry has been working for a number of years on a telecommunications architecture called the asynchronous transfer mode (ATM). This technique breaks digital data into small units of fixed size called *cells*. The goal is to create an integrated services architecture to enable the integrated handling of computer data, voice, and video data. In the computer industry, the integrated handling of these various forms of data (including content origination, editing, storage, and internetworked sharing, conferencing, and collaboration) is frequently referred to as *multimedia*. The basic exchange protocol in the Internet remains the IP. Some have expressed the view that "Broadband ISDN is happening — except it's spelled IP" (W. David Sincoskie in Ref. 13). The CSTB has published a report dealing with the role of the Internet in the development of the national information infrastructure that also deals with interoperability requirements (14).

It seems clear that as integrated services architectures are developed in the converging computer and telecommunications industries, interoperability of transmission techniques and protocols and interoperability of network protocols, as well as seamless network operation across different underlying transmission media, will be issues.

International Issues

In the international arena, the coordination and development of information policy are rapidly evolving. For information systems that cross boundaries seamlessly, the coordination and protection of national interests remain very important. However, the advantages of easy information exchange and the possibility of the enablement of international commerce via electronic means on the Internet bring a strong international dimension to any regulation and standard setting, either on a formal or on a voluntary basis.

Federal Role

There has been extensive discussion in the United States regarding the role of the federal government in furthering the Internet development activity (15). There is general agreement that the role of the government is to provide for "seeding" or advanced development activity of high-performance networking while not interfering with the private sector's role in developing, marketing, and operating the national information infrastructure. The federal role in oversight of standards processes and registration activity for the Internet is less well defined and not widely agreed on.

Broadband Access

The wealth of information on the Internet, from entertainment and games, business statistics, airline and other travel reservations and information, health systems and information, government information in diverse databases and agencies at local state and national levels, to on-line education opportunities, has brought the industry to a call for ubiquitous high-bandwidth Internet access. The questions are: How to get high bandwidth to residences and small businesses? How to achieve seamless, tetherless access? Competitive approaches are all undergoing serious development. A survey of architectural trends for broadband access is available in Ref. 16.

Cable Access

Cable television distribution systems pass near a large percentage of the population in urban and suburban settings. If these systems can be upgraded to function in a two-way configuration, as in trials in several cities recently, then there is a significant opportunity to provide high-bandwidth Internet service, perhaps including telephone service, via the cable system. These systems currently operate utilizing "cable modems" at the user location, along with substantial supporting infrastructure on the part of the provider, to provide Internet access downstream at about 10 megabits per second and upstream at about 1 megabit per second. Most cable access systems fall into the hybrid fiber coaxial (HFC) cable category, with fiber-based distribution complemented by coaxial cable drops into the final user facility.

Wireless Access

Several different approaches to wireless access systems are under development (16,17). The advent of the cellular telephone has popularized wireless telephony. Cellular wireless systems make effective use of the radio-frequency spectrum by dividing the service area into geographically differentiated cells. By controlling and coordinating power levels, the radio-frequency spectrum can be reused in cells with sufficient geographic separation. These systems are fundamentally interference-limited systems — users can be added only up to an acceptable level of interference. Early systems, called Advanced Mobile Phone System (AMPS), were analog systems. Upgrades to digital service systems are beginning to occur, with use of one of several digital standards: time division multiple access (TDMA); Global System for Mobile Communication (GSM) or Interim Standard 54 (IS-54); or code division multiple access (CDMA), IS-95 for example. Services are now available in the initial spectrum band of 900 megahertz (MHz) and in the more recent personal communications service (PCS) band of 1900 MHz. For voice applications, the acceptable level of interference is quite high compared to data applications.

Digital television is now being provided via satellite in several countries, and standard setting for high-definition digital television is making substantial progress.

Wireless, fixed local voice telephony is a very effective means to provide coverage in areas where wire-line plant is not available or in a situation for which bypass of the existing wire-line plant is desirable for competitive economic reasons. AT&T has just announced the development of a wireless local access system for fixed users.

Trials of several approaches to wireless distribution of television broadcasts have been made. Multipoint, multichannel distribution service (MMDS) at 2.1 to 2.7 gigahertz (GHz) and local multipoint distribution service (LMDS) at 27 to 30 GHz are systems undergoing extensive testing currently. Some companies have recently introduced spread-spectrum wireless Internet access systems operating in the unlicensed regions of the radio frequency spectrum. These systems, such as the one by Metricom, offer low-cost alternative access systems that may be particularly attractive for certain settings.

Wireless alternatives to the wire-line local-area network are also available.

For global coverage, several competing satellite-based systems are being developed. Several competing low-earth-orbit satellite systems (LEOS) are being developed, chosen to minimize problems due to delay and power constraints.

Wireless Internet access as a mobile and/or fixed access system is in a state of intense development, including field trials of some systems. Seamless interoperation with emerging wire-line systems is a requirement. Wireless techniques will clearly be put forward both as alternatives for wire-line systems in fixed installations and as complements to wire-line systems in mobile and portable installations.

Advanced Wire-Line Access

There is renewed interest in wire-line access via the twisted-pair telephone connections as well. The speed of access via twisted-pair lines has, under the intense pressure of data users and with the careful attention of communications engineers, advanced from 75 b/s through 300 b/s and, by means of several major breakthroughs in areas previously thought to be well known, then advanced to 9600 b/s, to 28.8 kb/s, and now to 56 kb/s on most dial-up, voice-grade circuits. For several years, in some locations Integrated Services Digital Network (ISDN) service at multiples of 64 kb/s has been available. Currently, there is a great deal of interest in a family of x-digital subscriber line (xDSL) technologies for getting several megabits per second of data downstream and upstream on top of a voice circuit over moderate local loop distances. Some of these techniques are asymmetrical in rate (asymmetric DSL, ADSL) and some (high-rate DSL, HDSL) are symmetrical.

On the Horizon

Next-Generation Internet

In late October 1996, the White House announced its intentions to support a "next-generation Internet" (NGI) initiative aimed at research universities and federal research organizations in order to develop next-generation applications of computer networking. As of February 1997, the next-generation Internet initiative has been included in the president's budget request to Congress ($100 million requested over each of the next three years). The funds are to be distributed among five federal agencies and managed by a tightly coupled interagency program team.

This initiative has three goals:

1. Connect universities and national laboratories with high-speed networks that are 100–1000 times faster than the Internet of today. These new networks will connect about 100 institutions at 100 times today's capabilities and about 10 universities at 1000 times today's capabilities.
2. Promote experimentation with the next generation of networking technologies, for example, real-time services such as high-quality videoteleconferencing and the provision of integrated services Internet architectures.
3. Demonstrate new advanced applications that meet important national goals and missions. These applications will include, for example, health care, national security, distance education, energy research, biomedical research, and environmental monitoring.

Internet 2

In early October 1996, a number of research universities agreed to create an organization named *Internet 2* to pursue high-performance networking to support the deployment of novel research applications of computer networking. As of February 1997, the Internet 2 organization had grown to about 100 research universities. The NSF-supported vBNS is being viewed as "an Internet 2 network." Whether or not the vBNS will be the only Internet 2 network is a subject of discussion within the university community.

The Internet 2 group states these goals:

1. A leading-edge network capability for the national research community will be created and sustained.
2. Network development efforts will be directed to enabling a new generation of applications that fully exploit the capabilities of broadband networks.
3. The Internet 2 project will be integrated with ongoing efforts to improve Internet services for all members of the academic community.

The first is the most important of the goals and will be the focus of activity; participation is anticipated from 100 universities, federal agencies, and leading computer and telecommunications companies.

Clearly, there is congruence between the primary goals of the next-generation Internet initiative and the Internet 2 project. The role of the vBNS will be come clearer as time progresses. About 30 awards have been made for high-performance connections to the vBNS, and 50 to 60 more sites have applied for connections to date. There is, in exploiting the intersection of these three activities, a major opportunity to advance substantially the state of the art of internetworking. The beneficiaries will range from digital libraries (18) and collaborative research to sporting events (19).

Local or campus area networks will have to be upgraded to handle the advanced applications and services as well.

Helpful Resources

There are many resources for information on the Internet. One of the greatest features of the Internet is that (as one might expect) much of the available information is on line on various servers. Some good locations from which to begin to access this information are www.cise.nsf.gov/ncri/, www.vbns.net, www.internet2.edu, and rs.internic.net. These resources are linked to many other locations.

References

1. Measuring the Return on Investment in University Based Research, Alumni Profile: Lawrence G. Roberts, President, ATM Systems, *MIT Research Laboratory of Electronics Currents*, 8:14–15 (Fall 1996).
2. Leiner, B. M., Cerf, V. G., Clark, D. D., Kahn, R. E., Kleinrock, L., Lynch, D. C., Postel, J., Roberts, L. G., and Wolff, S. S., The Past and Future History of INTERNET, *Commun. ACM*, 40:102–108 (February 1997).
3. Messerschmidt, D. G., The Convergence of Telecommunications and Computing: What Are the Implications Today? *Proc. IEEE*, 84:1167–1187 (August 1996).
4. Heart, F. E., Kahn, R. E., Ornstein, S. M., Crowther, W. R., and Walden, D. C., The Interface Message Processor for the ARPA Computer Network, *1972 Spring Joint Computer Conf. AFIPS Conf. Proc.*, 40:551–567 (1972).
5. *NSFNET: A Partnership for High-Speed Networking. Final Report 1987-1995*, Merit Network, Ann Arbor, MI, 1995.
6. Cerf, V. G., and Kahn, R. E., A Protocol for Packet Network Interconnection, *IEEE Trans. Commun. Tech.*, 22:627–641 (May 1974).
7. Jamison, J., and Wilder, R., vBNS: The Internet Fast Lane for Research and Education, *IEEE Commun.*, 35:61–63 (January 1997).
8. The Web Maestro: An Interview with Tim Berners-Lee, *Tech. Rev.*, 32–42 (July 1996).
9. Fernandes, T., Beyond the Browser, *Networker*, 1:47–53 (March/April 1997).
10. *Cryptography's Role in Securing the Information Society. Computer Science and Telecommunications Board*, National Research Council, National Academy Press, Washington, DC, 1996.
11. Neuman, B. C., Security, Payment, and Privacy for Network Commerce, *IEEE J. Sel. Areas Commun.*, 13:1523–1531 (October 1995).
12. Brasil, J. T., Low, S., Maxemchuk, N. F., and O'Gorman, L., Electronic Marking and Identification Techniques to Discourage Copying, *IEEE J. Sel. Areas Commun.*, 13:1495–1504 (October 1995).

13. Bell, T. E., and Riezenman, M. J., Technology 1997 Analysis and Forecast: Communications, *IEEE Spectrum*, 34:27 37 (January 1997).

14. *Realizing the Information Future the Internet and Beyond. Report of the Renaissance Committee*, Computer Science and Telecommunications Board, National Research Council, National Academy Press, Washington, DC, 1994.

15. Press, L., Seeding Networks: The Federal Role, *Commun. ACM*, 39:11–17 (October 1996).

16. Fenton, F. M., and Sipes, J. D., Architectural and Technological Trends in Access: An Overview, *Bell Labs Tech. J.*, 1:3–10 (Summer 1996).

17. Honcharenko, W., Kruys, J. P., Lee, D. Y., and Shah, N. J., Broadband Wireless Access, *IEEE Commun. Mag.*, 35:20–26 (January 1997).

18. Digital Library Initiative, Theme Issue, *IEEE Computer Mag.*, 29:22–78 (May 1996).

19. *Meeting the Olympic Challenge, Supplement to Telephony* (July 22, 1996).

AUBREY M. BUSH

Steinmetz, Charles Proteus

One of the most extraordinary geniuses in the history of American engineering (with 195 patents in his name and some of the most important discoveries in electronic transmission), Charles Proteus Steinmetz is also one of the most affectionately remembered for his brilliance, kindness, and passionate love of the outdoors.

He was born Carl August Rudolph Steinmetz on April 9, 1865, in Breslau, Germany, a city rich in commerce, industry, and newly developing technology. He was the son of Caroline Neubert and Carl Heinrich Steinmetz, a lithographer and amateur scientist. Steinmetz's mother, who was the widow of Carl Heinrich Steinmetz's brother, survived the boy's birth by only one year. The boy's earliest childhood was guided by a grandmother and an aunt, but as he was about to start kindergarten, they left Breslau, and his half sister was given responsibility for raising him.

From his father, young Carl August inherited both his dwarfish, humpbacked physical deformity and, much more important to his life, a consuming passion for machines. His father made a determined habit of staying current with all the new knowledge of science, which was occurring with rapidity, and of learning every detail of the operations of the latest mechanical devices.

As a very young child, Steinmetz had three years of elementary education, which is when he confronted, and had to struggle to develop a proficiency in, arithmetic. He was regarded as being a little bit dull, and his difficulty in doing something as basic as memorizing multiplication tables did nothing to discourage his teachers from believing this.

After elementary school, at age 8½ years, came the gymnasium. Latin was added to his studies immediately, and in the second year French was required. As school became more intense, so did the situation at home. When Steinmetz was 9, his father moved another widow and her children into the home, apparently without benefit of marriage.

During his third year at gymnasium, Greek was added to Steinmetz's curriculum. None of these studies were treated casually; many years later, Steinmetz could still recite long passages and would do so to entertain friends and students. In addition, his studies included Polish, Hebrew, philosophy, and as his skills developed, higher math. During his later student years at the gymnasium, Steinmetz had progressed to the point that he often tutored other students in mathematics.

Though known for his evident poverty and for being somewhat embittered, Steinmetz also developed a reputation among other students as someone who was particularly witty, kind, and interesting if students could break through his shell. In addition, he became known for his card tricks—and for his practical jokes—which would continue for the rest of his life.

When it came time to leave the gymnasium,

Steinmetz scored the highest marks in his graduating class on the written tests in mathematics and science and was exempted from taking the oral exams in these fields. The testing commission noted that his knowledge of mathematics far surpassed that taught at the school and that he was able to solve the most complicated problems. They also reported that his knowledge of physics was equally extensive and certain. (1, p. 5)

In 1883, he enrolled at the University of Breslau, where the boy who had once found it extremely difficult to memorize multiplication tables became the man who specialized in math and physics. It was while he was at the University that Carl August first became known as "Proteus." Although having decided to become a chemist, he had joined the mathematics club, which provided each of its members a nickname. He became Proteus, for the Greek mythological god of the sea. The name had implications of his deformity, but really was intended to signify his extraordinary brilliance and the breadth of his knowledge. (It was years later, after his move to the United States, that he legally changed his name to Charles Proteus Steinmetz.)

Steinmetz was also extremely active in the local Socialist party. Despite the fact that the party was under severe legal restrictions, the young man made his controversial views known. He wrote tracts, made speeches, and was generally visible as a proponent of Socialist causes.

By May 1887, Steinmetz had completed four years of study and was awarded a stipend so that he could work toward a doctorate. But his very visible activity as a Socialist had attracted the attention of both university officials and local law enforcement. During summer 1887, he failed to complete the paperwork for two of his courses, though he completed all the others. As a result, there was an excuse for the university to reconsider whether Steinmetz could stay on as a student.

A year later, after battling the system and attempting to complete his education, Steinmetz fled Breslau during the night. In later years he would tell many different versions of that story, but they seem to boil down to a need to escape rather than be arrested for his Socialist activities. He had been active in a group that planned to start an Icarian (Socialist Utopian) community, probably in the U.S. midwest—though the group later learned of the failure of the Iowa communities and determined against that idea. By the spring of 1888, the police were known to be preparing legal proceedings against the group.

Steinmetz fled to Zurich, where socialism was flourishing. He worked there as a math tutor but was also commissioned to write a series of monthly articles for the local newspaper. The wide variety of topics he covered attracted a lot of attention among both general readers and scientists. He also published a book on astronomy, but money was extremely tight.

Up to this point, Steinmetz was still planning to become a chemist, but in Zurich he began to develop an interest in engineering and took courses in mechanical engineering. As his interest in electricity grew, he studied on his own and soon began developing both new theories and practical applications in the new field. His particular interests at that time were the electrical circuit and the transformer, and he provided a unique perspective on their engineering requirements because he looked at them from a mathematical point of view rather than that of traditional engineering.

In March 1889, the Breslau police issued a warrant for his arrest, and his options for returning home were lost. Concerned about his safety in Zurich,

Steinmetz's friends convinced him to join another student who was about to embark for the United States. His friend, Oskar Asmussen, was quite wealthy and offered to pay Steinmetz's passage and help him financially until the two could find jobs.

On May 20, 1889, they arrived in New York City, and the physically deformed Steinmetz was almost turned back. Asmussen again came to his rescue and agreed to take personal and financial responsibility for him. The two young engineers found an apartment in Harlem, where they kept bachelor's quarters and were known for their Bohemian social gatherings.

Asmussen was educated as a mechanical engineer though he had not previously worked in the field. He found a job quickly. Steinmetz, however, applied at the Edison Electric Company and was turned down. Looking for other opportunities, he discovered a company owned and run by German socialist émigré Rudolf Eickemeyer. Steinmetz began work as a draftsman June 10, 1889.

Eickemeyer was something of a Renaissance man, and the two got along quite well. Eickemeyer took Steinmetz under his wing, encouraging him to turn his theoretical knowledge to practical applications. Among those efforts was a redesign of the alternating current motor, which would later be patented in Eickemeyer's name but which was acknowledged by him as being developed by the two men together.

In 1891, emotionally committed to both Eickemeyer, the Socialist friend, and to his new home in the United States, Steinmetz applied for U.S. citizenship.

On January 19, 1892, Steinmetz delivered the first of many papers on hysteresis before the American Institute of Electrical Engineers.

During his tenure at Eickemeyer's, Steinmetz published approximately 50 articles, most on alternating current and magnetic theory. This gained significant exposure for the young engineer, and his visibility increased still more when he joined the American Institute of Electrical Engineers and presented many papers before those practicing engineers. The work on hysteresis, which would occupy him in different forms throughout his life, was the source of much of his early recognition.

The problem of hysteresis — the tendency of some material to resist changes to its magnetic nature, that is, for magnetic materials to resist becoming nonmagnetic and vice-versa — provided a great deal of difficulty for everyone involved in the new science of electricity. The problems cause by hysteresis invariably resulted in drastically overheated machinery and was equally problematical with both alternating current and direct current (DC) hardware.

Steinmetz observed the phenomenon extensively, gathered a lot of data, and got himself into serious trouble by at first claiming that he had found a natural law innate to hysteresis. He eventually reversed himself and acknowledged hysteresis as merely a phenomenon, not a law of physics. Reconsidering, he applied complex mathematical models of analysis to the problem, and the eventual results would become known as the Steinmetz coefficient, Steinmetz curve, Steinmetz exponent, and, eventually, the Steinmetz law.

During 1892, Edison's company, General Electric, purchased the Eickemeyer organization, which set Steinmetz on the path that would occupy the rest of his life. (In Ref. 1, Kline suggests the merger was accomplished specifically to allow General Electric to compete with Westinghouse; pp. 66–68.) Steinmetz's genius was recognized right away, and he was assigned to perform complex calculations in the "Computations Department."

Within a year, Steinmetz had accomplished the first three of his patents. The 1893 filings dealt with methods for reducing transmission line inductance, new methods for using capacitors, and methods for operating synchronous motors. It was this work that led him to focus on the development of a synchronous condenser. His unique method for analyzing the problem required that he develop complex mathematical tools based on imaginary numbers—but his obscure and esoteric theories led, in the first of many successes, to the development of very practical solutions to problems of electrical transmission.

During January 1894, Steinmetz was transferred to the General Electric facility in Schenectady, New York. E. J. Berg, one of his acquaintances from the Lynn, Massachusetts, plant was there also, and the two became fast friends as they explored the new town. The abundance of water nearby was of particular interest to them, and they designed a boat together. Steinmetz later bought out Berg's share, but the two spent a great deal of time exploring the nearby waterways.

That summer during one of their boat rides, the two found themselves aground at a bluff above Viele's Creek, a tributary of the Mohawk River. Discovering a small campsite there, they were much taken by the view and returned often. Soon, Steinmetz acquired the land, which would become both his second home and the site of a laboratory, and had a small cabin built there. Dubbed Camp Mohawk, the lodge was the site of many of Steinmetz's most important research efforts. He made himself a desk of boards that fit into the canoe, and he would do a great deal of his work while floating along the Mohawk River at his camp. The camp also became legendary for its weekend parties, which Steinmetz was notorious for hosting while wearing a ragged red bathing suit.

Concurrently, Steinmetz and Berg rented a house in Schenectady, just a couple of blocks from the General Electric plant. They dubbed the place "Liberty Hall" and hired a staff of servants.

Steinmetz's extraordinary kindness and gentleness extended to animals, as well as people. Liberty Hall became home to a veritable zoo. Among the pets were Steinmetz's favorites, a pair of crows named John and Mary who flew about and took feed directly from his hands. After their deaths, Steinmetz had them stuffed and provided them a permanent perch atop a bookcase. The household collection also included cranes, eagles, owls, and a monkey named Jenny, plus miscellaneous squirrels, raccoons, and dogs. There was also an alligator. The men were known for inviting groups of schoolchildren to wander through Liberty Hall and to meet the residents of their little zoo.

Berg's brother Eskil was also an engineer. When he was hired by General Electric, the odd duo invited him to Liberty Hall, and the three lived there comfortably for years. Berg and Steinmetz had once formed something called the "Society for the Equalization of Salaries," which "met" as Saturday night social gatherings for years.

Steinmetz's work was an integral part of this friendly setting. His laboratory began in his bedroom, but was soon moved to the larger quarters of the stable, though some equipment was kept in each space. His many hobbies also flourished at Liberty Hall. His collection of rare cacti, ferns, and, in later years, orchids flourished in the conservatory he had built adjoining the house. Hammond says that Steinmetz's collection of cacti was eventually said to be second in the world only to the great British collection at Kew Gardens (2, p. 262).

During the winter of 1895–1896, Clara Steinmetz visited her brother at Liberty Hall. Beginning with this visit, she arrived periodically to assist with the housekeeping. She was apparently somewhat intrusive as a guest, challenging such things as the need for all the animals and the expense of pet food. Clara was an artist, and one of her portraits of her brother also hung in the house, but she spent only winter with the men, taking a New York apartment so that she could pursue her art.

During their life at Liberty Hall, Steinmetz and the Bergs created what they called the Mohawk River Aerial Navigation, Transportation, and Exploration Company, Unlimited, to build gliding aircraft. The men had several of the aircraft built but were never successful in getting one to maintain any real flight. Steinmetz, whose hobbies also included photography, took pictures of some of the planes on the ground, then retouched them so the planes seemed to be flying. The photos were widely shown around, and the reactions provided much delight for the two engineers.

By 1898, Steinmetz was devoting a significant portion of his time to a subject that would occupy him for the rest of his life—research into the behavior of lightning and its effects on electrical transmission. As a result of one of those experiments, there was a fire that destroyed the entire stable, convincing Steinmetz that pursuing his research on rental property was not a very good idea. During the autumn of 1900, the Liberty Hall residents began to move away, Eskil first, and later E. J. Berg. With the laboratory gone and his friends moving away, Steinmetz began looking for alternatives via General Electric's recently formed real estate division. He began to explore a variety of options but made no move yet.

During this period, friends continued to drop by Liberty Hall with great regularity. During the winter of 1901, some of those friends brought along a young mathematician, Joseph LeRoy Hayden (eventually adopted legally as Steinmetz's son). Hayden became one of the many young scientists and engineers who made a habit of dropping by Liberty Hall and discussing their research, as well as Steinmetz's many other interests. Gradually, Hayden became a close companion to Steinmetz and was, like many other young scientists, a frequent guest at Camp Mohawk, where he participated in many chores around the camp, including the construction of a large dam ordered by Steinmetz so that he could canoe more easily on the river.

Throughout 1900, Steinmetz had been working to develop significant improvements in understanding the ways that lightning worked. Simultaneously, he was working on the transmission of electrical power. As a result of those investigations, he eventually invented the magnetite arc lamp, which was noted for two things: it gave off significantly better light than that given off by the carbon arc lamp, and it would operate only via direct current.

Eventually, he would convince General Electric to build a small DC power plant on his property, where he could experiment with the new lighting system. He then convinced the local city powers to allow him to extend lighting from that plant and ran 25 pole-mounted lamps from the facility. Next, Steinmetz explored the idea of mercury lamps, which could be used closer to the ground, and designed them also to be operated via DC, which enabled the city to begin lighting its parks and pathways.

By spring 1901, Steinmetz had grown determined to build a home for himself. As might be expected, his plans for the project were somewhat unconventional. He determined that all construction in areas that would house the plants and animals must be done first to ensure that they were safe and protected. He convinced General Electric to build a laboratory for him on the same site at significantly reduced cost. His passion for water was demonstrated when he required that the house have a small indoor swimming pool, located in the conservatory. It started, however, as a fish pond containing fish and terrapins.

It was Hayden, by then an expert on arc lighting, who supervised the installation of lighting systems that Steinmetz designed. To make it simpler for the young man to find quarters, Steinmetz offered Hayden the opportunity to sleep in a spare room beside the laboratory and a chance to take his meals with Steinmetz (who by then was without servants) as well. As their friendship grew, Hayden took to calling the older scientist "Dad," and Steinmetz eventually adopted Hayden legally as his son.

During 1901, Steinmetz was elected president of the American Institute of Electrical Engineers. In June 1902, Steinmetz received an M.A. from Harvard in recognition of his work as an electrical engineer, and in 1903 Union College, Schenectady, conferred that long-elusive Ph.D., inviting him to become a professor of electrical engineering.

The man who had been a distant, socially uncomfortable student in Breslau immersed himself in college life in the United States. While continuing to work for General Electric, he became an active part-time instructor. He joined Phi Gamma Delta and occasionally dropped by the fraternity house, where he was a popular visitor. Later, he joined Sigma Xi and Tau Beta Pi, both honorary scientific fraternities, as well as Eta Kappa Nu, the engineering fraternity. He was also known for his financial support of athletics and was known to the students as "Steiny." His lectures were notoriously tough, but extremely popular. He also became a popular public speaker at local events, such as those sponsored by the YMCA.

In May 1903, Hayden moved away, marrying a young lady who had frequently visited Camp Mohawk. The young couple lived away for a short while, but after only a few months, they moved back in with "Daddy Charles," and the family thrived. In the later years, Hayden's sons, Billy and Joe, would spend their summers at Camp Mohawk, where summers and weekends were still hosted by Steinmetz himself. Throughout this era, Steinmetz's lifelong affection for practical jokes continued, and there is a story that he once presented Hayden's new bride with a box of chocolates, delicately wired to give her a startling, though harmless, electrical shock.

During the first decade of the 20th century, Steinmetz continued to address the problems that were getting in the way of the general transmission of electrical power. He spent a great deal of time developing a mechanism for stepping up or stepping down energy during transmission without substantial loss of power. By 1903, he had focused part of his effort on the problems of working with high-tension transmission lines and eventually found a way to make them function, "which made it possible to perfect the far-reaching electrical systems, and to utilize to the full the transformer invented by William Stanley" (2, p. 295).

As the decade continued, he solved still more of the pressing problems involved in the expansion of electrical power. During 1907, he presented the results of a systematic study of the general equation of the electric current before the American Institute of Electrical Engineers. During 1908, he and several of his colleagues received the Certificate of Merit from the Franklin Institute. The award recognized his invention of the magnetite arc lamp. In 1909, the first edition of Steinmetz's *Theory and Calculation of Transient Electrical Phenomena and Oscillation* was published (3).

Despite his extremely busy schedule, Steinmetz also made a foray into American politics during 1911. He had long been a quiet member of the American Socialist party, but that year Schenectady elected George R. Lunn, the Socialist candidate, as its mayor. Steinmetz offered his services, and the Mayor appointed him to the Board of Education. The popular Steinmetz was soon elected chair of the board and used his authority to press for improvements in elementary education and for the inclusion of classics in all levels of education.

During 1913, Steinmetz was awarded the Elliott Cresson gold medal for his successful mathematical solutions to so many problems of electrical engineering.

During the years from 1912 to 1918, he applied his engineering ideas (and ideals) to still another arena. Steinmetz began to explore seriously the idea of converting an automobile into a truck—something that had been little discussed as automobiles were still very much a luxury item. As might be expected, he wanted to develop his trucks to be powered by electricity. After some serious study, he concluded that electric motors were better for the trucks, but only for short-distance hauling.

By the late 1910s, he had been persuaded to design the Steinmetz electric truck, a large hauling machine with a large number of batteries. A prototype was built and performed satisfactorily, causing a group of promoters to consider developing the truck commercially. According to Hammond (Ref. 2, pp. 389–390) and others, the business was poorly managed and soon failed. This did not, however, diminish Steinmetz's belief in the future of electrically powered transportation.

As the years went on, his accomplishments continued to mount. In 1915, he was elected president of the National Association of Corporation Schools. That same year, he also became president of the Illuminating Engineering Society. In 1917, he published a paper on advanced investigation of the general theory of transients, a subject he addressed again in several papers published in 1919.

During 1920, lightning struck at Camp Mohawk. It first took out a tree and then forked, one part taking out a fence post and one taking out a window, striking lighting gear, and then spreading throughout the building. Others might have been distraught because of the destruction left in the wake of the strike, but Steinmetz found it a wonderful opportunity to study lightning. Among other things destroyed by the strike was a large mirror, and he collected the pieces, turning them into a jigsaw puzzle and tediously reassembling the mirror so that he could study the way it had shattered. Eventually, he succeeded in putting the shards together, sealed it within plate glass layers, and shipped it to the laboratory in Schenectady for further analysis. The timing was excellent, as he, Hayden, and engineer N. A. Lougee were focusing almost entirely on lightning-related research for General Electric.

As a result of those experiments, Steinmetz became known to the public as "the man who tamed the lightning." He had explored lightning-related questions all of his life, including building a very effective lightning generator (note its relation to his passion for hysteresis and for the mathematics of alternating current). He had learned very early that the damage was not due directly to the lightning, but to the power surges that resulted from it.

Among his more important inventions were lightning arresters (not lightning rods). They were the result of several years of research. The lightning arrester works by shunting off lightning energy via a "line of least resistance," thereby diverting it from the transmission current. (His examination of transient phenomena continued throughout the last two decades of his life and resulted in many papers and presentations.)

Among his investigations into lightning, Steinmetz made a serious effort to find a way to capture and utilize the energy of lightning for power generation, but concluded it was not a viable concept because the power did not last long enough to be readily captured and utilized.

Throughout this period, he continued to explore concepts of electrically powered transportation. He was a long-established proponent of the potential of electric trains to replace steam-powered ones. He reminded proponents of steam that their engines must slow down on extreme grades and that electric locomotives had no such limitations. It was one more example of his determination to minimize waste in all aspects of energy development and distribution.

In the later years of his life, he also was a regular and popular guest on radio and achieved significant popularity as a writer in general interest magazines. He was known for his ability to explain complex scientific issues effectively for the general public. Because of his ongoing work with lightning, he gradually became known as the "Jove of Schenectady."

In 1922, Steinmetz found himself involved once again in politics. He was nominated for state engineer and surveyor of the commonwealth of New York and ran on the Socialist and Farmer-Labor ticket. (He was not elected.) He chose to make no active campaign for the office and held only a limited number of meetings and newspaper interviews. The topic he pressed might surprise many: he felt that the state of New York was greatly underusing its water resources for power; he was not a proponent of the continued use of fossil power. Among his ideas were the capture of the power of Niagara Falls—without jeopardizing its natural beauty. His idea would require capturing the entire motion and flow of the falls during the week and then rereleasing its natural flow for tourists to watch on the weekends and holidays. His rationale was that the amount of power captured would be of immense usefulness to everyone.

By the spring of 1923, Steinmetz was working intensely on the lightning generator effort and had about three years of work planned for himself and his team. During an early experiment, his machine captured such a forcible strike of lightning that it sent a huge surge through the entire building, leading to plans to build a special laboratory for the lightning work. Construction of that room began during early summer.

With everything under way, Steinmetz announced to his family that he wanted to fulfill a long-time dream and go to the Pacific Coast of the United States, which he had never seen. As word of his plans got around, Steinmetz was invited to speak at the October meeting of the Pacific Coast chapter of the American Institute of Electrical Engineers and at various other organizational events along the route. He insisted the entire family, including the three Hayden children, accompany him, and the group set out via railroad on what would be a six-week journey.

Along the way, he spoke in Denver in an auditorium with seating for 900 people, but with a huge overflow crowd. They went next to Colorado Springs, where the passionate outdoorsman could visit many of the natural sites he had always dreamed about, and then on to the Grand Canyon and to Los Angeles (with trips to view the equipment at various motion picture studios), with speaking engagements all along the way.

By September 23, when they arrived in San Francisco, he was apparently tiring, but spent several days in meetings and making presentations. A week later, they went to Del Monte for the Pacific Coast conference of the American Institute of Electrical Engineers, and he delivered a paper on "High Voltage Insulation," which he had coauthored with Hayden. On October 6, they started back East, with many talks still scheduled along the way. Arriving home on October 12, Steinmetz admitted to being worn out by the tour. His physician recommended bed rest, which the Haydens found difficult to enforce but were able to encourage him to read in his room or in bed.

On October 26, 1928, after a restless night, Charles Proteus Steinmetz died within moments of a brief conversation with Hayden. The body first lay in state at home, then was buried at a huge funeral attended by a group that clearly represented his impact on the world. His family was there, of course, but there were also powerful administrators from General Electric, scientists and engineers who had been his colleagues, important scientific minds from around the world, and the local folks, who had learned so much from "the man who tamed lightning."

Bibliography

Steinmetz, C. P., *America and the New Epoch*, Harper & Brothers, New York, 1916.
Steinmetz, C. P., In Favor of Railroad Electrification, *Heart's International Magazine* (May, 1923).

References

1. Kline, R. R., *Steinmetz: Engineer and Socialist*, Johns Hopkins University Press, Baltimore, MD, 1992.

2. Hammond, J. W., *Charles Proteus Steinmetz: A Biography*, Century Co., New York, 1924.
3. Steinmetz, C. P., *Theory and Calculation of Transient Electrical Phenomena and Oscillation*, McGraw-Hill, New York, 1909.

MARYLIN K. SHEDDAN

Submarine Cable Systems

History

Telegraph

Although the first several attempts in the 1850s to install submarine telegraph cables ended in failure, so great was the need for such communications that efforts continued unabated. In 1850, the first telegraph cable laid across the English Channel failed after less than one day of operation, cut by a fisherman. The first two attempts to complete a transatlantic cable, between Ireland and Newfoundland in 1857 and 1858, also failed when the cable broke during installation and could not be recovered. The second 1859 attempt was successful and was greeted with great public adulation for the feat and its prime mover, Cyrus Field, an American entrepreneur. Unfortunately, this cable failed after about a month of operation. Nevertheless, by 1862, 44 submarine telegraph systems totaling some 9000 miles had been installed in less challenging locations than the Atlantic. Finally, after a lengthy scientific and engineering effort was brought to bear to identify and overcome the root causes of the early transoceanic failures, two transatlantic systems were successfully installed in 1866.

Telegraph cable consisted of several copper wires, separately covered by gutta percha insulation,* twisted together with tarred hemp strings filling the interstices; the whole was wrapped by a covering of the same material. This structure was covered by helically wound galvanized iron or steel wires, usually tarred and then wrapped with jute yarn before a final coating of tar was applied. The steel wires provided the needed tensile strength against breaking and protection against seabed hazards like fishing activities and sharp rocks.

Technological advances in both terminal apparatus and the cable itself continued over the next several decades to improve the reliability and speed (capacity). A major improvement in cable transmission, first recognized by Oliver Heaviside in the mid-1880s, resulted from increasing its per-unit-length inductance to compensate for its large capacitance. The improvement was lower attenuation and less phase distortion (phase velocity less dependent on frequency, allowing higher speed operation). Installation of submarine telegraph cables continued worldwide even into the 1930s, although competition with high-frequency radio, starting in the 1920s, significantly reduced its growth. The end of the telegraph cable era came with the introduction, in the late 1940s, of coaxial cable capable of carrying analog telephone traffic.

Coaxial Analog Telephone Systems

Overseas radio telephony, while offering considerable speed/capacity advantages over telegraph cables, nevertheless suffered from fade-outs and interfer-

---

*Gutta percha is the coagulated latex obtained from certain trees of the Sapotaceae family that is mainly indigenous to the Malay Peninsula; it has a chemical composition similar to rubber.

ence, making it less reliable. Moreover, limited frequency allocations would not allow significant traffic growth. In the 1930s, Bell Laboratories initiated a research and development (R&D) effort to design an electron tube and other components to achieve sufficient reliability to be usable in an undersea repeater (amplifier) that, combined with coaxial cable, could support multichannel telephony over transoceanic distances. The repeaters would be placed periodically in the cable to compensate for the signal attenuation in the cable. After being delayed by World War II, this effort culminated in a commercial system installation in 1950 between Florida and Cuba that used a separate cable containing three repeaters for each of the two directions of transmission. The cable structure was similar to the earlier telegraph cables except for the addition of an outer conductor and the use of polyethylene as the dielectric, which has less conductance loss and a lower dielectric constant than gutta percha. To power the undersea electronics, a constant direct current was supplied over the center conductor of the cable from the land terminal stations. The capacity was 24 voice-frequency channels, achieved by use of single-sideband, suppressed-carrier multiplexing equipment in the terminals and repeater gain carefully tailored to match the undersea cable loss over a frequency band extending to a little above 100 kilohertz (kHz).* The repeater components were assembled into a long, semiflexible tube only slightly larger in diameter than the cable and overarmored along with the cable. In this way, existing shipboard cable-handling machinery, which was used with telegraph cable, could be used for installation and repair.

The first transatlantic telephone cable (TAT-1), which was actually two repeatered cable systems in tandem, commenced service in 1956. The lengthier, deeper segment between Scotland and Newfoundland was similar in design to the Cuba cable, but with the top frequency extended to 164 kHz. The second segment, between Newfoundland and Nova Scotia, was patterned after short, shallow water systems installed earlier in the North Sea. It used a single cable and carried the two directions of transmission in separate frequency bands. This single-cable, band-splitting scheme became standard in future designs of undersea coaxial telephone cables.† TAT-1 was a major technical and commercial success.

In the next four years, while R&D efforts were under way to design the next-generation, higher-capacity system, four new TAT-1-like systems were installed, California–Hawaii and Washington–Alaska in the Pacific, Florida–Puerto Rico in the Caribbean, and New Jersey–France (TAT-2) in the Atlantic.

Technology advances were needed to keep pace economically with exponential traffic growth, which in the Atlantic has averaged about 20–25% per year since the installation of TAT-1. Table 1 lists the major technical advances, the year they were introduced, and their effect.

Undersea telephone coaxial cable systems with ever-increasing bandwidth (traffic capacity) continued to be developed and installed worldwide into the 1980s, in total length well over 100,000 nautical miles (186,000 kilometers

---

*This system remained in service well into the 1980s.
†The return path for DC powering a single cable is through the ocean.

TABLE 1   Major Technical Advances in Coaxial Systems

Technical Advances	Introduced	Effect
High-efficiency channel multiplexing in the terminal	1959	Increased the voice-channel traffic capacity by 33%
TASI (time-assignment speech interpolation)	1959	Increased the voice traffic capacity by approximately 80%
Lightweight (armorless) cable, with strength member a part of the center conductor	1962	Allowed for larger diameter cable, yielding lower attenuation and easier shipboard handling (loading, laying, and recovery)
Form-factor change in repeater housing (long and thin to short and fat)	1963	Allowed for "directional filters" for bidirectional transmission through a single wideband amplifier
New shipboard cable-handling machinery and stowage facilities	1963	Permitted high steady-state payout speeds during cable laying with large diameter repeaters
Sea Plow development	1967	Used to bury cable in shallow water (< 500 fathoms) while laying from a ship, to protect against cable damage from fishing activities, anchors, and the like
Germanium transistors, sold-state diodes	1968	Lowered repeater voltage and increased achievable operating bandwidth
Improved Sea Plow	1968	Achieved greater burial depth with improved operating controls
Silicon transistors	1976	Further increased operating bandwidth
Remotely controlled underwater vehicle for surveillance, burial, and assisting repair of cable on continental shelves	1981	Provided functions previously achievable with divers, but extended to depths well beyond practical diver capability

[km]).* Because of the high cost and long time to repair undersea faults, ultra-high reliability was foremost in the design and implementation of these systems. The achieved component reliability record is still remarkable by present-day standards. The repeater component failure rate was typically less than 15 FITs (a FIT is one failure in a billion hours) (1). As of this writing, very few of these systems remain in service, having been replaced by higher capacity cables using fiber-optic digital technology.

*The last major installation was TAT-7, commissioned in 1983, which contained 659 undersea repeaters and had a top frequency of 30 megahertz (MHz) (corresponding to a capacity, without TASI, of 4200 two-way voice-frequency channels).

Light-Wave Systems with Undersea Signal (Electronic) Regeneration

In the 1980s, about the time that submarine analog coaxial cable technology was "running out of steam" in its ability to meet the ever-increasing traffic capacity demands of overseas telephony efficiently, two new technologies emerged that were critical in overcoming this problem. One was low-attenuation glass fiber to act as the transmission medium, and the other was the semiconductor injection laser.

Earlier, in the 1970s, repeated terrestrial telephone transmission systems were deployed that used cabled multimode glass fiber, semiconductor positive-negative (PN) junction photodiodes (positive-intrinsic-negative [PIN] and avalanche photodiode [APD]) and light-emitting diodes (LEDs; the material system is AlGaAs) operating at 0.89-micron wavelength. In the repeater, the photodiode receiver converts photons (light) to electrons (electricity), and the LED transmitter does just the opposite. These building blocks are ideally suited for digital lightwave regenerative transmission: digital because at any particular time a light pulse is either present or not (1 or 0), lightwave because the transmitted signal is pulses of light, and regenerative because at each repeater between cable sections the presence of a light pulse is detected and regenerated (not amplified) and retransmitted into the next cable section.*

Though offering great promise, much development work was required to bring this technology to long-haul submarine cable applications. To achieve the high bit rate needed to meet the overseas traffic demands efficiently, it was essential to change to single-mode† fiber. Such fiber has a loss minimum (about 0.4 decibels per kilometer [dB/km]) and zero chromatic dispersion‡ near 1.3 microns. However, tensile strength had to be increased substantially for deep-sea cable laying and recovery, and a methodology to achieve low-loss, high-reliability fiber splices, usable aboard ship as well as in a factory, had to be developed.

Another essential change was from LED transmitters to injection lasers operating at 1.3 microns, which can efficiently couple light into the single-mode fiber and have a much narrower wavelength spectrum. This combination kept the pulse-broadening effect of chromatic dispersion from limiting the bit rate. In addition, a new material system had to be developed to realize a high-sensitivity photodiode that could operate at the 1.3-micron wavelength.§

Cables for optical-fiber systems were developed by modifying the designs used for coaxial analog systems. In most cases, the central steel-wire core of the coaxial designs was replaced by a similar, but fiber-containing, core structure that provided a protected, benign environment for the glass fibers. But, some

---

*Telephone traffic, including voice-frequency, television, and data-set (modem) analog signals, is converted to numbers (*digitized*) before being sent over a digital transmission system and reconverted at the receive end.

†A single-mode glass fiber has a much smaller core diameter than multimode fiber, which limits light transmission to a single fundamental waveguide mode.

‡Chromatic dispersion occurs when different wavelengths travel at different speeds. At electronic frequencies, this phenomenon is called delay distortion.

§Sensitivity of a receiver is a measure of the input optical power required to achieve a specified bit error ratio (usually 10^{-9}).

manufacturers began with central fiber structures and developed new strength-member configurations to match them. Such cable can accommodate many fibers, and all are placed near the center of the cable to limit fiber stress due to bending of the cable during coiling for factory and shipboard storage, cable laying, and repair operations.

Because cable insulation no longer served as a dielectric for signal transmission, as it had in coaxial cables, the dimensions were determined solely by voltage breakdown requirements. This allowed the fiber cables to be significantly smaller than their coaxial predecessors. Armor-wire protection designed for the fiber cables was very similar to that used for coaxial cables, with the dimensions scaled down appropriately. Moreover, optical-fiber cable encouraged development of a new undersea component, the *branching unit*, a three-legged body allowing traffic on different fiber pairs to be split apart.

In moving to digital lightwave technology for undersea cable applications, the challenge was to achieve the desired performance and prove in the required ultrahigh reliability of new-technology components that had no "track record." This was a major departure from the undersea coaxial design philosophy of not using components unless there was well-established terrestrial experience.

Exploratory development of undersea lightwave systems began in the early 1980s, and the first major application was TAT-8, installed in 1988. It operates at a line rate of 295 megabits per second (Mb/s). Two fibers in each direction carry traffic, each providing 4000 voice-frequency (64 kilobits per second [kb/s]) channels. Speech compression and digital time-assignment speech interpolation (TASI) equipment in the terrestrial extensions almost pentupled the effective voice circuit capacity.* TAT-8 was the first submarine cable to use an undersea branching unit (BU), which incorporated signal regeneration, allowed optical switching to be controlled from a terminal, and provided for separate cable landings in the United Kingdom and France. Other technical advances used to realize the TAT-8 system design and their effects are summarized in Table 2.

TAT-8 so successfully demonstrated the viability of digital undersea lightwave technology that by 1993 more than 125,000 km of such systems were in service worldwide, almost matching the total length of analog submarine coaxial installations, but accomplished in 5 rather than 30 years. A second (and, in one instance, a third) generation of regenerated undersea lightwave systems was developed, with repeater spacing as far apart as 150 km and bit rates as high as 2.5 Gb/s. These designs changed to even narrower output spectrum laser transmitters† that allowed operation in the 1.55-micron wavelength range without significant dispersion penalty, where another fiber loss minimum exists (≈ 0.2 dB/km). Use of branching units became the norm. TAT-9, for example, has five landing points, two in North America and three in Europe. It contains three branching units with regeneration and multiplexing catering to 140 and 45 Mb/s exchanges, respectively among fiber pairs.‡

---

*Land-based equipment also mitigated the differences in voice-frequency digital coding schemes (A law v. μ law) and digital multiplexing hierarchies between the United States and Europe.

†The material system is the same as for 1.3 microns, but employed distributed feedback (DFB) to narrow the spectrum.

‡These units were named underwater branching multiplexers (UBMs).

TABLE 2   Major Technical Advances for TAT-8

Technical Advance	Effect
Low-loss, depressed-cladding, single-mode glass fiber	With other factors, accommodated long repeater spacing (67 km for TAT-8)
Buried, double-heterostructure InAgAsP laser transmitter operating at a wavelength of 1.3 microns	Provided sufficiently narrow band spectrum to accommodate the required bit rate
Receiver technology	Provided reliable, high-sensitivity operation at 1.3 microns
Non-return-to-zero signal format	Provided maximum light per pulse and minimum electrical bandwidth
High proof-strength glass fibers	Permitted a simple, tightly coupled cable design such that fibers did not require further protection during cable burial, deep-sea laying, and recovery
Silicon microwave integrated regenerator circuitry	Provided reliable, high-speed signal processing capability
Regenerator section redundancy (spare fiber and regenerators)	Relaxed component FIT requirements
Laser transmitter redundancy (one active transmitter and one or more standby transmitters)	Relaxed laser FIT requirements
In-service monitoring and control of regenerators from the terminals	Provided computer-controlled, on-demand measurement from the terminal of laser-bias current span sensitivity margin parity error count and control of redundancy switching for regenerator sections and transmitters
Optical and electrical switches in the repeater and branching unit	Permitted use of redundant laser transmitters and regenerator sections and rerouting of traffic among terminals

Lightwave Systems with Undersea Signal (Optical) Amplification

Introduction

The latest technological change in submarine cable systems is from electrical regeneration in the undersea repeaters to optical amplification using an erbium-doped fiber amplifier (EDFA). When pumped by energy near either 980 or 1480 nanometers (nm), an EDFA provides gain to signals in approximately the 1550–1560-nm band. Major advantages of this technology over undersea regeneration are:

- increased bit-rate capability (not limited by speed of electronic components)
- lower repeater-component count (improved reliability)
- undersea transmission not restricted to a particular bit rate (changes in terminal equipment can upgrade the system traffic-carrying capacity)
- automatic gain adjustment in the undersea system (a property of the EDFA that is explained below)

But, as might be expected, this new technology introduced new challenges to be overcome. In particular, the long transmission distances without regeneration brought into prominence the cumulative degrading effects of dispersion, fiber nonlinearities, and light polarization phenomena that were not significant when regeneration occurred in each repeater. Further, the erbium-doped amplifiers add amplified spontaneous emission (ASE) noise to the signal path (ASE noise is analogous to thermal noise encountered with electronic amplifiers). How these challenges are handled is discussed in the remainder of this section.

The first submarine cable system using optical amplification in the repeaters, Americas-1 North, was installed in 1994 between Florida and the Virgin Islands. It provides two fiber pairs, each operating at 2.5 Gb/s at a single wavelength of 1558.5 nm, and contains 26 repeaters spaced nominally 80 km apart. More installations quickly followed, most operating at twice this bit rate, most notably TAT-12/13 across the Atlantic and TPC-5 across the Pacific. Systems currently being developed will operate with more than one signal wavelength (*channel*) to increase the net bit rate. This technique is referred to as wavelength division multiplexing (WDM).

Transmission

General

High-speed transmission of information over optical submarine cable systems involves three basic processes: propagation, compensation, and terminal signal processing. With light-wave systems, whether the undersea repeaters use electronic regeneration or optical amplification, the information is conveyed in synchronous digital form by pulses of light. The presence of a pulse in a time slot indicates a one, absence indicates a zero. For systems using undersea optical amplification, encoding and decoding (regeneration) occur only in the terminals and not in the undersea repeaters. While other signal formats have been proposed and studied, a non-return-to-zero (NRZ) pulse format is presently the one of choice for undersea use because the light pulse occupies the entire time slot (the duration of a time slot is the reciprocal of the bit rate) and requires the minimum electrical bandwidth, thereby optimizing the signal-to-noise ratio (S/N).*

*The propagation of waves in optical fibers is described by the Schroedinger or nonlinear wave equation. In general, solutions can be found only by numerical means. A class of solutions does, however, exist for certain ranges of pulse power and positive dispersion. These are called *solitons* and, once formed, can propagate as wave packets of unchanging or periodically changing shape. Solitons are a class of return-to-zero (RZ) pulses that promise the ultimate capacity on optical-fiber transmission systems. Their realization requires close control of dispersion and nonlinearity (i.e., power level). To date, the capacity required of ocean cable systems has been achieved with less elegant, although much simpler, negative-dispersion NRZ techniques.

Propagation and Compensation

To achieve the extremely low bit error rate (BER) required by users of digital submarine cable systems, the waveforms (in time) of the light pulses should not be seriously distorted as they travel along the fiber, and at the receiver the pulse power needs to be sufficiently large compared to the noise. Propagation of the optical signal spectrum in glass fibers is affected by both linear and nonlinear processes. The former include attenuation and both chromatic and polarization-mode dispersion. Chromatic dispersion was mentioned above in connection with regenerative light-wave systems.

Polarization-mode dispersion (PMD) can occur if the two orthogonal sets of transverse electric and magnetic fields (polarization modes) that coexist in the light pulses in a so-called single-mode fiber do not remain in step. PMD arises from asymmetry in the fiber cross-section (e.g., if it is elliptical instead of circular or if the core is not centered) and can cause waveform distortion. PMD cannot be compensated; rather, it must be controlled to an acceptable level through care in fiber and cable manufacturing. The dominant nonlinearity is the Kerr effect, by which the index of refraction in the glass core of the fiber increases with light intensity.* The degrading effects of both dispersion and nonlinearities can be significant for optical amplifier systems because of the long distances traversed without regeneration.

Attenuation in the fiber is compensated periodically by the EDF amplifiers in the repeaters, which are typically spaced from 40 to 90 km apart, depending on system length. Amplifier gain is a function of both pump and input signal power (see Fig. 1).

Referring to Fig. 1, when operated in compression (e.g., with input power in the −15 to −5-dBm range), the gain of an EDFA increases when its input power decreases and decreases when its input power increases. The time constant of gain change is about 1 millisecond (ms). Thus, automatic gain regulation is achieved by the system without distorting the shape of individual light pulses. The normalized gain spectrum of a representative type of EDF is shown in Fig. 2.

With many such amplifiers in tandem, the 3-dB bandwidth of a transoceanic system without the addition of gain-shaping equalization is only about 1 to 2 nm. Although this is adequate for operation with up to possibly three WDM channels, many more wavelengths can be applied when gain-shaping equalization is provided to broaden the band shape.

One scheme for accomplishing this is to place loss equalizers every so many repeater sections. The equalizer loss shape is complementary over a specified wavelength range to the gain shape provided by the cable-repeater combination. These equalizers can be fabricated using thin dielectric film technology with a Fabry–Perot-type transmission characteristic or using fiber gratings with a more general shape. The former presently has less flat loss, and a single design can effectively equalize up to about 10 nm over transoceanic distances. To compensate for the flat loss of these equalizers, the cable section into which they are placed is shortened.

---

*A secondary nonlinearity is the Raman effect, which transfers power (gain) from shorter to longer wavelengths. This effect is very small in the current designs of submarine systems with optical amplification in repeaters.

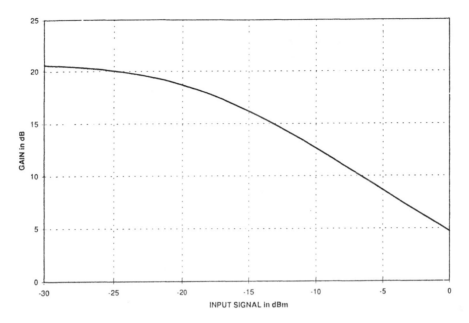

FIG. 1 Effect of input signal power on erbium-doped fiber amplifier (EDFA) gain.

The relative effect of fiber nonlinearities increases with the number of optical carriers and the system length. The optical Kerr effect transfers power between all spectral components with wavelengths that are related by

$$\lambda_1 + \lambda_2 = \lambda_3 + \lambda_4$$

through a process called four-wave mixing (4-WM). A single carrier also interacts with itself (self-phase modulation), with other carriers (cross-phase modula-

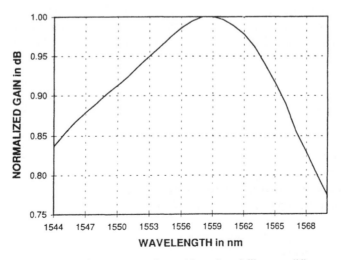

FIG. 2 Gain spectrum of an erbium-doped fiber amplifier.

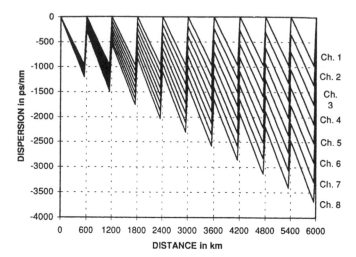

FIG. 3   Ideal dispersion map for an eight-channel wavelength division multiplexing system.

tion), and with ASE noise.* Although the net power transfer in the process is zero, the effect causes waveform distortion and increased noise in all channels. On long systems, best performance occurs when the light intensity of the signal is such that the power per channel introduced by ASE noise and nonlinear distortion is approximately equal. The nonlinear effects are proportional to the product of the power of the interacting waves and the inverse square of the local dispersion. For this reason, long WDM systems are operated at low power, less than 1 milliwatt (mW) per carrier at the amplifier output, using fiber with large negative local dispersion. Such fiber has a nominal zero-dispersion wavelength (ZDW) greater than 20 nm above the longest wavelength channel and is referred to as high dispersion fiber (HDF).

Chromatic dispersion, if not compensated, can extend the duration of the light pulses to longer than a time slot, resulting in errored transmission. The resolution of the two seemingly conflicting aspects of chromatic dispersion, that is, providing large local dispersion to minimize the accumulation of nonlinear distortion and providing no net dispersion in the system, is to use HDF in the cable for most of the system and then to compensate periodically with relatively short cable lengths with fiber that exhibits large positive dispersion in the signal band (ZDW ≈ 1300 nm).† The length of the cable containing compensating fiber is selected to return the dispersion at the longest wavelength channel to approximately zero. An example of an idealized dispersion profile of an eight-channel WDM system, called a *dispersion map*, is shown in Fig. 3, with Channel 1 at the longest wavelength. The final individual dispersion compensation for each channel is provided in the receiving terminal. Dispersion maps with excursions into the positive range are also effective.

---

*Four-wave, cross-phase, and self-phase modulation are similar to what at electronic transmission frequencies is referred to as intermodulation distortion.
†This is the fiber type used in the earlier submarine cable systems using undersea regeneration.

TABLE 3 Typical Transmission Parameters of
Dispersion-Shifted Fibers

Effective area	$> 50 \text{ mm}^2$
Nonlinear coefficient n_2	$2.6 \times 10^{-20} \text{ m}^2/\text{W}$
Polarization-mode dispersion (PMD)	$< 0.10 \text{ ps/km}^{1/2}$
Loss	0.21 dB/km
Loss slope	-0.00022 dB/km-nm
Average zero-dispersion wavelength (for WDM applications)	> 1580 nm
Dispersion slope	$0.075 \text{ ps/nm}^2\text{-km}$

Table 3 lists typical values of transmission-related parameters of dispersion-shifted fiber.

Terminal Signal Processing

Optical Signal Processing. A typical terminal transmitter (shown in Fig. 4) includes several functions. The digital forward error correcting (FEC) encoder is a Reed-Solomon 15/14 type. When coupled with a corresponding decoder in the receiving terminal, an internal BER as high as 10^{-5} in the undersea line is corrected to an external-delivered BER of 10^{-12}. This reduces the required signal-to-noise ratio in the undersea system and allows operation at an increased capacity with the same repeater spacing or at the same capacity with longer repeater spacing. Outputs of two or four FECs can be interleaved to produce nominal 5- or 10-Gb/s line rates.

A highly stable and spectrally pure single-mode laser is externally modulated by a lithium niobate or electroabsorptive modulator. It can also be internally bias-current modulated by a low-frequency line monitoring signal (discussed below). A lithium niobate polarization modulator depolarizes the signal, thus reducing the effect of polarization hole burning in the EDFAs (i.e., the buildup of ASE noise in the polarization state orthogonal to that of the signal).

In WDM systems, the launch power of each carrier signal can be adjusted

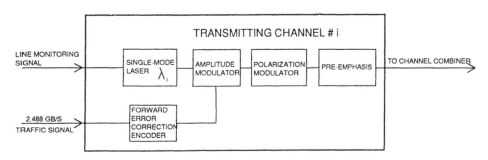

FIG. 4 Transmit terminal signal processing.

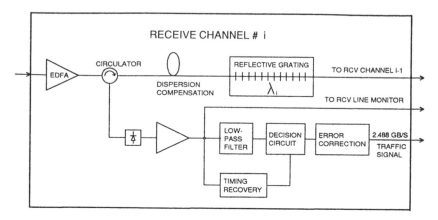

FIG. 5   Receive terminal signal processing.

(preemphasized) in the transmit terminal to achieve equal BER performance on all channels at the receiver. The WDM carriers are combined, amplified, and launched onto an input fiber of the submarine cable.

Figure 5 shows an example of a receive channel of the series-connected type. The output of the receive terminal amplifier (EDFA) passes through a circulator to a length of dispersion-compensating fiber (DCF)* and to a reflective grating that selectively reflects the spectrum of the longest wavelength incoming channel. All other carriers pass through the grating and to the input of the next channel receiver. The reflected signal passes back through the DCF and through the circulator to the photodetector.

The detected signal is amplified and recovered in a conventional terminal regenerator. Errors introduced after the FEC encoder are detected and corrected in the error-correction circuit. A linear sample of the received channel is sent on to the receive line-monitoring equipment (LME). If the nominal line rate is 5 or 10 Gb/s, the bit stream is deinterleaved between the decision circuit and the error corrector. A block diagram of the terminal showing the arrangement of transmit and receive channel equipment is shown in Fig. 6.

Interface to Terrestrial Facilities.   The interface to terrestrial transmission facilities can be at any standard asynchronous or synchronous digital rate as specified in ITU-T (International Telecommunication Union-Transmission) standards. The rates may be as low as 2.048 Mb/s (E-1) or as high as 2.488 Gb/s (synchronous transport mode-16), with higher rates expected in the future. Currently, the most common interface rate is 140 Mb/s.

Standard line muldex (multiplexer-demultiplexer) equipment is generally used in the submarine cable system terminal station; however, special equipment is sometimes needed to accommodate particular functionality requested by the

---

*This length of fiber restores cumulative chromatic dispersion at the highest wavelength to zero. Compensation of the other wavelengths is accomplished farther "downstream" in the terminal equipment.

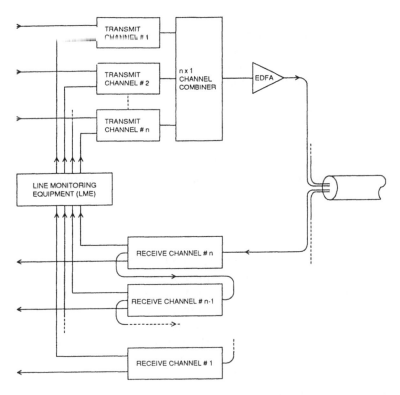

FIG. 6 Transmit and receive channel equipment arrangement.

system owners. To minimize loss of traffic caused by a failure in the terminal, the equipment is often duplicated with automatic transfer switching triggered by the failure. For WDM systems, protection switching may be provided on a wavelength or optical-line-pair basis. When the line-terminating equipment includes forward error correction or any form of time division multiplexing, optical line error monitoring is possible, as well as terminal-to-terminal voice-frequency channels (order wire or speaker circuits) to support maintenance activities. Otherwise, these functions are provided by the highest level muldex equipment.

Optical Line Monitoring and Fault Location. The approach taken to optical line monitoring and fault location in an amplifier-based submarine system is fundamentally different from that in a system with undersea regeneration. Because even a modest transmission degradation of a single repeater section in a regenerative system can result in the failure of that line, it is necessary to monitor the behavior of individual repeater sections to assess overall system performance. Amplifier-based systems are not as dependent on individual repeater section performance because the noise and waveform distortion contributed by each section is such a small amount of the end-to-end total. Moreover, the automatic gain regulation of EDFAs further limits the net effect of a local degradation. A single section must degrade by a very large amount before there is any measurable effect on the overall system performance. This significantly reduces the monitoring accuracy required in amplifier-based systems.

One technique for optical monitoring and fault location for such systems makes use of high-loss optical coupling between the two directions of transmission within the repeater. Figure 7 illustrates a loopback and OTDR (optical time-domain reflectometer) path between amplifier pairs. (The same paths exist for signals-coming and returning in the opposite directions.) At each fiber pair in a cable, this high-loss loopback path couples a small amount of transmitted signal from the outgoing fiber to the incoming fiber. It is thus possible to measure the loop loss from either terminal to/from each amplifier pair in every repeater. Such measurements allow fault localization to one repeater section, and comparison of recent loop loss data with a previous baseline measurement allows one to identify and locate even modest transmission changes whose effects on end-to-end performance are negligible.

One method for measuring loop loss is enabled by amplitude modulating the outgoing optical traffic with a low-frequency pseudorandom supervisory signal. Because the round-trip delay to/from each repeater is different, the supervisory signal coupled back at each repeater can be identified and measured separately. Though weak, these delayed copies of the outgoing supervisory signal are recovered using sophisticated techniques, including cancellation of the incoming traffic signal, narrowband filtering, and signal correlation. This monitoring method can be used in service (by maintaining a low percentage modulation of the traffic signal) or out of service (using a large percentage modulation to reduce the correlation time to achieve a particular measurement accuracy).

The OTDR path comes into play for out-of-service measurements when the traffic signal is replaced by one from a coherent optical time-domain reflectometer (COTDR) test set. That signal is constant in power to maintain stable gain regulation by the EDFAs in the outgoing line, but periodically shifts to another wavelength for short durations. Because detection of the signal returned to the

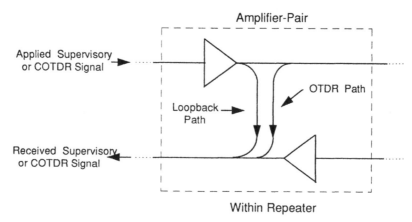

FIG. 7 Loopback and optical time-domain refractometer (OTDR) paths in amplifier pairs (COTDR = coherent optical time-domain reflectometer).

COTDR is restricted to the second wavelength, a traditional pulsed OTDR operation is achieved.

Although the COTDR is also able to measure loop loss via the loopback path in each amplifier pair, the primary measurement is of Raleigh scattering returned from the outgoing fiber following each amplifier and coupled onto the incoming line via the OTDR path. In the receiving portion of the COTDR, this weak returned signal is separated from the background noise and unwanted signals by use of coherent detection, catering to very narrowband filtering, and other sophisticated signal processing steps. As a result, the COTDR can provide a traditional OTDR loss-versus-distance trace for each outgoing fiber in each repeater section. (For a more detailed discussion of traditional OTDR operation, refer to the section describing fault location in repeaterless submarine cable systems.)

Measurement from both terminals provides this coverage for every fiber span in the system. Currently, the finest resolution one can obtain with the COTDR for locating a fiber span fault is 300 meters. Measurement time needed to achieve this resolution can be many tens of hours, but lesser resolutions are considerably faster (e.g., at 10 km the measurement is virtually in real time).

An alternative means of optical line monitoring and fault locating is also in use. It involves a command-and-response channel to/from each repeater, with electronic sensing and encoding circuits within the repeater. In that way, one can "interrogate" a repeater to determine the input and output power of each amplifier. This approach is patterned after a similar technique used with regenerative submarine systems and can localize a fault to a repeater section.

Powering

For repeatered submarine cable systems, the undersea electronics are powered by high-voltage, constant-current sources (power feed equipment [PFE]) that are located in the shore-end terminal stations. Distribution of this current to the undersea repeaters and branching units is via a copper conductor in the undersea cable, with the ocean itself used as the return path. The power system configuration depicted in Fig. 8 is for a point-to-point system. This is the simplest of configurations and forms the operational basis for more complex system architectures. The PFEs at each end of the system are operated in a series-aiding mode (one providing positive voltage and one providing negative voltage) to obtain up to 15,000 volts at nominal current in the undersea cable.*

The system voltage required to provide the constant current depends on the cable resistance, the number of electronic elements in the system, and the magnitude of the geomagnetic earth potential.† Thus, to the first order, the required system voltage is proportional to the length of the cable. The use of

---

*Modern-day submarine cable systems operate at 0.92 or 1.6 amperes.
†Various points on the surface of the earth can be at significantly different electrical potentials. Part of the difference is caused by the interaction of the earth's magnetic field with the moving conductor formed by ocean water currents. Other contributors are sunspots and magnetic "storms" that occur in space, causing a variation in the distribution of charge in the upper atmosphere and thus the earth's magnetic field.

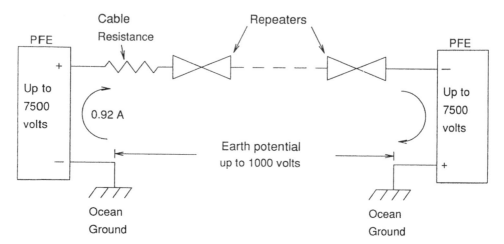

FIG. 8   Powering arrangement for repeatered systems (PFE = power feed equipment).

constant-current sources in conjunction with local regulators at each undersea repeater allows a single conductor in the cable to distribute power equally to each of the repeaters irrespective of relative location and prevailing geomagnetic potential. The power processing units, or converters, within the PFE have the characteristic of a negative slope resistance. That is, as current increases, the output voltage decreases. This feature allows current-regulated converters and PFEs to be operated in series without the need for master-slave control.

A cable system having a total required voltage less than the rating of a single PFE can be single-end fed, but standard practice is to provide a PFE at each end of the cable. This has several advantages. Service can be maintained even if one PFE suffers a failure or needs to be shut down for maintenance, and service can be maintained even if there is a shunt (insulation) fault in the system power path.* The last is achieved by adjusting the voltage of the two PFEs to place the virtual ground (0 Volts) at the site of the fault. For systems having a total required voltage greater than that of a single PFE, it is standard practice to equip each PFE with the maximum rating (rather than one-half the required voltage) in order to provide virtual ground positioning capability for as much of the cable length as possible.

For branched system configurations, high-voltage relays in the branching unit power-path circuitry provide the means for establishing the main power path from the positive PFE normally connected to the trunk to a negative PFE on one of the two branches. These relays also connect the power path of the remaining branch to an ocean electrode at the branching unit, which enables it to be powered independently. The high-voltage relays may be of either the power-switched type, which are controlled by the sequence of power application, or of the supervisory-switched type, which are controlled by a command signal sent from one of the terminal stations.

---

*A "shunt fault" occurs when the power conductor of the cable is shorted to sea ground, usually due to external aggression by fishing equipment or anchors.

The safety of operating personnel is of utmost importance. The input battery power is hazardous to work with, but the PFE high-voltage output, up to 7500 Volts DC (direct current), can cause death on contact. The primary protection against access to cable potential is the use of captive interlocks. These devices preclude entry into high-voltage areas unless the units providing power to these areas are de-energized and connected to the test load, or else the cable is connected to station ground. The use of a grounding probe can provide additional assurance that residual charges* are drained prior to maintenance activity. Similarly, the captive interlocks preclude power up or connection to the cable unless all compartments are closed and locked. Isolated controls and meters provide protection for operators, as do emergency shutdown switches, operable by non-technical personnel in the unlikely event of an equipment malfunction.

Protection for the undersea plant is provided by redundant automatic shutdowns for both excessive voltage and excessive current. In addition, alarms, generated at levels well below those that could cause damage, warn of potential malfunction. Training of operating personnel is a prerequisite to safety, and all PFE operations are carried out under the direction of a resident power safety officer.

Powering availability (reliability) is obviously critical to system performance. Powering availability is maximized by providing redundancy in both the PFE and in the system powering design. Within the PFE, sufficient redundancy is provided to account for the failure of any single element or subsystem, and means are provided to allow for routine maintenance and repair without having an impact on service. Nonredundant elements have proven reliability, and their numbers are minimized. To the greatest degree possible, system powering design accounts for all potential failures, be they PFE equipment failures or faults in the undersea system.

Repeaters and Branching Units

Amplifiers

Fiber amplifiers are the primary building blocks for both line repeaters and the latest technology branching units. Erbium is at present the most commonly used dopant in fiber amplifiers because, when incorporated into the glass core, it efficiently absorbs energy at 980 or 1480 nm and emits energy in the 1550–1560 nm range. When pumped sufficiently at either of these wavelengths, the population of energy states becomes inverted, meaning that the number of excited states exceeds the number of unexcited or ground states. Gain can occur only in the presence of inversion. That energy radiated in response to stimulation from an incoming photon is called *stimulated emission*; that which occurs randomly is called *spontaneous emission*. The former produces useful gain, while the latter produces noise that, when amplified in the fiber, is called ASE noise.

Part of the optimization of EDFAs is in minimizing the ASE for a given gain

*Various elements in the PFE, as well as the cable itself, have capacitance that can store significant amounts of electrical charge.

and output power. Each amplifier adds ASE noise power that is proportional to its gain. Because amplifiers tend to operate at a constant output power, the added noise also reduces signal power. Thus, longer systems generally require shorter spans to maintain a given level of performance (BER). The spontaneous emission occurs equally in two orthogonal polarization modes. The individual erbium ions are polarized and oriented randomly with respect to the fiber axis and the signal polarization. The signal selectively depletes those energy states with its corresponding polarization, thus reducing the signal gain relative to that in the orthogonal polarization state and increasing the gain for noise in the orthogonal state. The process is called *polarization hole burning*, and without countermeasures can cause noise saturation of the amplifiers due to buildup of this orthogonal component of ASE noise. The effect can essentially be eliminated by modulating the polarization of the transmitted signal at a rate that is fast compared to the gain response of the EDFA.

As shown in Fig. 9, the simplest EDFA contains only two components besides the erbium-doped fiber and the laser pump. A wavelength selective coupler (WSC) efficiently couples the pump power into the EDF while adding minimal loss in the transmission path at the signal wavelengths. The coupler does, however, slightly affect the gain shape of the amplifier. Amplification is provided for both forward- and backward-propagating light waves. An isolator at the output prevents backward-propagating light, such as reflections, from being amplified and consuming pump power.

Line Repeaters

Simplicity in support functions is key to realizing the full advantages (high reliability and performance at low cost) of fiber amplifier technology in the line repeater. The two functions are pump laser sparing and control and support of line monitoring. Figure 10 is a block diagram of a line repeater containing four EDFAs. Shown is an example of passive implementation of line monitoring via high-loss loopback paths, which was discussed above.

One type of implementation of the pump manifold function is shown in Fig.

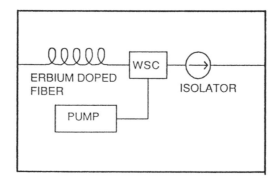

FIG. 9   Simple erbium-doped fiber amplifier (WSC = wavelength selective coupler).

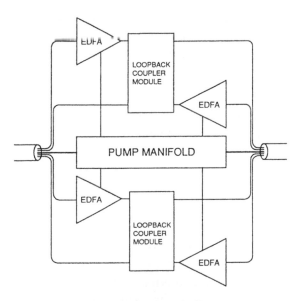

FIG. 10   Block diagram of a line repeater.

11. To achieve the full reliability advantage of redundancy with the fewest pump lasers, four lasers are coupled equally to four amplifiers. This requires only four interconnected equal-ratio couplers. Isolators on the pump output ports reduce interactions between lasers and the effects associated with laser-mode locking. In the powering scheme shown here, the laser diodes are powered in series with line current. The line current can be adjusted at any time, if required, to achieve a given average repeater launch power. This capability is beneficial in maintaining the average pump power required for optimum system performance.

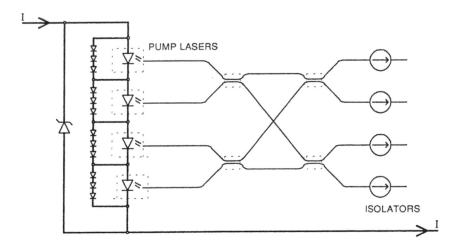

FIG. 11   Optical pump manifold detail (heavy lines are electrical connectors; light lines are optical fibers; I = constant direct current).

Other parts of the pump manifold include series-connected diodes in parallel with each laser, which conduct only if a laser fails in an open circuit mode. A Zener diode connected across the lasers protects against power surges or inadvertent polarity reversals of line current. The line current in such a system equals the nominal laser current, which is normally in the range of 200 to 400 milliamperes, with a voltage drop of less than 6.5 Volts.

Other types of pump manifolds can be used. One common arrangement includes two shared pumps, each of which is operated at constant power by an electronic circuit that adjusts the laser current to maintain a constant and factory preset value of laser power, usually sampled at the back face of the laser.

Branching Units

Branching units allow more than two terminals to interconnect over one undersea system. Their functionality varies with each application. The simplest is a passive unit in which fibers are internally and permanently routed among the three ports. Optical switching of fiber within a BU is also possible, with the states of the optical switches directly dependent on the states of the high-voltage power-switching relays.

With the advent of WDM technology, branching units have become more complex, adding optical amplification and being able to redirect selected wavelengths (channels) to/from another fiber. Such capability is known as *optical add/drop*. Figure 12 is a block diagram of an optical add/drop branching

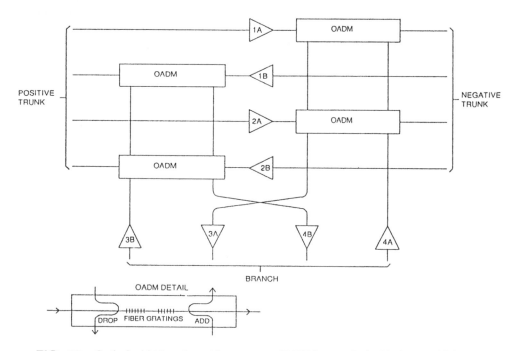

FIG. 12 Optical add/drop branching repeater (OADM = optical add/drop multiplexer).

repeater. It is capable of dropping and adding to a branch up to two optical channels on each of two fiber pairs. In this example, the technology used is reflective fiber gratings and couplers. The EDFAs compensate for the loss of the adjacent cable sections. The net loss of such a BU, which can be as low as 4–5 dB, is made up by shortening the adjacent cable sections. Powering of branching units was discussed above.

Physical Design

The optical and electronic components of undersea line repeaters are placed into pressure housings that permit them to operate at near atmospheric pressure independent of the external sea pressure. Some manufacturers use a copper-beryllium alloy for the pressure housing, which does not corrode significantly in seawater, while others use a steel housing in conjunction with various corrosion-protection systems. A typical repeater, being launched into the ocean, is shown in Fig. 13. The pressure housing in this example is approximately 30 centimeters (cm) in diameter and 80 cm long.

At the repeaters and branching units, the optical fibers and the electrical power conductor from the cable are terminated in a joint box, which is essentially the same as a cable joint except that on the end opposite from the cable there is a polyethylene-insulated metal tube (called a pigtail) instead of another

FIG. 13   Repeater housing, including bend limiters on cables.

cable. This pigtail tube carries electrical power, along with the optical fibers, into the amplifier housing through a high-pressure seal assembly. Tension and torque from the cable are transferred by the pressure housing to the cable on the other end.

A branching unit pressure housing is similar to that of a line repeater, except it is about 50% larger in diameter. This permits two branch cable connections to be made on one end and a trunk cable connection at the other. Also located at the trunk cable connection is the sea-ground electrode, which is required for powering one of the two branch cable legs. While line repeaters can be handled through shipboard cable machinery much the way cable is handled, branching repeaters cannot. They are installed slowly and carefully, using two cable engines, in accordance with strict procedures.

Cables and Joints

Submarine cables can be divided into two categories, lightweight and protected. The lightweight design is used directly in deep water and in many cases is used as the starting point for construction of protected designs. Some manufacturers make a special core for protected cables that is not used by itself.

Lightweight Cables

Cables from all manufacturers are similar because the operational requirements are the same for all and the materials that will survive for long times (25 years) in seawater (and which are economically viable) are very limited. The principal requirements for a lightweight, deep-water cable are

- provide strength to allow installation and recovery in water up to 8000 meters deep
- provide a benign and protected environment for the optical fibers
- provide DC-power-carrying and insulation capacity sufficient to power systems up to 9000 km long (typically 1 ampere at 7000 volts)
- provide sufficient abrasion protection to allow handling, installation, and, if necessary, recovery and reinstallation
- accommodate the fiber types required by the transmission design

In all designs commonly used today, strength is provided by high-strength steel wires arranged around a central core that contains the fibers. In addition to providing strength, the array of steel wires forms a pressure vessel that isolates the fibers from sea pressure. The fibers are placed in the center to reduce as much as possible tensile strain in the fibers resulting from cable bending. The DC conductivity is provided by a copper member, usually a tube, either inside or outside the steel wires. In addition to providing conductivity, the copper tube is usually made hermetic to protect the fibers from water and from environmental hazards such as hydrogen, which can react with the glass to increase fiber attenuation. The copper-steel structure is filled with a polymer to prevent ingress of water in the event that the cable is broken in service. Polyethylene insulation over the copper-steel structure provides both electrical insulation and abrasion resistance.

Fiber structures used in submarine cables are of two types, tightly buffered and loose tube. In the tightly buffered designs, the fibers are encased in an elastomeric solid that holds the fibers firmly, so the fiber strain and the cable strain are equal. The maximum allowable fiber strain determines the maximum load-carrying capacity of the cable, which is usually less than the breaking load.

In loose-tube designs, the fibers are placed inside one or more tubes. Each tube is filled with a very viscous liquid that holds the fibers approximately in place, but allows some slow movement under load. About 0.2% excess fiber length is placed in the tube, so that the fiber experiences no average strain until the cable strain exceeds that value. A diagram of a typical lightweight cable design, showing both kinds of fiber structures, is shown in Fig. 14.

Core Cables

Cables used only as a starting point for construction of protected cables are similar in many ways to lightweight cables, but in general do not have strength members and consist only of a fiber structure, a hermetic conductor tube, and polyethylene insulation. Strength is provided by the outer protection. Since these are used only in shallow water, the conductor tube provides pressure isolation for the fibers.

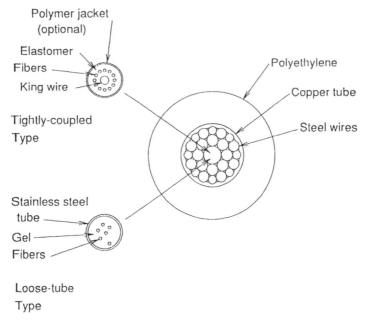

FIG. 14 Lightweight cable design.

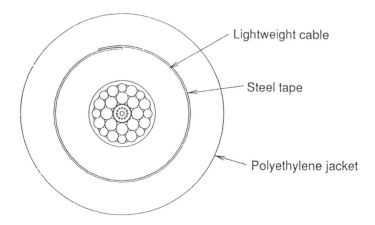

FIG. 15   Steel-tape-protected cable.

Protected Cables

The lowest level of added protection used in submarine cables consists of a layer of steel tape applied over lightweight cable, with a layer of polyethylene applied over that. This provides significant improvement in abrasion resistance and protects against shark bite.* A typical cross-section is shown in Fig. 15.

Higher levels of protection are provided by wire armor structures. Light, single-armored cable is used for deeper water and for burial. Heavy, single-armored cable is used in intermediate depths when cable is laid on the bottom rather than buried; it is often used for repair of light, single-armored cable since the need for repair indicates the need for added protection. Some suppliers provide three grades of single-armored cable to accommodate the various depth ranges and sea-bottom conditions. Figure 16 shows double-armored cable, used in very shallow water and up onto the beach. Single-armored cables are very similar in design, simply having the outer layer of armor and its underlying bedding omitted.

A special protected design called *rock-armored cable* has high crush resistance and is used where rock slides are a danger and in locations where icebergs may settle onto the cable. Rock-armored cable is similar in design to double-armored cable except that the outer layer of armor consists typically of six to eight large-diameter wires applied with very short lay length.

Another special cable design has been developed for use in bringing submarine cables to offshore floating oil production platforms. These cables, called *dynamic risers*, are suspended essentially vertically in the water from the sea bottom to the deck of the oil platform. They must withstand a service life under static tension from their own weight, as well as over 5 million superimposed tension cycles per year caused by motion of the floating platform due to weather, waves, and currents. Dynamic riser cables have multiple layers of

---

*Sharks have attacked cables in some areas. It is known that they find food by sensing electric fields from other fish; apparently they sense the electric field of the cable and mistake it for prey.

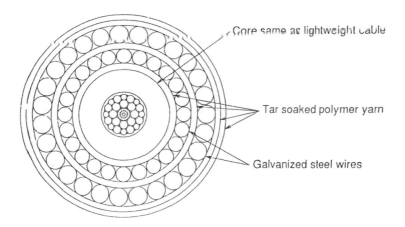

Core same as lightweight cable

Tar soaked polymer yarn

Galvanized steel wires

FIG. 16 Double-armored cable.

armor wires, with the dimensions and lay directions arranged so that tension in the cable produces essentially no torque. Termination of such cables at the platform presents significant design challenges in preventing failure from tension and bending fatigue.

Cable Joints

Joints are provided in order to connect two pieces of cable together. A universal design, the lightweight version of which is shown in Fig. 17, has been developed by a consortium of American, British, French, and Japanese system suppliers. This jointing system is capable of connecting cables from all major manufacturers. The steel wires of the cables are terminated into a steel structure that provides tension and torque transfer across the joint, as well as electrical continuity. The fibers are spliced and coiled into an internal protective chamber. The assembled joint is covered with a cylinder that forms a pressure vessel, and finally the joint is encased in polyethylene, which bonds to the polyethylene in the cables and provides electrical insulation, as well as a water-tight jacket.

An armored cable joint, shown in Fig. 18, consists of a lightweight joint connecting the central cables plus a means for terminating armor wires and transferring the armor wire load and torque across the joint.

Cable joints have a breaking strength of at least 90% of the breaking strength of the cables and are capable of being bent, under full operational load, over the same machinery surfaces as the cable.

End seals are used to terminate cable ends for storage, for connection to test equipment, and for pulling shore-end cables through underground ducts into beach joint chambers. One version allows the end of the cable to be sealed so that it can safely be placed in the sea temporarily during installation and repair operations. Another version provides electrical continuity between the cable power conductor and the sea so that the system can be powered for testing.

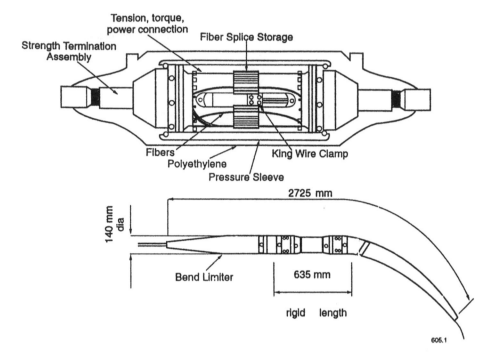

FIG. 17 Lightweight cable joint, universal type. (Courtesy of the Universal Jointing Consortium, Southampton, England.)

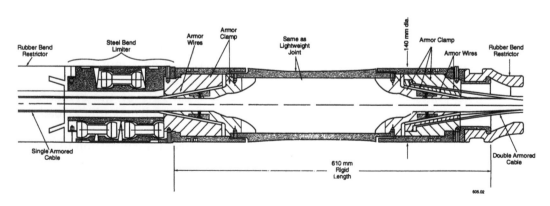

FIG. 18 Single-to-double-armored cable joint, universal type. (Courtesy of the Universal Jointing Consortium, Southampton, England.)

Reliability

Components

Repairs to a submarine system may take weeks to complete. Because modern optical systems carry a tremendous density of telecommunications traffic, system owners face large financial penalties during a repair due to the cost of restoration capacity on another system. In addition, system users are concerned about dropped telephone calls and other possible interruptions in their communications links. Faults in a system that require an at-sea repair can be caused by an internal failure (e.g., a broken fiber or failed component in a repeater) or by external aggression (e.g., a broken cable caused by fishing activity or a ship's anchor). Steps taken to minimizing external aggression faults are discussed separately. Here, the emphasis is on steps taken during the design, manufacture, installation, and operating phases to assure the quality of the submarine system against internal failures.

Purchasers of a submarine system typically specify that the expected number of ship repairs due to internal failure during a 25-year service life be less than three (for a transoceanic, repeatered system). To meet this requirement with a system composed of many thousands of electronic and optical components, optical-fiber splices, mechanical connections, and so on requires robust design of components, as well as freedom from defects and damage during the manufacturing, assembly, installation, and operating phases of the project. The outline below indicates some of the key approaches taken to meet these objectives:

Robust Design
  Components
  • Use technologies and designs with reliability that has been established by experience in other applications.
  • Rigorously qualify each design for undersea use. Typically, these programs involve theoretical analysis, stress testing, and aging tests designed to reveal possible modes of failure or degradation.
  System
  • Avoid active devices where possible, use redundancy where they are required. These typically have higher power densities, higher operating temperatures, and increased complexity. For example, in some optical amplifier systems, the semiconductor pump lasers are paired such that failure of one device does not cause a system failure.
  • In general, minimize the number and complexity of components and repeaters.
  • Use self-compensation to make the system robust to changes in individual components.
Avoid defects and damage
  Manufacturing
  • Perform qualification testing on devices made on the same manufacturing line, with the same processes, to be used for making product devices.
  • Screen every device at stresses equal to or exceeding worst-case use conditions. Device acceptance is conditional and depends on the device meeting

performance and stability requirements and absence of untypical behavior.
- Maintain production facilities to assure consistent processes.
- Require formal training and testing of all personnel working in manufacturing.

Assembly
- Perform periodic tests to look for any performance anomaly.
- Maintain facilities to assure consistent processes.
- Require formal training and testing of all personnel working in system
assembly.

Operation
- Use surveillance systems to monitor the performance of the system and
detect possible problems or changes in transmission characteristics before
the function of the system is seriously affected.

While the challenge of making large, high-capacity, reliable, undersea systems is indeed great, experience has shown that a rigorous, comprehensive, and
well-executed approach like that outlined above can result in nearly trouble-free
system operation. The high demand for undersea links throughout the world is
due in large part to the satisfaction that owners and customers have had with
the performance and reliability of these systems.

Cable Selection and Placement

Strategies to minimize cable failures from external aggression, which occur
almost exclusively in water depths less than 1500 meters, include use of protected cables, burial of cable under the sea bed, and careful selection of the
cable route from the shore to this depth. Land cables between the terminal
station and the beach joint, located very near the water's edge, are typically
placed in underground ducts. (Often, separate cables for DC power and terrestrial optical transmission are used between the terminal and the beach joint.)

Cable protected by double armoring is typically used from the beach joint to
a water depth of approximately 100 meters (up to 200 meters if required by
severe currents or bottom conditions) to provide maximum protection from
abrasion in the surf and tidal current areas, where the bottom soil may shift and
erode. This cable type is commonly buried in the sea bottom, either directly or
in ducts, at least through the surf zone and often even farther out. After this,
cable burial to depths ranging from 0.6 to 1.0 meters can be carried out with
plows to a water depth of between 1000 and 1400 meters, depending on the
bottom conditions. Burial in deeper water is possible with jetting equipment
attached to remotely operated vehicles, but the process is slow and costly. In
shallow water when the bottom is soft and where large ship anchors can be
encountered, much deeper burial, up to 3 meters, is sometimes done.

Buried cable is usually single armored. If bottom conditions do not permit
cable burial, heavily protected cable types are used to minimize damage from
abrasion and fishing activity. Proper route selection can often avoid areas of
heavy fishing activity and strong water currents.

For all cables types, the maximum deployment depth is determined by considerations of tension in the cable due to cable weight, water drag, and ship motion during recovery operations in the event that a repair is required.

Ring Networks

Even with the remarkable record of component reliability achieved in submarine cable systems, failures generally from external aggression still occur. Protection against loss of service and even short service interruptions can be gained by configuring cable systems in ring networks.

One example of ring architecture is two transoceanic cables terminating at separate (but generally nearby) landing points. Ocean cables or land links are used to connect the ends of the transoceanic cables on each side, forming the ring. Half of the primary traffic is carried on each cable. If one cable fails for any reason, the traffic is automatically switched to the other cable, and thus service is maintained. Because the reliability of submarine cable systems is very high, the ring is in fault condition for only a very small portion of its life. The rest of the time, lower priority traffic, which can be interrupted without notice, is carried on both cables in the protection channels. The TAT-12/13 and TPC-5 systems are ring networks of this type.

Another example of ring network architecture is a single trunk cable completely encircling a land mass, with a multiplicity of branching units providing a leg to connect to the land mass. This configuration has been proposed for the African continent.

System Manufacturing and Testing

Repeaters

Within the constraints imposed by the ocean environment, the submarine repeater is designed to house and protect a set of optical amplifier units, one for each fiber, from the sea pressure and the forces associated with installation and recovery. Stringent reliability objectives require careful attention to mechanical design for shock, vibration, and temperature performance.

Optical amplifier units are usually mounted on metallic frames in pairs, one for each direction of transmission. In addition to the optical components, which operate directly on the transmitted signal, there are electronic components for distributing power within the repeater to the pumps and the pump circuitry itself. Multiple optical amplifier pairs can be mounted in a single pressure housing. The mounting arrangement provides shock and vibration isolation while accommodating the dimensional changes of the housing due to sea pressure. Moreover, it provides a low thermal impedance path to the ocean from heat-generating elements such as pump lasers. Because the repeaters are powered in series over the cable, the internal frames on which the components are mounted can be at potentials of as much as 7500 volts with respect to the sea. Therefore, a reliable method for high-voltage insulation between the frames and the pressure housing is included in the design.

All components are thoroughly tested before being incorporated into repeater assemblies. The assembly operations take place with careful monitoring and inspection in clean rooms. Thorough testing is carried out at various stages to assure that the unit continues to meet performance requirements. When a full set of amplifier assemblies is completed, they are placed into the pressure housing, along with other power and control equipment, and connections are made to the flexible pigtail cable that carries the fibers and power outside the pressure housing. The covers are then welded in place, and the assembly is given a final set of tests.

Cable

Optical cable manufacturing begins with the manufacture of fibers to exacting specifications for physical dimensions and optical characteristics of attenuation, chromatic dispersion, and polarization-mode dispersion. While PMD normally can be controlled adequately, it is not possible to draw continuous fiber at cable-section lengths that meet, at acceptable cost, the attenuation and chromatic dispersion requirements. Consequently, a typical cabled fiber consists of three or four separate fibers spliced together, selected to produce the overall transmission properties required. In addition to requirements for each fiber, there are uniformity requirements that must be met among all fibers in a cable at any point along the length. The total length of a fiber to be cabled is typically 50 to 70 km.

While the details vary considerably depending on the particular cable design, a number of manufacturing processes are reasonably common among all standard designs. The fibers, the strength members, and some form of hermetic barrier are assembled to form the central portion of the cable. Fibers are placed at or near the center of the cable, with careful control over tension and placement, in the case of tightly coupled designs, and over slack, in the case of loose-tube designs. The strength members, usually round wires, are helically stranded around the fiber core with the tension and placement being carefully controlled.

A Japanese design includes, in addition to round wires, three 120° arc-shaped wires laid straight to form a cylinder around the fiber core to provide strength and pressure protection. The hermetic barrier can be formed inside or outside the wires. It usually consists of a stainless steel tube (in the case of loose-tube cables) or a copper tube continuously formed, seam welded, and drawn down to size. The weld, as much as 70 km in length, is continuously inspected for flaws, which are repaired. All vacant space in the assembly is filled with a polymer compound to prevent the ingress of water in the event that a cable is broken in the sea.

The insulation layer is extruded over the metallic core and is normally bonded to it by chemical means. Much attention is paid to the cleanliness and dimensional stability of the insulation in order to guarantee satisfactory long-term, high-voltage performance in powered cables. The extruded cable is continuously inspected for voids and other flaws in the insulation, which are repaired.

The description above summarizes the production of deep-water cable. To produce tape-protected cable, the deep water cable is wrapped with a metallic layer formed from strip creating an overlapped straight seam. A protective jacket, usually high-density polyethylene, is then extruded over the metal layer. The metal layer is chemically bonded to the adjacent polyethylene on both sides.

To manufacture armored cable, galvanized armor wires are wrapped helically around the core cable and flooded under and over with an asphalt or tar compound to prevent corrosion. The wires are continuously preformed to their required helical shape before being applied to the cable to reduce stored elastic energy and improve handling properties. In some designs, a bedding of polymer yarn is used under the wires; in others, it is omitted. For double-armored cable, a single layer of polymer yarn is wrapped over the first armor layer, then a second asphalt-flooded wire layer is applied. At least one, in most cases two, layers of polymer yarn are wrapped over the outer layer of armor wires to provide abrasion protection.

Fiber attenuation is measured at various points in the manufacturing process. In the deep-water or core stage, the cable is brought to a known temperature and the full set of attenuation and dispersion properties is measured. These values are compared to specification requirements to determine the applicability of the cable for a particular system. At this stage, the DC resistance of the inner metallic assembly also is measured, and any required high-voltage withstand capability tests are performed.

System Assembly

Unlike earlier coaxial systems in which cable sections and repeaters were loaded separately aboard the cable ship and spliced together there, the cable-repeater assembly of lightwave systems occurs mostly in the cable factory. The longer time and more complex jointing operations for fiber splicing and handling dictate this change. Typically, a few joints must be left to be done on the ship to accommodate the laying scenario (e.g., a planned bow-to-stern cable transfer or deployment of a branching unit) and/or multiple load lines (to reduce the time required for ship loading). Cable-to-repeater jointing operations are very similar to those used for cable-to-cable jointing, discussed above.

Prior to loading the ship, the assembled system is powered and tested to confirm correct repeater-cable connections and to qualify transmission performance before proceeding with the installation.* Tests on each fiber path include a determination of

- the gain-versus-wavelength characteristic
- net chromatic dispersion
- a measure of optical signal-to-noise ratio

---

*To attain the longest possible system length for testing, temporary connections are usually made in the factory at the locations where shipboard joints will later be made.

For systems operating at a single wavelength, a simple receive ASE spectrum measurement with no signal applied is all that is required to determine the system's gain peak. For multiple-wavelength (WDM) operation, it is necessary because of hole-burning effects in the erbium amplifiers to apply all the signal wavelengths simultaneously and measure the gain at each one. Such measurements are necessary to confirm the effectiveness of the gain equalization process. Measurement of net chromatic dispersion is the final step and allows one calculation of a nominal value, after correction for temperature, for the amount of dispersion compensation fiber that will need to be placed in the receiving terminal when the system is installed. Signal-to-noise ratio is the "bottom line" of transmission performance. A number of different measurement techniques can be used. One is the measurement of Q factor (2).

Installation

Land Cable

The land portions of submarine cable systems, between the point the cable is brought ashore and the terminal station, can be up to 30 km in length, though they are commonly only a few kilometers long. These segments are usually installed in a conduit and manhole infrastructure, but can also be directly buried. Instead of extending the submarine cable over land to the terminal station, two land cables are often used, one an optical-fiber cable and the other a high-potential DC power cable. The reasons are to accommodate cable installation in restrictive or crowded conduit systems, to provide robust cable designed for the particular environmental hazards found in terrestrial segments, and to allow for relatively simple repair procedures in the event of cable damage. A third power cable is also used to connect the return current path for the PFE to the power ground bed.

Power Ground Bed

A power ground bed is associated with the power feed equipment to provide a low-resistance return path for the DC current from the ocean. This bed is usually located near the beach joint. The return cable to the station is run close and parallel to the power (and optical) cables, thereby minimizing induction from commercial alternating current (AC) power distribution facilities that might be in the vicinity. Located in this way, inadvertent corrosion of underground metallic structures between the ocean and the terminal station is avoided. To achieve a low resistance to the earth (usually less than 2 ohms), the ground bed consists of an anode field containing no less than six durichlor rods surrounded by a carbon backfill. The mass of the rods is designed to ensure that electrolytic corrosion does not reduce the efficiency of the ground bed over the life of the system. The rule of thumb for corrosion of anodes is one pound per ampere per year. If several power feed systems share a single ground bed, the net ground bed amperage is the algebraic sum of the individual system amperages. In the common practice of sharing one ground bed among several systems, opposite polarity systems effectively reduce the net current flow.

Beach Joint

The transition between the land and ocean cables occurs at the beach joint. In addition, it is the place where the ocean cable is physically anchored by terminating the cable armor wires in a clamp. The clamp in turn is attached to the beach manhole or to a buried earth anchor to prevent the cable from being pulled offshore. The beach often contains an enclosure that is partitioned to allow the high-voltage DC to be transferred to the land cable in an isolation box safely away from the splicing shelf where the fiber splices are stored.

Cable Laying

After cable systems are assembled and tested in the factory, they are loaded aboard installation vessels. All but very short systems are installed by cable ships, which have large storage tanks, specially designed cable engines to lay the cable under precise control, large overboarding sheaves, instrumentation to monitor cable payout speed and tension continuously, and transmission test rooms to monitor fiber performance continuously during system installation. Short systems in shallow water are often laid by small ships chartered for the project and outfitted temporarily with portable cable-handling and instrumentation equipment.

Normally, one shore end is installed first, with the cable paid out by the ship and brought ashore to the beach joint. Figure 19 illustrates one method for landing a shore end. The cable is brought to the shore supported by a series of foamed floats, which are removed after the cable is secured to the beach joint. The ship is in the background. Machinery on the beach is used to pull the cable in and to handle it to the beach joint. The cable is then laid by the ship to a predetermined point, where it is sealed and laid on the bottom with a line leading up to a buoy. The ending point is normally chosen beyond any double-armored cable that may be used and at a location convenient for later jointing work.

For a point-to-point system, the other shore end is then installed in the same manner as the first, but in this case the ship continues to lay the entire system back to the buoy on the first shore end. The buoy and the first shore end are then brought aboard the ship. The fibers are joined, and end-to-end transmission measurements are made by the terminals to verify proper system performance. When this is done, the joint is completed and carefully lowered to the bottom.

For more complicated branched systems, one branch segment for each branching unit is installed from its beach landing to the planned location of its branching unit, where it is sealed off and buoyed. The trunk segment, containing the branching units, is then laid like a point-to-point system, except that along the route the branch segments are picked up and joined in. Transmission

FIG. 19   View of shore landing.

testing is carried out continually to assure that the installed system is satisfactory.

Commissioning and Acceptance

When installation of the terminal equipment and undersea plant is complete, the system as a whole is tested to

- verify that its performance and operational features meet expectations
- obtain reference information to aid in future maintenance and fault location activities
- build confidence in continuing satisfactory operation

Terminal-to-terminal pre-service testing is called *commissioning*. Before performance and features are measured and demonstrated, line-up adjustments are made (e.g., setting optimum signal wavelength and power levels at the terminal-undersea cable interface, setting alarm thresholds, installing dispersion compensation fiber in the receiving line, and the like). The primary measure of transmission performance is the BER. The problem from the commissioning standpoint is that typically there are so few errors with an optical amplifier system that the BER cannot be determined in a practical measurement time.*

However, because the performance might degrade over the system life (caused by such things as component aging or extra loss added as a result of undersea repair operations), system operators would like to have the system margin measured at commissioning. That is, they would like to determine how much the S/N could decrease and still maintain the specified BER requirement. The Q-factor measurement, mentioned above, has been the method used to determine the margin at commissioning, but it is not without its own problems.†

Other means of characterizing effective system margin are being considered. The fundamental problem is that the BER performance of these systems is almost perfect. Measuring degrees of perfection is understandably difficult.

To enhance the ability to detect and locate small transmission degradations in the undersea plant, baseline measurement runs are made with the line-monitoring equipment in the terminals. During the system life, small transmission changes can be recognized more easily as deviations from a baseline. Similarly, with the system unpowered, baseline DC and low-frequency power-path measurements are made to improve the fault location accuracy of these measurement techniques.

The last phase of commissioning is usually the confidence trial. The system is operated continuously over a period of one or two weeks, but sometimes longer, using a pseudorandom test pattern in the traffic bands, while recording any transmission anomalies (bit errors, out of frames, and the like) that occur. Such testing has the advantage of detecting an intermittent problem that might otherwise be missed.

After the system is in operation, steps must be taken to guard against aggression damage from commercial and other activity taking place near the cable. Fishermen and others working near the system are provided with information about the precise location of the cable so they can avoid it. When activity is observed along the cable route, ships or airplanes are often sent out to patrol the area and to warn away anyone who might pose a danger to the system. Close liaison is maintained with fisheries and others working on or near the seabed in the vicinity of the cable to minimize the risk of damage. Most coastal nations have established legislation protecting submarine cables and giving system owners recourse in the courts in the event of damage.

System Repair

Submarine systems require repair on occasion, most often because of damage by fishing gear or anchors, but sometimes due to other causes. In the segment between the shore and the first repeater, it is possible to locate fiber faults from the terminal with great precision using a conventional OTDR. In relatively shallow water, power-path faults can be located accurately by means of elec-

---

*It is not unusual at commissioning for no errors to occur for weeks at a time.
†Q-factor measurement uses the decision-threshold method. It has been shown to give incorrect (pessimistic) results when intersymbol interference (ISI) is an important contributor to the net S/N performance.

troding. With electroding, a low-frequency tone (usually 25 Hz) is transmitted into the power path along with the regular DC. The electric or magnetic field surrounding the cable that is created by this tone can be detected by special electric or magnetic probes. Some are designed to be hand held by divers and others to be towed from a repair vessel. Such detectors are also mounted on remotely controlled underwater vehicles that can be used to do post-burial bottom surveying, and assist in locating, recovering, and repairing submarine cables.

Line-monitoring measurements from the terminal can localize transmission and power faults anywhere in the system to a repeater section. On systems capable of using a COTDR, fiber faults can be localized with moderate accuracy within a repeater section. Shunt fault location using the PFEs can also often be made within a repeater section.

Faults, once located, are usually repaired in the following manner. The repair ship drags a cutting grapnel across the cable route to engage and sever the cable. The ship then moves approximately one depth of water to the side of the cut believed not to contain the fault and drags a holding grapnel to engage and pick up the cable end. The recovered cable end is tested to determine if the fault is in the recovered portion. If the fault is not found, the cable is cut cleanly, sealed, and placed on the sea bottom with a line running up to a buoy. The other cable end is then recovered with a holding grapnel, and the presence of the fault is verified. Cable is picked up until the fault is aboard the ship, then the cable beyond the fault is tested to be sure it is clear to the terminal. The end is joined to spare cable carried aboard the ship, which is laid back to the buoy. The spare cable is joined to the buoyed cable end, and terminal-to-terminal transmission tests are carried out to verify proper system operation. The final joint is then carefully installed, and the repair is complete.

Systems Without Undersea Repeaters

Introduction

Some applications for submarine cable systems are short enough not to require undersea repeaters. These quite naturally are referred to as *repeaterless* or *non-repeatered* systems. Besides the obvious, they differ from repeatered systems in a number of interesting ways:

- The cable and joints do not have to withstand high voltage, so the cable can be smaller and the joints simpler.
- Because of special techniques usable only in terminals, single spans in repeaterless systems can be much longer than maximum distances between repeaters.
- The number of fibers that can be used in a cable is not limited by the number of amplifiers that can be placed in a repeater housing, so quite high fiber count cables have been used and others are proposed (up to 100).
- No PFE or repeater-monitoring equipment is needed in the terminals.

- Small ships can be used for installation and repair since the cable storage volume is much less than for repeatered systems, and shipboard test equipment needs are minimal.

Transmission

Propagation and Compensation

Compared to repeatered systems, the distance traversed by the traffic signals over repeaterless links before regeneration is quite modest, a few hundred kilometers at most. Therefore, many of the waveform distortion problems of repeatered transmission are considerably reduced. In most, but not all, cases, attenuation in the fiber is the major transmission concern, and often lower loss pure silica fiber (Z fiber) is used. Much higher launch power can also be transmitted, even more than a tenth of a watt. With such signal power, one must be concerned with self-phase modulation and another fiber nonlinearity, called stimulated Brillouin scattering (SBS), which is analogous to overload in electronic transmission systems.

To avoid SBS at these signal powers, one can dither the signal to broaden its wavelength spectrum. Self-phase modulation can be controlled by using fiber with large negative dispersion in the first several tens of kilometers from the terminal in the transmit direction. The distance traversed without a repeater can be increased by many tens of kilometers by use of a technique called *remote pumping* (by which a length of EDF fiber is placed in the undersea cable, but is pumped from the terminal) and through the use of forward error correction coding in the terminals.

Remote Amplifiers

Erbium-doped fibers can be used to provide undersea amplification in nonrepeatered, as well as repeatered, systems. Instead of being combined with optical pumps in a repeater housings, the few tens of meters long EDF is either spliced to regular fiber and cabled or coiled into a cable joint housing. In either case, the optical pump power is supplied from the terminal, usually over the transmission fiber.* An alternative design is to use a separate fiber for the pump power, with an optical coupler located in the joint housing near the EDF to get the pump power to it. Transmission and available pump power considerations dictate that the EDF be located typically 50 to 80 kilometers from the receiving terminal, so pump power is supplied from this end of the system. The EDF location is a compromise between getting as much extra system length as possible (the reason for using the amplifier in the first place) and providing sufficient pump power to the amplifiers to achieve the needed gain.

---

*The high-power pump energy at 1480 nm has the added advantage of providing several decibels of gain in the fiber over which it is transmitted at the signal wavelength (1550 nm range) through a nonlinear process known as the Raman effect.

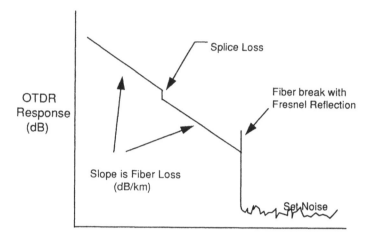

Distance (Kilometers)

FIG. 20   Typical optical time-domain reflectometer trace showing a fiber break.

Fault Locating

Undersea fault-locating methodologies for nonrepeated systems are simpler than those for repeated systems. The primary tool is the conventional OTDR, a test set that can detect and locate transmission anomalies in glass fiber. The OTDR launches repetitive, short-duration light pulses onto the fiber. That light is backscattered continuously along the length of the fiber (Raleigh scattering) and returns to the test set. With the constant speed of light in glass, time to return is converted to distance so that the OTDR displays loss versus distance on a cathode-ray tube (CRT) screen. For a normal fiber, this is a straight line with a slope that is the fiber loss characteristic (typically dB/km). A lossy splice, for example, is seen as a sudden drop in the trace at the splice site.* Of course, the trace ends at the site of a broken fiber. Often, at a fiber break a Fresnel reflection occurs that is stronger than the backscattered light and is seen as a "spike" on the screen trace. These OTDR trace features are illustrated schematically in Fig. 20.

The distance the OTDR can "see" into the fiber is limited by signal-to-noise considerations. Signal power can be increased by increasing the pulse duration, but the accuracy of spatially locating an anomaly is inversely proportional to pulse duration. These parameters are adjustable and can be optimized for a particular measurement. With an OTDR in each terminal, the complete fiber path for even the longest nonrepeated system can be seen.†

Though normally unused, the steel-copper strength member of the cable

---

*The drop in the trace at a splice is not necessarily an accurate measure of splice loss, especially so if the splice is between different fiber types. True splice loss can be determined with an OTDR by averaging the two splice losses seen by measuring from both ends of the fiber.
†Because the remote EDF amplifies bidirectionally, its gain can be seen in an OTDR trace, although narrowband filtering at the OTDR is needed because of the ASE noise.

can be used to carry an electroding signal. During at-sea repair operations or post-installation bottom surveys, low-frequency electroding signal detection can be a valuable tool for locating the cable from the ship, including finding the site of a broken cable.

Cable and Joints

Cable for Nonrepeatered Applications

Cable used in nonrepeatered systems is similar in many ways to the cable used in repeatered systems because it must provide many of the same features. The fibers need to be housed in a soft, protected environment; the cable must be strong enough to be installed and recovered; the structure must be able to withstand the sea pressure at the maximum depth of use; and the strength members need to be protected from corrosion. Although the cable does not have to carry electrical power to repeaters, it still must provide a conductive path insulated from the sea for purposes such as electroding. In addition, some system operators apply DC of a few milliamperes to the cable from the terminal in order to discover insulation faults that might exist but have not yet caused transmission impairment in the fibers.

Repeaterless systems are often installed in shallow water, so the strength and pressure requirements are less than for repeatered cable. This has led to families of lower cost cables that are essentially scaled-down versions of repeatered system cables, with fewer strength member wires, higher DC resistance, and less insulation. Figure 21 shows a typical lightweight cable of this type. The diameter of this cable is approximately half that of the corresponding repeatered system cable shown in Fig. 14. Abrasion-resisting jackets and fish-bite-protection metallic layers have been applied to these smaller cables to make them suitable for the ocean environment. Furthermore, single-armored, double-armored, and rock-armored versions of these cables have been developed that are similar in form to the corresponding types of repeatered cables. Because of their typical use in shallow water, the percentage of armored cable in repeaterless systems is

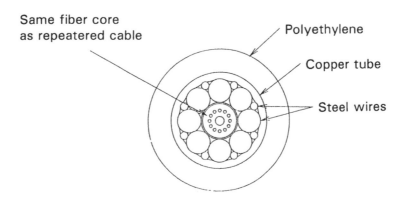

FIG. 21   Lightweight cable for repeaterless systems.

much higher than in repeatered systems. The use of smaller cables thus provides a significant cost advantage.

Joints

The tension-carrying, torque-carrying, and pressure-resisting portions of repeaterless cable joints are similar to those used for repeatered system joints, but the magnitudes of these forces are normally smaller. The lower voltage requirement for the repeaterless systems, however, allows different approaches to insulation of the joints. While injection molding of polyethylene is almost always used in high-voltage systems to maintain insulation continuity, low-voltage systems can use simpler, faster methods, such as adhesive-lined, heat-shrinkable polyolefin tubing, to withstand the voltage and to prevent seawater ingress.

Cable System Networks

Point-to-point and branched systems are widely used to connect points separated by open water, just as with repeatered systems. In addition, there are applications for repeaterless technology in which many landing points may be in the same country along a shoreline. This arises because in many cases connections by sea are less costly and simpler than adding land-based facilities, which can require construction work in heavily populated areas.

Two types of cable configurations are used for such applications (Fig. 22). The festoon type simply has cable loop at sea between two terminal stations. For connection to more terminals, such loops are continued in a daisy chain. This configuration results in two cables entering each terminal station. While

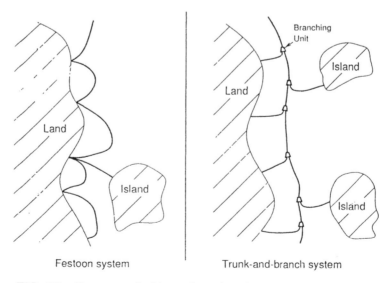

Festoon system   Trunk-and-branch system

FIG. 22   Two types of cable configurations for cable system networks.

this is simple to implement, it generally results in much cable being in shallow water, where vulnerability to damage by fishing equipment and anchors is greatest. Moreover, a single cable break interrupts transmission for more than the single faulted loop.

The second type is the trunk-and-branch configuration. Here, a cable is laid parallel to the shore, usually in deep water, and single branch cables are brought to each terminal from branching units. In addition to reducing the number of cables in vulnerable areas and the total amount of cable, it allows a wide variety of fiber interconnection arrangements among terminals without the need to deal with all of the fibers at each terminal. A fault in a branch cable only affects traffic at a single terminal.

Installation and Repair

The basic installation procedures for repeaterless systems are the same as for repeated systems. However, since repeaterless systems are shorter and are often installed in much shallower water than are repeated systems, ships can be used that have significantly less capacity than the large, purpose-built cable ships used for repeated systems. Most system installers have suites of portable cable machinery (storage tanks, cable engines, tension-monitoring instrumentation, cable troughing, etc.) that can be installed on "ships of opportunity" to turn them temporarily into cable ships. Ships of opportunity are relatively low-cost, smaller ships, usually used for coastal freight service or for servicing offshore installations such as oil platforms; these ships happen to be available when an installation ship is needed. Such ships must have adequate power to maneuver in moderately heavy weather and to hold station (stay in the same place without anchors) for long periods of time.

Repair of repeaterless systems is carried out in the same way as repair of repeated systems. The transmission test equipment needed on the ship is simpler, however, since there are no repeaters. Specially outfitted ships of opportunity can be used for repair, but the time necessary to equip them with cable machinery is a serious penalty since it adds to the time the system is out of service. Repair is usually done by fully equipped ships that are dedicated full time to repair duty for particular systems or for a particular area of the oceans.

Bibliography

Garratt, G. R. M., *One Hundred Years of Submarine Cables*, His Majesty's Stationery Office, London, 1950.

Dibner, B., *The Atlantic Cable*, Burndy Library, Norwalk CT, 1959.

Russel, W. H., *The Atlantic Telegraph (1865)*, 1865; reprint, David and Charles Reprints, Redwood Press, Towbridge, Wiltshire, England, 1972.

Transatlantic Submarine Cable System, *Bell Sys. Tech. J.*, 36(1):1–326 (January 1957).

SD Submarine Cable System, *Bell Sys. Tech. J.*, 43(4):1155–1479, Part 1 (July 1964).

SF Submarine Cable System, *Bell Sys. Tech. J.*, 49(5):601–783 (May–June 1970).
SG Submarine Cable System, *Bell Sys. Tech. J.*, 57(7):2313–2573, Part 1 (September 1978).
Ehrbar, R. D., Undersea Cables for Telephony, *IEEE Commun. Mag.* (August 1983).
Undersea Communications Technology, *AT&T Tech. J.*, 74(1):1–102 (January–February 1995).
Runge, P. K., and Trischitta, P. R. (eds.), *Undersea Lightwave Communications*, IEEE Press, New York, 1986.

References

1. Easton, R. L., and Marra, W. C., The Evolving Techniques for Achieving Undersea System Availability and Reliability, *L'Onde Électrique*, 73(1) (March–April 1993).
2. Bergano, N. S., Kerfoot, F. W., and Davidson, C. R., Margin Measurements in Optical Amplifier Systems, *IEEE Photonics Tech. Letters*, 5(3):304–306 (March 1993).

ROBERT F. GLEASON
CLEO D. ANDERSON
GREGORY M. BUBEL
ARMANDO A. CABRERA
ROBERT L. LYNCH
BRUCE O. REIN
WILLIAM F. SIROCKY
MARK D. TREMBLAY

Milton Keynes UK
Ingram Content Group UK Ltd.
UKHW052025071024
449327UK00027B/2430

9 780367 400835